BINARY STARS AS CRITICAL TOOLS AND TESTS IN CONTEMPORARY ASTROPHYSICS

IAU SYMPOSIUM No. 240

COVER ILLUSTRATION:

Night time view of the Prague Hradčany Castle (the seat of Czech kings and presidents) with the Charles Bridge, from the river bank of the Vltava. Photograph courtesy of Netzach Farbiash. Other photographs of Prague by Dr. Farbiash grace the session divider pages for Sessions 5, 7, 8, and 9, as well as the poster abstract pages for Sessions 6 and 8.

Speaker and other meeting photographs for all the session divider pages were provided by Miloslav Zejda. Additional "tourist" photos on these pages were provided by Brian Mason (Introductory Session), Robert Stencel (Session 3), Deborah Cline (Session 6), and William Hartkopf (Sessions 1, 2, and 6).

INTERNATIONAL ASTRONOMICAL UNION

UNION ASTRONOMIQUE INTERNATIONALE

BINARY STARS AS CRITICAL TOOLS AND TESTS IN CONTEMPORARY ASTROPHYSICS

PROCEEDINGS OF THE 240th SYMPOSIUM OF
THE INTERNATIONAL ASTRONOMICAL UNION
HELD IN PRAGUE, CZECH REPUBLIC
AUGUST 22–25, 2006

Edited by

WILLIAM I. HARTKOPF
United States Naval Observatory, Washington, DC, USA

EDWARD F. GUINAN
Villanova University, Villanova, PA, USA

and

PETR HARMANEC
Astronomical Institute of the Charles University, Prague, Czech Republic

CAMBRIDGE
UNIVERSITY PRESS

CAMBRIDGE UNIVERSITY PRESS
The Edinburgh Building, Cambridge CB2 8RU, United Kingdom
32 Avenue of the Americas, New York, NY 10013-2473, USA
477 Williamstown Road, Port Melbourne, VIC 3207, Australia
Ruiz de Alarcón 13, 28014 Madrid, spain
Dock House, The Waterfront, Cape Town 8001, South Africa

First published 2007

Printed in the United Kingdom at the University Press, Cambridge

Typeset in System LaTeX 2$_\varepsilon$

A catalogue record for this book is available from the British Library

Library of Congress Cataloguing in Publication data

ISBN-13 978 0521 86348 3 hardback
ISBN-10 0521 86348 1 hardback
ISSN 1743-9213

Table of Contents

Welcome and Dedications

New Observing Techniques and Reduction Methods

Session 1: Observing with High Angular and Spectral Resolution

Session 1 Poster Abstracts

Session 2: New Possibilities for Standard Observational Techniques

Session 2 Poster Abstracts

Session 3: Improved Methods of Data Analysis

Session 3 Poster Abstracts

Session 4: Observing in the Era of Large-Scale Surveys

Session 4 Poster Abstracts

Binary Stars as Critical Tools for Obtaining Direct and Reliable Information

Session 5: The Need to Improve Basic Calibrations; Increasing Possibilities of Classical Methods: a Few Examples

Session 5 Poster Abstracts

Session 6: Evolutionary Models for Binary and Multiple Stars

Session 6 Poster Abstracts

Binary Stars as Critical Tests for Studying Specific Phenomena

Session 7: Binary Stars as Probes of our Galaxy

Session 7 Poster Abstracts

Session 8: Asteroseismology; Stellar Activity

Session 8 Poster Abstracts

Summaries and Thanks

(Full posters are available at http://www.journals.cambridge.org/jid_IAU.
Note that page numbering of posters continues that of these proceedings;
these page numbers are shown in italics above, and also in the author and
object indexes. Full posters (pdf format only) for abstracts flagged by a † are
available as supplemental information at this same website.)

THE SCIENTIFIC ORGANIZING COMMITTEE

C. Allen (Mexico)
D. Bisikalo (Russia)
J. Davis (Australia)
E. Guinan (USA, co-chair)
P. Harmanec (Czech Republic, co-chair)
W. Hartkopf (USA, co-chair)
P. Lampens (Belgium)
J. Ling (Spain)
K. Oláh (Hungary)

T. Oswalt (USA)
K. Pavlovski (Croatia)
G. Peters (USA)
S. Rucinski (Canada)
C. Scarfe (Canada)
B. Warner (South Africa, ex officio)
M. Wolf (Czech Republic,
    National Organizing Committee)
H. Zinnecker (Germany)

Consultants

C. Aerts (Belgium)
A. Gimenéz (Spain/ESA)
H. McAlister (USA)

D. Pourbaix (Belgium)
V. Trimble (USA)

Acknowledgements

Symposium 240 was sponsored by:

- IAU Division IV *(Stars)*,
- Commission 26 *(Binary and Multiple Stars)*, and
- Commission 42 *(Close Binary Stars)*.

Supporting organizations included:

- IAU Division V *(Variable Stars)*,
- Commission 8 *(Astrometry)*,
- Commission 27 *(Variable Stars)*,
- Commission 30 *(Radial Velocities)*,
- Commission 33 *(Structure and Dynamics of the Galactic System)*,
- Commission 35 *(Stellar Constitution)*,
- Commission 37 *(Star Clusters and Associations)*,
- Commission 45 *(Stellar Classification)*,
- Working Group on *Massive Stars*,
- WG on *Catalog of Orbital Elements of Spectroscopic Binary Systems*,
- WG on *Optical/Infrared Interferometry*, and
- WG on *Binary and Multiple System Nomenclature*.

Financial support for some Symposium 240 participants was provided through a travel grant by the **International Astronomical Union**, as well as a matching grant by the **United States Office of Naval Research Global**. The organizers wish to express their gratitude to both organizations for their generosity.

Participants

Abolmasov, Pavel	Russian Federation	**Castermans**, Thierry	Belgium
Abt, Helmut	USA	**Caswell**, James	Australia
Aerts, Conny	Belgium	**Catalán**, Silvia	Spain
Ahmad, Amir	United Kingdom	**Celebre**, Cynthia	Philippines
Ahumada, Andrea	Argentina	**Chambliss**, Carlson R	USA
Albacete Colombo, Juan Facundo	Italy	**Chanamé**, Julio	USA
Allard, France	France	**Chandrasekhar**, Thyagarajan	India
Allard, Nicole	France	**Chappelle**, Richard	Czech Republic
Allen, Christine	Mexico	**Chatterjee**, Tapan K.	Mexico
Amado, Pedro J.	Spain	**Chatzidogiannaki**, Vasilea	Greece
Amores, Eduardo	Spain	**Chen**, Wen-Ping	China Taipei
Andersen, Johannes	Denmark	**Choi**, Chul-Sung	Korea, Rep. of
Andersen, Michael	Germany	**Christensen–Dalsgaard**, Jørgen	Denmark
Anglada-Escude, Guillem	Spain	**Chung**, HyunSoo	Korea, Rep. of
Arenou, Frédéric	France	**Clarke**, Cathie	United Kingdom
Argyle, Robert	United Kingdom	**Clausen**, Jens Viggo	Denmark
Arnett, David	USA	**Colome**, Josep	Spain
Aslan, Zeki	Turkey	**Corbally**, Christopher	Vatican City State
Asplund, Martin	Australia	**Corbin**, Brenda	USA
Atanackovic–Vukmanovic, Olga	Serbia	**Corbin**, Thomas	USA
Aufdenberg, Jason	USA	**Correia**, Serge	Germany
Awadalla, Nabil	Egypt	**Costa**, Joaquim	Brazil
Babina, Julia	Ukraine	**Costero**, Rafael	Mexico
Balega, Yuri	Russian Federation	**Cremaschini**, Claudio	Italy
Barkin, Yury	Spain	**Cropper**, Mark	United Kingdom
Barsunova, Olga	Russian Federation	**Cséki**, Attila	Serbia, Rep. of
Bartolini, Corrado	Italy	**Czart**, Krzysztof	Poland
Baruch, John	United Kingdom	**Dabringhausen**, Jörg	Germany
Batten, Alan	Canada	**Davis**, John	Australia
Baumgardt, Holger	Germany	**de Araújo**, Francisco Xavier	Brazil
Becker, Werner	Germany	**De Buizer**, James	USA
Bedding, Tim	Australia	**De Cat**, Peter	Belgium
Beers, Timothy	USA	**de Freitas Mourao**, Ronaldo R.	Brazil
Benedict, George	USA	**de Greve**, Jean-Pierre	Belgium
Berdyugina, Svetlana	Switzerland	**Demichev**, Vasily	Russian Federation
Berinde, Stefan	Romania	**Derekas**, Aliz	Australia
Berriman, Bruce	USA	**Deustua**, Susana	USA
Bisikalo, Dmitry	Russian Federation	**Devor**, Jonathan	USA
Blomqvist, Anna	Sweden	**Díaz**, Fabiola	Venezuela
Blondy, Pierre	USA	**Dieball**, Andrea	United Kingdom
Bobylev, Vadim	Russian Federation	**Dieckmann**, Mark Eric	Germany
Bochkarev, Nikolai	Russian Federation	**Donnison**, J. Richard	United Kingdom
Boffin, Henri	Germany	**Duncan**, Douglas	USA
Bonacic Marinovic, Axel	Netherlands	**Dworetsky**, Michael	United Kingdom
Bonanos, Alceste	USA	**Echevarría**, Juan	Mexico
Boss, Alan	USA	**Ederoclite**, Alessandro	Belgium
Bouy, Herve	USA	**Eenmäe**, Tnis	Estonia
Boyle, S.J., Richard P.	Vatican City State	**Eggleton**, Peter	USA
Bradley, Arthur	USA	**Elebert**, Patrick	Ireland
Breton, René	Canada	**Elias**, Nicholas	Germany
Briot, Danielle	France	**Errico**, Luigi	Italy
Brown, Anthony	Netherlands	**Evans**, Nancy R.	USA
Brown, John C	United Kingdom	**Eyer**, Laurent	Switzerland
Bruntt, Hans	Australia	**Fabrycky**, Daniel	USA
Bursa, Milan	Czech Republic	**Falceta-Gonçalves**, Diego	Brazil
Callanan, Paul	Ireland	**Farbiash**, Netzach	Israel

Fekel, Francis	USA	**Henry**, Todd	USA
Feltzing, Sofia	Sweden	**Hensberge**, Herman	Belgium
Fernández Lajús, Eduardo E.	Argentina	**Herdiwijaya**, Dhani	Indonesia
Firneis, Friedrich	Austria	**Hesser**, Jim	Canada
Fischer, Tobias	Switzerland	**Hestroffer**, Daniel	France
Foing, Bernard	Netherlands	**Hilditch**, Ron	United Kingdom
Franz, Otto	USA	**Hillberg**, Tomas	Sweden
Freyhammer, Lars M.	United Kingdom	**Hirai**, Masanori	Japan
Gabriel, Carlos	Spain	**Hojaev**, Alisher S.	Uzbekistan
Gagik, Tovmassian	Mexico	**Holley-Bockelmann**, Kelly	USA
Gális, Rudolf	Slovakia	**Hong**, Kyeong Soo	Korea, Rep. of
Garcia, Beatriz	Argentina	**Horniaková**, Erika	Slovakia
Garcia Lugo, Gabriela	Venezuela	**Hornoch**, Kamil	Czech Republic
Gaume, Ralph	USA	**Houziaux**, Leo	Belgium
Gavras, Panagiotis	Greece	**Howarth**, Ian	United Kingdom
Gavryuseva, Elena	Italy	**Hu**, Haili	Netherlands
Gazeas, Kosmas	Greece	**Hu**, Jian	China Nanjing
Gelderman, Richard	USA	**Hu**, Juei-Hwa	China Taipei
Gerbaldi, Michele	France	**Hudec**, René	Czech Republic
Gilmore, Alan	New Zealand	**Hurley**, Jarrod	Australia
Giménez, Alvaro	Netherlands	**Hut**, Piet	USA
Giridhar, Sunetra	India	**Imada**, Akira	Japan
Glushkova, Elena	Russian Federation	**Imaeda**, Yusuke	Japan
Goker, Umit Deniz	Turkey	**In-Ok**, Song	Korea, Rep. of
Golimowski, David	USA	**Iping**, Rosina	USA
Golovin, Alex	Ukraine	**Ishihara**, Daisuke	Japan
Gómez Maqueo Chew, Yilen	USA	**Ishioka**, Ryoko	USA
Gorshanov, Denis	Russian Federation	**Isik**, Emre	Germany
Gradoula, Georgia-Peristera	Greece	**Ivanova**, Natalia	Canada
Gráf, Tomáš	Czech Republic	**Ivantsov**, Anatoliy	Ukraine
Gräfener, Götz	Germany	**Izzard**, Robert	Netherlands
Granada, Anahí	Argentina	**Janík**, Jan	Czech Republic
Griffin, R. Elizabeth	Canada	**Janusz, S.J.**, Robert	Vatican City State
Griffin, Roger	United Kingdom	**Järvinen**, Silva	Germany
Groenewegen, Martin	Belgium	**Jassur**, Davoud	Iran, Islamic Rep. of
Groot, Paul	Netherlands	**Javadi**, Atefeh	Iran, Islamic Rep. of
Gros, Monique	France	**Jeon**, Young-Beom	Korea, Rep. of
Grosheva, Elena	Russian Federation	**Joss**, Paul C.	USA
Grygar, Jiří	Czech Republic	**Justham**, Stephen	United Kingdom
Guillard, Pierre	France	**Kafka**, Styliani	Chile
Guinan, Edward	USA	**Kaldalu**, Meelis	Estonia
Gull, Theodore	USA	**Kalinichenko**, Leonid	Russian Federation
Gustafson, Bo	USA	**Kalomeni**, Belinda	Turkey
Gvaramadze, Vasilii	Russian Federation	**Kang**, Young-Woon	Korea, Rep. of
Hadrava, Petr	Czech Republic	**Karitskaya**, Eugenia	Russian Federation
Halbwachs, Jean-Louis	France	**Karpov**, Sergey	Russian Federation
Halevin, Alexander	Ukraine	**Kawabata**, Koji	Japan
Han, Zhanwen	China Nanjing	**Kawka**, Adela	Czech Republic
Hao, JingFang	China Nanjing	**Kembhavi**, Ajit	India
Harmanec, Petr	Czech Republic	**Kharinov**, Mikhail	Russian Federation
Hartkopf, William	USA	**Kholtygin**, Alexander	Russian Federation
Hasan, Saiyid Sirajul	India	**Kim**, Dong-Woo	USA
Hashimoto, Osamu	Japan	**Kim**, Ho-il	Korea, Rep. of
Heacox, William	USA	**Kinugasa**, Kenzo	Japan
Hearnshaw, John	New Zealand	**Kiraga**, Marcin	Poland
Heggie, Douglas	United Kingdom	**Kiselev**, Alexey	Russian Federation
Heiter, Ulrike	Sweden	**Kiss**, Zoltán T.	Hungary
Hejna, Ladislav	Czech Republic	**Kisseleva–Eggleton**, Ludmila	USA
Henden, Arne	USA	**Kiyaeva**, Olga	Russian Federation

Klioner, Sergei	Germany	Manimanis, Vassilios	Greece
Klochkova, Valentina	Russian Federation	Marschall, Laurence	USA
Klutsch, Alexis	France	Marsden, Stephen	Switzerland
Koch, David	USA	Marshalov, Dmitry	Russian Federation
Kochukhov, Oleg	Sweden	Martioli, Eder	Brazil
Köhler, Rainer	Netherlands	Marvel, Kevin	USA
Komonjinda, Siramas	New Zealand	Maschberger, Thomas	Germany
Korčáková, Daniela	Czech Republic	Mason, Brian	USA
Korhonen, Heidi	Germany	Mathys, Gautier	Chile
Kosovichev, Alexander	USA	Matsuda, Takuya	Japan
Kotnik–Karuza, Dubravka	Croatia	Matthews, Jaymie	Canada
Kouwenhoven, Thijs	Netherlands	Matveyenko, Leonid	Russian Federation
Kövári, Zsolt	Hungary	Mayer, Pavel	Czech Republic
Kovetz, Attay	Israel	Mazeh, Tsevi	Israel
Kramer, Michael	United Kingdom	McAlister, Harold	USA
Kroupa, Pavel	Germany	McDavid, David	USA
Krtička, Jiří	Czech Republic	McMillan, Robert	USA
Kučerová, Blanka	Czech Republic	McMillan, Stephen	USA
Kuepper, Andreas	Germany	McSwain, M. Virginia	USA
Kuznyetsova, Yuliana	Ukraine	Menzies, John	South Africa
Lampens, Patricia	Belgium	Michalska, Gabriela	Poland
Landolt, Arlo	USA	Mickaelian, Areg	Armenia
Lane, Benjamin	USA	Mikulášek, Zdeněk	Czech Republic
LaSala, Jerry	USA	Mink, de, Selma	Netherlands
Latham, David	USA	Mitchell, Deborah	United Kingdom
Latković, Olivera	Serbia, Rep. of	Moffat, Anthony	Canada
Lattanzio, John	Australia	Moldon, Francisco Javier	Spain
Lebreton, Yvelin	France	Monier, Richard	France
Lee, ChungUk	Korea, Rep. of	Montalban, Josefina	Belgium
Lee, Jae Woo	Korea, Rep. of	Montes, David	Spain
Lee, Jeong-Ju	Korea, Rep. of	Morales–Rueda, Luisa	Netherlands
Lee, Ki-Won	Korea, Rep. of	Muñoz–Darias, Teodoro	Spain
Lee, Myung Gyoon	Korea, Rep. of	Nelemans, Gijs	Netherlands
Lee, Tse-Lin	China Taipei	Nesslinger, Stefan	Germany
Leedjärv, Laurits	Estonia	Netolický, Martin	Czech Republic
Leibowitz, Elia	Israel	Neuhäuser, Ralph	Germany
Leung, Kam-Ching	USA	Niarchos, Panagiotis	Greece
Levato, Hugo	Argentina	Nicol, Marie-Helene	Germany
Lewis, Brian	USA	Nogami, Daisaku	Japan
Li, Xiangdong	China Nanjing	Nordström, Birgitta	Denmark
Ling, Josefina F.	Spain	Norman, Colin	USA
Lipunov, Vladimir	Russian Federation	Nurmi, Pasi	Finland
Lloyd Evans, Tom	United Kingdom	Oblak, Edouard	France
Lo, Man Kit	China Nanjing	Okeke, Pius. N.	Nigeria
Löckmann, Ulf	Germany	Oláh, Katalin Ilona	Hungary
Lombardi, James	USA	Ondřich, David	Czech Republic
Longhitano, Marco	Switzerland	Orsatti, Ana Maria	Argentina
Lopez-Santiago, Javier	Italy	Osten, Rachel	USA
Lub, Jan	Netherlands	Oswalt, Terry	USA
Lugaro, Maria	Netherlands	Pan, Xiaopei	USA
Lyo, A-Ran	China Taipei	Pandey, S. K.	India
Mac Low, Mordecai-Mark	USA	Pandey, Vishambhar Nath	India
Maceroni, Carla	Italy	Paparo, Margit	Hungary
Macri, Lucas	USA	Parimucha, Stefan	Slovakia
Malaroda, Stella	Argentina	Patruno, Alessandro	Netherlands
Malekjani, Molayar	Iran, Islamic Rep. of	Paura, Glavio	Brazil
Malkov, Oleg	Russian Federation	Pavlenko, Elena	Ukraine
Malofeev, Valery	Russian Federation	Pavlovski, Krešimir	Croatia
Mamajek, Eric	USA	Pendharkar, Jayant Kumar	India

Peters, Geraldine	USA	**Sonneborn**, George	USA
Peterson, Bruce	Australia	**Soonthornthum**, Boonrucksar	Thailand
Petr-Gotzens, Monika	Germany	**Soulié**, Edgar	France
Phelps, Randy	USA	**Southworth**, John	United Kingdom
Philip, A. G. Davis	USA	**Spite**, Francois	France
Pigulski, Andrzej	Poland	**Stanghellini**, Letizia	USA
Pilachowski, Catherine	USA	**Stappers**, Benjamin	Netherlands
Pireaux, Sophie	France	**Stencel**, Robert	USA
Platais, Imants	USA	**Sterken**, Chris	Belgium
Plavec, Mirek J.	USA, Czech Rep.	**Strassmeier**, Klaus	Germany
Pols, Onno	Netherlands	**Stringfellow**, Guy	USA
Popa, Lucia Aurelia	Romania	**Stumpf**, Michaela B.	Germany
Portegies Zwart, Simon	Netherlands	**Swings**, Jean-Pierre	Belgium
Pourbaix, Dimitri	Belgium	**Szeidl**, Bela	Hungary
Poveda, Arcadio	Mexico	**Tanaka**, Yasuo	Germany
Prša, Andrej	Slovenia	**Tango**, William	Australia
Quirrenbach, Andreas	Germany	**Taracchini**, Andrea	Italy
Raghavan, Deepak	USA	**Tarter**, Jill	USA
Reipurth, Bo	USA	**Tata**, Ramarao	USA
Rey, Soo-Chang	Korea, Rep. of	**Tayal**, Swaraj	USA
Ribas, Ignasi	Spain	**ten Brummelaar**, Theo	USA
Richards, Mercedes	USA	**Terlevich**, Elena	Mexico
Richer, Harvey	Canada	**Terlevich**, Roberto	Mexico
Roelofs, Gijs	Netherlands	**Thomas**, Sandrine	USA
Rolland, Loïc	France	**Tiplady**, Adrian	South Africa
Rosenzweig, Patricia	Venezuela	**Tokovinin**, Andrei	Chile
Rowan, Sheila	United Kingdom	**Topinka**, Martin	Czech Republic
Royer, Frédéric	France	**Toropina**, Olga	Russian Federation
Rubio-Herrera, Eduardo	Netherlands	**Torra Roca**, Jordi	Spain
Rucinski, Slavek	Canada	**Torres**, Andrea Fabiana	Argentina
Ruiz-Lapuente, Pilar	Spain	**Torres**, Guillermo	USA
Ryabov, Yuri	Russian Federation	**Torres**, Kelly	Belgium
Saar, Steven	USA	**Trimble**, Virginia	USA
Samec, Ronald	USA	**Tsuji**, Takashi	Japan
Samus, Nikolay	Russian Federation	**Tuvikene**, Taavi	Belgium
Sarazin, Craig	USA	**Udry**, Stephane	Switzerland
Satterthwaite, Gilbert	United Kingdom	**Uemura**, Makoto	Japan
Savos, Raoul	Romania	**Ulas**, Burak	Turkey
Scarfe, Colin	Canada	**Urošević**, Dejan	Serbia, Rep. of
Scholz, Ralf-Dieter	Germany	**Usov**, Vladimir	Israel
Schuster, William J.	Mexico	**Valls-Gabaud**, David	France
Serabyn, Eugene	USA	**van Altena**, William	USA
Sergeeva, Tetyana	Ukraine	**van Belle**, Gerard	USA
Serlemitsos, Peter	USA	**van den Besselaar**, Else	Netherlands
Shakht, Natalia	Russian Federation	**van der Hucht**, Karel A.	Netherlands
Shao, Michael	USA	**van Hamme**, Walter	USA
Shengbang, Qian	China Nanjing	**van Rensbergen**, Walter	Belgium
Šíma, Zdislav	Czech Republic	**Vanbeveren**, Dany	Belgium
Simon, Michal	USA	**Vaz**, Luiz Paulo	Brazil
Skjaeraasen, Olaf	Norway	**Verbunt**, Frank	Netherlands
Škoda, Petr	Czech Republic	**Vergne**, Maria Marcela	Argentina
Šlechta, Miroslav	Czech Republic	**Vesperini**, Enrico	USA
Smart, Richard	Italy	**Viallet**, Maxime	France
Šmelcer, Ladislav	Czech Republic	**Vida**, Krisztián	Hungary
Smith, Martin C.	Netherlands	**Vidal Safor**, Erick Dennis	Peru
Smits, Derck	South Africa	**Vidal-Madjar**, Alfred	France
Šolcová, Alena	Czech Republic	**Vidojević**, Sonja	Serbia, Rep. of
Soleri, Paolo	Netherlands	**Vittone**, Alberto Angelo	Italy
Solheim, Jan-Erik	Norway	**Voges**, Wolfgang	Germany

Vrba, Frederick	USA	**Wolf**, Marek	Czech Republic
Wanas, Mamdouh	Egypt	**Wood**, Peter	Australia
Wang, Hongchi	China Nanjing	**Yakut**, Kadri	Turkey
Wang, Na	China Nanjing	**Yamaoka**, Hitoshi	Japan
Warner, Brian	South Africa	**Yaqoob**, Tahir	USA
Webb, Natalie	France	**Yishamuding**, Aili	China Nanjing
Weis, Edward	USA	**Yungelson**, Lev	Russian Federation
Weiss, Achim	Germany	**Zacs**, Laimons	Latvia
Weiss, Werner	Austria	**Zahn**, Jean-Paul	France
Weltevrede, Patrick	Netherlands	**Zejda**, Miloslav	Czech Republic
Whitelock, Patricia	South Africa	**Zhang**, Shuang Nan	China Nanjing
Whiting, Alan	United Kingdom	**Zhou**, Jianfeng	China Nanjing
Whitworth, Anthony	United Kingdom	**Zhu**, Li-Ying	China Nanjing
Williams, Iwan	United Kingdom	**Zhu**, Yongtian	China Nanjing
Williams, Robert	USA	**Zinnecker**, Hans	Germany
Wilson, Robert E.	USA	**Zwitter**, Tomaž	Slovenia
Wing, Robert	USA		

1

Welcome and Dedication

Figure 1. (top) Petr Harmanec, giving welcoming address. (middle) Dedicatee Mirek Plavec, receiving the applause of the audience. (bottom) Ed Guinan, giving symposium overview.

Figure 2. (top) Petr Harmanec, with poster honoring Mirek Plavec. (bottom left) St. Franciscus Seraphicus church, near the Old Town end of the Charles Bridge; the statue of Emperor Charles IV was erected in 1848, on the occasion of the 400th anniversary of the Prague University. (bottom right) Astronomical Clock (from 1410) at the Old Town Hall on Staroměstské Square.

Binary Stars as Critical Tools & Tests
in Contemporary Astrophysics
Proceedings IAU Symposium No. 240, 2006
W.I. Hartkopf, E.F. Guinan & P. Harmanec, eds.

Welcoming Address

Petr Harmanec†

Astronomical Institute of the Charles University, Faculty of Mathematics and Physics,
V Holešovičkách 2, CZ-180 00 Praha 8, Czech Republic
email: hec@sirrah.troja.mff.cuni.cz
and
Astronomical Institute, Academy of Sciences of the Czech Republic, CZ-251 65 Ondřejov,
Czech Republic
email: hec@sunstel.asu.cas.cz

Dear colleagues and friends, ladies and gentlemen,

It is great privilege for me to welcome so many of you to Prague, or Praha as we call our capital in Czech.

We have been witnessing a very rapid development of observational techniques over the past few decades, leading to many exciting discoveries, quite often in the field of extragalactic research. I sometimes sense the tendency to consider the studies of binaries and stars in general as exhausted topics. However, we all know that the better our real and deep understanding of the principal sources of radiative energy in the Universe becomes, the safer will be the ground for studies of all higher hierarchical systems.

I am therefore very pleased to see how great an interest this symposium attracted and I am especially happy to see not only old friends but also many young colleagues in the audience.

As most of you know, there were originally two proposals for this meeting and I find it great that it resulted in this joint effort. Interferometry and other new techniques are gradually removing the distinction between astrometric and spectroscopic binaries, and I am personally very keen to see the progress in this area. In particular, this will give us the chance to derive certain basic properties of stars like the limb- and gravity-darkening of rapidly rotating early-type stars simply via comparison of observed radiative properties of stars of the same mass seen under different aspect angle, from equator-on to pole-on.

Let me also express our appreciation for the help given to us by all members of our Scientific Organizing Committee and our advisers, who were active in all phases of preparation for this meeting. Our apologies at the same time to all of you who were not given the chance to present your favourite topics. Please, consider that we were put into a very uneasy position. We were first promised to have a five-day meeting and while the initially-suggested program was taking shape, the news came that all symposia can go on for only 3.5 days.

Some practical information:

1. It is now obligatory to record the discussions following all invited talks. Our assistants will distribute discussion sheets to everybody who asks a question after a talk and then to the speakers to add their answers. We hope this might naturally regulate your anxiety to discuss! Note also that no discussion will be allowed after the short oral presentations of selected posters, due to time constraints.

2. Thursday evening starting at 6 p.m. Marek Wolf, Miloslav Zejda and I have organized a public outreach: a discussion between professional and amateur astronomers

† Present address: Astronomical Institute of the Charles University, V Holešovičkách 2, CZ-180 00 Praha 8, Czech Republic

about binary-star observations by amateurs. This discussion will be held at the building of the faculty of Mathematics and Physics of the Charles University, located at the other side of Nuselský bridge. It is within walking distance of here, and we invite all interested people to join us.

Let me end this welcome by saying the following: From the very beginning of my career in astronomy, I always had a very pleasant feeling that the students of binaries all over the world represent a very friendly community joined by their love of binaries and multiple systems. I heartily hope that we all will share the same feeling during our meeting which I now declare open.

Binary Stars as Critical Tools & Tests
in Contemporary Astrophysics
Proceedings IAU Symposium No. 240, 2006
W.I. Hartkopf, E.F. Guinan & P. Harmanec, eds.

Introduction & Overview to Symposium 240: Binary Stars as Critical Tools and Tests in Contemporary Astrophysics

Edward F. Guinan[1], Petr Harmanec[2] and William Hartkopf[3]

[1]Department of Astronomy & Astrophysics, Villanova University, Villanova, PA, 19085, USA
e-mail: edward.guinan@villanova.edu
[2]Astronomical Institute of the Academy of Sciences of the Czech Republic, 251 65 Ondrejov, Czech Republic
[3]Astrometry Dept., U.S. Naval Observatory, Washington, DC, 20392, USA

Abstract. An overview is presented of the many new and exciting developments in binary and multiple star studies that were discussed at IAU Symposium 240. Impacts on binary and multiple star studies from new technologies, techniques, instruments, missions and theory are highlighted. It is crucial to study binary and multiple stars because the vast majority of stars (>60%) in our Galaxy and in other galaxies consist, not of single stars, but of double and multiple star systems. To understand galaxies we need to understand stars, but since most are members of binary and multiple star systems, we need to study and understand binary stars. The major advances in technology, instrumentation, computers, and theory have revolutionized what we know (and also *don't* know) about binary and multiple star systems. Data now available from interferometry (with milliarcsecond [mas] and sub-mas precisions), high-precision radial velocities (∼1-2 m/s) and high precision photometry (<1–2 milli-mag) as well as the wealth of new data that are pouring in from panoramic optical and infrared surveys (e.g., >10,000 new binaries found since 1995), have led to a renaissance in binary star and multiple star studies. For example, advances have lead to the discovery of new classes of binary systems with planet and brown dwarf components (over 200 systems). Also, extremely valuable data about binary stars are available across the entire electromagnetic spectrum — from gamma-ray to IR space missions and from the ground using increasingly more powerful and plentiful optical and radio telescopes as well as robotic telescopes. In the immediate future, spectral coverage could even be extended beyond the radio to the first detection of gravity waves from interacting close binaries. Also, both the quality and quantity of data now available on binary and multiple stars are making it possible to gain unprecedented new insights into the structure, and formation and evolution of binary stars, as well as providing valuable astrophysical information (like precise stellar masses, radii, ages, luminosities and distances) to test and constrain current astrophysical theory. These major advances permit tests of current theories and ideas in stellar astrophysics and provide the foundations for the next steps in modeling and improvements in theory to be taken.

1. Introduction

IAU Symposium 240, *Binary Stars as Critical Tools & Tests in Contemporary Astrophysics*, took place 22-25 August 2006, during the second week of the XXVIth IAU General Assembly in Prague, Czech Republic. The meeting lasted for 3-1/2 days and consisted of a mix of invited, contributed oral contributions and over 180 posters. This symposium was the result of a merger of two binary star symposia proposals developed for the General Assembly. One proposal, originated from Commission 26 (Binary and Multiple Stars) and led by Hartkopf, focused mainly on wider systems. The other proposal was organized by Commission 42 (Close Binaries) and led by Harmanec and Guinan, and it focused primarily on close/interacting binaries. The IAU Executive Committee recommended that the two proposals be merged, and after hundreds of e-mail exchanges between the wide and close binary groups and their respective organizing committees and related commissions, the present symposium was organized with a broader program that includes all varieties and flavors of binary stars — both close and wide as well as near and far binaries and multiple stars systems. Most of the different types of binary systems discussed at the symposium are given in Figures 1 – 3. The various classes of binary systems composed of non-degenerate stars are shown in Figure 1, while Figure 2 shows binaries containing at least one degenerate component (white dwarf, neutron star or black hole). Figure 3 shows recently discovered binary systems with planetary or brown dwarf components. Note that our solar system would be included in Figure 3.

The symposium brought together ∼500 astronomers from 54 countries who are involved in all aspects of binary and multiple star research, from very long-period, common-proper-motion pairs and other "fragile" binaries to short-period contact binaries, short-period binaries with degenerate components, as well as star/brown-dwarf/planet systems, with the aim of exploring interests common to all binary star researchers. Both the observational and theoretical aspects of binary and multiple star research are represented, but the main themes of the program are the new information and physical insights gleamed from the recent advances in instrumentation and techniques. The meeting also attracted those interested in the observational and theoretical aspects of modern stellar astrophysics that depend very strongly on the fundamental properties of stars found primarily from binary and multiple stars. Sponsored by Division IV [Stars] and Commissions 26 and 42, the symposium has also received strong support from Division V [Binary & Variable Stars], as well as from seven other commissions and three working groups. We thank these Divisions, Commissions, and Working Groups for their support.

The resulting program benefited greatly from the infusion of new ideas and differing perspectives from the two binary star communities, which resulted in a broader and more comprehensive program and most of all a more vibrant and interesting meeting. The format for the symposium was a mix of invited oral review presentations (∼30 min) and more narrowly-focused topical (∼15–20 min) presentations. There were also over twenty short oral/poster presentations (5–10 min) that were selected by the SOC from >180 submitted posters. These are also included in this volume, as are abstracts of all the poster presentations posted to the publisher's website.

1.1. *Venue and dedication of the symposium*

It is fitting that this symposium was held in the Czech Republic, because much of the pioneering work on binary and variable stars has been carried out in Central and Eastern Europe for over a century. The meeting was also fittingly held in the city of Prague, home to both Tycho Brahe and Johannes Kepler. It is noteworthy that Tycho carried out the first accurate astrometric measurements, while Kepler's laws of orbital motion play such an important role in binary star studies. As the first joint meeting of the "close and wide" binary communities in recent memory, it was also appropriate to jointly dedicate the symposium to outstanding representatives of both of those "close" and "wide" worlds of binary and multiple star research: Mirek Plavec of the Czech Republic and the late Charles Worley of the United States. Both gentlemen attended the last General Assembly held in Prague in 1967, and Plavec also attended the opening session of the 2006 meeting. Poster papers about both astronomers, displayed at the meeting and included here in these proceedings, provide brief descriptions of their professional careers as well as their many important contributions to binary and multiple star research.

Also, the reader should take note of the poster paper by Augensen, Mason & Hartkopf (see publisher's website) discussing the life and work of Dr. Wulff Heintz, who passed

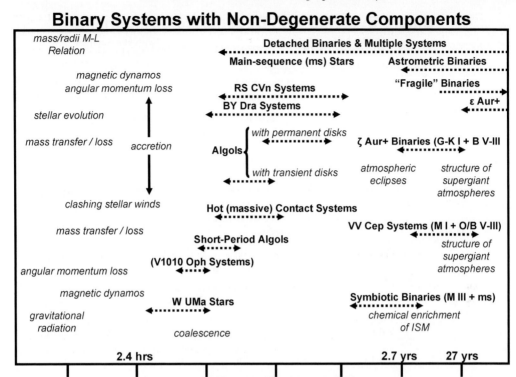

Figure 1. The period range and astrophysical interests of binary systems with non-degenerate components.

away in June 2006, shortly before this symposium was held. Noteworthy among Wulff's many contributions to binary and multiple star research is that he made a total of 54,000 micrometer measures of double stars and discovered over 900 new pairs. Wulff is second only to Willem van den Bos in the total number of astrometric measures. Interesting enough, Charles Worley is in third place in this category, with over 40,000 individual astrometric measures and the discovery of 39 new faint, cool stellar companions.

2. Background: Major Advances in Instrumentations and Techniques

In 1992, IAU Colloquium 135, "Complementary Approaches to Double and Multiple Star Research," was held at Callaway Gardens (near Atlanta, Georgia), and was meant to emphasize the expanding overlap of observational opportunities offered to binary star researchers by advances in precise radial velocity techniques, interferometry, etc. Additional topics included recent advances (for those times) in our knowledge of duplicity for young stars and pre-main sequence stars, the latest theories of binary and multiple star formation, and the tantalizing first results from HST and Hipparcos. The meeting successfully brought together nearly one hundred astronomers with a diversity of expertise until then rarely found at a single meeting on binary stars.

Much has changed in the ensuing 14 years! HST is now approaching the end of its life, and its larger successor, the James Webb Space Telescope (JWST) is under construction. Hipparcos results have been published and well-studied, and the next generation of astrometry satellites (such as the Space interferometry Mission [SIM] and Gaia) now

Binary Systems with Degenerate Components

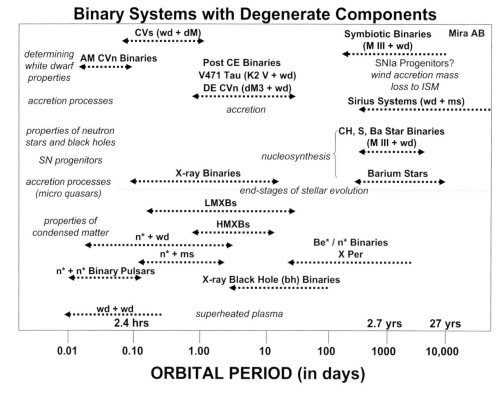

Figure 2. The period range and astrophysical interests of binary systems with degenerate components.

under development promise results orders of magnitude greater in both precision and quantity. Speckle interferometry and adaptive optics have replaced visual micrometry as the routine methods of measuring visual binaries, while long-baseline interferometry has now produced a significant body of results for closer binaries, with the promise of an outburst of activity in the resolution of spectroscopic binary systems during the next few years. From astrometric and spectroscopic orbits, masses and other important stellar properties are determinable. Interferometric results from the HST Fine Guidance Sensors (FGS) have also become more plentiful. As discussed in several papers at the meeting and in this volume, there have been dramatic changes in the way we conduct astrometry of binaries and profound increases in the reliability of the results.

Similar tremendous advances continue in other areas as well. Variability-induced motion and other color-based detection methods are being used to mine the SDSS database for new binaries and may be used with Gaia (http://www.rssd.esa.int/index.php?project =GAIA&page=index) and SIM (http://planetquest.jpl.nasa.gov/SIM/sim_index.cfm) data as well. Superior infrared detectors and techniques, as well as the recent availability of infrared spectroscopy and imaging from space with the Spitzer Space Telescope (SST), are revolutionizing the study of pre-main sequence binaries and binaries with cool star and brown dwarf components. Ever more accurate radial velocity techniques have yielded long-period orbits that further blur the distinction between the traditional spectroscopic and visual separation/period regimes and yield accurate stellar masses across the spectral range – not to mention the new short-period systems and 200+ exoplanets discovered over the last decade! Within the next several years hundreds

Binary Systems with Planetary/Brown Dwarf Components

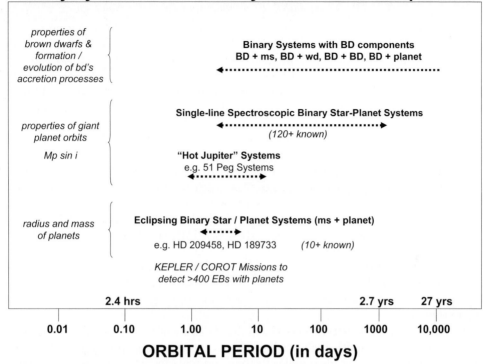

Figure 3. The period range and astrophysical interests of binary systems with brown dwarf and planetary components.

of additional exoplanets (some with diameters as small as the Earth) are expected to be discovered by the CoRot and Kepler missions.

2.1. *Organization of the papers*

The papers in this volume are organized into three major themes, as discussed in Sections 3–5 below. "New Observing Techniques and Reduction Methods" includes (3.1) Observing with high angular and spectral resolution; (3.2) New possibilities for standard observational techniques; (3.3) Improved methods of data analysis; and Observing in the era of large scale surveys. "Binary Stars as Critical Tools (for obtaining direct and reliable Astrophysical Information)" consists of (4.1) The need to improve basic calibrations and increasing possibilities of classical methods and (4.2) Evolutionary models for binary and multiple stars. Finally, "Binary Stars as Critical Tests" focuses on (5.1) Binary stars as probes of our Galaxy and (5.2) Asteroseismology and Stellar Activity. The volume concludes with excellent summaries on the observational aspects (given by Colin Scarfe) and on the theoretical aspects (given by Virginia Trimble) and a few closing comments from Hartkopf. The papers (both oral presentations and posters) presented at this symposium and included in this volume and on the publisher's website demonstrate that these are indeed very exciting times for the studies of binary and multiple star systems.

3. New Observing Techniques and Reduction Methods

3.1. *Observing with high angular and spectral resolution*

Recent results from powerful long-baseline interferometers both in the northern and southern hemispheres are discussed by Hal McAlister and John Davis. These instruments are yielding unprecedented high angular resolution measures, and refined astrometric orbits, of an increasing number of binaries with smaller and smaller separations. These papers show that astrometric measures are now closing the gap between wide binaries and shorter-period spectroscopic binaries and are providing fundamental properties of the component stars not possible previously except for eclipsing systems. In many cases additional components have been discovered to single stars and doubles. As discussed by Yuri Balega, interferometry using single telescopes (mostly from speckle interferometry), is discovering many new binaries and multiple star systems, and these observations are routinely yielding high precision orbits. Also with interferometric and adaptive optics techniques, many single and binary stars are being found to harbor fainter stellar and, in some cases, brown-dwarf and planetary-size companions. As discussed by Mike Shao and Xiaopei Pan, the Space Interferometry Mission (SIM), currently planned for 2015, will be capable of ultra-high angular precision, microarcsecond (μ-sec) astrometry that has the capability of astrometrically detecting components with masses as small as Earth. In a contributed oral/poster presentation, Frank Fekel & Jocelyn Tomkin discuss several spectroscopic binary candidates that are most suitable for ground-base multiple aperture interferometry.

A decade ago, most astronomers (including most members of Commission 42) would have been very happy with stellar radial velocity measures with precisions of $\approx$1–3 km/s. However, over the last decade radial velocity measures using, for example, iodine gas cells, are routinely being obtained with precisions of better than 5 m/s. More recently, radial velocity measures are being accomplished with precisions reaching as low as $\sim$1 m/s. As discussed at the meeting by Artie Hatzes (but not included in the book) and others, over 200 extrasolar planets have so far been discovered, mostly from these high-precision radial-velocity programs (see, e.g., `http://exoplanets.org`). It appears that $\sim$5% of nearby solar-type (i.e., dG and dK) stars are accompanied by Jupiter-size planets, many of these (known as "Hot Jupiters") are orbiting very close to their host stars. We have not yet reached the needed observational precision to detect Earth-size planets from high-precision spectroscopy but, as reported by several papers here, Earth-size planets should soon be discovered with SIM astrometry. These discoveries may come more immediately from high-precision photometry missions such as CoRot and Kepler, however, as discussed later in the book. Excellent summaries of high-precision radial velocity work are given by Marcy & Butler (`http://exoplanets.org/`) and Mayor (`http://obswww.unige.ch/~udry/planet/`). For the most part, the high-precision radial velocity work is focused on finding planets. However, in the near future, these measures will also be secured for double-line eclipsing binaries to obtain very accurate fundamental properties.

3.2. *New possibilities for standard observational techniques*

It is commonly thought by many stellar astronomers that observing time on large telescopes is impossible to get because this precious time is monopolized by extragalactic astronomers. As discussed in papers by Ignasi Ribas and Alceste Bonanos, however, this is clearly not the case. It is still possible to get observing proposals on binary stars passed by the scrutinizing eyes of TACs guarding the time on large telescopes. Ribas focuses mainly on spectroscopic observations of faint, but astrophysically important, eclipsing binaries that are members of Local Group galaxies (mostly O/B binaries) and intrinsically faint low-mass red-dwarf and brown-dwarf eclipsing systems. Bonanos discusses observations of O/B eclipsing binaries in the Local Group and how these stars are providing independent, accurate distances to the Large and Small Magellanic Clouds, M31 and her recent work on determining an accurate distance to the Triangulum spiral (M33). There are a number of posters on these topics presented at the meeting.

Brian Mason discusses the results of (in some cases) over a hundred years of standard (or classical) astrometric observations of visual and multiple stars and presents some interesting results and statistics. As demonstrated by previous papers, many additional fainter members of binary and multiple star systems are being discovered in increasing numbers, thanks to the advances in instrumentation and technologies (mostly from interferometry, adaptive optics and high precision spectroscopy). Hartkopf reports on efforts to meet this problem of success in his paper about the nomenclature of multiple stars in the Washington Multiplicity Catalog (WMC; http://ad.usno.navy.mil/wds/wmc.html).

Also included here are several interesting oral/poster contributions. These papers focus on a wide range of topics that include a new study and mass determination of the Polaris multiple star system (Nancy Evans *et al.*), a new and very interesting eclipsing binary (DE CVn) consisting of a dM3 star and a cool white dwarf that must have evolved through the common-envelope stage of binary star evolution (Else van den Besselaar *et al.*), a new study of the eclipsing triple system U Oph (Luiz Paulo Vaz *et al.*) and a new study of the very young binary stars in the Orion Nebula Cluster (Rainer Köehler *et al.*). Additionally, there is a related poster paper on DE CVn by Ümit Goker & Günay Tas. Also here, Petr Hadrava discusses new trends in disentangling the spectra of multiple stars using methods that he pioneered.

3.3. *Improved methods of data analysis*

Powerful and improved methods for the analysis of binary and multiple stars, such as spectra disentangling, Doppler tomography, Doppler imaging and eclipse mapping, as well as other innovative modeling procedures are the primary themes of this section. Herman Hensberge & Krešimir Pavlovski present detailed discussions of the modern analysis techniques now being applied mainly to spectroscopic binaries. They discuss the advantages and disadvantages of the different methods to extract precise radial velocity information from spectroscopic binaries. Most of the emphasis is on spectra disentangling and the information gained. Mercedes Richards discusses Doppler Tomography as applied to mapping streams and disks in Algol systems, while Klaus Strassmeier discusses Doppler Imaging and constructing eclipse maps of spots and other stellar surface features on chromospherically active components of close binary systems — such as RS CVn stars.

In a thoughtful paper by the developer of the most widely used program for analyzing and modeling eclipsing binary light curve data, Bob Wilson focuses mainly on advances and innovations in light curve modeling developed over the last three years. He also looks to the future and discusses methods for the modeling of the tens of thousands of light curves of eclipsing binaries expected over the next several years. Wilson advocates publication of future lightcurves of eclipsing binaries not in delta-mags, but rather in physical units such as ergs/sec/cm^2. This is easier said than done, but would allow direct distance estimations within the light and radial velocity curve modeling.

In the same section, Theo ten Brummelaar discusses the practicalities of reducing multi-aperture interferometric data. Applications of improved methods of binary star analysis are presented by Geraldine Peters in her study of the bipolar jets, hot interacting regions and colliding winds in OB interacting binaries, and by Styliani Kafka on the evidence for chromospheric activity on the cool, low mass secondaries of cataclysmic variable binaries.

In an oral-only presentation, Dimitri Pourbaix discussed the use of large-scale surveys for discovering and studying new binary stars. A good example of this is found in the multi-band astrometry detection of binaries from the Sloan Digital Sky Survey (SDSS). These binaries are detected from the astrometric color-induced displacement technique (CID) in which the displacement of the photocenter between different bandpasses arises from a varying contribution of differently colored components to the total light. (see Pourbaix *et al.* 2004, *A&A*, 423, 755 for more details.)

Several interesting contributed papers appearing are the contributions of Diego Falceta-Gonćalves *et al.* on the massive binary/multiple star system η Carinae and the paper by Bob Stencel about another favorite binary — the 27-yr eclipsing (F2I + disk) binary, ϵ Aurigae. It is noteworthy that ϵ Aur is scheduled to start its next $\sim$2 year-long eclipse in 2009. Another famous (infamous?) binary to join η Car and ϵ Aur in this section is β Lyr. Pavel Chadima and company analyzed hundreds of spectra for this perplexing binary and, using the KOREL program, found strong evidence that supports mass accretion and the presence of a thick accretion disk with bi-polar jets, as well as a circumbinary envelope. It will be interesting in the future to image β Lyr with the next generation of multi-aperture interferometers. This session has a great number of interesting posters worth visiting.

As discussed in the papers by Wilson, Ribas and others in this section, it will not be long before high-precision radial velocity data will be combined with high precision ($\sim$1 mmag) photometric light curves of selected double-line eclipsing binaries, yielding the masses and radii of their component stars with uncertainties better than $\sim$0.2%. Now most stellar masses and radii are known typically to no better than $\sim$5%. Examples of the high precision light curves of eclipsing binaries that are now even possible using the star tracker of the WIRE satellite are given by Hans Bruntt *et al.* The Canadian MOST micro-satellite (http://www.astro.ubc.ca/MOST/) also currently has the capability of milli-mag photometry, but only a few eclipsing binaries have been observed so far. As discussed in the next section, light curves, with even higher precisions (better than 0.1 mmag) are expected very soon with the CoRot and Kepler space missions.

3.4. *Observing in the era of large-scale surveys*

Another important development discussed in a number of papers and posters at this meeting is the bonanza of new (mainly eclipsing) binaries discovered over the last decade from photometric survey and monitoring programs. Most of these binaries have been serendipitously discovered from micro-lensing photometry programs such as EROS, OGLE, and MACHO. So far several thousand new eclipsing binaries have been found, mostly in the rich star fields of the Galactic Bulge and the Magellanic Clouds. Also, nearly a thousand additional eclipsing binaries have been discovered by the DIRECT program (see Bonanos' paper) and by Ribas and collaborators in the M31 and M33 spiral galaxies. Some of these newly discovered extragalactic eclipsing binaries have been used to secure accurate distances to the Local Group galaxies (see also several posters in this section). A brief inventory of the eclipsing binaries found so far in Local Group galaxies is given in Figure 4.

But in the near future, thanks to planned ground-based panoramic photometric surveys such as Pan-STARRS (http://www.pan-starrs.ifa.hawaii.edu/) and the Large Synoptic Survey Telescope (LSST; http://www.lsst.org./) and several other programs, the numbers of eclipsing binaries are expected to swell to over 10^5 systems. How do we cope with such an abundance of riches? Several groups are trying to find the solution to

ECLIPSING BINARIES IN THE LOCAL GROUP GALAXIES

Galaxy	# of known EBs	Notes
The Milky Way	~4000 (Many in Bulge)	Traditional photometry & OGLE
LMC	5000+	EROS, OGLE,
SMC	2000+	MACHO & MOA
M31	400+	DIRECT Project (see Bonanos – this volume) (also Ribas – this volume)
M33	40+	>10^5 expected at m~24th mag (Bonanos – this volume)
Fornax	~15	Measured from 1/8 CCD chips of mosaic; (extrapolated value from mosaic: ~300) (Clementini – priv. comm.)
Leo I	~10+	Down to 24th mag. Only short period (P < 2 days) discovered so far.
NGC 6822	6+	Only 25% of galaxy covered (Clementini & collaborators)
Carina (dSp)	5+	Carina Project (Monelli & collaborators)

Figure 4. Eclipsing binaries in local group galaxies.

this by developing automated methods to analyzed thousands of light curves at a time. Andrej Prša & Tomaž Zwitter discuss a very promising pipeline reduction method of analyzing large numbers of light curves, while Tsevi Mazeh and colleagues discuss their automatic approach and initial results when applied to the study of early-type binaries in the Magellanic Clouds. Their preliminary results indicate that, at least among B-type stars, binaries may be far less frequent in the LMC (<10%) as compared to similar stars in the solar neighborhood. There are posters and related papers at the conference by Jonathan Devor & David Charbonneau and by Aliz Derekas *et al.*

In the same session there are fascinating papers by David Koch (and the Kepler team) about the ultra-high precision photometry expected from the Kepler mission and a related paper by Carla Maceroni and Ribas on the impact of very high precision photometry on close binary stars expected shortly from the CoRot space experiment (http://www.esa.int/science/corot/). These missions are expected to discover hundreds of additional planetary transit systems. Currently, about a dozen eclipsing planet-star systems have been uncovered. Also there are short papers presented by Panos Niarchos *et al.* on the Gaia mission's impact on binary star studies. Other interesting papers include the results of multiplicity studies of very young Herbig Ae/Be stars (Sandrine Thomas *et al.*) and the results of a multiplicity study in nearby solar-type stars (Deepak Raghavan *et al.*), as well as a contributed paper on the feasibility of detecting planets in wide binaries from ground-based Adaptive Optics (AO) systems (Ralph Neuhäuser *et al.*).

4. Binary Stars as Critical Tools

In this section many examples are provided of using binary and multiple stars as critical tools for studying a wide variety of important astrophysical problems. There are a number of interesting examples discussed in the papers included here.

4.1. *The need to improve basic calibrations using binary stars and increasing the possibilities of classical methods*

Jason Aufdenberg and colleagues report on the recent images of the rotationally distorted and gravity darkened bright stars made with the longest baselines of the CHARA Array and the Fiber-Linked Unit for Optical Recombination (FLUOR). They discuss their previous result on imaging the pole-on rapid rotator Vega, in which the star is clearly resolved and imaged. Here they also discuss the most recent results on imaging the components of the 4-day, non-eclipsing, double-line binary Spica. For Spica, in addition to resolving and imaging the individual components of the close binary, they discuss improvements in determining its apsidal motion by combining their observations with previous interferometry measures. In the same session Alvaro Giménez discusses the apsidal motion studies of eccentric eclipsing systems as a powerful means of studying the interiors of the stellar components. Except for a few outstanding cases, the internal structure determined from apsidal motion studies agrees very well with stellar evolution and stellar structure theory. In another important talk, David Valls-Gabaud discusses the results of using different methods to secure a reliable distance to the Pleiades Cluster. In summary, binary stars are yielding the most reliable Pleiades distance of about 134±3 pc. (It should be recalled that several years ago Hipparcos astrometry returned a distance of only 118±3 pc for the Pleiades.) Thanks to binary stars, the distance to such an important cluster is now reliably known.

Todd Henry focuses on a decade-long endeavor to calibrate the Mass-Luminosity Relation (MLR) for the numerous low-mass and low-luminosity red dwarfs and brown dwarfs that make up over 70% of the stars in our Galaxy. To get an accurate measure of these faint but numerous stars and substars, they have to be members of binary systems. He focuses on efforts to calibrate the MLR for dM stars as well as T and L dwarfs. In this talk (the abstract of which is included in this volume) he gives these dim, low-mass stars the respect they deserve, since these stars may be the dominant contributor to baryonic mass in the Universe. Ben Lane discussed recent developments in the study of extrasolar planet systems, including work in nulling interferometry and the Palomar High-precision Astrometric Search for Exoplanet Systems (PHASES); unfortunately his talk is not included in these proceedings.

In another interesting and thought-provoking paper, Terry Oswalt discusses the use of the loosely bound "fragile" binary stars to investigate several astrophysical problems. One interesting aspect featured in this paper is using the old fragile binaries to probe the dynamics and early history of our Galaxy and also to use white dwarf + main sequence star binaries to better calibrate the cooling times of white dwarfs or use the cooling ages of white dwarfs to study the properties of its companion.

Other interesting contributed papers in this session are papers dealing with the diverse topics that include the direct measures of tidal dissipation in highly eccentric binaries (Andrei Tokovinin), a study of the famous cataclysmic binary U Gem (Juan Echevarría-Roman), the study of "runaway" stars (M. Virginia McSwain *et al.*) and an interesting paper on "mining" catalogs of common proper motion binaries (Julio Chanamé). Also not to be missed are over a dozen related poster papers.

4.2. *Evolutionary models for binary and multiple stars*

Cathy Clarke opens this session with a thorough review of the major advances in the theories of binary star formation that have been made over the last several years. She concludes that results of current binary star formation theories and simulations, in which star formation occur in high interactive modes, appear to fit the observed binary and multiple star characteristics and frequencies fairly well. However, there are still problems with an agreement between theory and observations when it comes to low mass stars and in the tendency for models not to produce enough binaries with low mass ratios. In a related paper Peter Eggleton and colleagues discuss the incidence of multiple stars among bright stars found in bright star catalogs. They have identified over 4500 such systems and subdivided them in terms of frequencies of multiplicity and distributions of periods and sub-periods. In this study they discuss the results of their Monte-Carlo simulations to generate systems with roughly the observed multiplicities and orbital properties of the stars in this sample.

In another interesting paper Hans Zinnecker presents the study of young binary stars as tests of pre-main sequence evolution tracks. Some fascinating examples of various type of binaries with PMS components are included in the paper. He finds that with a sufficiently large sample of different masses and ages of resolved PMS SB2 systems, that most of the parameter space of pre-main sequence evolution tracks can be tested. In an interesting theoretical study Dmitry Bisikalo & Takuya Matsuda discuss recent progress in the theory and modeling of mass exchange in close binaries. In this paper they present and discuss the 3-D gas dynamic models used in the gas flow and exchange simulations and reach some important conclusions. Also included in this session are contributed poster/oral papers on the study of eclipsing binaries in the Large Magellanic Cloud and the astrophysical quantities that can be determined from them (Derekas *et al.*) and a study of the initial-final, mass relationship of white dwarfs in common proper motion pairs and in open clusters (Silvia Catalán *et al.*). Also included is a short contribution about the detectability of planets in wide binaries using ground-based relative astrometry with Adaptive Optics techniques (Neuhäuser *et al.*). They conclude that it is feasible to detect the small astrometric wobble due to a planet using their method. They have carried tests of the technique using the VLT and obtained astrometric precisions as small as 50 μas!

5. Binary Stars as Critical Tests

It is well known that binary and multiple stars provide critical tests of the structural and evolutionary theories of stars but, as shown by the papers here, binary stars are also important as tests of galactic evolution and dynamics, as well as providing testbeds for stellar pulsation theory & asteroseismology, along with studying magnetic activity in close binaries with cool star components.

5.1. *Binary and multiple stars as probes of our Galaxy*

This section contains a number of very interesting and diverse papers and posters covering a wide range of phenomena that include binary stars as probes of our Galaxy, asteroseismology, stellar activity and magnetism. Included are papers on general themes of "Binary Stars of our Galaxy" in which Dany Vanbeveren and Erwin De Donder discuss how mass loss from the final stages of some close binary stars plays a major role in the chemical enrichment and evolution of our Galaxy (and presumably in other galaxies as well). On the related subject of binary stars as probes of the Galaxy are the papers by

Christine Allen, Arcadio Poveda and Helmut Abt that focus primarily on using the orbital properties of wide binaries to study binary star evolution, as well as the dynamical history and evolution of our Galaxy. When statistically analyzed, the frequency of these wide binaries with differing gravitational bindings and ages yield important clues about the early dynamical history of our Galaxy.

5.2. *Asteroseismology and stellar activity in binary stars*

In a comprehensive review paper by Conny Aerts, an overview of the current status of asteroseismology is provided. She points out the advantages of having a pulsating star as a member of a close binary — in particular as a member of an eclipsing binary. The important new things that can be learned from the study of pulsating components in close binaries are discussed. For example, if the pulsating star is a member of a double-line eclipsing binary, then the physical properties of the pulsating star (radii and masses) are determinable from the traditional analyses of the light and radial velocity curves. Securing reliable physical properties of pulsating stars (such as masses, radii, and luminosities) are critical for the calibration and interpretation of asteroseismic observations expected very shortly from CoRot. Several contributed poster papers also discuss the related issues of stellar pulsation and asteroseismology. For example, Andrzej Pigulski & Grzegorz Pojmański discuss the discovery of four β Cephei variables in eclipsing binary systems. As pointed out by the authors these binaries are very attractive objects for much-needed follow-up spectroscopic and multicolor standard photometry which will yield accurate physical properties of the stars. Once the masses and radii of these stars are reliably known, then their interiors can be probed by means of asteroseismology methods. The reader should check the poster papers on related topics.

Another important topic covered here is the magnetic activity frequently found in close binary systems with cool stars (i.e., stars with outer convective zones). Katalin Oláh's paper discusses the influence of binarity on stellar activity — such as star spots and chromospheric emissions — and finds evidence that some stars in close binaries show stronger manifestations of magnetic activity when compared to single stars with similar physical properties and rotations. On the related subject of stellar magnetic activity, Heidi Korhonen & Silvia Järvinen discuss magnetically-active stellar longitudes and the rapid changes (flip/flops) that are commonly seen. This behavior may be related to the reversal of the Sun's magnetic field over its $\sim$22 year activity cycle. There are a number of interesting poster papers on the topic of stellar activity and stellar magnetism in close binaries that should not be overlooked in the poster papers.

Acknowledgements

This work was sponsored in part by grants from the U.S. National Science Foundation and NASA, which we gratefully acknowledge. We are also grateful to the U.S. Office of Naval Research Global (grant N00014-06-1-1054) for the support of travel costs for some 25 symposium participants, including WIH. EFG would like to acknowledge travel support from a National Science Foundation grant to the American Astronomical Society, and we would like to thank Scott G. Engle of Villanova University for his help in the preparation of this paper.

Binary Stars as Critical Tools & Tests
in Contemporary Astrophysics
Proceedings IAU Symposium No. 240, 2006
W.I. Hartkopf, E.F. Guinan & P. Harmanec, eds.

© 2007 International Astronomical Union
doi:10.1017/S1743921307003742

Professor Mirek J. Plavec

Petr Harmanec[1]†, Jiří Grygar[2], Alan H. Batten[3],
Geraldine J. Peters[4], Albert P. Linnell[5],
Ivan Hubeny[6] and Edward F. Guinan[7]

[1]Astronomical Institute of the Charles University, Faculty of Mathematics and Physics,
V Holešovičkách 2, CZ-180 00 Praha 8, Czech Republic
email: hec@sirrah.troja.mff.cuni.cz
and
Astronomical Institute, Academy of Sciences of the Czech Republic, CZ-251 65 Ondřejov,
Czech Republic
email: hec@sunstel.asu.cas.cz

[2]Institute of Physics, Academy of Sciences of the Czech Republic, Na Slovance 2,
CZ-18221 Prague 8, Czech Republic

[3]Herzberg Institute of Astrophysics, NRC of Canada/DAO, Victoria, BC V9E 2E7, Canada

[4]Space Sciences Center and Department of Physics & Astronomy, University of Southern
California, Los Angeles, CA 90089-1341, USA

[5]Department of Physics and Astronomy, Michigan State University, E. Lansing, MI, 48824
and
Department of Astronomy, University of Washington, Seattle, WA 98195, USA

[6]Department of Astronomy and Steward Observatory, University of Arizona, Tucson, AZ
85721 USA

[7]Department of Astronomy, Villanova University, Villanova, PA 19085, USA

Abstract. In keeping with its co-sponsorship by members of both the "close" and "wide"
binary star communities, IAU Symposium 240 has been jointly dedicated to the honor of Czech
astronomer Mirek J. Plavec and the memory of U.S. astronomer Charles E. Worley.

1. A few bibliographical facts

Miroslav Plavec was born on October 7, 1925 in Sedlčany, Czech Republic. During
World War II, in 1942, his father died in a Nazi concentration camp. The next year, his
mother moved with him and his brother Jaroslav to Ondřejov and Miroslav had the first
chance to visit astronomical observatory and even to participate in some research there.
From 1945 to 1949 he studied astronomy at the Charles University and graduated there
in 1949. In 1950, he married his whole-life partner Zdenka.

In 1955, he obtained PhD in Astronomy and two years later he moved again to
Ondřejov, this time with his own family. In 1958, the Stellar Department was founded in
the Astronomical Institute of the Czechoslovak Academy of Sciences under the leader-
ship of Dr. Luboš Perek. At that time, Miroslav switched from meteor studies to stellar
astrophysics and became the deputy of Perek.

He visited Manchester at the invitation of Prof. Zdeněk Kopal and later the Dominion
Astrophysical Observatory (DAO) to collaborate with Dr. Alan H. Batten in the late
fifties and sixties. Since 1959, when the Czechoslovak government approved the purchase
of the 2-m Zeiss Jena reflector to be installed in Ondřejov, he, Luboš Perek and other

† Present address: Astronomical Institute of the Charles University, V Holešovičkách 2,
CZ-180 00 Praha 8, Czech Republic.

Figure 1. Foto Martin Šolc.

colleagues in the department were preparing the site for the construction of the dome and installation of the telescope.

In 1968 he obtained DSc. degree in astronomy at the Charles University of Prague and started to teach students there. In the years 1968 to 1970, he served as the Head of the Stellar Department after Perek was appointed the Director of the whole Institute.

In 1969, after the Soviet-led invasion to Czechoslovakia, he and his family moved to Canada and then to the USA; after shorter stays at the DAO, University of Pennsylvania and University of Ohio, he got a permanent position at the UCLA and served also as the Head of the Department of Astronomy there.

He has been living in Pacific Palisades, a suburb of Los Angeles since then. He also simplified his Christian name Miroslav, which was hard to pronounce correctly in English, to shorter and easier Mirek (which is a shorter version of Miroslav, very often used in spoken Czech).

In 1987, he for the first time after 1969 visited Czechoslovakia officially and had an invited review during the 10th European Regional Meeting of the IAU.

In October 2000, a very successful conference *Interacting astronomers: A symposium on Mirek Plavec's favorite stars* was held at the UCLA to celebrate Mirek's 75th birthday.

2. Reminiscences of several of Mirek's students, colleagues and friends

2.1. *Jiří Grygar*

In the late fifties Mirek Plavec was a household word in Czechoslovakia because he was an enthusiastic popularizer of astronomy in the public radio, eloquent lecturer and author of many popular articles and books. His scientific profile was even more attractive. From 1949 to 1959 he finished alone 14 papers and two monographies on meteors. However, in 1959 (the year of my graduation from Charles U.) he switched his interest to stellar astronomy and published his first two papers on close binaries (W Del and RS CVn). My own evolution followed a similar track: firstly meteors, then popularization but my dream was to study stars. Thus it was almost inevitable that I asked Mirek to join his tiny group of three (observer Z. Pěkný, computing assistant M. Smetanová) in Ondřejov. Soon he became the clandestine supervisor of my PhD Thesis; its theme (limb darkening in eclipsing binaries) was suggested to me by Z. Kopal during his private visit to Prague in 1960. I believe that Kopal bored a hole in Iron Curtain for Mirek and soon after Mirek published many papers on orbital changes of the eclipsing variables and their apsidal motion. However, Mirek was already deeply interested in the solution of the Algol paradox and together with his next students (S. Kříž, J. Horn, P. Kratochvíl/Harmanec) he found — independently of Polish and German groups — the solution: huge mass overflow over the Roche lobe. I was much surprised that this paradigm was never accepted by Kopal who subsequently detached himself from further so-productive applications of his own original concept of close binaries. The early model calculations were done on the first available digital computer ZUSE located in the confiscated Prague monastery; thus the fine-tuning of the codes were done in the silent seclusion of the monks' cells while the machine room was filled with a terrible high-pitched noise of the swiftly-rotating internal drum memory. The breakthrough was announced simultaneously by all

Figure 2. Shown at left are Mirek and his first Škoda car (1961). At right are members of the Stellar Department in 1963; Mirek is third from right.

three groups in Brussells in 1966. Soon after we experienced the IAU G.A. in Prague (1967), Prague Spring and Soviet invasion (1968). In 1969 I followed Mirek and his family to the D.A.O. in Victoria, B.C. but then our tracks divided for two decades. Mirek accepted positions at Ohio State and then UCLA while I returned back behind the Iron Curtain to Ondřejov. Needless to say we stayed in touch, although our contacts were difficult. All correspondence was censored and for some years we were not allowed to quote Mirek's papers in our publications! Fortunately, the fairy tales occur in real world, too. After the Velvet Revolution Mirek was elected the Honorary Fellow of the Czechoslovak Astronomical Society in 1992, Honorary Fellow of the Learned Society of the Czech Republic in 1995 and he is the second recipient of the prestigious Nušl Prize of the Czech Astronomical Society (2000). Minor planet #6076 bears his name for many millions of years to come.

2.2. *Alan Batten*

A few years ago, many of us here gathered in Los Angeles to celebrate the seventy-fifth birthday of Mirek Plavec. As befitted someone with Mirek's wonderful sense of humour, that was an occasion for light-hearted banter on all sides. Today, as I join in honouring someone who has been a good friend for half a century, I would like to strike a more serious note.

Mirek has been a leader in our field of the study of interacting binary stars. When we first met, each of us was heavily influenced by the late Zdeněk Kopal. The origin of the Algol systems was still a mystery, even though Crawford had published his suggestion of what we now call mass-transfer. I suspect that the younger astronomers here find it hard to believe how speculative that idea seemed in those days; Kopal argued against it consistently until the end of his life. At first I, certainly, and Mirek also, I believe, shared that skepticism. Ten years later, however, all was to change at a meeting held in Brussels in the summer of 1966 — a meeting that Mirek, Charles Worley and I all attended. I have written about that meeting in other places and will not repeat myself here. Suffice it to say that Mirek and his many students, first in what was then Czechoslovakia and later in Los Angeles, were exceedingly active in producing computations of mass-transfer and models of systems, both of which led the rest of us to a deeper understanding of the objects of our study. We owe much to those pioneers in our field even if not everything they did has stood the test of time and further observation.

Mirek is not just a scientist, however, he is also a remarkable human being who has overcome many difficulties and disappointments that would have embittered lesser mortals. Unlike many of you who will be at this symposium, I was never formally Mirek's pupil; but I look on him as a mentor, not only in science, but also in life. My long friendship with him is a privilege for which I am thankful, and one of the most significant in my life. He and I each know that, in the nature of things, we have now lived the greater part of our allotted spans. For whatever is left and for whatever lies beyond, I wish Mirek and Zdenka well.

2.3. *Gerrie Peters*

I met Mirek Plavec in the early autumn of 1970, shortly after he arrived at UCLA. I was a graduate student studying chemical abundances in B stars with Lawrence Aller and had recently discovered some unusual spectral activity in the Be star HR 2142. After being introduced to me, Mirek commented "I am interested in your star HR 2142. I think it is an interacting binary." This statement, along with a scientific collaboration that followed, opened up an entirely new line of research for me that has persisted to the present. Thank you very much Mirek! I continued to discuss new spectroscopic observations of

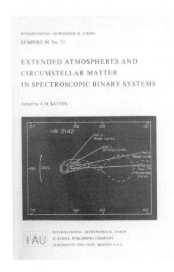

Figure 3. (left) Cover of the proceedings from IAUS 51, showing our model for HR 2142. (right) Mirek, my husband Phil, and me on the grounds of the Island Hall Hotel in Parksville, where IAUS 51 was held.

Figure 4. At the home of Mirek and Zdenka Plavec in 1976. Shown are Tony Keyes, myself, Phil, Mirek, and Mike Wright (Ron Polidan, photographer).

HR 2142 with Mirek and Ron Polidan, and these led to my first model for the system that Mirek encouraged me to present at IAU Symposium No. 51 held in Parksville, BC in September of 1972. This was the first IAU-sponsored meeting that I attended. In fact, I met many long-term colleagues there, some of whom are attending this IAU symposium. The Parksville experience was certainly a high point in my life. When the proceedings of IAUS 51 came out, I was quite surprised to find that our model had been used as a cover graphic (see Figure 3). Collaboration on interacting binary stars that display emission from circumstellar material continued, and in 1975 we assembled at another IAU symposium, IAUS 70 devoted to Be and shell stars. Along with Petr Harmanec, Mirek, Ron, and I put forth the bold hypothesis that all Be stars are interacting binaries. This created quite a stir among an audience hard-wired to the Struve model that explained the Be phenomenon as a result of critical rotation. While the current consensus does not support our 1975 idea, and the causes for the mass loss in Be stars are still unknown, the disks and activity in some Be stars are most certainly a result of close binary interaction. We enjoyed many great times with Mirek in the mid-early 1970s, and especially liked the social events at meetings. I recall one instance when Mirek stepped up to a booth to pay for our admission to the Desert Museum in Tucson, he said "one adult and four children (Phil, Ron, Tony, and me). But you probably don't believe they are children, so five adults!" Before we left UCLA for Cambridge, MA in 1976, Mirek and Zdenka hosted a fine dinner for us at their home (see Figure 4). It has been a long time, but Mirek's scientific reasoning and commitment to excellence and perfection have left an indelible imprint on my own scientific character.

2.4. *Al Linnell*

I had only a few mail exchanges with Mirek before he and Zdenka came to the US in May 1969. He stopped in East Lansing for a visit, on his way to Ohio State; I had been at Michigan State since 1966. Mirek, Zdenka, my wife and I had a great get-acquainted boat trip on the Red Cedar river, and I quickly became aware of his wonderful sense of humor and store of stories. It was while Mirek was at Ohio State that he published his well-known paper on binary star evolution (1970, PASP, 82, 957).

In 1979, Mirek introduced his proposed binary star category of W Serpentis stars (IAU Symposium 88, p. 251). The 1988 Algols conference in Sidney, B.C., Canada (1989, Space Sci. Rev., 50, 1-382) devoted seven of the contributed papers to W Serpentis stars. The category included β Lyrae, and Mirek subsequently urged me to work on that exotic object.

In 1987, Mirek put me in touch with Ivan Hubeny, who had recently come to the US. That connection began the most important collaboration of my career. Ivan and I collaborated on a paper (1994, ApJ, 434, 738) describing spectrum synthesis procedures for ordinary binary systems. Two years later (1996, ApJ, 471, 958) we extended the model to include binary systems with accretion disks. Dobias and Plavec (1985, AJ, 90, 773) had determined important parameters for β Lyrae, and subsequently Hubeny and Plavec (1991, AJ, 102, 1156) proposed a disk model to represent β Lyrae. This system seemed like an interesting subject for our new spectrum synthesis program, and with the possible feature of a geometrically thick accretion disk hiding the mass gainer. Petr Harmanec has had a longstanding interest in β Lyrae, and he joined our collaboration. Our study led to a new model (1998, ApJ, 509, 379). Three further papers on β Lyrae have appeared (2000, MNRAS, 319, 255; 2002, MNRAS, 332, 21; 2002, MNRAS, 334, 963).

Most of my interaction with Mirek has been on a professional level, and has been conducted by email. While visiting my older daughter in Los Angeles I have been a guest in Mirek's home. The 2001 Los Angeles symposium honoring Mirek (Proc. Astr. Inst. Acad. Sci. Czech. Rep., 89) was a wonderful event celebrating many of Mirek's interests in stellar astronomy, and I was honored to have been included among the participants. I want to take this opportunity to salute my friend Mirek and offer congratulations on an illustrious career in astronomy.

2.5. *Ivan Hubeny*

I have been interacting with Mirek for many years, 39 to be exact. Indeed, our first interaction occurred in 1967, but this time it was only one-sided. I was my first year at the university and was spending a few weeks at the Ondrejov Observatory as a summer student. I of course noticed Mirek because he was already quite famous; from my childhood I have read many of his semi-popular books on stars, history of astronomy, and even comets and meteors. Mirek was busy, and I was just one student out of several, so he did not notice me at all.

Then in 1969 he didn't return from his trip to Canada and USA, and thus ceased to officially exist in the normalization-period Czechoslovakia, and consequently no contact with him was officially allowed. Nevertheless, my colleagues from the Stellar Department, where I have in between begun working, were occasionally meeting him on conferences abroad, and were bringing news about how Mirek is successful in his research as well as teaching at the University of California, Los Angeles (UCLA), where he became one of the most popular professors on campus. (Much later, one colleague who was then a graduate student at UCLA, told me "When Mirek came to the Department, it was like a breath of fresh air".)

One day in 1986 I got an air mail letter from UCLA sent by an "Undergraduate Adviser". I was quite puzzled, feeling already a bit too old for an undergraduate student, but when I opened it I saw that it was sent by Mirek, who has concealed his name on the envelope in order not to cause me any problems at the Institute, as well as to prevent a possible confiscation of the letter by our watchful authorities. It was a very nice and encouraging letter: he has congratulated me for our recently published paper with Svatopluk Kříž on theoretical models of accretion disks, and expressed a keen interest to somehow collaborate with me on this topic. It seemed of course pretty difficult to arrange a collaboration with Mirek, but by an interesting twist of fate a few months after I got Mirek's letter I have finally managed to escape from Czechoslovakia, and after spending almost a year in Austria waiting for a proper authorization, moved to the United States. We have been corresponding with Mirek, this time quite freely, already from Austria, and Mirek grew very anxious to finally start our actual collaboration.

The first three years I worked in Boulder, CO, which is just about one and half hour flight from LA, so I have been visiting Mirek frequently there. This was a beautiful time. I have implemented all my computer programs for modeling stellar atmospheres (in particular, my general-use code TLUSTY), accretion disks, computing stellar spectra, etc., on Mirek's computers, and we started to work at the earnest. I have been staying with Mirek and his wife Zdenka in their house in Pacific Palisades (a beautiful suburb of LA), and during our regular walks with their legendary German sheppard named Shepinka, Mirek taught me a lot about the life in the US, about the university system, but also telling many stories and historical anecdotes about Czech and Slovak astronomers whom I only knew by name. At first I was commuting with Mirek and Zdenka (who also worked at UCLA) to the Department, but later, when I started to feel more confident with driving in Southern California, I got assigned one of their cars to be able to drive there myself on weekends or during the days they didn't go to the Department. In other words, I was quite pampered by them. There is a metaphor in Czech to express having a truly great time which goes like "To enjoy oneself like a pig in a field of rye"; so I have proposed an upgrade of it to express the extreme version of it by saying "To enjoy oneself like Ivan in Palisades".

But pampering aside, I really enjoyed working with Mirek very much. We were a sort of complementary. Mirek loves wild and exotic objects in the Universe, like interacting binaries (the more complex the better), symbiotic stars, Be stars, etc., while he loves and enjoys order and gentleness in everyday life, and in his interaction with Nature. I, on the other hand, love adventure, wilderness, mountain climbing in the Nature and life, but preferred to model well behaved objects like stellar photospheres (white dwarfs being the most friendly variety here), and reasonable accretion disks (after some time, however, I came to a realization that no such things exist). So, I had to tone down a little Mirek's excitement about interacting binaries with lots and lots of emission lines, which I could hardly model at all.

After several trials we settled on a good compromise, the famous β Lyrae. The object is enigmatic enough that it caught Mirek's attention already long ago, and he had already published a number of papers on it. However, after a concentrated effort during the history (β Lyrae is said to have more published papers on than any other astronomical object), but its nature was still quite uncertain - one could not even be sure that there is an accretion disk or else something else that hides the mass-gaining star of the system. But that "something" is at least optically thick, and thus could be modeled by TLUSTY without feeling myself completely foolish (that is, just acceptably foolish).

That time I have already moved from Boulder to NASA in Greenbelt, MD, but I still kept coming to LA about twice a year. Nevertheless, a bulk of our collaboration was done by e-mail. We have explored a large number of possible models, and finally came out with a model that invoked an optically thick (but geometrically still not so terribly thick) accretion disk model whose rim indeed largely obscures the primary star, but a tiny crescent of it (an early B star) would still provide the observed UV flux. We were reasonably happy with the model, but, like many good models this one also later turned out to be wrong. (Well, not completely, but the idea of the crescent, which I admit was originally mine, didn't work out when later confronted with detailed UV spectral energy distribution).

Over the years we remained in close contact, and continued trying to understand Mirek's wild objects. It was always a pleasure. I remember once I came with a slogan, or my motto, in Czech but still translatable to English, which says "More than bottle of whiskey, I love accretion diskee." (Well, not the best one, I admit, but perhaps it conveys a good spirit of our interaction).

Later, my main research interests switched somewhat from the stars and (non-existent) well-behaved accretion disks. But I have suddenly realized that there is one more thing I learned from Mirek, and will thus be able to carry his torch further – I found myself also working on wild objects: extrasolar giant planets (with their cloud formation, convective updrafts of clouds, rainouts, day-night side interactions, etc.), and even supernova explosions.

I am thus happy to carry Mirek's spirit, and love of wild and exotic objects, for the years to come.

But, at the moment, Mirek, I salute you!

2.6. *Ed Guinan*

I first became acquainted with the work of Mirek Plavec while I was a graduate student at the University of Pennsylvania during the late 1960s. For one practical thing, I extensively used the tables published of the relative sizes of Roche Equipotentials from a paper published in 1964 by Mirek and Petr Kratochvíl (= Petr Harmanec) in the Bull. Astron. Inst. Czechosl. (BAC). A few years later I first met Mirek in Philadelphia during his brief visit to the University of Pennsylvania in 1970. As a recent PhD, I was strongly influenced by Mirek's contagious enthusiasm for the study of strongly interacting binaries and was sorry to see him leave Pennsylvania. I was very impressed with his work and the joy that he got from trying to figure out mass transfer and accretion processes in semi-detached binaries. He took-on in his research some of the most complicated and difficult close binaries known. Because of this interaction with him, I soon found myself carrying out photometry of some of Mirek's favorite stars that included β Lyr, W Ser, U Cep and several other nasty systems.

I personally consider some of Mirek's most important contributions to the understanding interacting close binaries to be his extensive ultraviolet spectrophotometry of very active, mass transferring and mass losing close binaries carried out with NASA's International Ultraviolet Explorer (IUE) satellite. This research was made during the 1980s and early 1990s while he was a professor at UCLA. During this time, he published an impressive number of important papers about this work on interacting binaries made with the IUE satellite. He also received observing time at Lick Observatory to obtain high dispersion optical spectra of challenging binary systems. Also, Mirek gave invited talks and colloquia on this research with the IUE satellite. I was lucky to hear some of these talks. His talks were always well organized and presented with great vigor and enthusiam. My only refereed publication with Mirek was during the mid-1980s in a paper about the X-ray properties of several interacting binaries made using the EINSTEIN X-ray satellite. This paper concluded that most of the X-ray emissions from these long period Algol-type systems originated from dynamo-driven magnetic coronal activity from the cooler, secondary components.

Several years ago, on the occasion of Mirek's 75th birthday at UCLA, I had the honor of presenting a paper on the future (my prejudiced view of course) of research on close binary stars and included in the talk several of Mirek's Favorite Stars. And now after all of these years of knowing Mirek, I point out that Mirek is (and will always remain) one of my favorite "superstars" in the study of binary stars. I am happy that this conference, held in Mirek's homeland, is dedicated to him. All my best to you Mirek!

2.7. *Petr Harmanec*

I already dreamed of becoming an astronomer when I was six or seven years old. It is therefore natural that I knew Mirek's name quite soon since he published several very fine popular books on astronomy and also wrote an astronomical comentary to the (at that time new) Czech edition of Jules Verne's novel about the travel on a comet. But I first met him personally at the spring of 1959 on the occasion of an educational knowledge competition, run by the Czechoslovak radio for secondary-school students. It was *The 23rd Radio University: To the Near and Distant Universe*. In that year, the Czechoslovak government approved the purchase of a 2-m Zeiss reflector as a national facility for Czech and Slovak astronomers. Mirek was at that time a member of the Stellar Department of the Astronomical Institute of the Czechoslovak Academy of Sciences in Ondřejov and the Head of the department, Dr. Luboš Perek, made him resposible for the preparation of the site for the telescope and also for the formation of a group of young people to operate and use the telescope for the research. It was then natural that Mirek approached those of us who passed the initial tests via letters and were invited to a radio studio and told us that if some of us had a serious interest in astronomy, he or she should contact him and ask for advice. I immediately wrote a letter to Mirek and, starting in the summer of 1959, I used to come to the Ondřejov Observatory every year as a summer student. I remember that in 1959, I was accompanying Mirek every clear evening, tranporting a tripod with a small telescope. At different locations around the Ondřejov Observatory, we installed the telescope, defocused it and Mirek was estimating scintillation to find a place where the seeing conditions would be best. Well, after all that, the telescope was built in a location which did not have the lowest scintillation but a solid rock to built the ground of the dome on it....

Already at that time, Mirek was very interested in semi-detached binaries and I also had the chance to participate in his early photoelectric observations of the binary light curves with a small 0.2-m refractor equipped with a photoelectric photometer and a galvanometer! Only a year later, a 0.65-m reflector, built in a collaboration with the

Astronomical Institute of the Charles University and equipped with a strip-chart recorder, was put into operation.

With Mirek's advice, I began my studies at the faculty of Mathematics and Physics of the Charles University and Mirek later became the supervisor of my diploma thesis and — when I joined the Observatory in 1964 — also of my PhD thesis. I even had the privilege to share the office with him all the time since the spring of 1966, when the stellar department moved to a new building near the construction site of the dome of the telescope.

Mirek was very good in creating a team of collaboratorators and focusing the whole group on one principal topic. We all learned a lot from him. Before the installation of the telescope was finished (in summer of 1967), Mirek became aware of the possibility to model mass transfer in binaries via modified evolutionary models. In retrospect, I found admirable how quickly Mirek realized the great potential of this approach. Consider that Mirek's interest in interacting binaries was partly inspired by Prof. Zdeněk Kopal, who was a mathematician by his education and who had strongly opposed the idea of a large-scale mass transfer in binaries. Mirek focused the whole group on the effort to develop our own computer program for modelling stellar evolution and mass transfer in binaries. It was an adventure at that time! While both competing groups, German one around Prof. R. Kippenhahn, and the Polish one around Dr. B. Paczynski, had already IBM 360 or 370 series computers at their disposal, we were using a Russian Minsk 22 which had just 4096, not Mb, not even kB but plain 4096 memory locations where both the program and data had to be stored! The computational time per one model was less than a minute for the IBM, and something like 20 minutes for us. However, the computer was a property of the Astronomical Institute and people from the Stellar Department (Mirek, S. Kříž, J. Horn and I) were the largest consumers of its computing capacity at those years. The important thing also was that Mirek had a very deep insight into the problem of semi-detached binaries and this way we were able to compete.

After the Soviet-led invasion to Czechoslovakia in August 1968, Mirek felt uncertain about the future fate of the country and took the opportunity of an invitation from his friend Dr. A.H. Batten to leave for a collaborative visit to the Dominion Astrophysical Observatory in Canada and then to Dr. R. Koch at the University of Pennsylvania. Sometimes at the end of 1969, Mirek and his family made the painful decision to remain in the USA. Already in 1968, when Dr. L. Perek became the Director of the Institute, Mirek was appointed as the Head of the Stellar Department. Before Mirek announced his decision early in 1970, my colleagues and I had very lively letter exchanges with him. Mirek was a vibrant personality of the Institute and I felt he should not give up to less competent and less generous people.

Mirek was then invited by Prof. D. Popper to come to the UCLA and soon became a very popular professor there. He always loved teaching and maybe thanks to this, he certainly also played a major role in making the theory of the mass transfer well known also in North America. He continued binary studies there and succesfully applied with several colleagues for observing time to study binaries with the OAO3 Copernicus and later with the IUE satellite. With R. Koch, he became known for the identification of a group of strongly interacting binaries with numerous emission lines in the far-UV spectra which they called W Ser stars. He later published a number of important detailed studies of these objects.

For some time, Mirek and I still tried to collaborate on joint projects but it became difficult since we could only communicate via personal letters which (due to censorship) travelled typically three weeks one way. We then only maintained regular personal corre-spondence and exchanged reprints of our papers. Later, we had several chances to meet on various meetings and also in Praha when Mirek received US citizenship and could come privately to visit his mother. When in 1972 P. Koubský, J. Krpata and I discovered that the Be star 88 Her is a 87-d binary, Mirek very kindly provided us with his Lick spectra which were instrumental to eliminate 1-d aliases of the orbital period. Ironically, we were not allowed to thank him in the Acknowledgements for this. So, I am doing that in written form at least now, after 34 years...

Many happy returns to Praha, Mirek!

Figure 5. Foto Martin Šolc.

Binary Stars as Critical Tools & Tests
in Contemporary Astrophysics
Proceedings IAU Symposium No. 240, 2006
W.I. Hartkopf, E.F. Guinan & P. Harmanec, eds.

Charles Edmund Worley (1935–1997)

Brian D. Mason[1], William I. Hartkopf[1], Thomas E. Corbin[2], and Geoffrey G. Douglass[3]

[1]Astrometry Department, U.S. Naval Observatory, 3450 Massachusetts Avenue, NW, Washington, DC 20392-5420, USA
email: (bdm,wih)@usno.navy.mil

[2]Retired.

[3]Deceased.

Abstract. In keeping with its co-sponsorship by members of both the "close" and "wide" binary star communities, IAU Symposium 240 has been jointly dedicated to the honor of Czech astronomer Mirek J. Plavec and the memory of U.S. astronomer Charles E. Worley.

Charles Worley, long-time astronomer at the U.S. Naval Observatory, was born on May 22, 1935, in Iowa City, Iowa, and grew up in Des Moines the son of an M.D., Charles L. Worley, and his wife Iona Cooney Worley, a homemaker. He became interested in astronomy at age nine.

His first observational work as an amateur astronomer was plotting and recording more than 10,000 meteors for the American Meteor Society. Continuing his love for astronomy he attended Swarthmore College, where he took part in the parallax program as an Observing Assistant. He also met the other love of his life, his wife, Jane Piper. They were married in 1956 next to Sproul Observatory on the Swarthmore campus. He obtained a B.A. in mathematics from San Jose State College in 1959. He worked for the Lick Observatory in California (1959–1961) as a Senior Assistant and Research Astronomer under a Naval Research grant to observe double stars. After arriving at the U.S. Naval Observatory in 1961, he was the motive force behind an extensive program of double star observation (being himself, a prolific observer), instrumental innovation, and double star cataloging. He quickly gained recognition as one of the world's leading experts in the field of double star astronomy. Charles died on New Year's Eve, 1997, two days before his scheduled retirement.

Keywords. binaries (including multiple): general, binaries: visual, catalogs, history and philosophy of astronomy

1. Micrometer Observer

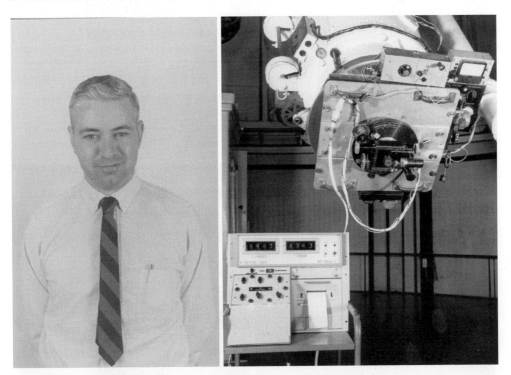

Figure 1. *Left:* Charles in 1964, a few years after his arrival at the USNO. *Right:* The micrometer, constructed by the USNO instrument shop and used for most of his observations, mounted on the 26″.

During his career Charles made over 40,000 measures of double and multiple stars using the USNO filar micrometer on telescopes in the northern and southern hemispheres, making him the 3rd most prolific double star observer ever. His special interest in nearby stars led to the discovery of 39 new, cool stellar companions. These companions, which are faint and difficult to observe, provide critical census information on the solar neighborhood. Apart from this foray into new binaries, his work was almost exclusively the measurement of close, fast moving known pairs where the scientific return was maximized. He was never interested in resolving new pairs for the sake of discovering more. His "mean positions" in the WDS were the result of many nights, each of many measurements per night, to ensure that his measurements were as accurate as possible. He was certainly gifted with the "double star eye," and was always able to push telescopes to their theoretical resolution limit when seeing permitted. Charles was introduced to double star work by Olin Eggen while at Lick Observatory. His micrometer wires were of etched tungsten rather than spider threads, and he was always eager to upgrade his technology, first with digital encoders and line printers, whenever possible.

2. Cataloger

In 1965 Charles arranged for the database of double star data, the *Index Catalogue of Visual Double Stars* (IDS), to be transferred from the Lick Observatory to the USNO and later renamed it the *Washington Double Star Catalog* (WDS)†.

This database became a truly comprehensive resource under his guidance, and is formally recognized as the international source of double star data by the International Astronomical Union (IAU). He updated the database on a continual basis, adding 290,400 observational records to the original 179,000 and increasing the original 64,000 systems by an additional 17,100 through careful literature searches and extensive communication with other double star observers throughout the world.

In collaboration with William Finsen and later Wulff Heintz, Charles produced two *Catalogs of Orbits of Visual Binary Stars*, the most recent published in 1983. At the time of his death he was preparing what would have been a new version. He was known for exacting standards and high quality best typified by his paper challenging all other double star observers; "Is This Orbit Really Necessary?"

Figure 2. Collecting data in the USNO Library for inclusion in the WDS.

† Charles was justifiably proud of the WDS and said that his catalog was "the oldest continuously maintained dataset in astronomy." He was right that sunspots were not continuously a maintained dataset. I think I won by saying the calendar was older, but he argued, in a typically curmudgeonly fashion, that counting sunrises was not data or at least not astronomy. *Anecdote courtesy of Demetrios Matsakis.*

3. IAU involvement

Charles first attended a General Assembly in 1961, and with the exception of R.G. Aitken, the first Commission 26 President, was personally acquainted with every Commission 26 President. He was a frequent correspondent with many double star luminaries of the day and his postal exchanges with, for example Finsen at the time of the closing of Republic Observatory, are a treasure trove of personal anecdotes and insights into selected events. He vociferously advocated for the reorganization of the stagnant administration of the Double Star Commission in the 1970s. In 1994 he became president of Commission 26.

Figure 3. *Left:* At the Santiago de Compostella Double Star Meeting (1996) with Jose Docobo (current Vice President of the Commission). *Right:* IAU Colloquium 100: Fundamentals of Astrometry (Belgrade, Yugoslavia, 1987) with his wife, Jane.

4. Speckle Observer

Recognizing the accuracy and precision afforded by speckle interferometry over micrometry, Charles advocated obtaining a speckle interferometer for the USNO. This addition, obtained well into the autumn of his career, was to improve his double star measurements. During the last seven years of his life, he oversaw improvements in both instrumentation and software implementation that resulted in making the USNO the world's second largest producer of double star observations using a speckle interferometer. Under Charles' direction more than 9,200 observations were made with the speckle interferometer on 1,100 systems down to separations of one-fifth of an arcsecond, the theoretical limit of the 26-inch refractor.

Acknowledgements

B.D.M. is supported by NASA under JPL task order # NMO710776. W.I.H. is supported by the Office of Naval Research, Global by grant N00014-06-1-1054. We gratefully acknowledge this support.

Figure 4. *Left:* Charles visiting speckle observers at Lowell. *Right:* With the USNO camera on the 26″.

Session 1: New Observing Techniques and Reduction Methods:

Observing with high angular and spectral resolution

Figure 1. (top left to bottom right) Speakers Harold McAlister, John Davis, Mike Shao, Yuri Balega, Artie Hatzes, and Frank Fekel.

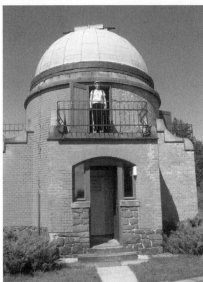

Figure 2. (top) Marek Wolf talking to Deniz Göker. (bottom left) One of the Ondřejov brick domes built around 1910, which hosted originally an 8-in Clark telescope and in the 1960s and 70s the 0.65-m reflector of the Astronomical Institute of the Charles University. It was jointly used for *UBV* photoelectric photometry by people from both institutions, especially by Pavel Mayer (AU UK), who was the main constructor of the telescope, and by Mirek Plavec and collaborators. Now an astronomical museum. (bottom right) Neo-Gothic Habsburg monument (1850) on the Smetana Riverside devoted to Franciscus II (reigned 1792–1835); this reactionary Emperor was known for his hatred of both the French Revolution and science: "The so called genii and scholars are good for nothing, they will all know better and obstruct only my administrative which they dislike. Common sense and good buttocks, this is the best. I need good citizens, not the scholars!"

Binary Stars as Critical Tools & Tests
in Contemporary Astrophysics
Proceedings IAU Symposium No. 240, 2006
W.I. Hartkopf, E.F. Guinan & P. Harmanec, eds.

Overview of Multiple–Aperture Interferometry Binary Star Results from the Northern Hemisphere

Harold A. McAlister

Center for High Angular Resolution Astronomy, Georgia State University, Atlanta, GA, USA
email: hal@chara.gsu.edu

Abstract. Long-baseline optical interferometry (LBI) can nearly close the gap in selection space between astrometric and spectroscopic detection of binary star systems, bringing the complementary powers of astrometry and spectroscopy to bear on a complete dynamical understanding of such systems, particularly including the determination of the masses of the individual stellar components. In the case of double-lined spectroscopic systems, their resolution by long-baseline interferometry also yields the orbital parallax and hence the luminosities of the individual stars. In some of these cases, the angular diameters of one or more components are accessible, and so a complete specification of a star in terms of its mass, radius and luminosity is made.

The northern hemisphere is now equipped with several interferometers of unprecedented capability in terms of their baseline sizes, numbers of telescopes and telescope apertures. These instruments, most notably the Palomar Testbed Interferometer at Mt. Palomar Observatory, have produced very significant results of a number of interesting systems fulfilling interferometry's promise to produce fundamental astrophysical data at levels of accuracy that challenge or confirm astrophysical theory.

This paper presents basic principles of long-baseline interferometric study of binary stars and summarizes results from northern interferometers with specific examples of their broad impact on binary star astronomy.

Keywords. instrumentation: high angular resolution, instrumentation: interferometers, stars: binary, stars: fundamental parameters.

1. Some Relevant Interferometry Basics

The simplest analog to a long-baseline interferometer employed in the observation of binary stars is the classic Young's double slit experiment in which fringes are detected from two point sources illuminating the slits at a wavefront tilt angle of θ corresponding to the angular separation of a binary system as seen on the sky. Each "star" gives rise to a sinusoid with peaks (or "fringes") separated by λ/B where B is the slit spacing or interferometric "baseline". The two fringe patterns exactly cancel when θ equals one half the fringe spacing, which occurs at $\lambda/2B$, $3\lambda/2B$, $5\lambda/2B$, etc. This forms a convenient definition for the limiting resolution of a two-element binary star interferometer as $\theta_{\lim} = \lambda/2B$. Notice that this compares favorably with the limiting resolution of full-aperture speckle interferometry as given by the Rayleigh criterion $\theta_{\lim} = 1.22\lambda/B$ where, in this case, B is the telescope aperture.

An interferometer operating within a finite spectral bandwidth will yield a fringe packet whose width is related to the coherence length $\lambda^2/\Delta\lambda$. The basic observable of an interferometer is the fringe "visibility" as originally defined by A. Michelson to be:

$$V = \frac{I_{\max} - I_{\min}}{I_{\max} + I_{\min}}$$

(1.1)

where I_{max} and I_{min} are the maximum and minimum values of the fringe packet. A typical experimental setup involves the simultaneous detection of fringe packets on the two sides of a beam splitter within which the interference occurs. Because those packets are 180° out of phase with each other, their difference forms the fringe signal and the visibility is given simply by the fringe amplitude once the fringe is normalized to its mean and subtracted by unity. For an excellent tutorial on interferometry, see the proceedings of the 1999 Michelson Summer School (published by the Jet Propulsion Laboratory and edited by P. Lawson) which are available at olbin.jpl.naxa.gov/intro/.

The interferometric visibility obtained for a binary star is a function of the binary star parameters (position angle θ, angular separation ρ, magnitude difference Δm), the angular diameters of the component stars (Θ_1, Θ_2), and the instrumental parameters (projected baseline length B, wavelength of observed passband λ, position angle on sky of projected baseline ψ). The analytic expression for visibility is:

$$V^2 = (1 + \beta)^{-2}(\beta^2 V_1{}^2 + V_2{}^2 + 2\beta V_1 V_2 cos(\frac{2\pi B}{\lambda}\rho cos(\theta - \psi)))$$ (1.2)

where

$$\beta = 10^{0.4\Delta m}$$ (1.3)

and

$$V_{1,2} = 2\frac{J_1(\pi\Theta_{1,2}B/\lambda)}{\pi\Theta_{1,2}B/\lambda}.$$ (1.4)

J_1 is the Bessel function of first order. Clearly, interferometers observing binary stars by means of visibility measurements do not measure the simple binary parameters (ρ, θ) directly and so cannot contribute in the usual way to catalogs such as the Washington Double Star (WDS) catalog.

Systems with separations exceeding about 10 milliarcseconds (mas) will typically no longer have overlapping fringe packets at long baselines and so visibility measurements of their separations become meaningless. Instead, separation can be measured astrometrically from the displacement of the separated fringe packets. The ability to measure large Δm systems is limited by the precision with which visibility can be measured since the visibility curve approaches that of a single star as Δm increases. With these limitations in mind, the current interferometric arrays are capable of resolving binaries with separations down to about 0.2 mas, a limit strictly determined by the longest baseline and shortest operating wavelength. Interferometers with beam combiners that spatially filter interfering beams with fibers or pinholes can to push to Δm's as large as 4 to 5 magnitudes at the expense of sensitivity.

2. Pioneering Work at Mt. Wilson

2.1. *Aperture Masking the 100-inch Telescope*

Inspired by Michelson's successful application of stellar interferometry, George Ellery Hale invited Michelson to try out his technique on the 60- and 100-inch telescopes. Using simple masks, Michelson easily demonstrated fringes, and Mt. Wilson astronomer John Anderson expanded Michelson's approach by designing a visual interferometer with variable slit spacing and orientation that was suitable to the accurate measurement of stellar diameters and binary star geometry. With this instrument, Anderson resolved the double-lined spectroscopic binary Capella during the winter of 1919-20 and employed six observations to calculate the visual elements of the 104-day system. In this process, Anderson (1920) determined the individual masses of the giant components and the orbital parallax of the system. This historic combination of the complementary capabilities of spectroscopy and interferometry when applied to binary stars can only now be widely exploited by the current generation of long-baseline arrays.

Anderson's Mt. Wilson colleague Paul Merrill obtained additional observations of Capella and resolved κ UMa and ν^2 Boo for the first time in a seven-month series of observations begun in the fall of 1920 (Merrill 1922). He inspected an additional 85 stars and marginally suspected duplicity in five cases (δ Cnc, 10 LMi, o Leo, ϵ UMa and υ Sgr). Of these five, three were subsequently inspected by speckle interferometry and found to be single, o Leo was resolved by long-baseline interferometry at a separation too close for the 100-inch experiment, but a companion was found by speckle interferometry in the case of δ Cnc that would have demonstrated variable visibility to Merrill. While Merrill expressed doubt as to the "reality of the changes" in visibility, he went on to encourage others to observe them "as the double star work at the Mount Wilson Observatory has been discontinued." Thus ended, after a very promising start, high angular resolution measurements of binary stars at the 100-inch telescope where these objects would be ignored for 60 years until speckle observers came along. However, the elegant success of Anderson and Merrill's work did inspire the long-term and highly productive program of visual interferometry with an "eyepiece interferometer" by W.S. Finsen at the Union/Johannesburg Observatory in South Africa. Finsen's program continued nearly to the advent of speckle interferometry (Finsen 1964). Since neither of those techniques is "multiple aperture" in nature, nothing further will be said about them here.

2.2. *The Mount Wilson "Long-Baseline" Interferometers*

With intuition that the maximum baseline available from a 100-inch aperture was just on the verge of providing diameters of giant and supergiant stars, Michelson and F.G. Pease collaborated on the design and use of the "20-ft interferometer", a beam that fed light from two movable mirrors into the optical train of the Hooker telescope. The only application to binary stars with this remarkable instrument was related by Pease in brief notes describing the determination of the orbit of Mizar from eight observations obtained with angular separations as small as $0\rlap{.}''008$ (Pease 1925, 1927). Interestingly, Pease credited "Professor Russell" as contributing the orbital analysis, although Henry Norris Russell was not listed as a co-author on the paper. The 20-ft was retired shortly after the second publication appeared, wrapped in heavy canvas and stored in the rafters of the 100-inch telescope building. At the instigation of this writer, the instrument was "exhumed" on 20 May 1999 in preparation for its display in CHARA's interferometry exhibit hall on Mt. Wilson where it can now be seen mounted atop the original prime focus cage of the telescope.

The irresistible desire for higher resolution led Pease to build a 50-ft stand-alone interferometer. Sadly, that instrument never fulfilled expectation, no doubt the result of extrapolating beyond the then accessible engineering performance limits. One can only wonder, had the 50-ft been a success, how it might have impacted binary star astronomy as a substantial number of spectroscopic binaries would have yielded to its very high limiting resolution.

3. The Modern Contributing Interferometers

Long-baseline, binary star interferometry came of age as recently as the late 1980's when the Mark III interferometer, a joint venture of the U.S. Naval Observatory, the Naval Research Laboratory, the Smithsonian Astrophysical Observatory and the Massachusetts Institute of Technology, became operational on Mt. Wilson. Working at visible

Table 1. Northern Interferometers that have Contributed to Binary Star Astronomy

Facility	Site	No. of Elements	Element Aperture	Max. Baseline	Operating Wavelength	Operational Status
Mark III	Mt. Wilson	2(4)	15 cm	32 m	0.45-0.80 μm	1987-1992
COAST	Cambridge, UK	4	40	100	0.4-0.95 + 2.2	since 1991
IOTA	Mt. Hopkins	3	45	38	0.5-2.2	1993-2006
NPOI	Anderson Mesa	6	60	(435)	0.45-0.85	since 1995
PTI	Mt. Palomar	3	40	110	1.5-2.4	since 1995
CHARA	Mt. Wilson	6	100	331	1.5-2.4	since 1999
KI	Mauna Kea	2	1,000	85	1.25-10.0	since 2001

wavelengths, the Mark III's 32-m longest baseline was more than adequate to reaching into the domain of spectroscopic binaries. Many of the problems unique to long-baseline interferometry were definitively solved by the Mark III, and its successors have drawn heavily on this highly successful, if short-lived, instrument, which was closed in 1992 in favor of the construction by the Navy of a new much larger interferometer near Flagstaff, Arizona, the Navy Prototype Optical Interferometer (NPOI). NPOI, which became operational in 1995, will ultimately have baselines almost 15 times longer than those of the Mark III. Before the Mark III was closed, Cambridge University opened its Cambridge Optical Aperture Synthesis Telescope (COAST) and successfully achieved the goal of producing the first optical aperture synthesis image in 1996.

The Smithsonian's Infrared Optical Telescope Array (IOTA) was commissioned on Mt. Hopkins in 1993 and, regrettably, closed in 2006. While IOTA was not used extensively for binary star studies, it rather naturally turned to binaries to demonstrate its imaging capability in 2004 after a third telescope was added to the previously two-telescope interferometer. In 1995, Caltech and the Jet Propulsion Laboratory began observations with their Palomar Testbed Interferometer (PTI), a facility with a 110-m longest baseline that has yielded very fine analyses of a number of resolved spectroscopic binaries to be described in more detail below. Georgia State University's CHARA Array saw first fringes from Mt. Wilson in 1999 but only became routinely scheduled for science operations in 2004. The CHARA Array currently possesses the longest operational baselines in the world and is capable of resolving a very large fraction of cataloged SB's. Last, and clearly not least, is the Keck Interferometer (KI), which is now limited to the 85-m baseline separating the Keck I and Keck II telescopes. The long-anticipated addition of the "Outrigger" telescopes was thwarted in 2005 when NASA canceled that effort.

Additional interferometers in the northern hemisphere have been built in France, where Antoine Labeyrie and others were among those responsible for the rebirth of the field, but those instruments are not described here as they had limited or no application to binary star astronomy. One example of an important negative result was the inspection by Harmanec *et al.* (1996) of the complex system comprising β Lyr, which was found to be unresolved by the GI2T, located on the Calern plateau in southern France and operated until recently by the Nice Observatory. The negative result from the north-south baseline provided evidence for an east-west orientation of the orbital plane as indicated by polarimetric measurements. The facilities listed in chronological order of their "first fringes" date in Table 1 were selected because of their relevance to the topic at hand.

A survey of the literature is summarized in Table 2, which gives an account of the refereed papers dealing with binary stars that have appeared as of early 2006 from these interferometers. As will be emphasized below, a total of 32 papers describing some 40

Table 2. Binary Star Output of Northern Interferometers

Facility	No. of Papers	No. of Systems	Emphasis
Mark III	11	20	SB' with P > 100 days
COAST	1	1	First optical aperture synthesis images (Capella)
IOTA	2	3	Closure phase imaging
NPOI	5	5	SB's and first 6-telescope imagery
PTI	11	11	SB's and ultra-precise astrometry of binaries
CHARA	1	1	Precise astrometry of 12 Per - more to come!
KI	1	1	PMS binary

binary star systems is a very modest contribution considering the potential LBI has in this field. The majority of the published results is from two instruments: the Mark III and Palomar Testbed Interferometers. While the CHARA Array has not yet contributed much in this area, the reader can be assured that binary stars will receive considerable attention from Mt. Wilson in coming years.

4. Some Example Results

4.1. *Mark III Results*

The baselines of the Mark III interferometer made it ideally suited to the resolution of spectroscopic binaries with intermediate periods, and the instrument was productively used in that domain, clearly demonstrating the effectiveness of LBI in complementing the spectroscopy to yield three-dimensional orbit solutions and component masses. The first of a series of such studies was that of α And (Pan *et al.* 1992) in which the sub-milliarcsecond precision of LBI was demonstrated through the calculation of a semi-major axis of 24.15 ± 0.13 mas. Similarly, Armstrong *et al.* (1992a) resolved the SB2 binary ϕ Cyg, which had previously been resolved by speckle interferometry (McAlister 1982), and improved the semi-major axis determination by a factor of five. Just as Capella had played a role in the first interferometers on Mt. Wilson, it was ideally suited to a definitive orbit determination by the Mark III (Hummel *et al.* 1994). Again, compared with speckle interferometry, the higher resolution of the Mark III naturally led to significant improvement of the orbital elements. Interestingly, the Anderson and Merrill observations of Capella show separations consistently too small by a few mas, most likely due to the difficulty in determining the effective wavelength of the visual interferometry process.

Other Mark III studies were completed for: β Ari (Pan *et al.* 1990); the AB,C system within Algol (Pan *et al.* 1993); the K4 Ib + B5V spectroscopic and eclipsing binary ζ Aur (Bennett *et al.* 1996); the G8III system η And (Hummel *et al.* 1993); ζ^1 UMa and η Peg in combined Mark III and NPOI analyses (Hummel *et al.* 1998); and, most recently, in another joint Mark III / NPOI venture targeting the Hyades binary θ^2 Tau (Armstrong *et al.* 2006). In a single paper (Hummel *et al.* 1995), the orbits of seven spectroscopic binaries were determined (π And, θ Aql, ζ^1 UMa, 93 Leo, 113 Her, β Tri and δ Tri). A three-way partnership of data exploiting the Mark III, NPOI and PTI facilities led to a solution of the orbit of o Leo (Hummel *et al.* 2001). The legacy of the Mark III interferometer is a powerful one in that the instrument not only brought interferometry into the modern world but it set an excellent standard for the application of this technique to binary stars.

4.2. *Interferometric Imaging of Binaries*

Binary stars serve as a natural target for optical aperture synthesis imaging, which was first achieved on interferometry's old friend Capella by Baldwin *et al.* (1996) in a beautiful demonstration of orbital motion over a ten-day period in the fall of 1995. This achievement was soon followed up at NPOI by the first multi-spectral channel binary images with good orbital coverage of the 20.54-day system ζ^1 UMa (Benson *et al.* 1997). Hummel *et al.* (2000) employed imaging in their discovery of a new companion to the massive O star ζ Ori A, which had originally been suspected from observations at the Narrabri Intensity Interferometer by Hanbury Brown *et al.* (1974). In a *tour de force* demonstration, (Hummel *et al.* 2003) employed all six NPOI light-collecting telescopes simultaneously to produce images of the 71-day and 13-yr components comprising the triple star system η Vir. Subsequent images have been produced at IOTA in its three-telescope configuration in the case of λ Vir (Monnier *et al.* 2004) and, yet again, for Capella (Kraus *et al.* 2005). Experiments in imaging binaries and triple systems is presently underway at the CHARA Array using the University of Michigan infrared beam combiner now capable of simultaneous four-way beam combination.

4.3. *Palomar Testbed Interferometer Results*

Constructed essentially by the same group that designed and built the Mark III interferometer, the Palomar Testbed Interferometer rather naturally followed in the scientific footsteps of its predecessor and immediately embarked on a productive program of binary star studies. The first effort was an important negative result by Boden *et al.* (1998) in which a search for a stellar companion to the exoplanet host star 51 Peg yielded no such companion. There quickly followed analyses of the RS CVn system TZ Tri by Koresko *et al.* (1998) and the SB2 systems ι Peg (Boden *et al.* 1999a) and 64 Psc (Boden *et al.* 1999b).

In an analysis of the equal mass system 12 Boo, Boden, Creech-Eakman & Queloz (2000) came to the surprising conclusion that one component was significantly more luminous than the other. This finding was confirmed five years later after the incorporation of additional visibilities and new radial velocities by Boden *et al.* (2005) with the determination that $M_1 = 1.44 \pm 0.02$, $M_2 = 1.41 \pm 0.02$ $M_\odot$ and the luminosity difference is 0.50 ± 0.09 magnitudes. This system has apparently been caught at that instant at which the slightly more massive component is entering into its red giant phase and evolving rapidly from the main sequence, temporarily leaving behind its companion. In a related analysis, Boden *et al.* (2006) found the high proper motion binary HD 9939 traversing the Hertzsprung gap enabling them to date the system at 9.12 ± 0.25 Gyr.

Other PTI work on spectroscopic binaries includes: the thick disk old system HD 195987 (Torres *et al.* 2002); HD 6118 and HD 27483 (Konacki & Lane 2004) for which the latter possesses the smallest semi-major axis yet determined by LBI (1.2 mas); and, the Pleiades SB2 Atlas (HR 1178) (Pan, Shao & Kulkarni 2004) for which an orbital parallax was determined that resolved the apparent discrepancy between the Hipparcos parallax and stellar models in favor of the models.

PTI was specifically designed to perform differential astrometry in a "narrow angle" mode employing two delay lines per telescope. This has permitted the instrument to undertake a program of very precise astrometry of binaries that are otherwise too wide for LBI. Thus, the Palomar High-precision Astrometric Search for Exoplanet Systems (PHASES) is monitoring binaries that fall in the separation regime of speckle interferometry. Lane & Muterspaugh (2004) attained an accuracy of ± 16 μas for the system HD 171779 Accuracies approaching that quality have subsequently been attained for δ Equ (Muterspaugh *et al.* 2005) and κ Peg (Muterspaugh *et al.* 2006a), and the relative inclination of the orbital planes in the triple system V819 Her have recently been determined by Muterspaugh *et al.* (2006b).

Finally, members of the PTI collaboration have teamed with others to combine observations from the Keck Interferometer and the Hubble Space Telescope Fine Guidance Sensors to determine the masses to about 10% accuracy of the SB2 system comprising the B component of the quadruple pre-main-sequence star HD 98800 B (Boden *et al.* 2005).

4.4. *The CHARA Array Binary Star Program*

One of the major motivating factors for the CHARA Array was an extrapolation of CHARA's long-term binary star program from the regime of separation accessible to speckle interferometry down to angular separations two orders of magnitude smaller. CHARA's facility on Mt. Wilson only became routinely operational in 2005 and most of its initial work has not been on binary stars. The exception to date is the astrometric study of the speckle binary 12 Per by Bagnuolo *et al.* (2006) in which an accuracy of ± 25 μas was utilized to refine the orbital parameters for the system.

This study is based on the analysis of separated fringe packets arising from binaries that are too wide to have overlapping packets that are amenable to standard visibility analysis but are sufficiently close so as to be encompassed within a fringe scan. This allows us to search for companions in the range of about 7 to 70 mas, giving overlap into the speckle regime. The approach is also being used to look for binaries in other selected samples of stars and also to employ one component as the visibility calibrator for the other in the case of a triple system.

CHARA is collaborating with F. Fekel (Tennessee State University) and J. Tomkin (University of Texas) on combining visibilities and velocities for selected SB2 systems. As a result of this symposium, we are also exploring a similar collaboration with P. Harmanec and P. Koubsky of the Ondrejov Observatory.

5. Prospects

The advantages of resolving spectroscopic binaries and thereby determining their orbits three-dimensionally are well-known with the most complete results emanating from a resolved SB2. In that case, knowledge of the individual masses is joined by the determination of the "orbital parallax" (with the potential for greater accuracy than Hipparcos). If the magnitude difference is also available, as it is in the case of LBI, then the individual luminosities fall out of the solution as well. In some cases, LBI will also yield the angular diameter of one or both components. Thus, LBI has the potential for contributing to the fundamental stellar mass/luminosity and mass/radius relations.

The potential for resolving known SB's has been explored in the context of parallaxes by Vinter Hansen (1942) and resolution by speckle interferometry (McAlister 1976). In order to better understand the regime accessible by LBI, CHARA initiated a bibliographic update (Taylor, Harvin & McAlister 2003) of the known SB's to create an input catalog for observational planning. In the case of the longest baseline of the CHARA Array (331 m), we find that at the K-band infrared some 370 or 43% of SB1's and 250 (49%) of SB2's are potentially resolvable. In the V-band, these numbers increase to 500 (57%) SB1's and 360 (70%) SB2's. Thus, it is clear that only a small fraction of the potential has yet been tapped by LBI observers for extracting fundamental stellar parameters from binary stars. In many of these cases, however, modern radial velocity studies are needed to improve the accuracy of the spectroscopic orbits so that the most accurate stellar parameters are obtainable. The results to date, particularly those from the PTI, are showing the benefits of such active collaborations.

6. Conclusions

Long-baseline interferometry from the northern hemisphere, particularly resulting from the Mark III and Palomar Testbed interferometers, has made important contributions to extracting fundamental astrophysical parameters from binary stars with accuracies sufficient to challenge astrophysical theory. But, the technique has really just begun to realize the full extent of its potential for resolving spectroscopic binary systems whose exploitation by interferometry can best be achieved in partnership with modern, high-precision radial velocity and quantitative spectroscopy programs.

Acknowledgements

The author wishes to thank the SOC for its invitation to participate in this Symposium and Georgia State University and the U.S. National Science Foundation for the support of the design and construction of the CHARA Array and its ongoing scientific work. Ellyn Baines kindly assisted in preparing this document.

References

Anderson, J.A. 1920, *ApJ* 51, 263

Armstrong, J.T., Mozurkewich, D., Vivekanand, M., Simon, R.S., Denison, C.S., Johnston, K.J., Pan, X.P., Shao, M., & Colavita, M.M. 1992, *AJ* 104, 241

Armstrong, J.T., Hummel, C.A., Quirrenbach, A., Buscher, D.F., Mozurkewich, D., Vivekanand, M., Simon, R.S., Denison, C.S., & Johnston, K.J. 1992, *AJ* 104, 2217

Armstrong, J.T., Mozurkewich, D., Hajian, A.R., Johnston, K.J., Thessin, R.N., Peterson, D.M., Hummel, C.A., & Gilbreath, G.C. 2006, *AJ* 131, 2463

Bagnuolo, W.G., Taylor, S.F., McAlister, H.A., ten Brummelaar, T.A., Gies, D.R., Ridgway, S.T., Sturmann, J. Sturmann, J., Turner, N.H., & Berger, D.H. 2006, *AJ* 131, 2695

Baldwin, J.E., Beckett, M.G., Boysen, R.C., Burns, D., Buscher, D.F., Cox, G.C., Haniff, C.A., Mackay, C.D., Nightingale, N.S., Rogers, J., Scheuer, P.A.G., Scott, T.R., Tuthill, P.G., Warner, P.J. Wilson, D.M.A., & Wilson, R.W. 1996, *A&A* 306, L13

Bennett, P.D., Harper, G.M., Brown, A., & Hummel, C.A. 1996, *ApJ* 471, 454

Benson, J.A., Hutter, D.J., Elias, N.M., Bowers, P.F., Johnston, K.J., Hajian, A.R., Armstrong, J.T., Mozurkewich, D., Pauls, T.A., Rickard, L.J., Hummel, C.A., White, N.M., Black, D., & Denison, C.S. 1997, *AJ* 114, 1221

Boden, A.F., van Belle, G.T., Colavita, M.M., Dumont, P.J., Gubler, J., Koresko, C.D., Kulkarni, S.R., Lane, B.F., Mobley, D.W., Shao, M., & Wallace 1998, *ApJ* 504, L39

Boden, A.F., Koresko, C.D., van Belle, G.T., Colavita, M.M., Dumont, P.J., Gubler, J., Kulkarni, S.R., Lane, B.F., Mobley, D., Shao, M., & Wallace J.K. 1999, *ApJ* 527, 360

Boden, A.F., Lane, B.F., Creech-Eakman, M.J., Colavita, M.M., Dumont, P.J., Gubler, J., Koresko, C.D., Kuchner, M.J., Kulkarni, S.R., Mobley, D.W., Pan, X.P., Shao, M., van Belle, G.T., Wallace, J.K., & Oppenheimer, B.R. 1999, *ApJ* 527, 360

Boden, A.F., Creech-Eakman, M.J., & Queloz, D. 2000, *ApJ* 536, 880

Boden, A.F., Torres, G., & Hummel, C.A. 2005, *ApJ* 627, 464

Boden, A.F., Sargent, A.I.; Akeson, R.L., Carpenter, J.M., Torres, G., Latham, D.W., Soderblom, D.R., Nelan, E., Franz, O.G., & Wasserman, L.H. 2005, *ApJ* 635, 442

Boden, A.F., Torres, G., & Latham, D.W. 2006, *ApJ* 644, 1193

Finsen, W.S. 1964, *AJ* 69, 319

Hanbury Brown, R., Davis, J., & Allen, L.R. 1974, *MNRAS* 167, 121

Harmanec, P., Morand, F., Bonneau, D., Jiang, Y., Yang, S., Guinan, E.F., Hall, D.S., Mourard, D., Hadrava, P., Božić, H., Sterken, C., Tallon-Bose, I., Walker, G.A.H., McCook, G.P., Vakili, F., Stee, Ph., & Le Contel, J.M. 1996, *A&A* 312, 879

Hummel, C.A., Armstrong, J.T., Quirrenbach, A., Buscher, D.F., Mozurkewich, D., Simon, R.S., & Johnston, K.J. 1993, *AJ* 106, 2486

Hummel, C.A., Armstrong, J.T., Quirrenbach, A., Buscher, D.F., Mozurkewich, D., & Elias, N.M. 1994, *AJ* 107, 1859

Hummel, C.A., Armstrong, J.T., Buscher, D.F., Mozurkewich, D., Quirrenbach, A., & Vivekanand, M. 1995, *AJ* 110, 376

Hummel, C.A., Mozurkewich, D., Armstrong, J.T., Hajian, A.R., Elisa, N.M., & Hutter, D.J. 1998, *AJ* 116, 2536

Hummel, C.A., White, N.M., Elias, N.M., Hajian, A.R. & Nordgren, T.E. 2000, *ApJ* 549 L93

Hummel, C.A., Carquillat, J.-M., Ginestet, N., Griffin, R.F., Boden, A.F., Hajian, A.R., Mozurkewich, D., & Nordgren, T.E. 2001, *AJ* 121, 1623

Hummel, C.A., Benson, J.A., Hutter, D.J., Johnston, K.J., Mozurkewich, D., Armstrong, J.T., Hindsley, R.B., Gilbreath, G.C., Rickard, L.J., & White, N.M. 2003, *AJ* 125, 2630

Konacki, M., & Lane, B.F. 2004, *ApJ* 610, 443

Koresko, C.D., van Belle, G.T., Boden, A.F., Colavita, M.M., Creech-Eakman, M.J., Dumont, P.J., Gubler, J., Kulkarni, S.R., Lane, B.F., Mobley, D.W., Pan, X.P., Shao, M., & Wallace, J.K. 1998 *ApJ* 509, L45

Kraus, S., Schloerb, F.P., Traub, W.A., Carleton, N.P., Lacasse, M., Pearlman, M., Monnier, J.D., Millan-Gabet, R., Berger, J.-P., Hanuenauer, P., Perraut, K., Kern, P., Malbet, F., & Labeye, P. 2005, *AJ* 130, 246

Lane, B.F., & Muterspaugh, M.W. 2004, *ApJ* 601, 1129

McAlister, H.A. 1976, *PASP* 88, 317

McAlister, H.A. 1982, *AJ* 87, 563

Merrill. P.W. 1922, *ApJ* 56, 40

Monnier, J.D., Traub, W.A., Schloerb, F.P., Millan-Gabet, R., Berger, J.-P., Pedretti, E., Carleton, N.P., Kraus, S., Lacasse, M.G., Brewer, M., Ragland, S., Ahearn, A., Coldwell, C., Haguenauer, P., Kern, P., Labeye, P., Lagny. L., Malbet, F., Malin, D., Maymounkov, P., Morel, S., Papaliolios, C., Perraut, K., Perlman, M., Porro, I.L., Schanen, I., Souccar, K., Torres, G., & Wallace, G. 2004, *ApJ* 602, L57

Muterspaugh, M.W., Lane, B.F., Konacki, M., Wiktorowicz, S., Burke, B.F., Colavita, M.M., Kulkarni, S.R., & Shao, M. 2005, *AJ* 130, 2866

Muterspaugh, M.W., Lane, B.F., Konacki, M., Wiktorowicz, S., Burke, B.F., Colavita, M.M., Kulkarni, S.R., & Shao, M. 2006a, *ApJ* 636, 1020

Muterspaugh, M.W., Lane, B.F., Konacki, M., Wiktorowicz, S., Burke, B.F., Colavita, M.M., Kulkarni, S.R., & Shao, M. 2006b, *A&A* 446, 723

Pan, X.P., Shao, M., Colavita, M.M., Mozurkewich, D., Simon, R.S., & Johnston, K.J. 1990, *ApJ* 356, 641

Pan, X., Shao, M., Colavita, M.M., Armstrong, J.T., Mozurkewich, D., Vivekanand, M., Denison, C.S., Simon, R.S., & Johnston, K.J. 1992 *ApJ* 384, 624

Pan, X., Shao, M., & Colavita, M.M. 1993, *ApJ* 413, L129

Pan, X.P., Shao, M., & Kulkarni, S.R. 2004, *Nature* 427, 326

Pease, F.G. 1925, *PASP* 37, 155

Pease, F.G. 1927, *PASP* 39, 313

Taylor, S.F., Harvin, J.A., & McAlister, H.A. 2003, *PASP* 115, 609

Torres, G., Boden, A.F., Latham, D.W., Pan, M., & Stefanik, R.P. 2002, *AJ* 124, 1717

Vinter-Hansen, J.M. 1942, *PASP* 54, 137

Discussion

ROBERT WILSON: What about the narrowness of filters used by the various groups - are they mainly the same or not? Are they Johnson filters or Strömgren or what? Can you comment on tradeoffs (limiting magnitude, good definition of λ_{eff}, etc.)?

MCALISTER: The various groups do use standard filters. In CHARA's case, we currently use near-infrared filters at the H and K bands, where we presently have a limiting magnitude of +6.5 or 7.0, depending on seeing conditions. We expect these limits to improve. We can measure the effective filter wavelengths by observing in an FTS mode.

Binary Stars as Critical Tools & Tests
in Contemporary Astrophysics
Proceedings IAU Symposium No. 240, 2006
W.I. Hartkopf, E.F. Guinan & P. Harmanec, eds.

© 2007 International Astronomical Union
doi:10.1017/S174392130700378X

Overview of Multiple–Aperture Interferometry Binary Star Results from the Southern Hemisphere

John Davis

School of Physics, University of Sydney,
NSW 2006, Australia
email: j.davis@physics.usyd.edu.au

Abstract. The first multiple-aperture interferometric study of a binary system, in which the power of combining interferometric and spectroscopic data was demonstrated, was made from the Southern Hemisphere. The observations of α Vir with the Narrabri Stellar Intensity Interferometer (NSII) were combined with spectroscopic and photometric data to yield the mass, radius and luminosity of the primary as well as an accurate distance to the system. The NSII also revealed a number of stars, previously thought to be single, to be binary systems. Several of these systems have subsequently been shown to be spectroscopic binaries.

The Sydney University Stellar Interferometer (SUSI) and the European Southern Observatory's Very Large Telescope Interferometer (VLTI) are the two current Southern Hemisphere multiple aperture interferometers. SUSI is being used to determine interferometric orbits for some of the binary systems discovered with the NSII including β Cen and λ Sco and, in combination with spectroscopy, to determine accurate masses for early-type stars and accurate dynamical parallaxes for the systems.

The VLTI has operated with three beam-combining instruments, namely VINCI, MIDI and AMBER. The few observations of binary systems that have been made so far are summarised and, while in general they are of a preliminary nature, they demonstrate the potential of the VLTI for binary star studies.

One double-lined spectroscopic binary that has been observed with all three Southern Hemisphere instruments is γ^2 Vel, which has the brightest Wolf-Rayet star in the sky as its secondary. The observations and preliminary results for the masses of the O-type primary and WC8 secondary and for the distance to the system are summarised.

Keywords. techniques: interferometric, techniques: spectroscopic, stars: binaries, stars: fundamental parameters, stars: distances

1. Introduction

The Narrabri Stellar Intensity Interferometer (NSII) (Hanbury Brown, Davis & Allen 1967) was the first of the three Southern Hemisphere multiple-aperture optical interferometers. It represents an important milestone in the study of binary stars as it was the first long-baseline optical interferometer with independently controlled apertures to be used to determine the orbital parameters of a binary system and to demonstrate the power of combining interferometric and spectroscopic data.

The other two Southern Hemisphere multiple-aperture interferometers are the Sydney University Stellar Interferometer (SUSI) (Davis *et al.* 1999) and the European Southern Observatory's Very Large Telescope Interferometer (VLTI) (Glindemann *et al.* 2003). Each of the three instruments and the binary star results obtained with them will be discussed in turn as they generally differ in nature or in the types of system studied.

Table 1. Parameters determined for α Vir from the combination of interferometric, spectroscopic and photometric data (Herbison-Evans *et al.* 1971). The symbols are defined in the text.

Interferometry		Plus Spectroscopy	
i	$= 65.9 \pm 1.8\,\mathrm{deg}$	a	$= (1.93 \pm 0.06) \times 10^7\,\mathrm{km}$
θ_a	$= 1.54 \pm 0.05\,\mathrm{mas}$	d	$= 84 \pm 4\,\mathrm{pc}$
θ_LD1	$= 0.90 \pm 0.04\,\mathrm{mas}$	R_1	$= 8.1 \pm 0.5\,R_\odot$
β	$= 6.4 \pm 1.0$	M_1	$= 10.9 \pm 0.9\,M_\odot$
		M_2	$= 6.8 \pm 0.7\,M_\odot$
Plus Spectrophotometry		$\log L_1/L_\odot$	$= 4.17 \pm 0.10$
T_e1	$= 22400 \pm 1000\,\mathrm{K}$	$\log g_1$	$= 3.7 \pm 0.1$

An exception is the case of γ^2 Vel since this system has been observed with all three instruments and the results will be discussed together in Section 5.

2. The Narrabri Stellar Intensity interferometer

The NSII (1965-1972) was used to measure the angular diameters of 32 early-type stars (Hanbury Brown, Davis & Allen 1974), to study the double-lined spectroscopic binaries α Vir (Section 2.1) and γ^2 Vel (Section 5), and to discover previously unsuspected binary systems (Section 2.2).

2.1. α Virginis

The double-lined spectroscopic binary α Vir (B1 III-IV+B3 V) was observed in 1966 and 1970 with the NSII at $\lambda442\,\mathrm{nm}$ to determine an interferometric orbit for the 4.015 day period system (Herbison-Evans *et al.* 1971). The results were combined with spectroscopically determined parameters for the system based on observations by Struve & Ebbighausen (1934) and Struve *et al.* (1958) to demonstrate, for the first time, the power of combining interferometric and spectroscopic observations of double-lined spectroscopic binaries for the determination of fundamental properties of stars.

The two techniques allow the determination of some orbital parameters in common, namely the period P, the longitude of periastron ω, and the eccentricity e. However, there are also complementary parameters: interferometry provides the inclination of the orbit i, the angular size of the semi-major axis θ_a, the brightness ratio of the two components β, and the angular size of at least the primary component of the system θ_LD1 (limb darkened), whereas spectroscopy provides $a \sin i$, $M_1 \sin^3 i$, and $M_2 \sin^3 i$ where a is the semi-major axis of the system and M_1 and M_2 are the masses of the component stars.

Full details have been given by Herbison-Evans *et al.* (1971) and some of the results are given in Table 1 to illustrate the importance of the combination of the two techniques. The addition of measurements of the spectral flux distribution allows the effective temperature T_e1 and luminosity L_1 of the primary to be determined.

The Hipparcos distance for α Vir is $80.4 \pm 5.5\,\mathrm{pc}$ in good agreement with the NSII value of $84 \pm 4\,\mathrm{pc}$.

2.2. Discovered Binaries

A by-product of the NSII programme was the discovery that several stars, previously thought to be single, were in fact binary systems (Hanbury Brown, Davis & Allen 1974) but the instrument lacked the capability for determining their orbits. These included β Cen, λ Sco, δ Vel, ζ Ori, β Cru, σ Sgr and δ Sco. Each of these systems has subsequently had its binary nature confirmed by spectroscopy and/or interferometry and the first three will be discussed in more detail below.

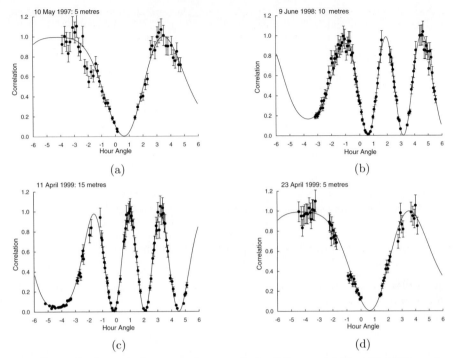

Figure 1. Normalised correlation versus hour angle for β Cen on (a) May 10, 1997 at a baseline of 5 m, (b) June 9, 1998 at 10 m, (c) April 11, 1999 at 15 m, and (d) April 23, 1999 at 5 m. In each case the filled circles represent the measurements and the line is the fit to determine the vector separation of the components.

3. The Sydney University Stellar Interferometer

SUSI is a long baseline optical interferometer with a 640 m long North-South array of input siderostats, the light from any two of which can be combined at a time. Baselines from 5-160 m are fully operational and the spectral range of the instrument is 430-950 nm. It has been used to observe a number of binary systems including β Cen and λ Sco discussed below, both discovered to be binaries with the NSII, and γ^2 Vel to be discussed in Section 5.

3.1. β Centauri

The bright southern binary β Cen was observed with SUSI at $\lambda442$ nm from 1997-2002 to determine an interferometric orbit (Davis *et al.* 2005). The component stars are of almost equal brightness (Δm at $\lambda442$ nm is 0.15$\pm$0.02) and of spectral type B1 III. Observations were made over a range in hour angle for each observing night to enable vector separations to be determined. Four examples are shown in Figure 1 and inspection shows that the plots for 10 May 1997 and 23 April 1999, both obtained with a baseline of 5 m, are almost identical. This is to be expected since they are separated in time by 1.9973$\pm$0.0004 orbital periods. The orbit fitted to the vector separations is shown in Figure 2.

The interferometric results were combined by Davis *et al.* (2005) with a revision of the spectroscopic results of Ausseloos *et al.* (2002) and the results are summarised in Table 2. There are two points to note about the parameters listed in Table 2. Firstly, Ausseloos *et al.* (2006) have improved the disentangling of the spectral lines of the component stars of β Cen, resulting in a revision of the parameters in the table. Using the same approach as Davis *et al.* (2005) with the revised spectroscopic data led to masses of

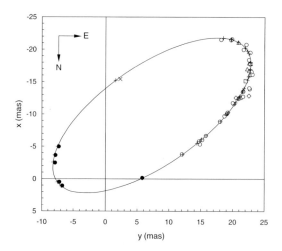

Figure 2. The orbit of β Centauri. The data points are SUSI results except for the MAPPIT point which is from Robertson *et al.* (1998). SUSI observations: 1997 $\Diamond$; 1998 $\triangle$; 1999 $\bigcirc$; 2000 $\bullet$; 2002 $\square$. MAPPIT observation: 1995 $\times$. The points on the fitted orbit that correspond to the observational points are marked with $+$.

$M_1 = 11.2 \pm 0.7\,M_\odot$ and $M_2 = 9.8 \pm 0.7\,M_\odot$ with a reversal of the identification of primary and secondary. The dynamical parallax was revised to $\pi_{\rm d} = 9.3 \pm 0.3$ mas. The uncertainties have been doubled in a conservative approach because the uncertainty in the disentangling process could not be propagated directly. The mass estimates were further refined using additional observational information and the extensive database of main-sequence stellar models by Ausseloos *et al.* (2004) to give $M_1 = 10.7 \pm 0.1\,M_\odot$ and $M_2 = 10.3 \pm 0.1\,M_\odot$. The second point to note is that the dynamical parallax differs significantly from the Hipparcos value of $\pi = 6.21 \pm 0.56$ mas. It is thought to be due to the fact that the binary nature of β Cen, with its relatively large variation in the projected angular separation of the two components on the sky, was not taken into account in determining the Hipparcos parallax.

Table 2. Parameters determined for β Cen taken from Davis *et al.* (2005). The three left-hand columns list values determined from spectroscopy, identified by S in the third column, and from interferometry, identified by I. The right-hand columns list parameters determined by the combination of the interferometric and spectroscopic data. P, a, a'' and i represent the period, semi-major axis, angular semi-major axis and inclination of the orbit respectively. β is the brightness ratio of the two components determined from the interferometric observations. π_d is the dynamical parallax.

Parameter	Value	Source	Parameter	Value
P (days)	357.00 ± 0.07	S	$\pi_{\rm d}$ (mas)	9.77 ± 0.15
$M_1 \sin^3 i\ (M_\odot)$	7.15 ± 0.24	S	$M_1\ (M_\odot)$	9.09 ± 0.31
$M_2 \sin^3 i\ (M_\odot)$	7.15 ± 0.23	S	$M_2\ (M_\odot)$	9.09 ± 0.30
$a \sin i$ (AU)	2.391 ± 0.032	S	M_{1V}	-3.85 ± 0.05
a''	0.02532 ± 0.00023	I	M_{2V}	-3.70 ± 0.05
i (deg)	67.4 ± 0.3	I		
β (at 442 nm)	0.868 ± 0.015	I		

Table 3. Parameters determined for λ Sco with SUSI (from Tango *et al.* 2006). The two left-hand columns list values determined interferometrically with SUSI and the right-hand columns list parameters determined by the combination of the interferometric and spectroscopic data. P, a'' and i represent the period, angular semi-major axis and inclination of the orbit respectively. The Δm values are the magnitude differences at the specified wavelengths determined from the interferometric observations. M_1 and M_2 are the masses of the components and π_d is the dynamical parallax.

Parameter	Value	Parameter	Value	Spectral Type
P (days)	1052.8 ± 1.2	M_1 ($M_\odot$)	10.4 ± 1.3	B1.5 IV
a''	0.0493 ± 0.0003	M_2 ($M_\odot$)	8.1 ± 1.0	B2 IV
i (deg)	77.2 ± 0.2	π_d (mas)	8.9 ± 0.4	
Δm_{442}	0.65 ± 0.10			
Δm_{700}	0.67 ± 0.10			

3.2. λ Scorpii

λ Sco is a well-known single-lined spectroscopic binary with a β Cep primary and a low mass companion orbiting it with a period of $\sim$6 days. The detection of a third component by the NSII has been confirmed by radial velocity measurements (Uytterhoeven *et al.* 2004) and by SUSI. The third component is a B star orbiting the primary with a period of $\sim$2.9 years and the interferometric orbit determined with SUSI (Tango *et al.* 2006) is shown in Figure 3. Tango *et al.* (2006) have combined the SUSI results with the spectroscopic results of Uytterhoeven *et al.* (2004), revised using the more accurate values for P, T, e and ω from the interferometry, and these are summarised in Table 3.

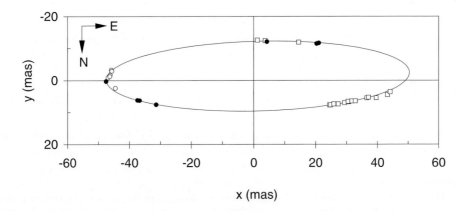

Figure 3. The orbit of λ Sco determined with SUSI. Key: ○ 1999 observations; □ 2000 and 2001 observations; ● 2004 and 2005 observations made during the next orbit. The 1999-2001 observations were made at a wavelength of 442 nm and the 2004-5 observations at 700 nm.

As in the case of β Cen, the dynamical parallax for λ Sco differs significantly from the Hipparcos value of $\pi = 4.64 \pm 0.90$ mas and it is thought to be for the same reason.

3.3. *SUSI Summary*

A major component of the SUSI observational programme is the observation of spectroscopic binary stars to determine stellar masses and dynamical parallaxes as demonstrated for β Cen and λ Sco. γ^2 Vel is discussed in Section 5. Several systems are in the programme including the 33 day period double-lined spectroscopic binary σ Sco, another system with a β Cep component.

4. ESO's Very Large Telescope Interferometer

ESO's VLTI (Glindemann *et al.* 2003) was commissioned with two test siderostats of 0.4 m diameter and baselines in the range 8-202 m. It also has the capability of using any of the four 8.2 m Unit Telescopes (UTs) with baselines in the range 47-130 m and, more recently, four 1.8 m Auxiliary Telescopes (ATs) with the same baseline range as the test siderostats. Three beam-combining instruments have been used with the VLTI to date, namely VINCI, MIDI and AMBER.

4.1. *VINCI*

VINCI (Kervella *et al.* 2003) was the commissioning instrument for the VLTI and worked in the K band (2.0-2.4 μm). It combined the light from either the two test siderostats or from two UTs. It was primarily a test instrument and many observations were made of single stars including Cepheids and a rapidly rotating star. However, the only binary star observations were of δ Vel.

The primary of δ Vel was found to be a binary system with the NSII (Hanbury Brown *et al.* 1974) and it has subsequently been found to be a 45 day period eclipsing system (Otero, Fieseler & Lloyd 2000). δ Vel is therefore a triple system in which the primary (Aa+Ab) is the eclipsing system. Aa and Ab are classified as A2 and A4 and the B component as F2/5. Observations of the eclipsing system were made at four orbital phases in 2003 with VINCI and at one orbital phase in 2005 with AMBER. It has been found that not all the VINCI data fit a simple model and there are significant discrepancies (A. Kellerer, private communication). The conclusion to be drawn is that more data are needed before a satisfactory solution for this system can be established.

4.2. *MIDI*

MIDI (Leinert *et al.* 2003) is a two aperture beam-combining instrument working in the N band (7.8-13.5 μm). Observations have been made of a number of binary systems containing T-Tauri stars (Ratzka & Leinert 2007). The emphasis has been on the study of disks associated with widely spaced component stars such as VV CrA (Ratzka *et al.* 2007). Essentially this has been a study of the disks rather than of the binaries themselves.

Vector separations have been measured for Z CMa (Ratzka & Leinert 2007) in good agreement with adaptive optics measurements, and for FU Ori (Quanz *et al.* 2006) in good agreement with earlier work . These measurements, although not providing new results, are a confirmation of the performance of the VLTI-MIDI combination and demonstrate its capability.

MIDI has also been used to observe the disks around the cores of planetary nebulae. Arguments have been advanced to suggest that the formation of the disks and their chemistry are caused by the presence of binary companions. Although binary companions have not been detected from the limited observations to date, the MIDI data and their interpretation have been discussed by Matsuura *et al.* (2006) and Chesneau *et al.* (2006).

Observations of some stars with MIDI have revealed circumstellar disks that are smaller than expected and this has led to the suspicion that the disks are being truncated by the presence of unseen companions – an example is the Be star α Ara (Chesneau *et al.* 2005).

To summarize the work so far with the relatively new MIDI instrument, most of the observational data have not been fully interpreted and in some cases more data are needed. A fair summary is to say that observations made with MIDI are mainly work in progress but that its potential for the study of young stars and disks in binary systems has been clearly demonstrated.

Table 4. Orbital parameters for γ^2 Vel: NSII values from Hanbury Brown *et al.* (1970), spectroscopic values from Schmutz *et al.* (1997), and SUSI values are preliminary results from work in progress. Values marked with an asterisk are the values adopted and fixed in the orbital solution for the interferometric data.

Parameter	NSII	Spectroscopy	SUSI
P (days)	78.5^*	78.53 ± 0.01	78.53 *
θ_a (mas)	4.3 ± 0.5		3.6
e	0.17 ± 0.03	0.326 ± 0.010	0.33
T (MJD)		50120 ± 2	50121
ω (deg)	$267 \pm 9^*$	248 ± 4	247.3
Ω (deg)			67.8
i (deg)	70^*	65 ± 8	65.5

4.3. *AMBER*

AMBER (Petrov *et al.* 2003) is a three aperture beam-combining instrument covering the J, H and K bands with the capability of recording spectra, dispersed fringes and closure phases. It is early in the life of the instrument and, apart from some early observations of δ Vel mentioned in Section 4.1, the only binary studied so far is γ^2 Vel and these observations will be discussed in Section 5.

Observations of η Car with AMBER, primarily aimed at studying the wavelength dependence of the optically thick wind region with high spatial and spectral resolution, have also been used to investigate the detectability of a hypothetical hot binary companion (T. Gull – private communication). While not directly detecting the companion, it is hoped that additional AMBER observations of higher accuracy will be sensitive enough to establish its existence.

Future plans for the VLTI-AMBER combination include observations of eclipsing binaries for distance determinations (A. Richichi – private communication).

5. γ^2 **Velorum**

γ^2 Vel is a double-lined spectroscopic binary system containing the brightest Wolf-Rayet star in the sky and it is therefore of particular interest. The primary is an O star with spectral classifications that range from O7 III to O9 I and the secondary is the WC8 Wolf-Rayet star. The system has been observed with the NSII, SUSI and VLTI-AMBER.

The first interferometric observations were made with the NSII in 1968 (Hanbury Brown *et al.* 1970). Observations were made in the continuum at 443 nm and in the C III-IV emission lines at 465 nm. At the time the Wolf-Rayet star was thought to be the primary and so the interpretation of the 443 nm results for the Wolf-Rayet star actually referred to the O star but the value determined for the angular semi-major axis of the orbit of 4.3 ± 0.5 mas is valid. The 465 nm observations were aimed at establishing the relative size of the C III-IV emission region around the Wolf-Rayet star but need to be repeated with the new knowledge of the system and this is planned for SUSI.

Observations with SUSI at 700 nm and a baseline of 80 m have been made to determine the orbit for γ^2 Vel and preliminary results are listed in Table 4. Also listed in Table 4 are the NSII results and spectroscopic results from Schmutz *et al.* (1997).

γ^2 Vel was observed in the dispersed K band on 25 December 2004 with the VLTI-AMBER combination using three UTs to give a range of interferometric and spectroscopic data (Petrov *et al.* 2007, Malbet *et al.* 2006, and R. Petrov, private communication). Although it yielded only a data set for a single night with limited u-v coverage it illustrated

Table 5. Fundamental quantities for γ^2 Velorum: NSII values from Hanbury Brown *et al.* (1970), SUSI values and uncertainties are preliminary results from work in progress, the AMBER distance is from Malbet *et al.* (2007) and the Hipparcos distance (d) is from Schaerer *et al.* (1997). The values in parentheses have been derived from the published data.

Parameter	NSII	SUSI	AMBER	Hipparcos
$M(\mathrm{O})/M_\odot$		28.6 ± 1.5		
$M(\mathrm{WR})/M_\odot$		9.0 ± 0.8		
π (mas)	(2.9 ± 0.4)	2.96 ± 0.16	(2.72 ± 0.37)	3.88 ± 0.53
d (pc)	350 ± 50	338 ± 20	368^{+38}_{-13}	258^{+41}_{-31}
$M_\mathrm{V}(\mathrm{O})$	-5.6	-5.7	(-5.9)	(-5.1)
$M_\mathrm{V}(\mathrm{WR})$	-4.6	-4.4	(-4.6)	(-3.8)

the potential of AMBER to provide spatial resolution for close binaries. In summary, one vector separation has been determined from the observations with the 180° ambiguity present for a two-aperture interferometer removed. Modelling of the spectra based on a simple geometrical binary model has been carried out giving a reasonable but not fully consistent fit with the observed data. The distance to the system has been determined from the AMBER data and this is listed in Table 5.

The distance has also been determined by combining the preliminary SUSI results with the spectroscopic results of Schmutz *et al.* (1997) to establish preliminary values for the masses of the component stars and the distance to the system and these are listed in Table 5 with the absolute visual magnitudes of the components.

In Table 5 the values shown in parentheses have been derived from the published results. The preliminary SUSI distance is in good agreement with the NSII and AMBER results and the following conclusions can be reached. The system is intrinsically brighter by a factor of ~1.7 than given by the Hipparcos parallax. The accuracy of the SUSI mass determinations, although not final values, shows that the accuracy is limited by the spectroscopic data. The separation of the spectra of the component stars and the determination of their relative flux contributions is a matter for further work.

6. Summary

The main contributions to binary star studies with multiple-aperture interferometry from the Southern Hemisphere have been by the NSII (α Vir) and currently by SUSI. SUSI is contributing mass determinations for early type stars including β Cep types in collaboration with Belgian spectroscopists. ESO's VLTI has yet to make a major contribution to binary star research but it has great potential as demonstrated by the observations of young binary systems, particularly those involving disks, with the N-band MIDI instrument and by early observations with the AMBER instrument in the K-band which, with its dispersed fringes, closure phase and imaging possibilities, has demonstrated its potential for studies of close binary systems. The planned programmes for SUSI and the VLTI are significantly different, particularly in the wavelength ranges covered but also, to a large extent, in the types of system observed. For these reasons the two instruments can be regarded as complementary for Southern Hemisphere interferometric binary star studies.

Acknowledgements

I would like to acknowledge my SUSI colleagues for the assistance they have given me in preparing this review and, in particular, Julian North for providing me with the preliminary SUSI results for γ^2 Vel. I also acknowledge information on the status of binary star observations with the VLTI that I have received from Olivier Chesneau, Aglaé Kellerer, Christoph Leinert, Romain Petrov, Thorsten Ratzka, Andrea Richichi and Markus Schöller.

References

Ausseloos, M., Aerts, C., Uytterhoeven, K., Schrijvers, C., Waelkens, C., & Cuypers, J. 2002, *A&A* 384, 209

Ausseloos, M., Aerts, C., Lefever, K., Davis, J., & Harmanec, P. 2006, *A&A* 455, 259

Ausseloos, M., Scuflaire, R., Thoul, A., & Aerts, C. 2004, *MNRAS* 355, 352

Chesneau, O. *et al.* 2005, *A&A* 435, 275

Chesneau, O. *et al.* 2006, *A&A* 455, 1009

Davis, J., Tango, W.J., Booth, A.J., ten Brummelaar, T.A., Minard, R.A., & Owens, S.M. 1999, *MNRAS* 303, 773

Davis, J., Mendez, A., Seneta, E.B., Tango, W.J., Booth, A.J., O'Byrne, J.W., Thorvaldson, E.D., Ausseloos, M., Aerts, C., & Uytterhoeven, K. 2005, *MNRAS* 356, 1362

Glindemann, A. *et al.* 2003, in: W.A. Traub (ed.) *Interferometry for Optical Astronomy II*, Proceedings of SPIE Vol. 4838, p. 89

Hanbury Brown, R., Davis, J., & Allen, L.R. *MNRAS* 1967, 137, 375

Hanbury Brown, R., Davis, J., & Allen, L.R. *MNRAS* 1974, 167, 121

Hanbury Brown, R., Davis, J., Herbison-Evans, D., & Allen, L.R. *MNRAS* 1970, 148, 103

Herbison-Evans, D., Hanbury Brown, R., Davis, J., & Allen, L.R. *MNRAS* 1971, 151, 161

Kervella, P. *et al.* 2003, in: W.A. Traub (ed.) *Interferometry for Optical Astronomy II*, Proceedings of SPIE Vol. 4838, p. 858

Leinert, C. *et al.* 2003, in: W.A. Traub (ed.) *Interferometry for Optical Astronomy II*, Proceedings of SPIE Vol. 4838, p. 893

Malbet, F., Petrov, R.G., Weigelt, G., Stee, P., Tatulli, E., Domiciano de Souza, A., & Millour, F. and the AMBER consortium 2006, in J.D. Monnier, M. Schöller, W.C. Danchi (eds.) *Advances in Stellar Interferometry*, Proceedings of SPIE Vol. 6268, CID 626802

Matsuura, M. *et al.* 2006, *ApJ* 646, 123

Otero, S.A., Fieseler, P.D., & Lloyd, C. 2000, *IAU Information Bulletin on Variable Stars*, No. 4999 (Konkoly Observatory, Budapest)

Petrov, R.G. *et al.* 2003, in: W.A. Traub (ed.) *Interferometry for Optical Astronomy II*, Proceedings of SPIE Vol. 4838, p. 924

Petrov, R. *et al.* 2007, in: F. Paresce & A. Richichi (eds.), *The Power of Optical/IR Interferometry*, Proc. ESO Workshop (Berlin: Springer-Verlag), in press

Quanz, S.P., Henning, Th., Bouwman, J., Ratzka, Th., & Leinert, Ch. 2006, *ApJ* 648, 472

Ratzka, Th., Leinert, Ch. 2007, in: F. Paresce & A. Richichi (eds.), *The Power of Optical/IR Interferometry*, Proc. ESO Workshop (Berlin: Springer-Verlag), in press

Ratzka, Th., Leinert, Ch., Przygodda, F., & Wolf, S. 2007, in: F. Paresce & A. Richichi (eds.), *The Power of Optical/IR Interferometry*, Proc. ESO Workshop (Berlin: Springer-Verlag), in press

Robertson, J.G., Bedding, T.R., Aerts, C., Waelkens, C., Marson, R.G., & Barton, J.R. *MNRAS* 1999, 302, 245

Schaerer, D., Schmutz, W., & Grenon, M. 1997, *ApJ* 484, L153

Schmutz, W., Schweickhardt, J., Stahl, O., Wolf, B., Dumm, T., Gäng, Th., Jankovics, I., Kaufer, A., Lehmann, H., Mandel, H., Peitz, J., & Rivinius, Th. 1997, *A&A* 328, 219

Struve, O., Ebbighausen, E. 1934, *ApJ* 80, 365

Struve, O., Sahade, J., Huang, S-S., Zebergs, V. 1958, *ApJ* 128, 310

Tango, W.J., Davis, J., Ireland, M.J., Aerts, C., Uytterhoeven, K., Jacob. A.P., Mendez, A., North, J.R., Seneta, E.B., & Tuthill, P.G. 2006, *MNRAS* 370, 884

Uytterhoeven, K., Willems, B., Lefever, K., Aerts, C., Telting, J.H., & Kolb, U. 2004, *A&A* 427, 581

Binary Stars as Critical Tools & Tests
in Contemporary Astrophysics
Proceedings IAU Symposium No. 240, 2006
W.I. Hartkopf, E.F. Guinan, & P. Harmanec, eds.

Using SIM for Double Star Astronomy

Michael Shao

Jet Propulsions Laboratory, California Institute of Technology, Pasadena, CA, USA
email: mshao@huey.jpl.nasa.gov

Abstract. The SIM (Planet Quest) mission is a space-based long-baseline stellar interferometer designed for ultra-precise astrometry. This paper describes how SIM can be used for double star research. There are several regimes of operation. For binary stars separated by more than $1\rlap{.}''5$, SIM treats these as distinct objects. Double stars less than ∼10 milliarcsec in separation are seen as a single object and SIM measures the photocenter of the composite object. Between 10 mas and $1\rlap{.}''15$, SIM is able to see the double star as two distinct objects, but because photons from both stars are detected there is the possibility of increased noise and measurement bias. This paper describes how double stars are observed with SIM and what information can be derived.

1. Introduction

The SIM mission (Figure 1) is a long-baseline interferometer designed to measure the positions of stars (and other point-like astronomical objects) with great precision. When the object is a double star, however, the observational procedure and data analysis needs to be slightly different.

We consider three types of double star observations:

(*a*) *wide binaries* >$1\rlap{.}''5$, where only one of the two stars can be observed at a time.
(*b*) *unresolved binaries*, where the separation is ≲10 mas.
(*c*) *narrow binaries*, whose separation is resolved by the interferometer but both stars' photons are detected by the fringe detector.

2. Cases 1 and 2

These two cases are the "trivial" situations. Binaries whose separations are wider than $1\rlap{.}''5$ are simply observed as two distinct objects. Their relative positions can be measured with 1 μas accuracy (at each epoch) and their absolute positions, proper motions, and parallax can be measured to ∼4 μas for position and parallax and ∼3 μas/yr for proper motions.

For Case 2, binaries <10 mas are unresolved by the interferometer and SIM just measures the photocenter of the binary.

3. Case 3

The third case is the one that merits more thought. The bottom line is that for binary stars whose separations are greater than ∼20 mas SIM can measure the position of both objects simultaneously with just two roughly orthogonal baseline orientations. This is possible because SIM measures the white light fringe in ∼80 spectral channels, and this information can be used to "synthesize" an image of the double star.

For angular separation between 10 mas and 20 mas the synthesized image must be done by rotational aperture synthesis.

Figure 1. The SIM PlanetQuest satellite.

SIM Fringe Detector

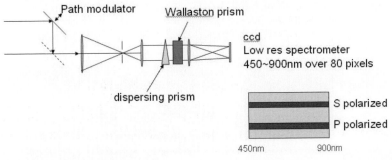

Figure 2. Schematic of the SIM fringe detector.

The beam combiner for SIM is shown in Figure 2. After the light is combined at a beam splitter, it goes through a 3″ diameter field stop, and then is dispersed; light in the 450 – 950 nm wavelength range is spread over 80 ccd pixels.

For double star observations, one would measure the fringe visibility and phase of the double star at 80 spectral channels. Before discussing double stars, we first illustrate the data processing procedure for a single star. In white light, the fringe pattern of a star (intensity vs. delay) is shown in Figure 3. Figure 4 shows the fringe pattern in two narrow spectral bandpasses (of the 80). We can synthesize an image of the star using

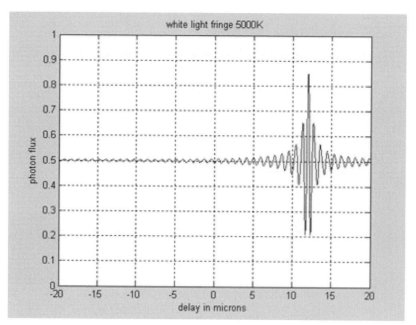

Figure 3. White-light fringe pattern of a single star.

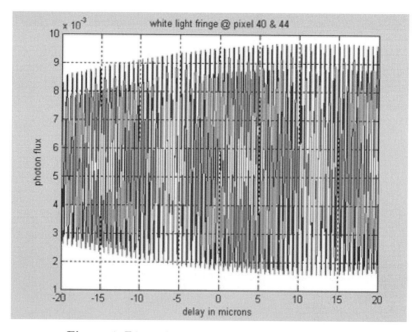

Figure 4. Fringes in two narrow spectral bandpasses.

the group delay estimator, which is simply the square of the Fourier transform of the complex visibilities. This is shown in Figure 5.

For a double star, this observation and analysis process leads to an image of the double star, as shown in Figures 6 and 7. Figure 6 is the fringe pattern of the double star and Figure 7 is the synthesized image of the double star.

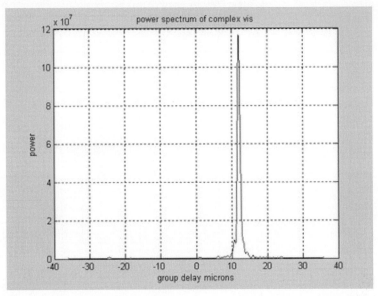

The Fourier transform of the complex Vis(1/λ)

Figure 5. Fourier transform of the complex visibilities.

4. Obtaining quantitative information

The synthetic image generated by taking the Fourier transform of the complex visibilities is only used to identify the location of the dimmer star of the binary. Quantitative information is obtained by a non-linear least squares fit of a double star model to the complex visibilities at the 80 spectral channels. That model should include:

- Angular separation of the stars
- Position angle of the companion
- Brightnesses of the two stars (at 80 spectral channels)
- Angular diameters of the two stars

The baseline of SIM is only 9 m, so we can expect that the diameter measurement would have an uncertainty of a few percent of the resolution of a 9-m telescope: 2% of $\lambda/2 \times 9\text{m} \approx 150\ \mu$as. That is, if the star's diameter is 5 mas, its uncertainty would be $\sim$150 μas.

The astrometric accuracy of the double star measurement is the same as for a single star, with one exception. Because both stars are seen by the fringe detector, the position of each star has not just the photon noise from one star but from both. For example, it takes an integration time of $\sim$500 sec to get 1-μas astrometry on a 10th mag target. If the binary consists of a 10th mag and a 12.5 mag star, then their separation would have an uncertainty of 10 μas after a 500-sec integration. The 12.5 mag star has 1/10 the signal. If we were only observing a 12.5 mag star the noise would be $\sqrt{10}$ lower as well, and the accuracy would be $\sim$3 μas. But because the 12.5 mag star is being observed in the presence of a 10th mag star the noise is not lower. The 1/10 lower signal means 1/10 the SNR, and 10 times lower astrometric accuracy. If very accurate binary separations are desired for stars with large magnitude differences, the integration time must be increased to account for the photon noise of the primary on the secondary's position.

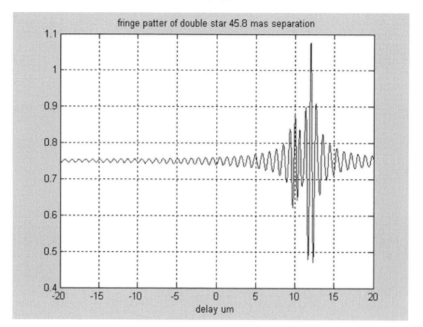

Figure 6. Double star fringe pattern.

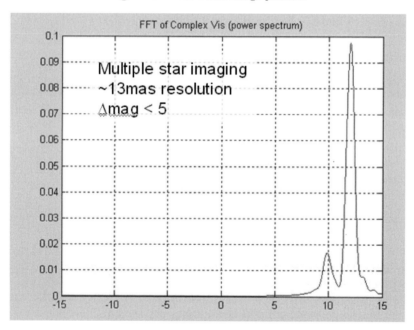

Figure 7. Synthesized double star image.

5. Summary

The SIM mission is capable of making very precise binary star measurements, both in a relative sense (separation and position angle) but also in an absolute sense, absolute parallax and reflex motion of the primary due to the secondary. In addition, SIM can measure the magnitude difference and colors of both binary components, as well as estimate the diameters of both stars.

Binary Stars as Critical Tools & Tests
in Contemporary Astrophysics
Proceedings IAU Symposium No. 240, 2006
W.I. Hartkopf, E.F. Guinan, & P. Harmanec, eds.

© 2007 International Astronomical Union
doi:10.1017/S1743921307003808

Spectroscopic Binary Candidates for Interferometers

Francis C. Fekel[1] and Jocelyn Tomkin[2]

[1]Center of Excellence in Information Systems, Tennessee State University,
Nashville, TN 37209, USA
email: fekel@evans.tsuniv.edu

[2]Department of Astronomy, University of Texas, Austin, TX 78712, USA
email: jt@alexis.as.utexas.edu

Abstract. We present a progress report on a program to improve the orbits of known spectroscopic binaries that are potential targets for ground-based optical interferometers, such as the CHARA array and PTI. The combination of such observations results in three-dimensional orbits that produce very accurate masses (uncertainties of less than 0.5%) as well as orbital parallaxes that are generally more precise than those from *Hipparcos*. After additional analyses determine other basic parameters of the stars, the components can be compared with theoretical evolutionary tracks. We are currently using three observatories to obtain high-resolution, red-wavelength spectra of 44 systems. Results from four years of observation are highlighted. Interferometric observations of some systems are in progress.

Keywords. binaries: spectroscopic, stars: fundamental parameters

1. Introduction

With routine operation of several ground-based interferometers such as the Palomar Testbed Interferometer (PTI) (Colavita *et al.* 1999), the Naval Prototype Optical Interferometer (NPOI) (Hummel *et al.* 2003), and the Center for High Angular Resolution in Astronomy (CHARA) array (ten Brummelaar *et al.* 2003), the period overlap of spectroscopic and visual binaries has been greatly increased. The interferometric observations, combined with precise radial velocities, provide an opportunity to determine very accurate masses plus orbital parallaxes more accurate than those from *Hipparcos* for a significant number of systems.

2. Binary sample

We searched The Eighth Catalogue of the Orbital Elements of Spectroscopic Binary Systems (SB8) (Batten *et al.* 1989) for double-lined binaries that might be resolved interferometrically and could profit significantly from spectroscopic observations with modern CCD detectors. The new velocities would produce more precise minimum masses as well as updated orbital ephemerides to facilitate interferometric observations at the most appropriate orbital phases. Our initial list included 83 double-lined binaries north of $-30°$ in declination, and generally brighter than $V = 7.5$ mag. In the hope of turning useless single-lined systems into valuable double-lined ones (e.g., Stockton & Fekel 1992), we added to our list 50 single-lined binaries with mass functions greater than 0.05 $M_\odot$. Also included were a few binaries from the more recent literature and systems with a giant component that we had previously been observing.

Of the 44 systems listed in Table 1, we have acquired at least 10 spectra for 37. In addition to the HD number, we give the V mag, spectral type of the primary, the

Table 1. Observing list

HD	V (mag)	Spectral Type	P (days)	Mass Ratio	Pri- ority	HD	V (mag)	Spectral Type	P (days)	Mass Ratio	Pri- ority
434	6.5	Am	34.3	0.84	2	120064	6.0	F6IV	36.0	0.85	2
8374	5.6	F1m	35.4	0.98	2	141458	6.8	A0V	28.9	0.91	2
9021	5.8	F6V	134.1	0.75	1	148367	4.6	A5Vm	27.2	0.85	3
9312	6.8	G5IV	36.6	0.75	1	157950	4.5	F2	26.3	0.63	1
15138	6.1	F4V	11.0	0.93	2	160922	4.8	F5V	5.3	0.82	1
20210	6.2	F0m	5.5	0.61	1	168913	5.6	F0m	5.5	0.85	2
24623	7.1	F2	19.7	0.98	3	170153	3.6	F7Vmw	280.6	0.72	1
30453	5.9	F0m	7.0	0.66	1	171653	6.6	Am	14.3	0.94	2
40084	5.9	G5III	219.1	0.98	1	171978	5.8	A2V	14.7	0.98	3
42083	6.2	A5III	106.0	0.96	1	178619	6.5	F5IV-V	4.8	1.00	3
61859	6.0	F7V	31.5	0.89	2	182490	6.2	A1III	7.4	0.74	1
82191	6.6	Am	9.0	0.85	3	185734	4.7	G8III	434.1	0.96	2
86146	5.1	F5V	9.3	0.80	1	188088	6.2	K3V	46.8	1.00	1
93903	5.8	Am	6.2	0.50	1	191747	5.5	A2IV	9.3	0.91	2
96511	7.1	G0IV	18.9	0.92	2	203439	6.0	A1IV	20.3	0.58	1
102713	5.7	F5IV	32.9	0.80	1	205539	6.3	F2IV	12.2	0.88	2
103578	5.5	A3IV	6.6	0.6	2	210027	3.8	F5V	10.2	0.62	2
106677	6.4	K0III	64.4	0.98	1	210763	6.4	F6V	42.4	0.81	2
108642	6.5	A2m	11.8	0.5	2	214686	6.9	F7V	21.7	0.99	3
110318	5.2	F5V	44.4	0.50	1	218527	5.4	G8III	920.1	0.99	1
112486	5.8	A8m	5.1	0.94	3	221950	5.7	F6V	45.5	0.94	3
120005	6.6	F5	39.3	0.58	1	224355	5.6	F6V	12.2	0.99	3

orbital period, the secondary to primary mass ratio (mostly from our own work), and our observational priority. The primaries have spectral classes from early-A to early-K and are mostly dwarfs, but some subgiants and giants are included. The periods range from 5 to 920 days. Highest priority, a value of 1, has been assigned to those binaries with mass ratios significantly less than unity and also to systems containing evolved components. Such systems will have the greatest leverage in a comparison of observational results with theoretical evolutionary tracks.

3. Spectroscopic observations

Extensive observing began in 2002 April, and we are currently obtaining spectrograms at three observatories. Our observations at McDonald Observatory are acquired with the 2.1 m telescope and the Sandiford Cassegrain echelle spectograph. Those spectra cover the wavelength range 5600 – 7000 Å and have a resolving power of 60,000. Additional spectroscopic observations are collected with the Kitt Peak National Observatory 0.9 m coudé feed telescope, coudé spectrograph, and a TI CCD. The observations are centered at 6430 Å, cover a wavelength range of about 80 Å, and have a resolution of over 30,000. Typical signal-to-noise ratios of the spectra are 250. Spectra of some binaries are being obtained at Fairborn Observatory with the 2 m automatic spectroscopic telescope and a fiber-fed echelle spectrograph with a SITe CCD. The wavelength range is 4920 – 7100 Å, and the resolving power is about 35,000.

4. Results

We have detected the secondaries of 10 binaries with previous single-lined orbits, HD 434, HD 9021, HD 9312, HD 20210 (Fig. 1), HD 96511, HD 102713, HD 103578, HD

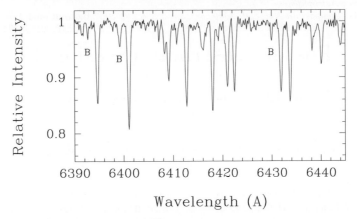

Figure 1. Spectrum of HD 20210. Several lines of component B are indicated.

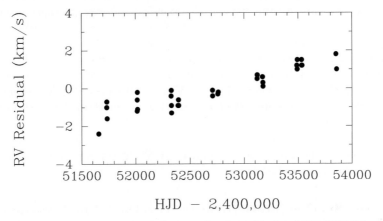

Figure 2. Velocity residuals from a circular orbit fit of HD 108642A.

108642, HD 110318, and HD 120005. Velocities for two of those systems, HD 103578 and HD 108642 (Fig. 2), have systematic residuals, so the systems are at least triple. Our spectra of HD 148367 have three sets of lines, so that system also is at least triple. Tomkin & Fekel (2006) published the first results of our program, new spectroscopic orbits for RR Lyn, 12 Boo, and HR 6169. Because RR Lyn is eclipsing and 12 Boo already has an interferometric orbit, we obtained masses with a precision better than 0.5% for the components. Our new orbit of HR 6169 produced significantly revised minimum masses. Many of our top priority systems are now being observed with the CHARA array.

Acknowledgements

This work is supported by NSF grant HRD-9706268 to Tennessee State University.

References

Batten, A.H., Fletcher, J.M., & MacCarthy, D.G. 1989, *Publ. Dom. Astrophys. Obs.* 17, 1
Colavita, M.M., *et al.* 1999, *ApJ* 510, 505
Hummel *et al.* 2003, *AJ* 125, 2630
Stockton, R.A., & Fekel, F.C. 1992, *MNRAS* 256, 575
ten Brummelaar, T.A., *et al.* 2003, *Proc. SPIE* 4838, 69
Tomkin, J., & Fekel, F.C. 2006, *AJ* 131, 2652

Binary Stars as Critical Tools & Tests
in Contemporary Astrophysics
Proceedings IAU Symposium No. 240, 2006
W.I. Hartkopf, E.F. Guinan & P. Harmanec, eds.

Poster Abstracts (Session 1)

(Full Posters are available at http://www.journals.cambridge.org/jid_IAU)

Comparison of Selected Methods of Radial Velocity Measurement

Stefan Parimucha[1] and Petr Skoda[2]
[1] *Safarik University, Kosice, Slovakia,*
[2] *Astronomical Institute of CAS, Ondrejov, Czech Republic*

The measurements of the radial velocity (RV) of spectral lines is a fundamental task of spectral analysis. Most of the current astronomical software packages use in addition to cross-correlation, fitting of line profiles with Gaussian, Lorentzian or Voigt functions. One of the less known and used method is a comparison of the direct and reverse images of the line profiles, so called "method of mirroring". It is successfully used in the SPEFO program developed at Astronomical Institute of the Academy of Sciences of the Czech Republic. Lately, it was implemented in the Virtual Observatory (VO) enabled package SPLAT-VO. Because of its simplicity and intuitivity, this method can still find application in the modern analysis of broad noisy and misshapen line profiles.

In this contribution we tried for an objective comparison of the accuracies of these methods. We show comparison of these methods on different types of lines and discuss their advancements and disadvantages.

Double Stars Speckle Interferometry with the 3.5 m Telescope at Calar Alto (Almeria, Spain)

J.A. Docobo[1], V.S. Tamazian[1], M. Andrade[1], J.F. Ling[1], Y.Y. Balega[2], J.F. Lahulla[3], A. Maximov[2], and J.R. Gonzalez-Romay[1]
[1] *Observatorio Astronómico Ramón María Aller, Santiago de Compostela, Spain,*
[2] *Special Astrophysical Observatory, Karachai-Cherkessia, Russia,*
[3] *Observatorio Astronómico Nacional, Madrid, Spain*

The first results of speckle interferometry carried out with the 3.5-m telescope of the C.A.H.A. (Almeria, Spain) during a run in July, 2005 are presented. Forty nine stars with separations between 0.058 and 2.1 arcsec were observed under good seeing conditions. On the basis of these observations three improved orbits are presented. The Time Allocation Committee's report with its high scientific qualification to our proposal is included. It confirms the relevance of binary and multiple star research in modern astronomy and the great significance of large telescopes in this kind of studies as well.

GL 569B: A Brown Dwarf Triple?

M. Simon[1], C. Bender [1], and L. Prato[2]
[1] *SUNY, Stony Brook, NY, United States,*
[2] *Lowell Observatory, Flagstaff, AZ, United States*

GL 569 is a multiple system comprised of GL 569 A, a M2.5V star (Henry & Kirkpatrick, ApJ 359, L29, 1990) of mass $M = 0.35 \pm 0.03$ $M_\odot$ (Gorlova *et al.*, ApJ 593, 1074, 2003) at distance 9.81 ± 0.16 pc, and GL 569 B, thought to be a binary brown dwarf with period ~ 876 days (Zapatero Osorio *et al.*, ApJ 615, 958, 2004). We began a study of GL 569 B in 2002 because the astrometric orbit is well determined and Ba and Bb are bright enough to be observed at high spectral resolution in the near-IR. Thus, GL 569 B is well suited for the fusion of visual and spectroscopic binary techniques to measure the mass of its components.

Here we describe the evidence that GL 569 B is actually a hierachical triple with components of roughly equal mass, $\sim 0.040 M_\odot$. Details, and evidence that at age about 100 Myr, the system is younger than had been reported earlier, are presented in Simon *et al.* (ApJ, 644, 1183, 2006).

Interferometric Investigations of Eclipsing Binaries as a Key to an Improved Distance Scale

K. Shabun[1], A. Richichi[1], U. Munary[2], A. Siviero[2], and B. Paczynski[3]
[1] *European Southern Observatory, Garching, Germany,*
[2] *INAF Osservatorio Astronomico, Padova, Italy,*
[3] *Princeton University, Princeton, NJ, United States*

Binary and multiple systems constitute one of the main tools to obtain fundamental stellar parameters, such as masses, radii, effective temperatures and distances. One especially fortunate, and at the same time rare, occurrence is that of double-lined eclipsing binaries with well-detached components. In this special case, it is possible to obtain a full solution of all orbital and stellar parameters, with the exception of the effective temperature of one star, which is normally estimated from spectral type, reddening-corrected photometric colors, or derived from atmospheric analysis of the spectrum. Long-baseline interferometry at facilities such as the ESO VLTI is beginning to have the capability to measure directly the angular separation and the angular diameter of some selected eclipsing binary systems, and we have proposed such observations with the AMBER instrument.

In particular, we aim at deriving directly the effective temperature of at least one of the components in each binary system by iterative convergence of the orbital solution and of the interferometric measurements, thereby avoiding any assumptions or required external calibration in the global solution through the Wilson-Devinney method. We will also obtain an independent check of the results of this latter method for what concerns the distance to the systems.This represents a first step towards a global calibration of eclipsing binaries as distance indicators. Our results will also contribute to the effective temperature scale for hot stars.The extension of this approach to a wider sample of eclipsing binaries could provide an independent method to assess the distance to the LMC.

Introduction to the 30 m Ring Interferometric Telescope

Z. Liu and S.-B. Qian

Yunnan Observatory, Kunming, China

For the demands of astronomical limitation observations, such as exploring extra-terrestrial planets or black hole accretion disks and jets in the near-infrared and optical wave band, extremely large telescopes (optical and infrared) have become the principal ground-based astronomical instrumentation. With the maturation of interferometric imaging theory, the borderline between new generation ground-based extremely large telescopes and interferometric arrays for aperture synthesis imaging is becoming increasingly blurred, and the differences in their technical methods and characteristics are also gradually disappearing. Based on the research results of interferometric imaging in Yunnan Observatory, we bring forward a new concept ground-based extremely large telescope — the 30m Ring Interferometric Telescope (30mRIT). It has the direct imaging ability and resolution of a single aperture telescope, and it also can image with high resolution like the aperture synthesis imaging mode. The 30m RIT has a ring spherical primary mirror with 90 segmented mirrors, the width of the ring is 1 meter and the F/D ratio is about 0.8. This report also introduces some high resolution astronomical observe results by a one-meter ring which is 1 m in diameter and 100 mm in width. The 30mRIT project is remarkably different from the conventional ground-base ELT and its pivotal techniques have received the support of CAS and China NSF.

V379 Cep: A Quadruple System of Two Binaries

P. Harmanec[1], P. Mayer[1], H. Bozic[3], P. Eenens[4], E.F. Guinan[5], G. McCook[5], P. Koubsky[2], D. Ruzdjak[3], D. Sudar[3], M. Slechta[2], M. Wolf[1], and S. Yang[6]

[1]*Astronomical Institute, Charles University, Praha, Czech Republic,*
[2]*Astronomical Institute, Academy of Sciences, Ondrejov, Czech Republic,*
[3]*Hvar Observatory, Faculty of Geodesy, Zagreb University, Zagreb, Croatia,*
[4]*Departamento de Astronomia, Universidad de Guanajuato, Guanajuato, Mexico,*
[5]*Department of Astronomy, University of Villanova, Villanova, United States,*
[6]*Physics & Astronomy Department, University of Victoria, Victoria, Canada*

Based on several published photometric and spectroscopic studies, V379 Cep (HR 7940) was identified as a 99.76-day eclipsing binary composed of two unusually-low-mass B type stars with projected rotational velocities of 15 and 55 km/s. Our investigation of new series of spectra and UBV observations from several observatories shows that the object is actually a quadruple system: the two observed sets of spectral lines belong to two primaries of two different binaries. The narrower lines belong to component Aa with a 99.76-day orbital period, while the broader ones belong to component Ba with a 159.3-day period. Both binaries revolve around a common centre of gravity with a period of about 6200 days; this motion is measurable via the change of the systemic velocities of both binaries.

Our result resolves the problem of anomalous masses. It is probable that the masses of all four bodies are quite normal for somewhat evolved stars. However, the ultimate test must come from interferometric resolution of the AB pair of binaries; it might be expected that the separation can reach about 40 mas. There is a very good chance that continuing systematic observations will permit accurate determination of all four individual masses.

Spitzer Observations of the Eclipsing Binary GU Bootes

G. van Belle[1], K. von Braun[1], D. Ciardi[1], D. Hoard[2], and S. Wachter[2]
[1] *Michelson Science Center, Pasadena, CA, United States,*
[2] *Spitzer Science Center, Pasadena, CA, United States*

We present a carefully controlled set of Spitzer MIPS time series observations of the newly discovered low-mass eclipsing binary star GU Bootes. These observations serve to characterize the MIPS-24 observing techniques of the spacecraft, precisely establishing the photometric repeatability of this instrument at the sub-percent level. The long wavelength characterization of this object's light curve allows for improved characterization of the primary and secondary component linear radii, in addition to other aspects of their surface morphology.

Tomography of the X-Ray Binary Cyg X-1 Based on High-Resolution Optical Spectroscopy

E.A. Karitskaya[1], M.I. Agafonov[2], N.G. Bochkarev[3], A.V. Bondar[2,4], and O.I. Sharova[2]
[1] *Astronomical Institute of RAS, Moscow, Russia,*
[2] *Radiophysical Research Institute (NIRFI), Nizhny Novgorod, Russia,*
[3] *Sternberg Astron. Institute, Moscow, Russia,*
[4] *IC AMER, Terskol, Russia*

We used optical spectra with resolution R=13000 obtained in the course of Cyg X-1 spectral monitoring over 2003-2004, carried out with the echelle spectrometer of the 2-m telescope of Peak Terskol Observatory (3100 m, Caucasus). The sequence of line profile variations with orbital phases is clearly pronounced. The Doppler images were reconstructed by an improved Doppler tomography method developed by Agafonov (2004) (radioastronomical approach) on the base of HeII λ4686Å profiles of 2003 ("soft" X-ray state) and 2004 ("hard" X-ray state). The main features of the reconstruction are: deconvolution in the image space with the introduction of the synthesized beam (equivalent summarized transfer function) and the removal of the distortions on the summarized image (after back projecting) caused by the sidelobes of this beam using the CLEAN algorithm. The method is developed specially for a small number of irregularly distributed observations.

The Doppler images and Roche lobe model allowed putting a limitation on the black hole to supergiant mass ratio $1/4 \leqslant M_X/M_O \leqslant 1/3$.

The emission may come from the accretion disk outer regions heated by the hot supergiant emission, from the "hot line" discussed by Kuznetsov *et al.* (2001), or/and from the accretion stream (focused stellar wind).

This work is supported by RFBR grant 04-02-16924.

The Orbit and Properties of the Spectroscopic-Eclipsing-Interferometric Triple System ξ Tauri

C.T. Bolton and J.H. Grunhut
David Dunlap Observatory, University of Toronto, Richmond Hill, ON, Canada,
e-mail: bolton@astro.utoronto.ca

ξ Tauri is a triple-lined spectroscopic binary consisting of a bright, broad-lined, main sequence B star in orbit about a close pair of sharp-lined A stars. We have used radial velocities measured from more than 100 years of spectrograms from the Lick, Perkins and David Dunlap Observatories to derive the spectroscopic orbital elements for this system. The ratio of the orbital periods, 145.1317 ± 0.0040 d to 7.1466440 ± 0.0000049 d is among the smallest known for triple systems. The inner orbit is circular, while the outer orbit has a modest eccentricity, $e = 0.149$. Photometric observations obtained during the HIPPARCOS mission show that ξ Tauri is an eclipsing binary with eclipses that are at least 0.1 mag deep. Based on the steepness of the ingress to primary eclipse,we believe that the eclipses of the inner pair are total, but additional photometry is required to prove this. In any event, the orbital inclination of the inner pair must be very close to $90°$. If we assume this is correct, then the masses of the inner pair are 2.21 ± 0.02 and 2.12 ± 0.02 M$_\odot$. When this is combined with the results of the outer orbit, we find that the inclination of this orbit is either $63° \pm 4°$ degrees or $116° \pm 4°$. This yields a primary mass of 3.12 ± 0.16 M$_\odot$. This system has also been observed as an interferometric binary by a number of groups, but as yet, no one has published an orbital solution. Both the depth of eclipses and the strengths of the A-star lines are inconsistent with some of the interferometric estimates of the magnitude difference between the B star and the close pair of A stars. We will present a combined solution for the astrometric and spectroscopic data and comment of the stability of the system.

Session 2: New Observing Techniques and Reduction Methods:

New possibilities for standard observational techniques

Figure 1. (top left to bottom right) Speakers Ignasi Ribas, Alceste Bonanos, Brian Mason, William Hartkopf, Nancy Evans, and E.J.M. van den Besselaar.

Figure 2. (top left to bottom) Speakers Luis Paolo R. Vaz, Petr Hadrava, and Rainer Köhler.

Figure 3. (left) Leopold gate (1678), south entrance gate to the Vyšehrad castle from the time when this area was changed to a military fortress. (right) Romanesque rotunda of St. Martin at Vyšehrad castle, from the 11th century.

Binary Stars as Critical Tools & Tests
in Contemporary Astrophysics
Proceedings IAU Symposium No. 240, 2006
W.I. Hartkopf, E.F. Guinan & P. Harmanec, eds.

The New Era of Eclipsing Binary Research with Large Telescopes

Ignasi Ribas

Institut de Ciències de l'Espai (CSIC–IEEC), Campus UAB, Facultat de Ciències, Torre
C5-parell, 2a planta, 08193 Bellaterra, Spain
email: iribas@ieec.uab.es

Abstract. The advent of larger telescopes and powerful instrumentation enables the exploration of new aspects of faint eclipsing binaries that are just now becoming accessible. An example of this are eclipsing binaries in Local Group galaxies such as the LMC, SMC, M31 and M33, whose study yields not only stellar properties of stars formed in different chemical environments (thus providing useful model tests) but also direct distance determinations to the host galaxies. In general this is also applicable to eclipsing binaries belonging to any stellar ensemble. Another example is the observation and study of eclipsing very-low mass stars, brown dwarfs and planets. Besides the need for large telescopes because of their faintness, these also benefit from improved observational capabilities in the infrared spectral windows. Here we discuss the prospects for eclipsing binary research using photometry and spectroscopy from large telescopes.

Keywords. binaries: eclipsing – stars: distances – stars: low-mass, brown dwarfs – stars: magnetic fields – distance scale

1. Introduction

Eclipsing binary (EB) research has made a strong impact on stellar astrophysics for over a century. The high-accuracy stellar properties that result from their study have greatly contributed to our understanding of stellar structure and evolution in a broad range of masses, evolutionary stages and chemical compositions. But, in addition, EBs have also played a role in many other aspects, such as close binary evolution, accretion physics, stellar atmospheres, etc (see Ribas 2006a for a review). For most of this research, observational data have been collected using small- to moderate-size telescopes because the EB systems studied are generally bright. In recent times, the increasing instrumental capabilities and telescope sizes have expanded the horizons to the study of faint EBs and made it possible to address new topics that had been beyond reach.

Classical EB work uses, in general, data in three different flavors: *1)* Multi-band time-series photometry in the form of light curves; *2)* Time-series spectroscopy to build a radial velocity curve; and *3)* Standard photometry or spectrophotometry. It is only from the combined analysis of these datasets that a full characterization of the EB system, including orbital and physical properties (M, R, L, $T_{\rm eff}$, $[Fe/H]$), is possible. In the case of time-series photometry, this was carried out mostly with photomultipliers until some 15 years ago, when CCDs became widely available and greatly increased the capabilities and efficiency of small telescopes. However, the accuracy reachable by CCDs was poorer than that of photoelectric detectors (typically 0.01–0.015 mag vs. a few mmag). This has remained true until recently, when the flawless cosmetics and improved stability of CCDs have greatly enhanced their performance. Also, sophisticated analysis techniques, such as optimal image subtraction (Alard & Lupton 1998), have made it possible for time-series CCD photometry to reach the 1 mmag level (see, e.g., Moutou *et al.* 2004).

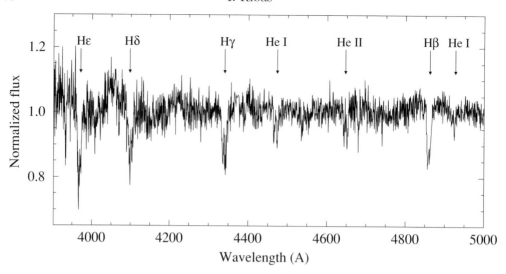

Figure 1. Spectrum of an EB target in M31 ($V = 19.3$ mag) obtained with the GMOS instrument of Gemini-N with a 1 hour integration. The average S/N of the spectrum is 15.

In general, EB photometry work is carried out with relatively small telescopes. To serve as a reference, good light curves of EBs of 14–15th mag can be obtained with a 1–1.5-m telescope, while 2.5–4-m telescopes can reach down to 19–20th mag. Obviously, these numbers depend on the length of the possible integrations and therefore on the orbital period of the binary. Much larger apertures are needed in the case of spectroscopy. Stars of 14–15th mag require 4–6-m telescopes and stars of 19–20th mag require today's largest 8–10-m telescopes, and even then, low spectral resolution has to be used to attain reasonable S/N values. The spectrum in Figure 1, obtained using a combination of a 8-m telescope (Gemini-N) and a high-throughput spectrograph (GMOS) illustrates this latter fact. A 1-hour exposure with a relatively low resolution of 3800 yielded a spectrum with a S/N of ∼15 for an EB target of visual magnitude 19.3.

In view of the numbers above, large telescopes are needed to study EBs in case these are intrinsically faint (such as low-mass stars or degenerates), are in highly absorbed lines of sight, are located at large distances or when there is some need for very high angular or time resolution. Also, it is true that new instrument developments and resources will be focused in years to come on large facilities. In spite of the strong competition, there are a number of EB projects that are being successful in obtaining precious large telescope time. In general, such success is related to the impact of the proposed research not only in understanding an individual object but in a broader area of Astrophysics. Studies making significant contributions to stellar structure and evolution, accretion physics or with cosmological implications stand the best chances. Also, it is important to point out that the instrumentation available in large telescope is not always perfectly suitable for stellar work (e.g., often only low resolution modes are available in spectrographs), a fact that requires adapting of the science goals and target selection to maximize the outcome.

Some examples of EB projects where the use of large telescopes make a difference will be discussed next. Basically, these are EBs in stellar ensembles (and, particularly, the Local Group galaxies) or EBs with components of very low mass, such as M-type stars, brown dwarfs or even planets.

2. Eclipsing Binaries in Stellar Ensembles

EBs provide simultaneous determinations of stellar masses and radii to an accuracy potentially better than a few percent, and these have been used for decades to better understand stellar structure and evolution (Pols *et al.* 1997; Ribas *et al.* 2000a). As a consequence of the full characterization of the system components, EBs provide determinations of the distances, which can be also as accurate as a few percent (see Clausen 2004 for a review). In case an EB system belongs to a stellar ensemble, its relevance is automatically enhanced. First, it provides much better and demanding tests of stellar models thanks to the age and chemical composition constraints, but also because of the added value of a distance measurement. Such stellar ensembles include open and globular clusters, star-forming regions, wide binaries, the Galactic bulge, and Local Group galaxies.

As discussed above, different datasets are needed to analyze EBs and, in particular, systems in stellar ensembles. Photometry requires arguably the larger investment of observing time, but can use relatively small telescopes. The single-target observing mode was the standard one until the mid/late 1990s, when the results of massive photometric surveys became available. Such surveys, consisting of calibrated time-series photometry, were designed to detect microlensing events and were focused on the Galactic bulge and the Large and Small Magellanic Clouds (LMC and SMC) (EROS: Grison *et al.* 1995; MACHO: Alcock *et al.* 1997; OGLE: Udalski *et al.* 1998, Wyrzykowski *et al.* 2003). But the legacy of the surveys goes beyond microlensing and they have provided phenomenal databases of photometry of variable stars, in particular EBs. As a result, the number of known EBs has increased ten-fold. However, it has to be noted that such observations were not tailored to be of use to EB research and therefore they have some shortcomings, most notably, the use of just one or two passbands. More recently, there have been specific surveys aimed at the detection of EBs in Local Group galaxies (Bonanos *et al.* 2003; Vilardell *et al.* 2006) and in a number of galactic open and globular clusters that have reported hundreds of new systems. It seems, therefore, that the availability of photometric data is not the limiting factor to the analysis of EBs in stellar ensembles.

The situation with spectroscopy is completely different. As we have mentioned before, the study of the faint targets detected by the microlensing surveys mostly requires large telescope time. This is certainly the bottleneck to a full determination of their properties, although new instrumentation with multi-object spectrographs are due to change the situation in the next several years. Note, however, that spectroscopy may not be the only way to extract scientifically valuable information from the survey EB data. Several groups (e.g., Tamuz *et al.* 2006; Devor & Charbonneau 2006) are developing codes for massive modeling of time-series photometry of EBs with very encouraging results.

Focusing on large telescopes, there are a number of examples illustrating their contribution to the characterization of EBs. Starting with Galactic systems, globular clusters pose very interesting cases to the study of their member EBs, from the point of view of distance and age determination and also to investigate stars formed in a very low metallicity environment. The CASE (Cluster AgeS Experiment) project illustrates the science case and the observational requirements of the study of EBs in globular clusters (see Kaluzny *et al.* 2005). Young open clusters also provide very interesting objects to analyze. Indeed, the the open cluster Westerlund 2 contains the Wolf-Rayet binary WR20a, whose component masses were determined to be 83 and 82 $M_\odot$ from spectroscopy using the NTT 3.5-m telescope and SOFI (Rauw *et al.* 2004). This binary holds the honor of being composed by the stars with the highest masses measured directly so far (Bonanos *et al.* 2004). Other stars are thought to have larger masses (such as the Pistol star or η Car), but these have been inferred through model comparisons.

A further step in distance leads to the Local Group galaxies. Our nearest neighbors, the LMC and SMC, with today several thousand EBs, have been the subject of spectroscopic studies for over a decade. Some of the brightest EBs can be measured spectroscopically with 1–2-m class telescopes (e.g., Niemela & Bassino 1994), but accurate work requires 3.5 to 4-m telescopes. This is the case of the LMC targets of Guinan *et al.* (1998), Fitzpatrick *et al.* (2002, 2003) and Ribas *et al.* (2002), whose magnitudes are between 14 and 15, and were observed with the 4-m Blanco telescope at CTIO. In contrast, the approach taken by Harris *et al.* (2003) and Hilditch *et al.* (2005) to study SMC EBs is quite different. They sacrificed spectral resolution but gained enormously in overall efficiency by using the 2dF spectrograph on the AAT 3.5-m telescope, which was indeed specially designed for Cosmology (galaxy redshifts). In this manner, the authors managed to measure radial velocity curves for a relatively large number of systems (50) with a quite modest investment of time.

The main driver behind studying EBs in Local Group galaxies has been the determination of the Cosmic Distance Scale. This was mainly motivated by the controversy in the late 1990s over the true distance to the LMC. Competing groups claimed distance values differing by over 20% (the so-called "long" and "short" distance scales) and this had direct consequences on the overall distance scale and, in particular, on the value of the Hubble constant, of clear cosmological implications. The analysis of EB systems permits the determination of the absolute radii of their components and also their temperatures, which, when combined with the fluxes observed from Earth, yield an accurate determination of the distance (e.g., Clausen 2004). The study of EBs, together with the improvement of other techniques, has helped to clear out the situation and converge towards an LMC distance value of 48.0–48.5 kpc (e.g., Macri et al. 2006), which is somewhat shorter than the canonical value of 50 kpc.

Recent further improvements in instrumental capabilities has allowed for the study of EBs in even more distant Local Group galaxies: M31 and M33. Photometric surveys with 2-m class telescopes (Bonanos *et al.* 2004; Vilardell *et al.* 2006) have provided hundreds of new EB systems in these galaxies. Spectroscopic studies, much more demanding for such objects of magnitudes 19–20, have also been carried out with success for two EB systems, one in M31 (Ribas *et al.* 2005) and one in M33 (Bonanos *et al.* 2006), using the GMOS instrument at the 8-m Gemini-N telescope. A more detailed discussion of EBs in Local Group galaxies and their implications to the Cosmic Distance Scale is provided by Bonanos (this volume). Challenges for the future include reaching to fainter magnitudes to avoid dealing with the hottest systems of each Local Group galaxy. As is well known, reliable determination of temperatures for the O- and early B-type stars is still an open issue. In the case of the LMC/SMC the solution is already in hand since late B-type EB pairs would have magnitudes of about 17–18 and these can be observed spectroscopically using 6–8-m class telescopes. The study of EBs with components of ~15000-K, whose temperature scale is considerably better defined, will provide a final check of the current results and the ultimate determination of the distances to the Magellanic Clouds.

But Local Group EB systems are much more valuable than just accurate distance indicators. Decades of work has shown that EBs constitute excellent benchmarks to test stellar structure and evolution models. The extension of this to binaries in the Local Group, which have been formed in environments with a chemical history differing from that of the solar neighborhood, is the true legacy of the efforts to study these objects. A conspicuous example is the LMC EB HV 2274, which was used to assess the impact of convective overshoot in the evolution of high-mass stars with sub-solar metal abundance

(Ribas *et al.* 2000b). The same idea could be applied to SMC EBs. In this case, their metallicity is about 1/10th solar and could act as laboratories to study high-mass stars that existed long ago in our Galaxy before the enrichment of the interstellar medium. Research in this direction should become quite active in the near future once accurate stellar properties are available for a sizable sample of EBs in Local Group galaxies.

Finally, the Local Group galaxies are also home to some particular types of EB systems that have no counterparts in the Milky Way. This is the case of EBs with component stars of type Cepheid or RR Lyr. Such systems are truly "Rosetta stones" for stellar Astrophysics. On the one hand, they could provide a direct calibration of the most widely used extragalactic distance indicators. And, on the other hand, their study would yield the fundamental properties of these variable stars and thus constrain their structure and evolution. The MACHO and OGLE projects identified three EB systems in the LMC with Cepheid components (Welch *et al.* 1999; Udalski et al. 1999; Alcock *et al.* 2002). Unfortunately, none of the Cepheid components seems to be a fundamental mode pulsator, being overtone, W Vir and unclassified. Further analysis of their light curves (Lepischak *et al.* 2004) has helped to clarify the nature of some of the companions, while spectroscopic observations taken with HST – and currently under analysis (Guinan *et al.* 2005) – should provide better constraints on the fundamental properties of these systems. In addition, the OGLE team has identified three RR Lyrae variables that are members of EB systems (Soszynski *et al.* 2003). Close inspection reveals that probably only one of them is a true EB system with the other ones being likely blends. Surprisingly, little has been done to further study these important objects and exploit their potential.

3. Eclipsing Binaries with Low-Mass Components

The other end of the Main Sequence, i.e., the realm of very low mass stars, also profits very much from the use of large telescopes. In spite of being the most numerous stellar population in the Galaxy, the faintness of these stars hinders careful study. Recent efforts, both observational and theoretical, have greatly contributed to a better understanding of the individual and overall properties of low-mass stars (see Henry, this volume). However, it is still important to carry out stringent tests to stellar structure models by constraining the maximum number of observables. In one of such efforts, Torres & Ribas (2002) analyzed the M-type EB YY Gem (Castor C). Tight constraints on the age and chemical composition of the system could be obtained from their more massive companions Castor A and B. Coming as a surprise, the detailed comparison of the fundamental properties of the components with the predictions of state-of-the-art stellar models revealed very significant differences. Models predicted stellar radii some 10–20% smaller than observed and effective temperatures some 3–5% hotter.

Stemming from the results for YY Gem, the interest in the field has raised and a number of surveys and detailed studies have contributed to increasing the sample of EBs with low-mass components, with the ultimate goal being the improvement of the statistical significance of the possible differences between observation and theory. It is revealing to note that until 2002 the number of EBs with late-K and M-type components and well-determined physical properties was only 2, in 2003 came the third, the number had gone up to 6 in 2005 and we now have almost a dozen EBs with such low-mass components (and more keep coming). The reason for this increase, apart from the growing interest, has been the availability of larger telescopes and more powerful instrumentation. Most of these EB systems have visual magnitudes in the range 13–16 and even fainter. Currently, the faintest system studied, with a visual magnitude of 19.3, is a member of the open cluster NGC 1647 and might contain the star with the lowest mass in a double-lined EB

(Hebb *et al.* 2006). Because of this faintness, some of the resulting physical properties still lack sufficient accuracy to perform a critical test to stellar models, but continuing efforts will decrease the current error bars.

Interestingly, planetary transit searches have also contributed to research on low-mass EBs. Follow-up using the 8.2-m VLT of OGLE planetary transit candidates has uncovered a number of eclipsing systems consisting of main sequence F-G stars with M dwarf companions (Bouchy *et al.* 2005; Pont *et al.* 2005). Because of selection effects, their light curves have shallow and flat-bottom eclipses corresponding to the transit of the M-type star (the occultation not observable). Also, only the lines of the F-G components are visible in the spectra due to the large contrast. These restrictions imply that the masses and radii of the M-type stars have to be determined through different assumptions (some of which are model dependent). For the 11 objects studied thus far the resulting accuracies are in the range 5–20% and therefore do not provide very stringent tests on models.

So, where are we now with respect to the contrast between observation and theory? As a matter of fact, the new additions to the list of low-mass stars with accurately determined physical properties has done nothing but confirm the mismatch observed for YY Gem. The mass-radius plot now covers almost the entire realm of late-K and M-type stars (0.8–0.2 $M_\odot$), as shown in Figure 2. Represented are the mass and radius measurements for 22 components of EB stars, which are: 2MASS J05162881+2607387 A&B (Bayless & Orosz 2006), V818 Tau B (Torres & Ribas 2002), RXJ0239.1-1028 A&B (López-Morales & Shaw 2006), GU Boo A&B (López-Morales & Ribas 2005), YY Gem AB (Torres & Ribas 2002), NSVS01031772 A&B (López-Morales *et al.* 2006), UNSW-TR-2 A&B (Young et al. 2006), TrES-Her0-07621 A&B (Creevey *et al.* 2005), 2MASS J04463285+1901432 A&B (Hebb *et al.* 2006), BW3 V38 A&B (Maceroni & Montalbán 2004), CU Cnc A&B (Ribas 2003), and CM Dra A&B (Lacy 1977; Metcalfe *et al.* 1996).

As can be seen in Figure 2, stars tend to fall systematically above the theoretical line, leaving no doubt that a significant discrepancy exists between models and observations with regards to stellar radii. On average, the observed values are some 10% larger than those predicted by theory. Other detailed comparisons have also shown that the stellar effective temperatures appear to be overestimated by $\sim5\%$. This, together with the good agreement in the mass-luminosity plot (see Henry, this volume), argues in favor of a scenario in which the stars have larger radius and cooler temperature than predicted by models but just in the right proportions to yield identical luminosities. A sensible hypothesis to explain the discrepancy is the effect of magnetic activity (Ribas 2006b). The close EB systems observed so far have orbital periods below 2.8 days and their components are forced to spin in orbital sync. The resulting high rotational velocities (10–60 km s^{-1}) give rise to a very efficient dynamo and thus enhance phenomena related to magnetic activity. The significant spot areal coverage observed in these eclipsing systems has the effect of lowering the overall photospheric temperature, which the star compensates by increasing its radius to conserve the total radiative flux. But also, the supposedly strong magnetic fields may change the heat transport efficiency in the stellar envelope and also alter the structure of the stars (Mullan & MacDonald 2001). Ongoing efforts, both observational and theoretical, are due to resolve this issue and yield a better understanding of these small but numerous stars.

Recently, EB research gave a further very important step down in mass. Stassun *et al.* (2006) reported the detection of a remarkable EB with brown dwarf components. The system is a member of the Orion Nebula star forming region and therefore has a very young age of about 1 Myr. Because of its youth, these 0.054 and 0.034-$M_\odot$ brown dwarfs have radii that are more akin to those of stars (0.67 and 0.51 $R_\odot$, respectively), which improves the chances for the occurrence of eclipses. Photometric observations of this system,

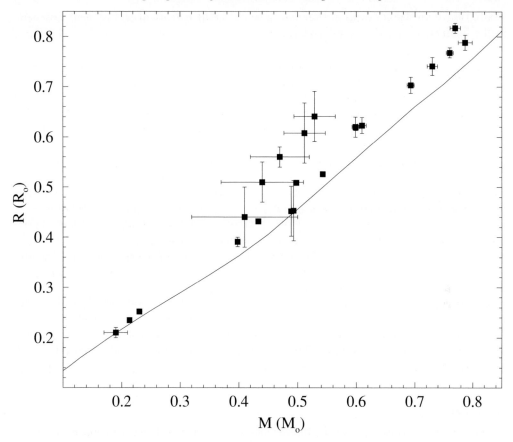

Figure 2. Mass-radius plot for EB stars with components of masses below 0.8 $M_\odot$. The solid line represents a theoretical 300 Myr isochrone calculated with the Baraffe *et al.* (1998) models.

with $V \approx 19.5$ mag, were carried out with 1-m class telescopes and the corresponding spectroscopy for radial velocities was obtained from the 8-m Gemini-S telescope. Detailed comparison of the observed properties with model predictions is still pending, but Stassun *et al.* (2006) find a reversed temperature ratio in the sense that the more massive component is also cooler. This would be a surprising result and in contradiction with theoretical expectations, but it still needs to be further confirmed.

Strictly speaking, transiting planets are just particular examples of EB systems and provide the final step down in mass. Some 12 planets have been found to date to transit the disk of their parent stars (see Bakos *et al.* 2007 for the full list). In all cases, confirmation of the planetary nature requires spectroscopic observations and these are obtained with the largest facilities available (8-10-m class telescopes). With their accurately measured masses and radii and the increasing statistical significance of the sample some surprises became evident. It was the first transiting exoplanet that already rose a red flag. Indeed, the mean density of the planet (which could be computed directly) turned out to be significantly smaller than that of Jupiter and in disagreement with the predictions of models. The additional 11 transiting planets detected have done nothing but confirm an unsuspected variety of physical properties in an otherwise quite homogeneous group of objects. The densities of the exoplanets detected cover values varying over a factor of 3 (see figure 6 in Bakos *et al.* 2007). Different scenarios involving irradiation, core sizes or evaporation are currently being investigated, but such dispersion in the intrinsic properties is still not understood.

4. Conclusions

Here we have shown several examples of studies related to EBs that require the use of large facilities. These were grouped in two classes, namely EBs in stellar ensembles, in which the faintness of the targets is due to the long distance, and EBs with low-mass components, in which the instrinsic low luminosity of the objects makes the use of large telescopes necessary. But there are other areas that will also have great importance. As opposed to the "classical" techniques discussed here, there are many aspects that are being and will be exploited, e.g., asteroseismology, Dopper tomography, high-precision/rapid photometry (e.g., eclipse mapping), interferometry (visual & spectroscopic binaries) and the opening of other windows such as the infrared. Some of these areas are discussed within this volume. It is also interesting to note that *all* examples shown here require the use of both large and small telescopes (spectroscopy vs. photometry). Thus, EB research constitutes a vivid example of the need to preserve also telescopes of smaller sizes to carry out high-quality and high-impact science.

The largest telescopes today are in the 10-m class but the next decade will witness the arrival of even larger telescopes in the 30–50-m range. The examples shown in this paper prove that EB research can be successful in competing against other scientific areas in obtaining large telescope time. The use of strong science cases, addressing topics of broad impact to Astrophysics and Cosmology, will ensure good prospects for research using EBs in an era in which most resources are likely to be put in large facilities.

Acknowledgements

E. Guinan, W. Hartkopf and P. Harmanec are thanked for the invitation and support to attend the Symposium. The author acknowledges support from the Spanish MEC through a Ramón y Cajal fellowship and from the Spanish MEC grant AyA2003-07736.

References

Alard, C., & Lupton, R. H. 1998, *ApJ*, 503, 325
Alcock, C., *et al.* 1997, *AJ*, 114, 326
Alcock, C., *et al.* 2002, *ApJ*, 573, 338
Bakos, G. A., *et al.* 2007, *ApJ*, 656, 552
Baraffe, I., Chabrier, G., Allard, F., & Hauschildt, P. H. 1998, *A&A*, 337, 403
Bayless, A. J., & Orosz, J. A. 2006, *ApJ*, 651, 1155
Bonanos, A. Z., Stanek, K. Z., Sasselov, D. D., Mochejska, B. J., Macri, L. M., & Kaluzny, J. 2003, *AJ*, 126, 175
Bonanos, A. Z., *et al.* 2004, *ApJ*, 611, L33
Bonanos, A. Z., *et al.* 2006, *ApJ*, 652, 313
Bouchy, F., Pont, F., Melo, C., Santos, N. C., Mayor, M., Queloz, D., & Udry, S. 2005, *A&A*, 431, 1105
Clausen, J. V. 2004, *NewAR*, 48, 679
Creevey, O. L., *et al.* 2005, *ApJ*, 625, L127
Devor, J., & Charbonneau, D. 2006, *ApJ*, 653, 647
Fitzpatrick, E. L., Ribas, I., Guinan, E. F., DeWarf, L. E., Maloney, F. P., & Massa, D. 2002, *ApJ*, 564, 260
Fitzpatrick, E. L., Ribas, I., Guinan, E. F., Maloney, F. P., & Claret, A. 2003, *ApJ*, 587, 685
Grison, P., *et al.* 1995, *A&AS*, 109, 447
Guinan, E. F, *et al.* 1998, *ApJ*, 509, L21

Guinan, E., Fitzpatrick, E., Ribas, I., Engle, S., Welch, D., & Lepischak, D. 2005, *BAAS*, 37, 1479

Harries, T. J., Hilditch, R. W., & Howarth, I. D. 2003, *MNRAS*, 339, 157

Hebb, L., Wyse, R. F. G., Gilmore, G., & Holtzman, J. 2006, *AJ*, 131, 555

Hilditch, R. W., Howarth, I. D., & Harries, T. J. 2005, *MNRAS*, 357, 304

Kaluzny, J., *et al.* 2005, in Stellar Astrophysics with the World's Largest Telescopes, Mikolajewska, J. and Olech, A. (eds)., *AIP Conf. Proc.*, 752, 70

Lacy, C. H. 1977, *ApJ*, 218, 444

Lepischak, D., Welch, D. L., & van Kooten, P. B. M. 2004, *ApJ*, 611, 1100

López-Morales, M., & Ribas, I. 2005, *ApJ*, 631, 1120

López-Morales, M., & Shaw, J. S. 2006, in 7th Pacific Rim Conference on Stellar Astrophysics, *ASP Conference Series*, in press (astro-ph/0603748)

López-Morales, M., Orosz, J. A., Shaw, J. S., Havelka, L., Arévalo, M. J., McIntyre, T., & Lázaro, C. 2006, *ApJ*, submitted (astro-ph/0610225)

Maceroni, C., & Montalbán, J. 2004, *A&A*, 426, 577

Macri, L. M., Stanek, K. Z., Bersier, D., Greenhill, L., & Reid, M. 2006, *ApJ*, 652, 1133

Metcalfe, T. S., Mathieu, R. D., Latham, D. W., & Torres, G. 1996, *ApJ*, 456, 356

Moutou, C., Pont, F., Bouchy, F., & Mayor, M. 2004, *A&A*, 424, L31

Mullan, D. J., & MacDonald, J. 2001, *ApJ*, 559, 353

Niemela, V. S., & Bassino, L. P. 1994, *ApJ*, 437, 332

Pols, O. R., Schröder, K.-P., Hurley, J.R., Tout, C. A., & Eggleton, P. P. 1998, *MNRAS*, 298, 525

Pont, F., Bouchy, F., Melo, C., Santos, N. C., Mayor, M., Queloz, D., & Udry, S. 2005, *A&A*, 438, 1123

Rauw, G., *et al.* 2004, *A&A*, 420, L9

Ribas, I. 2003, *A&A*, 398, 239

Ribas, I. 2006a, in Astrophysics of Variable Stars, Sterken, C. and Aerts, C. (eds)., *ASP Conf. Ser.*, 349, 55

Ribas, I. 2006b, *Ap&SS*, 304, 89

Ribas, I., Jordi, C., Torra, J., & Giménez, Á. 2000a, *MNRAS*, 313, 99

Ribas, I., *et al.* 2000b, *ApJ*, 528, 692

Ribas, I., Fitzpatrick, E. L., Maloney, F. P., Guinan, E. F., & Udalski, A. 2002, *ApJ*, 574, 771

Ribas, I., Jordi, C., Vilardell, F., Fitzpatrick, E. L., Hilditch, R. W., & Guinan, E. F. 2005, *ApJ*, 635, L37

Soszynski, I., *et al.* 2003, *AcA*, 53, 93

Stassun, K. G., Mathieu, R. D., & Valenti, J. A. 2006, *Nature*, 440, 311

Tamuz, O., Mazeh, T., & North, P. 2006, *MNRAS*, 367, 1521

Torres, G., & Ribas, I. 2002, *ApJ*, 567, 1140

Udalski, A., Soszynski, I., Szymanski, M., Kubiak, M., Pietrzynski, G., Wozniak, P., & Zebrun, K. 1998, *AcA*, 48, 563

Udalski, A., Soszynski, I., Szymanski, M., Kubiak, M., Pietrzynski, G., Wozniak, P., & Zebrun, K. 1999, *AcA*, 49, 223

Vilardell, F., Ribas, I., & Jordi, C. 2006, *A&A*, 459, 321

Welch, D. L., *et al.* 1999, New Views of the Magellanic Clouds, Chu, Y.-H. *et al.* (eds)., *IAU Symp.* 190, 513

Wyrzykowski, L., *et al.* 2003, *AcA*, 53, 1

Young, T. B., Hidas, M. G., Webb, J. K., Ashley, M. C. B., Christiansen, J. L., Derekas, A., & Nutto, C. 2006, *MNRAS*, 370, 1529

Discussion

JUAN MANUEL ECHEVARRÍA: YY Gem is a milestone on the lower end of the M-R diagram. Since the models are not in agreement with the observations, can there be any room for error in the observations?

RIBAS: The same disagreement between models and observations of YY Gem has also been found in all low-mass eclipsing binaries observed to date. In contrast, eclipsing binaries with components of higher mass are in good accord with the prediction of models. Since the techniques used to determine the masses and radii of the stars are independent of mass, systematic errors in the analysis could not explain the observed discrepancies.

JONATHAN DEVOR: To skim the best candidates one can find (for followup) software such as MECI to measure the masses of EB's components.

Binary Stars As Critical Tools & Tests
in Contemporary Astrophysics
Proceedings IAU Symposium No. 240, 2006
W.I. Hartkopf, E.F. Guinan & P. Harmanec, eds.

Eclipsing Binaries: Tools for Calibrating the Extragalactic Distance Scale

Alceste Z. Bonanos

Carnegie Institution of Washington, 5241 Broad Branch Road, Washington, DC 20015, USA
email: bonanos@dtm.ciw.edu

Abstract. In the last decade, over 7000 eclipsing binaries have been discovered in the Local Group through various variability surveys. Measuring fundamental parameters of these eclipsing binaries has become feasible with 8 meter class telescopes, making it possible to use eclipsing binaries as distance indicators. Distances with eclipsing binaries provide an independent method for calibrating the extragalactic distance scale and thus determining the Hubble constant. This method has been used for determining distances to eclipsing binaries in the Magellanic Clouds and the Andromeda Galaxy and most recently to a detached eclipsing binary in the Triangulum Galaxy by the DIRECT Project. The increasing number of eclipsing binaries found by microlensing and variability surveys also provide a rich database for advancing our understanding of star formation and evolution.

1. Introduction

The last decade has seen a dramatic increase in the number of extragalactic eclipsing binaries discovered. Most of these have been found as a side product of the microlensing surveys towards the Large and Small Magellanic Clouds (LMC and SMC). Starting in the early 1990s the EROS Experiment, the MACHO Project, the Microlensing Observations in Astrophysics (MOA) and the OGLE Project began monitoring the Magellanic Clouds with 1 meter telescopes in a search for dark matter in the form of massive compact halo objects. As a side product they have discovered thousands of variable stars, including many eclipsing binaries. The 75 EROS binaries in the LMC published by Grison *et al.* (1995) doubled the known binaries in the galaxy at the time. The Microlensing Observations in Astrophysics (MOA) group soon after released a catalog of 167 eclipsing binaries in the SMC (Bayne *et al.* 2002), followed by the extensive catalogs of the OGLE-II Project which comprise of 2580 eclipsing binaries in the LMC (Wyrzykowski *et al.* 2004) and 1350 in the SMC (Wyrzykowski *et al.* 2004). A catalog of the eclipsing binaries found by the MACHO project with ~ 4500 binaries in the LMC and 1500 in the SMC is underway. Note, that some of these binaries are foreground or galactic. The SuperMACHO project, using the CTIO Blanco 4-m telescope, has surveyed the Magellanic Clouds down to $VR \sim 23$ mag (Huber *et al.* 2005), extending the sample to include solar type stars, some of which will be W UMa variables. The first extragalactic W UMa detection was recently made by Kaluzny *et al.* (2006), with photometry from the 6.5-m Magellan telescopes at Las Campanas, Chile.

Fewer eclipsing binaries are known in more distant galaxies, partly due to their faintness and the necessity of large amounts of time on medium size telescopes (2-4 meters) to obtain good quality light curves. Two microlensing surveys are underway toward M31. The Wendelstein Calar Alto Pixellensing Project (WeCAPP, Fliri *et al.* 2006) discovered 31 eclipsing binaries in the bulge of M31. The variable star catalog released by the POINT-AGAPE Survey (An *et al.* 2004) has not been searched systematically for eclipsing binaries, but is bound to contain some among the 35000 variables.

The DIRECT Project (see Stanek *et al.* 1998; Bonanos *et al.* 2003) began monitoring M31 and M33 in 1996, specifically for Cepheids and detached eclipsing binaries (DEBs) with the 1.2-m telescope on Mt Hopkins, Arizona. In M31, a total of 89 eclipsing binaries were found in the 6 fields surveyed. A variability survey using the 2.5-m Isaac Newton telescope by Vilardell *et al.* (2006) has found 437 eclipsing binaries in M31, bringing the total to over 550 eclipsing binaries. In M33, the DIRECT Project has found 148 eclipsing binaries (see Mochejska *et al.* 2001, and references within). Eclipsing binaries have also been discovered in NGC 6822, a dwarf irregular galaxy in the Local Group, by the Araucaria Project (Mennickent *et al.* 2006).

It is worth mentioning the eclipsing binaries discovered beyond the Local Group. The first such discovery was made back in 1968 by Tammann & Sandage, who presented the light curve of a 6-day period binary in NGC 2403 (M81 group) with $B \sim 22$ mag. More recently, the Araucaria Project has discovered a binary in NGC 300 (Sculptor group). Mennickent *et al.* (2004) present the light curve of the $B \sim 21.5$ mag detached eclipsing binary in this galaxy.

Finally, the discovery of 3 Cepheid binaries in the LMC by Udalski *et al.* (1999), Alcock *et al.* (2002) provides a new way of calibrating the Cepheid period-luminosity relation and the extragalactic distance scale.

2. Eclipsing Binaries as Distance Indicators

Eclipsing binaries provide an accurate method of measuring distances to nearby galaxies with an unprecedented accuracy of 5% – a major step towards a very accurate and independent determination of the Hubble constant. Reviews and history of the method can be found in Andersen (1991) and Paczynski (1997). The method requires both photometry and spectroscopy of an eclipsing binary. From the light and radial velocity curve the fundamental parameters of the stars can be determined accurately. The light curve provides the fractional radii of the stars, which are then combined with the spectroscopy to yield the physical radii and effective temperatures. The velocity semi-amplitudes determine both the mass ratio and the sum of the masses, thus the individual masses can be solved for. Furthermore, by fitting synthetic spectra to the observed ones, one can infer the effective temperature, surface gravity and luminosity. Comparison of the luminosity of the stars and their observed brightness yields the reddening of the system and distance.

Measuring distances with eclipsing binaries is an essentially geometric method and thus accurate and independent of any intermediate calibration steps. With the advent of 8-m class telescopes, eclipsing binaries have been used to obtain accurate distance estimates to the LMC, SMC, M31 and M33; these results are presented below.

3. Eclipsing Binary Distances to the Magellanic Clouds and M31

The first extragalactic distance measurement using a detached eclipsing binary system was published by Guinan *et al.* (1998), demonstrating the importance of EBs as distance indicators. The detached system 14th mag system HV 2274 was observed with the Faint Object Spectrograph (FOS) onboard the *Hubble Space Telescope*. The UV/optical spectrophotometry was used to derive the radial velocity curve and reddening. The distance to HV 2274 was determined to be 47.0 ± 2.2 kpc (Guinan *et al.* 1998, Fitzpatrick *et al.* 2002). Distances to 3 more systems in the LMC have been determined: the detached 15th magnitude system HV 982 (Fitzpatrick *et al.* 2002) at a distance of 50.2 ± 1.2 kpc, the 15th magnitude detached system EROS 1044 (Ribas *et al.* 2002) at 47.5 ± 1.8 kpc and the 14th magnitude semi-detached system HV 5936 at 43.2 ± 1.8 kpc (Fitzpatrick *et al.* 2003). These support a "short" distance scale to the LMC, in contrast to the LMC distance of 50 kpc adopted by the Key Project (Freedman *et al.* 2001. The spread in the distances is most likely an indication of the intrinsic extent of the LMC along the line of sight.

Harries *et al.* (2003) and Hilditch *et al.* (2005) have conducted a systematic spectro-scopic survey of eclipsing binaries in the SMC, obtaining fundamental parameters and distances to 50 eclipsing binary systems. Their sample was selected from the OGLE-II database of SMC eclipsing binaries as the brightest systems ($B < 16$ mag) with short periods ($P_{orb} < 5$ days) to increase the efficiency of multi-fiber spectroscopy over a typical observing run. The mean true distance modulus from the whole sam-ple is $18.91 \pm 0.03(random) \pm 0.1(systematic)$ and the implied LMC distance is $18.41 \pm 0.04(random) \pm 0.1(systematic)$, again in support of the "short" distance scale.

M31 and M33, being the nearest spiral galaxies, are crucial stepping-stones in the extragalactic distance ladder. Ribas *et al.* (2005) have determined the first distance to a spiral galaxy, specifically to a semi-detached system ($V = 19.3$ mag) in M31. Note that at such distances, the location of the binary within the galaxy has an insignificant effect on the distance ($< 1\%$). Light curves for the system in M31 were obtained from the survey of Vilardell *et al.* (2006) with the 2.5-m Isaac Newton telescope and spectroscopy with the 8-m Gemini telescope using GMOS. Such stars are at the limit of current spectroscopic capabilities. The resulting distance is 772 ± 44 kpc and distance modulus is 24.44 ± 0.12 mag, in agreement with previous distance determinations to M31.

4. DIRECT Distance to a Detached Eclipsing Binary in M33

4.1. *Motivation*

The DIRECT Project (see Stanek *et al.* 1998; Bonanos *et al.* 2003) aims to measure distances to the nearby Andromeda (M31) and Triangulum (M33) galaxies with eclipsing binaries and the Baade-Wesselink method for Cepheids. It began surveying these galaxies in 1996 with 1-m class telescopes. The goal of the DIRECT Project is to replace the current anchor galaxy of the extragalactic distance scale, the LMC, with the more suitable spiral galaxies in the Local Group, M31 and M33. These are the nearest spiral galaxies to ours, yet more than ten times more distant than the LMC and therefore more difficult to observe stars in them. The Cepheid period-luminosity relation is used to measure distances to a few tens of Mpc, while Type Ia supernovae are used to probe distances out to a few hundred Mpc. Galaxies hosting both Cepheids and Type Ia supernovae become calibrators of the luminosities of supernovae, which are used to determine the Hubble constant, H_0.

How is the Cepheid period-luminosity calibrated? Benedict *et al.* (2002) in their Figure 8 show 84 recent measurements of the distance modulus of the LMC using 21 methods. The large spread in the different measurements is quite disturbing. There are several problems with using the LMC as the anchor of the distance scale, which demand its replacement. The zero point of the period-luminosity relation is not well determined and the dependence on metallicity remains controversial. There is increasing evidence for elongation of the LMC along the line of sight that complicates a distance measurement. One has to additionally include a model of the LMC when measuring distances, which introduces systematic errors. Finally, the reddening across the LMC has been shown to be variable (Nikolaev *et al.* 2004), which has to be carefully accounted for. These effects add up to a 10-15% error in the distance to the LMC, which in the era of precision cosmology is unacceptable. The replacement of the current anchor galaxy of the distance scale with a more suitable galaxy or galaxies is long overdue. Furthermore, the *Hubble Space Telescope* Key Project (Freedman *et al.* 2001) has measured the value of H_0 by calibrating Cepheids measured in spiral galaxies and secondary distance indicators and found $H_0 = 72 \pm 8$ km s^{-1} Mpc^{-1}. This result is heavily dependent on the distance modulus to the LMC they adopt (18.50 mag or 50 kpc).

4.2. *DIRECT Observations*

The DIRECT project involves three stages: surveying M31 and M33 in order to find detached eclipsing binaries and Cepheids; once discovered selecting and following up the best targets with medium size telescopes (2-4 m class) to obtain more accurate light curves and lastly, obtaining spectroscopy which requires 8-10 m class telescopes. DIRECT completed the survey stage in 1996-1999 with 200 full/partial nights on 1-m class telescopes in Arizona. Follow up observations of the 2 best eclipsing binaries were obtained in 1999 and 2001 using the Kitt Peak 2.1-m telescope in Arizona. The total number of eclipsing binaries found in M33 were 237, however only 4 are bright enough ($V_{\mathrm{max}} < 20$ mag) for distance determination with currently available telescopes. The criteria for selection include a detached configuration (stars are well within their Roche lobes) and deep eclipses, which remove degeneracies in the modeling and a short period (< 10 days) that makes follow up observations feasible.

Bonanos *et al.* (2006) presented the first distance determination to a detached eclipsing binary (DEB) in M33 that was found by Macri *et al.* (2001). D33J013346.2+304439.9 is located in the OB 66 association. Follow up optical data were obtained in order to improve the quality of the light curve and additional infrared observations were made using the 8-m Gemini telescope in order to better constrain the extinction to the system. Spectra of the DEB were obtained in 2002-2004 with the 10-m Keck-II telescope and 8-m Gemini telescope on Mauna Kea. Note that ~ 4 hours of observations per epoch were required for radial velocity measurements, a large investment of 8-10 m class telescope time. Absorption lines from both stars are clearly resolved in the spectrum, making it a double lined spectroscopic binary.

Careful modeling with non-local thermodynamic equilibrium model spectra yielded effective temperatures $T_{\mathrm{eff1}} = 37000 \pm 1500$ K and $T_{\mathrm{eff2}} = 35600 \pm 1500$ K. The primary star is defined as the hotter star eclipsed at phase zero. We measured radial velocities from the spectra and from the light and radial velocity curves derived the parameters of the DEB components. The $V-$band light curve model fit for the DEB is shown in Figure 1. Note that the deviation of the secondary eclipse from phase 0.5 is due to the eccentricity of the system. The radial velocity curve is presented in Figure 2. The rms residuals are 26.0 km s^{-1} for the primary and 28.0 km s^{-1} for the secondary star. We find the DEB components to be O7 type stars with masses: $M_1 = 33.4 \pm 3.5$ M$_\odot$, $M_2 = 30.0 \pm 3.3$ M$_\odot$ and radii $R_1 = 12.3 \pm 0.4$ R$_\odot$, $R_2 = 8.8 \pm 0.3$ R$_\odot$.

4.3. *Distance Determination*

Having measured the temperatures of the stars from the spectra, we computed fluxes and fit the optical and near-infrared $BVRJHKs$ photometry. The best fit that minimized the photometric error over the 6 photometric bands yielded a distance modulus to the DEB and thus M33 of 24.92 ± 0.12 mag (964 ± 54 kpc). The fit of the reddened model spectrum to the photometry is shown in Figure 3.

There are several avenues for improving the distance to M33 and M31 using eclipsing binaries. Wyithe & Wilson (2002) propose the use of semi-detached eclipsing binaries to be just as good or better distance indicators as detached eclipsing binaries, which

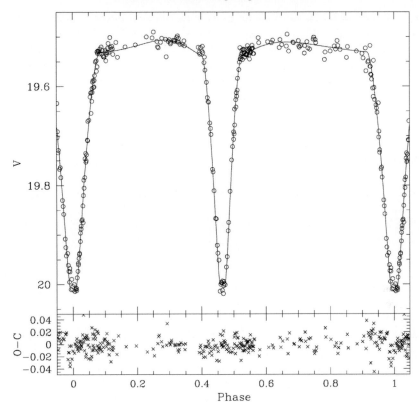

Figure 1. $V-$band light curve of the M33 DEB with model fit from the Wilson-Devinney program. Circles correspond to the 278 V-band observations and the solid line to the model; the rms is 0.01 mag (from Bonanos *et al.* 2006).

have been traditionally considered to be ideal. Semi-detached binaries provide other benefits: their orbits are tidally circularized and their Roche lobe filling configurations provide an extra constraint in the parameter space, especially for complete eclipses. Bright semi-detached binaries in M33 or M31 are not as rare as DEBs, and are easier to follow-up spectroscopically, as demonstrated by Ribas *et al.* (2005) in M31. Thus, for the determination of the distances to M33 and M31 to better than 5% we suggest both determining distances to other bright DEBs and to semi-detached systems found by DIRECT and other variability surveys. Additional spectroscopy of the DEB would also improve the current distance determination to M33, since the errors are dominated by the uncertainty in the radius or velocity semi-amplitude.

How does our M33 distance compare to previous determinations? Table 1 (adapted from Bonanos *et al.* 2006) presents a compilation of 13 recent distance determinations to M33 ranging from 24.32 to 24.92 mag, including the reddening values used. Our measurement although completely independent yields the largest distance with a small 6% error, thus is not consistent with some of the previous determinations. This possibly indicates unaccounted sources of systematic error in the calibration of certain distance indicators. Note the Freedman *et al.* (2001) distance to M33 is not consistent with the DIRECT measurement. This could be due to their ground based photometry which is likely affected by blending, but highlights the importance of securing the anchor of the

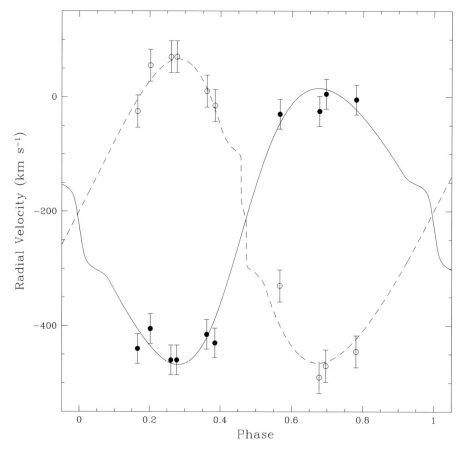

Figure 2. Radial velocities for the DEB measured by two-dimensional cross correlation with synthetic spectra. Model fit is from Wilson-Devinney program. Error bars correspond to the rms of the fit: 26.0 km s^{-1} for the primary (filled circles) and 28.0 km s^{-1} for the secondary (open circles).

extragalactic distance scale. The eclipsing binary distances to the LMC presented above indicate a shorter distance to the LMC. Combined with eclipsing binary distances to M31 and M33, we should soon be able to reduce the errors in the distance scale and thus the Hubble constant to 5% or better.

4.4. *Epilogue*

The accelerating rate of discovery of eclipsing binaries provides immense opportunities. With current spectroscopic capabilities it has become possible to measure distances to Local Group galaxies out to 1 Mpc, thus providing distances independent of the controversial LMC distance and the calibration of Cepheids, which most methods rely on. An independent calibration of the extragalactic distance scale has become possible. The recent distance determinations to the LMC, SMC, M31 and M33 are providing 6% distances to these galaxies that will improve over the next few years. Combined with geometric distances to the maser galaxy NGC 4258, the extragalactic distance scale will soon be anchored to several spiral galaxies (M31, M33, NGC 4258).

In addition to their use as distance indicators, eclipsing binaries provide many more opportunities to advance our understanding of star formation and evolution. In

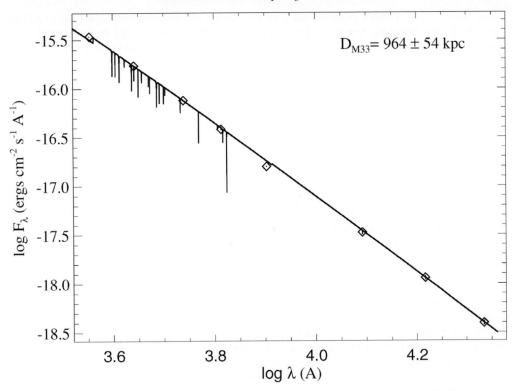

Figure 3. Fit of the reddened DEB model spectrum to the $BVRJHK_s$ ground-based photometry. Overplotted is the U and I photometry from Massey *et al.* (2006). The distance modulus to the DEB and thus M33 is found to be 24.92 ± 0.12 mag (964 ± 54 kpc).

Table 1. RECENT DISTANCE DETERMINATIONS TO M33

Study	Method[a]	Distance Modulus	Reddening
Bonanos *et al.* (2006)	DEB	24.92 ± 0.12	$E(B-V) = 0.09 \pm 0.01$
Sarajedini *et al.* (2006)	RR Lyrae	24.67 ± 0.08	$\sigma_{E(V-I)} = 0.30$
Brunthaler *et al.* (2005)	Water Masers	24.32 ± 0.45	—
Ciardullo *et al.* (2004)	PNe	$24.86^{+0.07}_{-0.11}$	$E(B-V) = 0.04$
Galleti *et al.* (2004)	TRGB	24.64 ± 0.15	$E(B-V) = 0.04$
McConnachie *et al.* (2004)	TRGB	24.50 ± 0.06	$E(B-V) = 0.042$
Tiede *et al.* (2004)	TRGB	24.69 ± 0.07	$E(B-V) = 0.06 \pm 0.02$
Kim *et al.* (2002)	TRGB	$24.81 \pm 0.04(r)^{+0.15}_{-0.11}(s)$	$E(B-V) = 0.04$
Kim *et al.* (2002)	RC	$24.80 \pm 0.04(r) \pm 0.05(s)$	$E(B-V) = 0.04$
Lee *et al.* (2002)	Cepheids	$24.52 \pm 0.14(r) \pm 0.13(s)$	$E(B-V) = 0.20 \pm 0.04$
Freedman *et al.* 2001	Cepheids	24.62 ± 0.15	$E(V-I) = 0.27$
Pierce *et al.* (2000)	LPVs	24.85 ± 0.13	$E(B-V) = 0.10$
Sarajedini *et al.* (2000)	HB	24.84 ± 0.16	$< E(V-I) >= 0.06 \pm 0.02$

†DEB: detached eclipsing binary; TRGB: tip of the red giant branch; PNe: planetary nebulae; RC: the red clump; LPVs: long period variables; HB: horizontal branch stars.

particular, they provide direct means of measuring masses, radii and luminosities of stars. For example, applications to the extremes of stellar mass ranges are underway in order to provide constraints to theoretical models of stellar atmospheres and evolution, such as for M-dwarfs and at the other extreme for very massive ($> 50\,M_\odot$) O-stars and

Wolf-Rayet stars. Future projects such as the wide field imaging surveys Pan-STARRS and the Large Synoptic Survey Telescope (LSST) will survey the sky down to 24th mag and yield thousands of binaries in the Galaxy, the Local Group and beyond.

References

Alcock, C., Allsman, R.A., Alves, D.R., *et al.* 2002, *ApJ*, 573, 338

An, J.H., Evans, N.W., Hewett, P., *et al.* 2004, *MNRAS* 351, 1071

Bayne, G., Tobin, W., Bond, I., *et al.* 2002, *MNRAS*, 331, 609

Benedict, G.F., McArthur, B.E., Fredrick, L.W., *et al.* 2002, *AJ*, 123, 473

Bonanos, A.Z., Stanek, K.Z., Sasselov, D.D., *et al.* 2003, *AJ*, 126, 175

Bonanos, A.Z., Stanek, K.Z., Kudritzki, R.P., *et al.* 2006, *ApJ*, 652, 313

Brunthaler, A., Reid, M.J., Falcke, H., *et al.* 2005, *Science*, 307, 1440

Ciardullo, R., Durrell, P.R., Laychak, M.B., *et al.* 2004, *ApJ*, 614, 167

Fitzpatrick, E.L., Ribas, I., Guinan, E.F., *et al.* 2002, *ApJ*, 564, 260

Fitzpatrick, E.L., Ribas, I., Guinan, E.F., *et al.* 2003, *ApJ*, 587, 685

Fliri, J., Riffeser, A., Seitz, S., & Bender, R. 2006, *A&A*, 445, 423

Freedman, W.L., Madore, B.F., Gibson, B.K., *et al.* 2001, *ApJ*, 553, 47

Galleti, S., Bellazzini, M., & Ferraro, F.R. 2004, *A&A*, 423, 925

Grison, P., Beaulieu, J.-P., Pritchard, J.D., *et al.* 1995, *A&AS*, 109, 447

Guinan, E.F., Fitzpatrick, E.L., Dewarf, L.E., *et al.* 1998, *ApJ*, 509, L21

Harries, T.J., Hilditch, R.W., & Howarth, I.D. 2003, *MNRAS*, 339, 157

Hilditch, R.W., Howarth, I.D., & Harries, T.J. 2005, *MNRAS*, 357, 304

Huber, M.E., Nikolaev, S., Cook, K.H., *et al.* 2005, *BAAS*, 207, 122.16

Kaluzny, J., Mochnacki, S., & Rucinski, S.M. 2006, *AJ*, 131, 407

Kim, M., Kim, E., Lee, M.G., *et al.* 2002, *AJ*, 123, 244

Lee, M.G., Kim, M., Sarajedini, A., *et al.* 2002, *ApJ*, 565, 959

Macri, L.M., Stanek, K.Z., Sasselov, D.D., *et al.* 2001, *AJ*, 121, 870

Massey, P., Olsen, K.A.G., Hodge, P.W., *et al.* 2006, *AJ*, 131, 2478

McConnachie, A.W., Irwin, M.J., Ferguson, A.M.N., *et al.* 2004, *MNRAS*, 350, 243

Mennickent, R.E., Pietrzynski, G., & Gieren, W. 2004, *MNRAS*, 350, 679

Mennickent, R.E., Gieren, W., Soszynski, I., & Pietrzynski, G. 2006, *A&A*, 450, 873

Mochejska, B.J., Kaluzny, J., Stanek, K.Z., *et al.* 2001, *AJ*, 122, 2477

Nikolaev, S., Drake, A.J., Keller, S.C., *et al.* 2004, *ApJ*, 601, 260

Paczynski, B. 1997 *The Extragalactic Distance Scale* 273

Pierce, M.J., Jurcevic, J.S., & Crabtree, D. 2000, *MNRAS*, 313, 271

Ribas, I., Fitzpatrick, E.L., Malonley, F.P., *et al.* 2002, *ApJ*, 574, 771

Ribas, I., Jordi, C., Vilardell, F., *et al.* 2005, *ApJ*, 635, L37

Sarajedini, A., Geisler, D., Schommer, R., *et al.* 2000, *AJ*, 120, 2437

Sarajedini, A., Barker, M., Geisler, D., *et al.* 2006, *AJ*, 132, 1361

Stanek, K.Z., Kaluzny, J., Krockenberger, M., *et al.* 1998, *AJ*, 115, 1894

Tammann, G.A., & Sandage, A. 1968, *ApJ*, 151, 825

Tiede, G.P., Sarajedini, A., & Barker, M.K. 2004, *AJ*, 128, 224

Udalski, A., Soszynski, I., Szymanski, M., *et al.* 1999, *AcA*, 49, 223

Vilardell, F., Ribas, I., & Jordi, C. 2006, *A&A*, 459, 321

Wyithe, J.S.B., & Wilson, R.E. 2002, *ApJ*, 571, 293

Wyrzykowski, L., Udalski, A., Kubiak, M., *et al.* 2003, *AcA*, 53, 1

Wyrzykowski, L., Udalski, A., Kubiak, M., *et al.* 2004, *AcA*, 54, 1

Discussion

CARLSON CHAMBLISS: A problem with very distant eclipsing binaries is that the component stars are necessarily very hot. This leads to substantial uncertainties in bolometric corrections since the photometry is *UBV*. Is this a major source of error?

BONANOS: We have used one of the best available non-LTE codes (from Rolf Kudritzki's group) to determine the temperatures of these stars. We do not use bolometric corrections, but fit synthetic photometry of the reddened model spectrum to observed optical and infrared photometry. Finally, we have adopted a conservative error in the temperature of 1500K.

Binary Stars as Critical Tools & Tests
in Contemporary Astrophysics
Proceedings IAU Symposium No. 240, 2006
W.I. Hartkopf, E.F. Guinan & P. Harmanec, eds.

Classical Observations of Visual Binary and Multiple Stars

Brian D. Mason

Astrometry Department, U.S. Naval Observatory, 3450 Massachusetts Avenue, NW,
Washington, DC 20392-5420, USA
email: bdm@usno.navy.mil

Abstract. Changes in the double star database are highlighted, describing various methods of observation (both historically and those of the past few years) and their effectiveness in different regimes of separation space. The various niches for wide- and narrow-field work as they apply to double and multiple stars are examined and the different types of information which each can provide are described. Despite the significant growth of the double star database, much can still be done, such as finding lost pairs, filling in missing parameters so that observing programs can select all stars appropriate to their capabilities, or providing at least gross kinematic descriptions. After more than 20 years of successful work, speckle interferometry and conventional CCD astrometry have replaced filar micrometry and photography as preferred classical techniques. Indeed, most work in filar micrometry is now being done by amateurs. Work on pairs described as neglected in the last major WDS data release (2001) is given as a specific example. Finally, the continued need to publish data in classical double star parameters is also discussed.

Keywords. binaries (including multiple): general, binaries: visual, catalogs

1. Growth of the WDS

The *Washington Double Star Catalog* (hereafter, WDS) has had significant growth in the past decade, both in the total number of mean positions (measures) and the number of systems. Due to the small number of measures per system, establishing which are true binaries (physical systems) is possible only for a subset of pairs. While the mean number of measures per system is 7.1, the median is only 3. In Figure 1 the growth of the WDS is plotted at top. The solid line and filled circles indicate the number of mean positions (left axis) for the major releases of the WDS, while the dashed line and open circles indicates the number of systems (right axis). At bottom is a histogram indicating the number of means per system. While the average number of means per system has climbed due to matching with various astrometric catalogs, determining the kinematic properties for the majority of systems remains an elusive task. Of all the pairs currently in the catalog, 1522 have orbits of varying quality, 354 have common parallax, and an unknown number have common proper motion, a known physicality rate of about 2%. About 1% are certainly optical: 1163 have rectilinear solutions computed and 174 have mutually exclusive parallax, leaving the vast majority (∼97%) unknown.

Table 1 gives a few statistics on data added to the WDS database in time for its various releases. For example, the line labelled "1984" describes all data included in the WDS database at the time of the first WDS release in 1984, "1996" the data added between 1984 and 1996, and so on. Measures added since the 2001 release have been divided into two categories: "old" measures (data published before 2001 but not added to the WDS until recently) and "new" measures (data published since 2001). As this break-down illustrates, a considerable fraction of the additions to the WDS in the 5.5 years

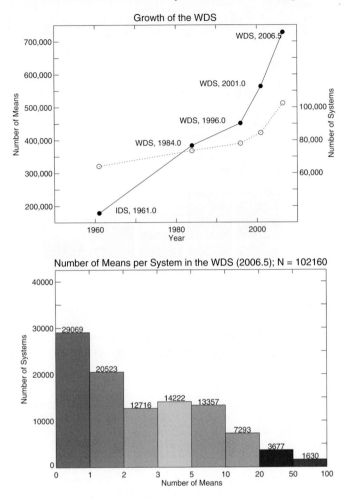

Figure 1. *Top:* Growth of the WDS. *Bottom:* Histogram of means per system.

since the last major release have come from data mining of astrometric catalogs. Improving WDS coordinates from arcminute to arcsecond precision allowed the matching of many historical astrometric catalogs against the WDS to add measures for these (mostly wide) pairs. While this can help considerably for wide pairs and aid in determining if they have the same proper motion, these data are generally not applicable to orbit pairs which require long-focus, and in some cases high angular resolution techniques. In Figure 2 is shown the normalized additions by various techniques to the major releases of the WDS. Micrometry has seen a steady decline. Wide-angle work made major contributions in the past few years, primarily from the AC and 2MASS (Wycoff *et al.* 2006) catalogs. High-resolution observations are primarily due to speckle interferometry, but also includes adaptive optics and similar techniques. Space-based measures include results from both the high-precision HST fine guidance sensors as well as the all-sky Hipparcos and Tycho satellite (hence the large contribution by this technique to the 2001 release).

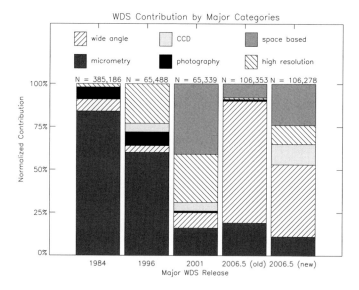

Figure 2. Sources of WDS data.

Table 1. Statistics of WDS additions

WDS Release	N means added	Mean ρ ($''$)	Median ρ ($''$)	% $< 1''$	N $<0''.1$	N $<0''.01$
1984	385,186	8.01	2.30	29.3	943 (0.2%)	0
1996	65,488	12.62	1.68	41.1	1670 (2.6%)	0
2001	65,339	10.71	1.87	32.2	1230 (2.0%)	255
2006.5 (old)	106,353	33.12	14.42	2.0	73 ($\ll$1%)	0
2006.5 (new)	106,278	19.92	7.62	16.8	301 (0.3%)	34

2. Physical Systems

Visual orbits can be of varying quality, with some that are well-observed and well-defined to others that are only marginally defined. However, even a curve indicating Keplerian motion with scant coverage can yield a solution where the important quantity $3\log(a) - 2\log(P)$ is often not grossly erroneous. Pairs with poor quality orbits or physical binaries with no orbits at all need continued vigilance at the appropriate observing cadence while others need observations with specific techniques and/or telescopes. Orbits of two well-observed pairs are displayed with their data in Figure 3. In these, and other figures like them, the calculated orbit of the secondary relative to the primary is indicated by the solid ellipse. Measures by different techniques (e.g., plus signs for micrometry, solid circles for speckle interferometry) are connected to their predicted positions along the orbit by solid $O - C$ lines. The scale in arcseconds is indicated on the axes and the direction of motion at the lower right. The broken line indicates the line of nodes. The pair at left, 70 Ophiuchi, is the most frequently observed binary in the WDS and has an orbital period of $\sim$88 years with data as far back as 1779, so is very well characterized. The plotted orbit is by Pourbaix (2000). At right is the relative orbit of the wider pair of ϵ Hydrae. While also well-characterized over the observed portion of its orbit, only about 20% of the orbit has been observed, despite having data since 1825. The nearly 1000-yr period orbit shown in the figure is by Heintz (1996). The closer pair in this system, with an orbital period of $\sim$15 years, is extremely well characterized, as well.

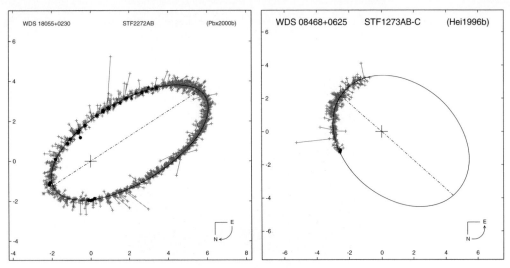

Figure 3. *Left:* Relative orbit of 70 Ophiuchi. *Right* Relative orbit of the wider pair of ϵ Hydrae.

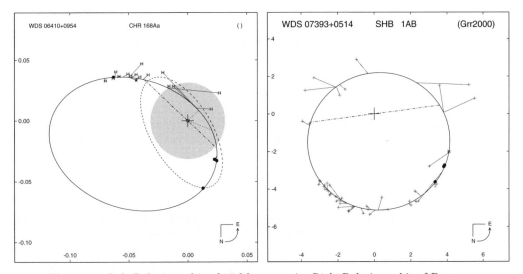

Figure 4. *Left:* Relative orbit of 15 Monocerotis. *Right* Relative orbit of Procyon.

While ϵ Hydrae just needs more time to complete its orbit, the pairs in Figure 4 need more specialized data acquisition. The massive and very important speckle/spectroscopic O star binary 15 Monocerotis is at left. A challenging target, this pair has only been split by speckle interferometry with a 4-m class telescope (open circle and $\star$), Hubble Space Telescope fine guidance sensors (HST-FGS; **H**), or long-baseline optical interferometry (filled circle). The resolution limit of a 4-m telescope is indicated by the lightly-shaded circle at the origin. This orbit is based on many unpublished observations. At right is plotted the relative orbit of Procyon by Girard *et al.* (2001). Although the secondary has been known for well over a century, the large Δm has continued to make resolving this pair quite challenging — no astronomer living has split this pair with the naked eye!

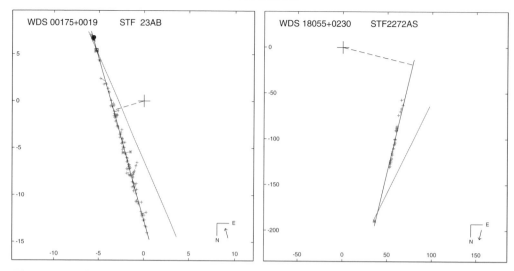

Figure 5. *Left:* This is the linear fit to the double STF 23. *Right* Linear fit to one of the wide components of 70 Ophiuchus.

3. Optical Systems

Even optical systems can be of value. While continuing to catalog these pairs can prevent their "re-discovery," the motion of one relative to the other due to differential proper motion can aid in a variety of applications. Due to the often longer timebase and frequently larger number of observations, the relative proper motion of optical double stars has been found to be more accurate than either high precision short- or long-period classical proper motion determination techniques (Kaplan & Snell 2001) and submotions to the linear fits can indicate the presence of closer pairs. Finally, these pairs can be exceptional calibrators of rotation and plate scale. In Figure 5 are examples of these two. At left is the linear fit to the double STF 23. Seen almost due North by F.G.W. Struve in 1825 (1837), the companion has moved southwest since then. The best relative proper motion here indicated is from Tycho-2 and deviates significantly from the historical measures. At right submotion seen in a more distant optical component can indicate a closer physical pair, in this case the known binary 70 Ophiuchus (see Figure 3).

4. Work Still to be Done

Filling in missing parameters in catalogs such as the WDS is not mere "stamp collecting." When all these parameters are complete — position, relative position, magnitude, proper motion, etc. — observing programs may be more efficient at observing stars appropriate to their capabilities, so that at least approximate kinematic descriptions can be assigned to them.

A large number of systems in the WDS may be characterized as "neglected." These include unconfirmed binaries as well as systems which have not been resolved for many years. The reasons for this neglect are varied: poor coordinates or large proper motion (so the systems are "lost"), erroneous magnitude or Δm estimates (so the systems are skipped over or misidentified), or true neglect (too many binaries and too few observers). While the veracity of some of these systems is certainly suspect, many (if not most) of these are *bona fide* double stars. Unconfirmed pairs need to be verified as real and if their

Table 2. Neglected Doubles by Catalog Release.

Release	easy	easy but close	unknown
2001	6307	6630	58,842
2006.5	946	4376	39,585

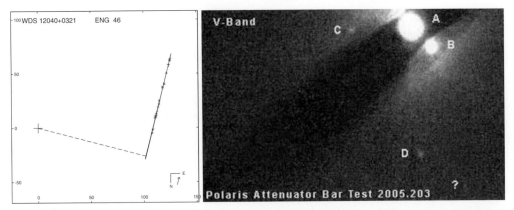

Figure 6. *Left:* This is the linear fit to the double ENG 46. *Right* CCD frame of Polaris (image courtesy of Jim Daley).

motion cannot be characterized, at least the appropriate re-observation cadence needs to be determined.

In Table 2, "easy" pairs are defined as those wider than 3″, with a Δm smaller than 3 magnitudes and both components brighter than 11th magnitude. Those characterized as "easy but close" have the same two latter discriminators but are closer than 3″. Any not matching these two sets, including pairs with one or more completely unknown characteristics, are in the largest, and most challenging, third set. Note that while the number of pairs (which start off as neglected as they are unconfirmed) has significantly increased the number of neglected pairs in all catagories has decreased. The five largest contributors, by number in the observation of these pairs is:

(a) USNO matching of 2MASS (39,580; Wycoff *et al.* 2006)
(b) Tycho-2, primarily the TDSC (8071; Fabricius *et al.* 2002)
(c) Washington Speckle Interferometry (6233; most recently Mason *et al.* 2006)
(d) T. Tobal (4256; http://ad.usno.navy.mil/wds/wdstext.html#unpublished)
(e) D. Arnold (2509; most recently Arnold 2006)

As enumerated above, much of the work on negected pairs has been done by amateurs. While quality control can be a concern, qualifying who is and who is not an amateur is at least as difficult a task. The work produced by these "financially uncompenated astronomers" has made and continues to make significant contributions. Examples of this are given in Figure 6. At left, historical measures of the AC, the AGK3, and others as well as more recent measures from 2MASS and Dave Arnold (2006) have allowed for this precise differential proper motion to be determined. At right, the C and D components of Polaris were observed by S.W. Burnham (1894) on the Lick 36-in at the close of the 19th Century and not split again until 2005 when Jim Daley (2006) recovered them. Many other large Δm systems still remain neglected.

Table 3. Major New Sources of Data.

Data Source	Number of Observations	Approximate Date
Hipparcos	19,146	Early 1990s
Tycho	40,277	Early 1990s
2MASS	42,078	Late 1990s
UCAC3	50,000+	Late 1990s

Other work could possibly be more beneficial to other areas of binary star astronomy. For example, third light components are detected in many eclipsing binary solutions. These can also show up as stationary contributors to the spectrograms of close binaries. A comprehensive listing of available information about these systems, such as the date of detection, an upper limit on ρ, and an approximate Δm, would be beneficial. It is possible that the magnitude difference may be more valuable than the separation for these objects.

5. Problems & Solutions

While specific large projects have made significant number of observations, and have found many pairs that were listed more than once, they are over a limited timebase. Although this can aid in the observation of wide pairs and in determining which ones have the same proper motion, the programs themselves are not appropriate for orbital solutions. They may be adequate for simple kinematics, determining common proper motion pairs, and finding missing doubles, but they are not adequate for complex dynamics. Characterizing these orbit pairs requires perseverance over an extended period of time.

These data sources have indicated systems appropriate for follow-up observation. For example, the resolution capability of Hipparcos is of the same order of magnitude as ground-based speckle interferometry and follow-up data for the fast-moving Hipparcos discoveries have produced numerous short-period orbits (see Balega *et al.* 2005, 2006). Continued observations of these pairs will keep ground based astronomers busy for many years (Horch *et al.* 2004, Mason *et al.* 2001).

6. LBOI

Of all techniques for resolved pairs the one with the greatest potential is long baseline optical interferometry (LBOI) which can observe many of the most interesting systems and have very meaningful synergy as the pairs resolved with it are preferentially spectroscopic binaries and can be eclipsing pairs as well.

However, rather than the more conventional separation (ρ) and position angle (θ), data are often presented in terms of the less intuitive baselines (B) and visibilities (V or V^2). The advantage of getting greater comprehension among colleagues and (perhaps as important), funding officials is very important. Visualizing ρ and θ on the sky can be easier than the more abstract B and V^2. While many LBOI targets are new resolutions, there exist many appropriate known systems as well. While the published data obtained using different techniques are of admittedly lower accuracy and/or precision than obtainable by LBOI, they may still provide other benefits. Probably the best example of this would be the pair known as "The Interferometrist's Friend," Capella.

First resolved in 1919 with the 20-foot beam inteferometer on Mount Wilson (Anderson 1920), Capella had completed almost 3 full revolutions by the time of Merrill's orbit (1922). It was not recovered again for nearly fifty years and by the time of the orbit

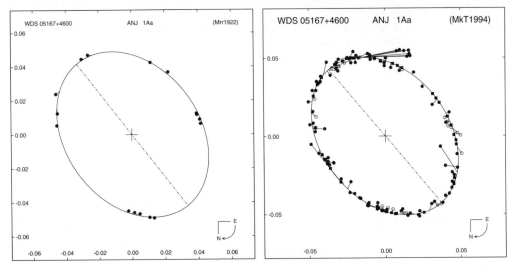

Figure 7. *Left:* Data and Merrill's 1922 orbit of Capella. *Right* Data and the 1994 orbit from the Mark III of Capella.

update of McAlister (1981), Capella had gone around over 217 times. While the accuracy and precision of speckle interferometry were important, the enormous lever arm afforded by the earlier astrometry allowed the period to be determined with far greater precision. Nearly 50 additional revolutions occurred before data from the Mark III, also on Mt. Wilson, were used to further refine the period (Hummel *et al.* 1994). This, coupled with the superior data from LBOI has given this orbit staying power.

If the interferometer has more than two elements and they are not in a line it should be possible to express parameters in the conventional ρ-θ space (see, for example, the NPOI measures of o Leo; Hummel *et al.* 2001). If the P $\gg$ 1 day the rotation of the Earth can help fill out the uv-plane (e.g., the SUSI measures of β Cen; Davis *et al.* 2005) and also allows observers to express parameters in ρ-θ space. The difficulty may be in rapid moving systems when only two-element interferometers are used (e.g., the PTI orbits of Boden *et al.* 1999). Are we in a regime where the B-V^2 are only a intermediate step and we'll eventually end up back in a ρ-θ way of thinking, or do we need a new paradigm and new orbit calculation tools? Catalogers, observers and orbit computers need to know, and the observables which should be cataloged remains an elusive question. Heed should be given to words of Ejnar Hertzsprung (Lippincott 1962), 45 years ago:

> *The debt to our ancestors for the observations they made to our benefit, we can pay only by doing the same for our ancestors.*

As we use observations made in the past, we need to ensure that the observations made today are not just for today, but speak to the future.

Acknowledgements

The author is supported by NASA under JPL task order # NMO710776. This support is gratefully acknowledged.

References

Anderson, J.A. 1920, *ApJ* 51, 263

Arnold, D. 2006, *JDSO* 2, 163

Balega, I.I., Balega, Y.Y., Hofmann, K.-H., Malogolovets, E.V., Schertl, D., Shkhagosheva, Z.U., & Weigelt, G. 2006, *A&A* 448, 703

Balega, I.I., Balega, Y.Y., Hofmann, K.-H., Pluzhnik, E.A., Schertl, D., Shkhagosheva, Z.U., & Weigelt, G. 2005, *A&A* 433, 591

Boden, A.F., Koresko, C.D., van Belle, G.T., Colavita, M.M., Dumont, P.J., Gubler, J., Kulkarni, S.R., Lane, B.F., Mobley, D., Shao, M., *et al.* 1999, *ApJ* 515, 356

Burnham, S.W. 1894, *Publ. Lick Obs.* 2

Daley, J.A. 2005, *JDSO* 2, 44

Davis, J., Mendez, A., Seneta, E.B., Tango, W.J., Booth, A.J., O'Byrne, J.W. Thorvaldson, E.D., Ausseloos, M., Aerts, C., & Uytterhoeven, K. 2005, *MNRAS* 356, 1362

Fabricius, C., Høg, E., Makarov, V.V., Mason, B.D., Wycoff, G.L., & Urban, S.E. 2002, *A&A* 384, 180

Girard, T.M., Wu, H., Lee, J.T., Dyson, S.E., van Altena, W.F., Horch, E.P., Gilliland, R.L., Schaefer, K.G., Bond, H.E., Ftaclas, C. *et al.* 2001, *AJ* 119, 2428

Heintz, W.D. 1996, *AJ* 111, 408

Hummel, C.A., Armstrong, J.T., Quirrenbach, A., Buscher, D.F., Mozurkewich, D. Elias, N.M., II, & Wilson, R.E. 1994, *AJ* 107, 1859

Hummel, C.A., Carquillat, J.-M., Ginestet, N., Griffin, R.F., Boden, A.F., Hajian, A.R., Mozurkewich, D., & Nordgren, T.E. 2001, *AJ* 121, 1623

Kaplan, G.H. & Snell, S.C. 2001, *BAAS* 33, 1493

Lippincott, L.S. 1962, *PASP* 74, 5

Mason, B.D., Hartkopf, W.I., Holdenried, E.R., & Rafferty, T.J. 2001, *AJ* 121, 3224

Mason, B.D., Hartkopf, W.I., Wycoff, G.L., & Rafferty, T.J. 2006, *AJ* 131, 2687

McAlister, H.A. 1981, *AJ* 86, 795

Merrill, P.W. 1922, *ApJ* 56, 40

Pourbaix, D. 2000, *A&AS* 145, 215

Struve, F.G.W. 1837, *Mensurae Micrometricae Petropoli*

Wycoff, G.L., Mason, B.D., & Urban, S.E. 2006, *AJ* 132, 50

Discussion

TOM CORBIN: Considering the large astrometric catalogs that are coming out and will come out (in particular Gaia), will the WDS become a catalog that is mostly stars where a common proper motion is all that is known about the pairs?

MASON: It may be appropriate to have an addition "faint stars annex" for many new Gaia common proper motion pairs. The amount of data it generates will present challenges.

RAGHAVAN: Here are some WDS stats — optical versus physical — for an FGK sample within 25pc. Some 204 out of 455 primaries have WDS entries. These 204 primaries have 448 entries, and of the 448 entries:

    271 (60%) are optical pairs

    125 (28%) are physical pairs (i.e., have orbits or are clearly CPM)

    52 (12%) are undetermined as of now

Also, contamination from field stars is expected to be higher for nearby stars and hence my sample will have a higher percentage of optical alignments.

Binary Stars as Critical Tools & Tests
in Contemporary Astrophysics
Proceedings IAU Symposium No. 240, 2006
W.I. Hartkopf, E.F. Guinan & P. Harmanec, eds.

Toward a Common Language: the Washington Multiplicity Catalog

William I. Hartkopf

Astrometry Department, U.S. Naval Observatory, Washington, DC, United States
email: wih@usno.navy.mil

Abstract. Due to improvements in technology (interferometers, precision radial-velocity techniques, etc.), the traditional separation/period regimes of "wide" and "close" binaries are witnessing increasing overlap. This is expected to lead to increasing confusion in component identification, since different observing techniques adopted their own rules for designation of these components. A quick overview will be given of the *Washington Multiplicity Catalog*, an effort to create a common nomenclature scheme for all types of stellar and sub-stellar companions.

Author's note: The following is an abbreviated version of remarks made at Symposium 240; see Hartkopf & Mason (2004) for a more detailed description of the WMC and its history.

1. Introduction

The introduction of new observing and reduction techniques in recent years — from long-baseline interferometers to iodine–cell spectrographs to millimag–precise photometry to Doppler tomography techniques — have revolutionized much of our field. New classes of companions (e.g., brown dwarfs, exoplanets) have been discovered, and the previously distinct classes of binaries (e.g., "visual" or "spectroscopic") are now observable by multiple techniques. The result is greater understanding for the scientist, but greater challenges for the cataloger; furthermore, future generations of interferometers, astrometric satellites, computers, etc. means the "problems" will only become worse.

So what *are* the problems?

(*a*) Different techniques mean different nomenclature! Visual binaries are traditionally given "discoverer designations" (e.g., Σ or STF 14, β or BU 96). Spectroscopic binaries are usually referred to by their HD numbers, eclipsing binaries by variable star designations, occultation binaries by SAO or ZC numbers. Discovery or analysis using multiple techniques results in multiple designations in the literature; cross–reference resources such as SIMBAD are great aids in reducing the ensuing confusion, but much useful research on a given object may remain unknown to another astronomer simply due to the use of different object designations.

(*b*) A bigger problem is component confusion! Practitioners of different observing techniques have adopted different designation schemes for identifying the primary, secondary, etc.: capital letters (**A/B**, **N/S** or **E/W**, etc.), lower–case letters (**a/b**, **p/s**, etc.), numbers, or some combination. Further, the definition of "primary" may differ (for example, the brightest star in *V*- or *B*- or *K*-band, the most massive star, the star with sharpest spectral lines, etc.), meaning one astronomer's **AB** pair may be another's **BA**. An additional component to a binary may be assigned a **C** designation by someone unaware that the system is already a known triple via discovery of a different component using a different technique.

The purpose of the *Washington Multiplicity Catalog* (WMC) is to attempt to address these problems by creating a single catalog of all types of binary and multiple stars and developing a consistent designation scheme for all components of those systems.

2. Addressing the Problem

Informal working groups dating back to 1999 began work on various schemes for a common component designation. Discussions were held at IAU Symposium 200 (Reipurth & Zinnecker 2000), and later during the XXIVth and XXVth General Assemblies, where a Working Group was formed, a nomenclature scheme was adopted, and a sample catalog was presented. The plan formulated at the Sydney GA was to have a completed all-sky version of the WMC ready in time for the XXVIth General Assembly in Prague. However, manpower shortages have delayed completion of the catalog.

The root of the WMC is the *Washington Double Star* (WDS; Mason *et al.* 2006) database, maintained at the U.S. Naval Observatory and updated nightly. At present it is comprised of over 725,000 measures of 100,000+ pairs; as the largest single database of double stars, it was logical to propose a scheme based on its component rules. Of course, the WDS lists only resolved systems (i.e., it excludes spectroscopic pairs, eclipsing binaries, occultation systems, etc.), but it was agreed that WDS nomenclature rules (with slight modification) could accommodate all types of double stars.

3. Designations and Hierarchies

The WMC is a **hierarchical** scheme. Rules of component designation are as follows:

- **Level 1** = capital letters (e.g., STF 1523 **A,B**)
- **Level 2** = lower case letters (e.g., FIN 347 **Aa,Ab**)
- **Level 3** = numbers (e.g., BNK 1 **Ba1,Ba2**)
- **Higher levels**: alternate lower case letters and numbers (no examples yet known)

System hierarchy is ideally based on orbital period and/or semi-major axis separation, with a ~3:1 ratio of semi-major axes generally followed. However, wider, long-period systems (including most visual doubles) typically have insufficent information to determine whether motion is Keplerian or rectilinear. In these cases, apparent separation is all that is available. Typically systems with separations $> 1''$ are given upper-case letters, as are known optical pairs. The *primary* of a system is defined as the most massive object in the system, if known. Otherwise, this designation is given to the brightest object (preferably in V) in the system.

Figure 1 (from Hartkopf & Mason 2004) illustrates an imaginary system, where a simple wide pair becomes more complex as an additional faint companion is found, then newer observing techniques resolve components into two or more objects. The most important thing to note is that definitions of existing components do not change as new objects are discovered. For example, when the **C** component of the system is resolved into two objects by interferometry, the individual stars comprising **C** are labelled **Ca** and **Cb**, but **C** still refers to that collection of material unresolved on those astrographic plates. In terminology developed by Andrei Tokovinin, **C** is described as the "parent" of **Ca** and **Cb**.

No designation scheme is perfect, of course. The imaginary system in Figure 1 is rather idealized, in that new components are discovered in such an order that their designations evolve logically. What if, for example, a level-3 hierarchical pair (such as a

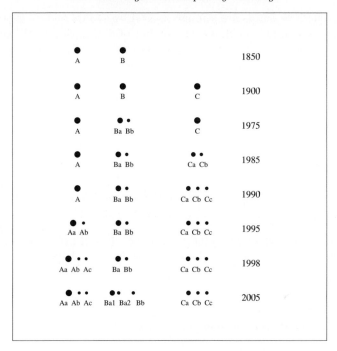

Figure 1. Nomenclature assignment as a (ficticious) system grows more complex:
    1850: visual pair is discovered
    1900: wide common proper motion companion is found on astrograph plates
    1975: **B** component is found to be spectroscopic binary
    1985: **C** component is split by speckle interferometry
    1990: additional speckle **C** component is resolved at a similar separation
    1995: planet is found orbitting the **A** component
    1998: second planet is found
    2005: primary of **B** is resolved by long-baseline interferometry

close spectroscopic pair) is discovered — and components are designated — before the level-2 hierarchical pair which includes it (e.g., a speckle binary) is resolved? What do we do if we know that one component of a close interferometric pair is an SB but we can't tell which one? What do we do with all the "incorrect" or mutually incompatible designations already in the literature? We of course cannot just ignore published designations and rename system components as we learn more about them. As in all aspects of science (and life), complications cannot be avoided.

The sample version of the WMC generated for the Sydney GA used data available from various double-star catalogs covering a half-hour band in RA. The techniques which contributed to that sample broke down as follows:

    95.8% visual binaries and optical pairs,
     1.7% spectroscopic binaries,
     1.4% cataclysmic variables or related objects,
     1.0% occultation binaries,
     0.3% astrometric binaries,
     0.2% each eclipsing binaries and X-ray binaries,
     0.1% each spectrum binaries and planets.

Please note that these totals are strongly biased by several factors (two examples: (1) exoplanet discovery is a new field, so the number of discoveries is still small; (2) visual binaries are cataloged upon discovery, while SB's are usually not cataloged until their orbits have been determined). These relative numbers are expected to change drastically in the future, as new surveys (e.g., Gaia) release their findings.

Individual pairs in this sample were then grouped into 1,465 multiple systems and designations were assigned to all components; the breakdown was as follows:

> 1,336 (91%)   simple binaries
> 80 (5.5%)   non-hierarchical triples
> 16 (1.1%)   non-hierarchical systems, > 3 components
> 25 (1.7%)   hierarchical triples
> 8 (0.5%)   hierarchical systems, > 3 components

Again, this breakdown will undoubtedly change as the optical/physical nature of wider pairs are determined and as the results of surveys for SBs, EBs, and other closer pairs are published.

4. WMC Contents

The WMC will include the following information:

- ⋆ J2000 arcsecond coordinates
- ⋆ component designation (A,B; Aa,Ab; etc.)
- ⋆ components' *parent*
- ⋆ identifiers:
  - − HD, DM, variable star designation, discoverer designation, etc.
  - − Tycho-2, GSC2, UCAC, etc.
  - − GJ or other nearby star designation, cluster ID, Bayer/Flamsteed, WDS, etc.
- ⋆ physicality (optical, physical, cpm, etc.) and method
- ⋆ magnitudes, spectral types, masses if known
- ⋆ orbital period and/or angular separation
- ⋆ references for all information

It is expected that the parent designations and extensive notes will be able to clarify most hierarchy problems that may arise due to the complications mentioned above.

5. So why this talk?

The reasons for giving this brief overview of the WMC to this audience are twofold:

1. To promote use of a consistent designation scheme
2. To urge all observers to send their data to databases

Much of the information in catalogs such as SB9 (Pourbaix *et al.* 2004) or the WDS has arrived there via literature searches by the authors of those databases. Both these projects are severly understaffed, and would benefit enormously if, as a matter of course, appropriate electronic data tables are sent to relevant catalogers by authors when their journal papers are accepted for publication†.

† By the way, if you wish your data to be of use to future researchers, *please* include the date of observation with your published binary star data.

Further information on the WMC and the catalogs used in its complilation may be found on the USNO's *Double Star Library* website:

(http://ad.usno.navy.mil/wds/dsl.html).

Your input or comments are welcome!

References

Hartkopf, W.I. & Mason, B.D. 2004, *RMxAC* 21, 83

Mason, B.D., Hartkopf, W.I., & Wycoff, G.L. 2006, *Washington Double Star Catalog*, http://ad.usno.navy.mil/wds/wds.html. See also 2001, *AJ*, 122, 3466

Pourbaix, D., Tokovinin, A.A., Batten, A.H., Fekel, F.C., Hartkopf, W.I., Levato, H., Morell, N.I., Torres, G., & Udry, S. 2006, *SB9: The Ninth Catalogue of Spectroscopic Binary Orbits*, http://sb9.astro.ulb.ac.be. See also 2004, *A&A* 424, 727

Reipurth, B. & Zinnecker, H. (eds.) 2000, IAU Symp. 200, *The Formation of Binary Stars*, (San Francisco: ASP)

Discussion

CARLSON CHAMBLISS: In eclipsing binaries with third components (e.g., V505 Sgr, 44 Boo, etc.) the variable pair should be referred to by its "classical" variable star name. What about the third component — either resolved on unresolved?

HARTKOPF: Individual systems will of course have to be examined on a case-by-case basis. In general, however, the components of close eclipsing pairs would probably be assigned component designations of **Aa** and **Ab** (assuming the eclipsing pair is brighter than the third body) and the third component would be labelled the **B** component. The more massive (or brighter if masses are unknown) star of the eclipsing pair would be assigned the **Aa** designation.

TERRY OSWALT: This is not really a question but a compliment to you for trying to make some sense out of a growing problem of nomenclature in binary star research **and** also to suggest that we address the equally challenging problem of multiple NAMES for binaries. Perhaps we should consider, as Luyten and others did so long ago, that the basic name of a binary be the earliest one given in the literature. Then the hierarchical scheme you proposed can be added as a suffix.

Binary Stars as Critical Tools & Tests
in Contemporary Astrophysics
Proceedings IAU Symposium No. 240, 2006
W.I. Hartkopf, E.F. Guinan & P. Harmanec, eds.

Polaris: Mass and Multiplicity

Nancy Remage Evans[1], Gail Schaefer[2], Howard E. Bond[2], Edmund Nelan[2], Giuseppe Bono[3], Margarita Karovska[1], Scott Wolk[1], Dimitar Sasselov[4], Edward Guinan[5], Scott Engle[5], Eric Schlegel[6] and Brian Mason[7]

[1]SAO, 60 Garden St., Cambridge MA 02138, USA
email: nevans@cfa.harvard.edu

[2]STScI, 3700 San Martin Dr., Baltimore, MD 21218, USA

[3]Univ. Roma, Rome, Italy

[4]Harvard University, 60 Garden St., Cambridge MA 02138, USA

[5]Villanova Univ., Dept. of Astronomy, Villanova, PA 19085 USA

[6]University of Texas, San Antonio, Dept. of Physics and Astronomy, 6900 N. Loop 1604 West, San Antonio TX 78249-0697 USA

[7]US Naval Observatory, 3450 Massachusetts Ave., NW, Washington, D.C. 20392-5420, USA

Abstract. Polaris, the nearest and brightest classical Cepheid, is a member of at least a triple system. It has a wide (18″) physical companion, the F-type dwarf Polaris B. Polaris itself is a single-lined spectroscopic binary with an orbital period of ∼30 years (Kamper 1996). By combining *Hipparcos* measurements of the instantaneous proper motion with long-term measurements and the Kamper radial-velocity orbit, Wielen *et al.* (2000) have predicted the astrometric orbit of the close companion. Using the *Hubble Space Telescope* and the Advanced Camera for Surveys' High-Resolution Channel with an ultraviolet (F220W) filter, we have now directly detected the close companion. Based on the Wielen *et al.* orbit, the *Hipparcos* parallax, and our measurement of the separation (0″.176 ± 0″.002), we find a preliminary mass of 5.0 ± 1.5 $M_\odot$ for the Cepheid and 1.38 ± 0.61 $M_\odot$ for the close companion. These values will be refined by additional *HST* observations scheduled for the next 3 years.

We have also obtained a *Chandra* ACIS-I image of the Polaris field. Two distant companions C and D are not X-rays sources and hence are not young enough to be physical companions of the Cepheid. There is one additional stellar X-ray source in the field, located 253″ from Polaris A, which is a possible companion. Further investigation of such a distant companion is valuable to confirm the full extent of the system.

Keywords. Cepheids, masses, multiplicity, *Chandra*, *HST*

1. Introduction

Polaris, like most massive stars, is a member of a multiple system. It is also a supergiant (F5 Ib) which is the nearest and brightest classical Cepheid. It is a low amplitude, somewhat quirky, Cepheid with a variable amplitude, which is pulsating in the first overtone mode, as shown by the *Hipparcos* parallax (Feast & Catchpole 1997)

The goals of this discussion are to present preliminary results on two topics. The first is a direct dynamical measurement of the mass, which is of particular interest because Polaris Aa is a Cepheid. The second is to explore how many physical companions belong to the system, which is of interest as the "footprints" of star formation.

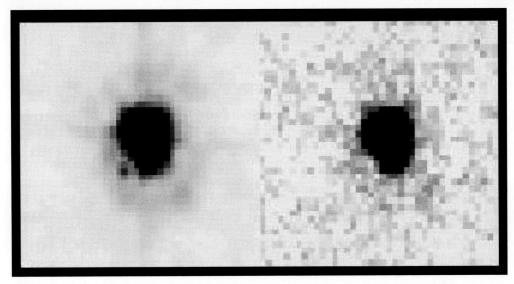

Figure 1. Left: Polaris Aa + Ab taken with the *HST* ACS HRC F220W filter (coadded images). The companion Ab can be seen at approximately 7 o'clock. The image is approximately $1''$ on a side. Right: Polaris B from the same image shown to the same scale.

2. Mass

In order to determine the mass, an orbit must be derived. Polaris is a member of a ~30-year spectroscopic system. We have used the orbit redetermined by Kamper (1996). He includes high-precision velocities, which is important since the orbital velocity amplitude is only 4 km sec^{-1}. Wielen *et al.* (2000) made an important breakthrough when they used *Hipparcos* proper motions to derive an inclination for the system.

However, no further information could be determined about the mass because the companion had never been detected, nor the separation measured. We have obtained a *Hubble Space Telescope* (*HST*) Advanced Camera for Surveys (ACS) High Resolution Channel (HRC) image on 2 August 2005 (Figure 1). The companion can be seen in Figure 1 (left) at about "7 o'clock".

Because the point-spread function (PSF) is not completely symmetric, and faint wings can be seen in Figure 1, we have performed several tests to confirm the companion detection. The companion Polaris B is also in the images, and is shown in Figure 1 (right). It is noisier than A since B is 6 mag fainter. However the PSF is very similar, and there is no artifact at the location of the companion of Polaris A = Polaris Ab. In addition, we have examined several images of single white dwarfs taken with the same instrumentation from the *HST* archive, and also find no indication of an artifact at the location of Ab. For a full discussion of the 2005 observation and a second image obtained in 2006, see Evans *et al.* (2007).

Combining the orbit with the separation measured from the *HST* images ($0''.176 \pm 0''.002$), we find a mass of 5.0 ± 1.5 M$_\odot$ for the Cepheid and 1.38 ± 0.61 M$_\odot$ for the close companion. These values are preliminary and will be refined through successive *HST* observations.

3. Companions

The second aspect of this study is an investigation of the number of members of the system. While there are many ways to detect a binary companion in a system, it is much more difficult to be certain that the full list of members has been identified.

In the Polaris system, there are 2 highly probable companions, Ab, the member of the spectroscopic system, and Polaris B. Polaris B (19″ distant) is a probable physical companion on velocity grounds (Kamper 1996). Two fainter, more distant stars (C and D) might be companions and would be dK stars if they are at the distance of Polaris.

Does a system with three stars more massive than the sun have a number of low-mass companions as would be predicted by the initial mass function? In order to investigate this, we used the following approach. Any low-mass companions (mid-F spectral type and later) as young as the Cepheid would produce X-rays. See, for instance, the study of the α Per cluster (Randich *et al.* 1996).

We have obtained a 10 ksec *Chandra* image to look for low-mass companions. Stars C and D do not appear on the X-ray image, and hence are not young companions. Their motion has also been found (by BDM) to be incompatible with the Polaris system. We do find X-rays at the location of Aa + Ab and are working to determine whether they come from the Cepheid or the companion. B was not detected in X-rays, but that is not surprising for an early F star. In addition there are a number of background AGNs in the image. They can be distinguished from stars since they do not have optical counterparts on the 2MASS images, whereas stars at the distance of Polaris do have counterparts. However, there is one X-ray source that does have a 2MASS counterpart consistent with an early M star. This source "E" becomes the one candidate for a further member of the system in the $16' \times 16'$ *Chandra* ACIS-I field.

In summary, only a late dM star would have remained undetected on the *Chandra* image. This means we have searched for companions to approximately a mass ratio of 0.1. The full content of the system is made up of the Cepheid and two probable companions with one possible additional low-mass companion.

Acknowledgements

Support for this work was provided by grants HST-GO-10593.01-A and NAS8-03060, and also Chandra grant GO6-7011A and Chandra X-ray Center NASA Contract NAS8-39073.

References

Evans, N.R., Schaefer, G., Bond, H.E., Nelan, E., Karovska, M., Bono, G., & Sasselov, D. 2007, in preparation
Feast, M.W., & Catchpole, R.M. 1997, *MNRAS*, 286, L1
Kamper, K.W. 1996, *JRASC*, 90, 140
Randich, S., Schmitt, J.H.M.M., Prosser, C.F., & Stauffer, J.R. 1996, *A&A*, 305, 785
Wielen, R., Jahreiss, H., Dettbarn, C., Lenhardt, H., & Schwan, H. 2000, *A&A*, 360, 399

Binary Stars as Critical Tools & Tests
in Contemporary Astrophysics
Proceedings IAU Symposium No. 240, 2006
W.I. Hartkopf, E.F. Guinan & P. Harmanec, eds.

© 2007 International Astronomical Union
doi:10.1017/S1743921307003882

DE CVn: A Bright, Eclipsing Red Dwarf – White Dwarf Binary

E. J. M. van den Besselaar[1], R. Greimel[2], L. Morales-Rueda[1],
G. Nelemans[1], J. R. Thorstensen[3], T. R. Marsh[4], V. S. Dhillon[5],
R. M. Robb[6], D. D. Balam[6], E. W. Guenther[7], J. Kemp[8],
T. Augusteijn[9] and P. J. Groot[1]

[1]Department of Astrophysics, IMAPP, Radboud University Nijmegen, PO Box 9010, 6500 GL
Nijmegen, The Netherlands email: [besselaar;lmr;nelemans;pgroot]@astro.ru.nl

[2]Isaac Newton Group of Telescopes, Apartado de correos 321, E-38700 Santa Cruz de la
Palma, Spain email: greimel@ing.iac.es

[3]Department of Physics and Astronomy, Dartmouth College, 6127 Wilder Laboratory
Hanover, NH 03755, USA email: thorsten@partita.dartmouth.edu

[4]Department of Physics, University of Warwick, Coventry CV4 7AL, UK
email: t.r.marsh@warwick.ac.uk

[5]Department of Physics and Astronomy, University of Sheffield, Sheffield S3 7RH, UK
email: vik.dhillon@sheffield.ac.uk

[6]Department of Physics and Astronomy, University of Victoria, Victoria, BC, V8W 3P6,
Canada email: robb@uvic.ca; cosmos@uvvm.uvic.ca

[7]Thüringer Landessternwarte Tautenburg, Sternwarte 5, D-07778 Tautenburg, Germany email:
guenther@tls-tautenburg.de

[8]Joint Astronomy Centre 660 N. A'ohoku Place University Park Hilo, Hawaii 96720, USA
email: j.kemp@jach.hawaii.edu

[9]Nordic Optical Telescope, Apartado 474, E-38700 Santa Cruz de La Palma, Spain
email: tau@not.iac.es

Abstract. DE CVn is a relatively unstudied eclipsing binary where one of the components
is an M dwarf and the other is a white dwarf. Its brightness makes it an ideal system for a
detailed study in the context of common-envelope evolution of a detached white dwarf - red
dwarf binary with a relatively short orbital period (∼8.7 hours). We present a detailed study of
the basic parameters (e.g. orbital period, components' masses and spectral types) for this system
from photometric and spectroscopic studies. The eclipses observed during several photometric
observing runs were used to derive the ephemeris. We have used spectroscopic data to derive the
radial velocity variations of the emission lines and these are used to determine the components'
masses and the orbital separation. The secondary component in DE CVn is an M3 main-sequence
star and the primary star, which only contributes to the blue continuum, is a cool white dwarf
with a temperature of ∼8000 K. From the photometry and spectroscopy together, we have set
a limit on the binary inclination. This system is a post-common-envelope system where the
progenitor of the present day white dwarf was a low-mass star ($M \leqslant 2M_\odot$). The time before DE
CVn becomes a semi-detached system is longer than the Hubble time.

Keywords. Stars: individual (DE CVn), Binaries: close, Binaries: eclipsing, Stars: late-type,
white dwarf, Stars: fundamental parameters

1. Introduction

DE CVn is a high proper motion object ($-0''.198 \pm 0''.002$ in RA, $-0''.178 \pm 0''.003$ in
Dec) which was first discovered as an X-ray source by ROSAT (Voges *et al.* 1999). It is a

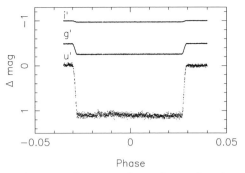

Figure 1. Primary eclipse observed with ULTRACAM.

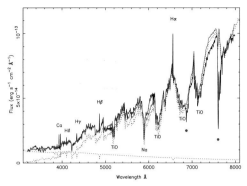

Figure 2. A combined ISIS spectrum of DE CVn (black line) together with the best model (grey/red). See Table 1 for information on the model.

detached white dwarf – red dwarf (WD+RD) binary with a relatively short ($\sim$8.7 hours) orbital period. All close WD+RD binaries must have gone through a common-envelope (CE) phase during their evolution. DE CVn's brightness and the presence of eclipses makes this system ideal for a more detailed study. Our aim is to derive its system parameters from a study of photometric and spectroscopic observations and ultimately to set limits on the physics of the CE phase.

2. Observations

We obtained photometry and spectroscopy of DE CVn on a number of telescopes and epochs. Simultaneous photometry was obtained in u', g' and i' bands with ULTRACAM on the WHT on May 24th, 2003. Additional photometry was obtained with the automatic 0.5-meter telescope of the Climenhage Observatory in Victoria, Canada (R, V and clear filters), with the 1.8-meter telescope of the Dominion Astrophysical Observatory (B filter) and with the 1.3-meter telescope of the Michigan-Dartmouth-MIT Observatory (MDM) in Arizona (B, $BG38$ filters) over a period of 10 years. The main spectroscopic observations are echelle observations with the 2-m telescope of the Thüringer Landessternwarte 'Karl Schwarzschild' in Tautenburg and long-slit spectroscopy with the MDM and WHT telescopes.

3. Photometry

By combining our photometry and the published times of mid-eclipse from Robb & Greimel (1997) and Tas *et al.* (2004), we obtain a new, more accurate ephemeris:

$$HJD_{\mathrm{min}} = 2452784.55337(2) + 0.36413945(4) \times E$$

with the uncertainty on the last digits in parentheses. Figure 1 shows the ULTRACAM photometry of the eclipse of DE CVn on the night of May 24th in 2003.

The eclipse depths in u', g' and i' are 1.11 ± 0.04, 0.235 ± 0.004 and 0.028 ± 0.004 magnitudes respectively.

4. Spectroscopy

We have used the low-resolution spectra obtained with ISIS on the WHT to derive the composition of DE CVn. The WD atmospheres were kindly provided by P. Bergeron

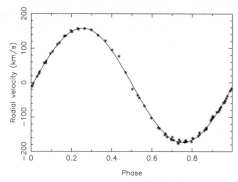

Figure 3. Radial velocity measurements for the Hα lines. The solid line is the best fit to these velocities. The uncertainties on the points are smaller than the symbols.

(with 1500 K < T < 17000 K). RD templates (M0V - M6V) have been obtained from Pickles (1998). Single templates are first scaled to 10 pc, then added and fitted to the observed spectrum, including a distance offset. The combination with the lowest χ^2 is taken as the best fit and is shown in Figure 2. From the echelle spectroscopy of the Hα line we derive the radial velocity curve of the RD as shown in Figure 3. The semi-amplitude is 166 ± 4 km s^{-1}.

5. Conclusion

The results of our photometric and spectroscopic analysis are listed in Table 1. DE CVn is an eclipsing binary consisting of a cool WD and an M3 main-sequence star that must have experienced a CE phase. From evolutionary tracks we derive a WD progenitor mass of M $\leqslant 2M_\odot$. The time remaining before the system becomes semi-detached is 1.7×10^{10} years. These kinds of systems will not contribute to the current sample of Cataclysmic Variables, unless the loss of angular momentum in the current detached phase is much higher than that given by magnetic braking alone.

Table 1. System parameters of DE CVn

Parameter	Value	Parameter	Value
WD temperature	8000 ± 2000 K	WD mass	$0.54 \pm 0.04 M_\odot$
WD $\log g$	7.5	WD radius	$0.0132 \pm 0.0006 R_\odot$
RD spectral type	M3V	RD mass	$0.40 \pm 0.05 M_\odot$
Semi-amplitude	166 ± 4 km s^{-1}	RD radius	$0.40 \pm 0.04 R_\odot$
Orbital separation	$2.10 \pm 0.06 R_\odot$	Inclination	$\geqslant 82°$
Distance	26 ± 3 pc		

Acknowledgements

We thank Pierre Bergeron for making his cool white dwarf models available to us.

EvdB, LMR and PJG are supported by NWO-VIDI grant 639.042.201 to P.J. Groot. GN is supported by NWO-VENI grant 638.041.405 to G. Nelemans. JRT thanks the U.S. National Science Foundation for support through grants AST-9987334 and AST-0307413. TRM was supported by a PPARC Senior Fellowship during the course of this work.

ULTRACAM is supported by PPARC grants PP/D002370/1 and PPA/G/S/2003/ 00058. The William Herschel Telescope is part of the Isaac Newton Group of Telescopes, operated on the island of La Palma by the Instítuto de Astrofísica de Canarias on behalf of the British PPARC and the Dutch NWO. We acknowledge the use of the 0.5-m telescope of the Cimenhage Observatory and the 1.8-m telescope of the Dominion Astrophysical Observatory located in Victoria, Canada. We acknowledge the use of the 1.3-m telescope and the 2.4-m Hiltner Telescope of the Michigan-Dartmouth-MIT observatory in Arizona and the 2-m Alfred–Jensch–Teleskop at the Thüringer Landessternwarte.

References

Pickles, A.J. 1998, *PASP*, 110, 863

Robb, R.M. & Greimel, R. 1997, *Informational Bulletin on Variable Stars*, 4486, 1

Tas, G., Sipahi, E., Dal, H.A., *et al.* 2004, *Information Bulletin on Variable Stars*, 5548, 1

Voges, W., Aschenback, B., Boller, T., *et al.* 1999, *A&A*, 349, 389

Binary Stars as Critical Tools & Tests
in Contemporary Astrophysics
Proceedings IAU Symposium No. 240, 2006
W.I. Hartkopf, E.F. Guinan & P. Harmanec, eds.

© 2007 International Astronomical Union
doi:10.1017/S1743921307003894

The Eclipsing Triple System
U Ophiuchi Revisited

Luiz Paulo R. Vaz[1], Johannes Andersen[2,3] and Antônio Claret[3]

[1]Departamento de Física, ICEx, UFMG, Brazil
email: lpv@fisica.ufmg.br

[2]Niels Bohr Institute, Astronomical Observatory, Copenhagen, Denmark
[3]Nordic Optical Telescope Scientific Association, E-38700 Santa Cruz de La Palma, Spain.
email: ja@astro.ku.dk

[4]Instituto de Astrofísica de Andalucía, CSIC, Granada, Spain
email: claret@iaa.es

Abstract. We have redetermined the absolute dimensions of the mid B-type eclipsing binary
U Oph from new light and radial-velocity curves, accounting for both the apsidal motion and
the light-time orbit around the third star. The stars in U Oph have masses of 5.27 and 4.74
$M_\odot$($\pm 1.5\%$) and are located in the middle of the main-sequence band for an an age of $\sim$50 Myr.
U Oph and three other systems (V760 Sco, MU Cas and DI Her) all have components within
10% of 5 $M_\odot$ and ages below 100 Myr; we find significant heavy-element abundance differences
between these young nearby stars.

Keywords. stars: binaries: close: eclipsing: evolution: individual (U Oph, MU Cas, V760 Sco,
DI Her)

1. Introduction

Accurate empirical stellar masses, radii, and luminosities are fundamental test data
for stellar models (Andersen 1991). Good data sets exist for stars below $\sim$4$M_\odot$, but
good determinations for higher-mass stars remain few. We have therefore reanalyzed the
detached, mid-B type eclipsing binary U Oph from new spectroscopic and photometric
data, using improved analysis techniques. Our primary results, including figures, are
given in the on-line version of the poster and in the research paper by Vaz *et al.* (2007).

2. Observations and Analysis

New light curves of U Oph in the Strömgren $uvby$Hβ system were obtained with the
Danish 0.5-m telescope at ESO, La Silla, Chile, in 1992-94; spectroscopic orbits were
determined from high-quality coudé spectrograms from the ESO 1.5 m telescope. Radial
velocities were measured from spectral lines without broad damping wings, which may
blend and cause errors in the derived masses (Andersen 1991).

Photometric and spectroscopic orbital elements were determined in a fully consistent analysis, taking into account both the exceptionally short-period apsidal motion
($\sim$21.2 yr) and the light-time orbit around a third star ($P_3 \sim$38.4 yr). A new combined
ephemeris for both orbits was determined from 353 published times of minimum. The
Wilson-Devinney (WD) light-curve analysis code was modified to treat apsidal and light-
time orbital motion effects together and used to analyse our $uvby$ light and radial-velocity
curves simultaneously.

Figures 1–3 of the poster show our observational data and the differences between our
new ephemeris and the previous best determination by Kämper (1986).

3. Results

We find the masses and radii of the components of U Oph to be 5.27±0.09 and 4.74±0.07 $M_\odot$, and 3.48±0.02 and 3.11±0.03 $R_\odot$, respectively. With $\log g \sim 4.1$, both stars are located near the middle of the main-sequence band. From the colour indices, we find $T_{\rm eff} \sim 16,000 K$ and derive $\log L/L_\odot \sim 2.8$ and a distance around 200 pc (details given in Table 2 of the poster).

We have compared our results for U Oph with stellar evolution models, both the latest series by Claret (2004) and the earlier models (Claret 1995, 1997), as well as with two generations of Padova models (Bressan *et al.* 1993, Girardi *et al.* 1996 and Girardi *et al.* 2000). In each case, we have considered models for both Z= 0.02 and Z= 0.01 in the diagrams of $\log g$ vs. $\log(M/M_\odot)$ and $\log T_{\rm eff}$.

In order to extend the comparison to stars of the same mass but different degrees of evolution within the main sequence, we have identified three other binary systems (MU Cas, DI Her, and V760 Sco) with components within 10% of 5 $M_\odot$ and with accurate data in the literature. Of these stars, DI Her is close to the ZAMS, MU Cas near the TAMS, and V760 Sco intermediate in age between DI Her and U Oph.

We find that the differences between different model calculations for the same chemical composition remain within the observational errors, given the small difference in mass and degree of evolution between the components in each system. When metallicity and age are free parameters, models can be selected that fit each system within the errors, and nothing significant is learned about the models themselves – underscoring that stellar models are best tested in binary systems with very different components and known metal abundances (Andersen 1991). For plausible compositions, the four systems are in the age range 0-100 Myr, with U Oph itself around 40-50 Myr.

The data are, however, accurate enough to show that not all four systems can be fit by models for a single value of Z; a range of about a factor 2 must be invoked. We emphasise that this result is obtained already from the $\log M - \log g$ diagram and is thus independent of uncertain temperature-colour relations. The chemical diversity of these young nearby stars presents a challenge to models of Galactic chemical evolution.

Acknowledgements

We thank Dr. Claud Lacy for kindly providing his ephemeris code. Support from ESO and the Danish Natural Science Research Council is gratefully acknowledged, as is the financial support of the Brazilian institutions CNPq, CAPES, FAPEMIG and FINEP (to LPRV). This research has made use of the Simbad database, operated at CDS, Strasbourg, France, and of NASA's Astrophysical Data System bibliographic services.

References

Andersen, J. 1991, *A&AR* 3, 91
Bressan A., Fagotto F., Bertelli G., & Chiosi C. 1993, *A&AS* 100, 647
Claret, A. 1995, *A&A* 109, 441
Claret, A. 1997, *A&A* 125, 439
Claret, A. 1998, *A&A* 330, 533
Claret, A. 2004, *A&A* 242, 919
Girardi L., Bressan A., Chiosi C., Bertelli G., & Nasi E. 1996, *A&AS* 117, 113
Girardi L., Bressan A., Bertelli G., & Chiosi C. 2000, *A&AS* 141, 371
Kämper, B.-C. 1986, *Ap&SS* 120, 167
Vaz, L.P.R., Andersen, J., & Claret, A. 2007, *A&A*, in press (accepted March 29, 2007)

Binary Stars as Critical Tools & Tests
in Contemporary Astrophysics
Proceedings IAU Symposium No. 240, 2006
W.I. Hartkopf, E.F. Guinan & P. Harmanec, eds.

New Trends in Disentangling
the Spectra of Multiple Stars

Petr Hadrava

Astronomický ústav, Akademie věd České republiky, CZ-251 65 Ondřejov, Czech Republic
email: had@sunstel.asu.cas.cz

Abstract. The method of spectra disentangling has been applied in many studies of different stellar systems up to the distance of Andromeda Galaxy. In some of these applications the underlying assumptions are not precisely satisfied. This is why new generalizations of the method are needed. Ways to overcome these problems are discussed here.

Keywords. techniques: spectroscopic, (stars:) binaries: spectroscopic

1. Introduction

The method of spectra disentangling developed using different approaches by Simon & Sturm (1994) and Hadrava (1995) enables the simultaneous determination of the orbital parameters of a system as well as the spectra of individual component stars. It proved to be very useful for various studies, including several presented in this volume. However, an improper application of the method as a black box without understanding to systems which do not satisfy the underlying assumptions is risky. A review of the method together with discussion of related techniques and practical hints is available at Hadrava (2004c). In this abbreviated version of the present contribution, only the basic, sometimes misunderstood, principles of disentangling are reviewed and recent developments towards a generalization of the method are summarized.

2. Disentangling = decomposition of spectra + fitting of parameters

In the classical treatment, orbital elements of spectroscopic binaries are obtained by subsequent measurement of radial velocities (RVs hereafter) and their fitting preferably together with light curves and other data (cf. Wilson 1979; Kallrath & Milone 1999; Hadrava 2004b, 2005). To measure the RVs for blended lines, special techniques (cf. Hill 1993, Zucker & Mazeh 1994, Rucinski 1992) requiring the knowledge of component spectra are needed. Special techniques have been developed (e.g. by Bagnuolo & Gies 1991) to separate the component spectra if RVs are known.

In disentangling, the observed spectra $I(x,t)$ are decomposed into spectra $I_j(x)$ of n component stars assuming that (in the logarithmic wavelengths $x = c \ln \lambda$) they are a linear combination

$$\sum_{j=1}^{n} I_j(x) * \delta(x - v_j(t,p)) = I(x,t), \qquad (2.1)$$

and simultaneously the orbital parameters p (or RVs $v_j(t)$) are fitted (cf. Figure 1 in the full version of this contribution).

3. Disentangling of spectra or Spectral disentangling?

Some users incorrectly denote the method of disentangling by adjective "spectral", what reveals misunderstanding of its principle. Spectral methods are generally a mathematical treatment of functions as a linear vector space, while for the spectra disentangling in astrophysics is essential that the treated functions represent some observed parts of electromagnetic spectrum radiated by the stellar system. The spectra decomposition can be performed either by the method of Singular Value Decomposition in the x- representation (Simon & Sturm 1994), or using the Fourier transform method, which reduces the set of Equations (2.1) into n- dimensional systems for each Fourier component (Hadrava 1995). Recently a kind of iterative procedure for the decomposition was also demonstrated by Gonzáles & Levato (2006).

4. Disentangling with line-profile variability

The standard disentangling does not impose any restriction on the component spectra I_j apart from their invariability in time. However, in real systems I_j may change because of proximity effects, eclipses and circumstellar matter, or from intrinsic line-profile variations arising from oscillations, rotation or secular changes of the component stars. In some systems, we can neglect these effects in the first approximation, and study them from $O - C$ residuals as higher-order perturbations only. But there is a danger of systematic errors e.g. in the disentangled orbital parameters. Moreover, the solution by standard disentangling may fail completely in extreme cases.

The relevant effects should thus be involved directly into the procedure of disentangling by generalization of Equation (2.1) to the form

$$\sum_{j,k} I_j^k(x) * \Delta_j^k(x,t,p) = I(x,t), \qquad (4.1)$$

where the spectrum of each component star j can be a superposition of several functions I_j^k corresponding e.g. to different limb-darkening modes, each one broadened by appropriate broadening function Δ_j^k (cf. Hadrava 1997, 2004c).

The first such step consists of involving line-strength factors $s_j(t)$ into Equation (2.1), as done in the KOREL04 code (Hadrava 1997, 2004c). This enables the measurement of eclipses of (not too fast rotating) stars from their spectra, but also to disentangle telluric lines (cf. Hadrava 2006a). More tricky possibilities are to disentangle the rotation of ellipsoidal components (Hadrava & Kubát 2003), rotational effects in eclipsing binaries (Hadrava 2006c), or oscillations and pulsations — either radial (Hadrava 2004a, 2004c) or non-radial. Other models of line-profile variability, like spots etc. could be included using a proper model of the stellar disc like in light-curve models (Hadrava 2005).

5. Disentangling with constraints

It is generally advantageous to solve all types of data (spectroscopy, photometry, astrometry etc.) simultaneously, as it is enabled, e.g., by FOTEL code (cf. Hadrava 2004b). The previous versions of KOREL provided RVs of individual components at each exposure, which could be subsequently included into FOTEL solution with other RVs (e.g., from the literature), photometry etc. However, the information available in the additional data is not then utilized in the disentangling itself, which can result in a false solution, or, at least, its convergence may be more difficult. In a case of disentangling with LPVs caused by eclipses, the solution of radii or inclination would be safer to perform simultaneously with light-curve solution (despite a possibility of different photospheric and chromospheric radii should be considered). Such constraints of parameters can be, in some cases, taken into account by their fixing to values found from the additional sources. However, more generally, we need to search for a minimum of $(O - C)^2$ bound by some conditions $F_k(p) = 0$ to a subspace of the parameter space. This can be performed by minimization of the sum

$$S = \sum_t \int |I - \sum_j I_j * \Delta_j|^2 dx + \sum \lambda_k F_k^2(p) \qquad (5.1)$$

with some Lagrange multiplicators λ. Because, in practice, the constraints obtained from the additional data have also some uncertainty, the additional terms F_k^2 can be the $(O - C)^2$ for those data and we thus arrive at the problem of simultaneous disentangling and solution of the other data with λs determining their relative weights, like it is done in FOTEL and other similar codes (cf. Holmgren 2004; Hadrava 2004b, 2006b).

Also, information about a component spectrum (e.g., from known spectral type or for the telluric lines) may help the convergence and to avoid instabilities (especially of low Fourier modes). The new version of KOREL thus permits the selection of some component spectra to be restricted by templates J_j given on input and to disentangle the others (Hadrava 2006b, cf. Figure 2), i.e., to solve the equation

$$\sum_{j=1}^{m} I_j(x) * \Delta_j(x, t, p) = I(x, t) - \sum_{j=m+1}^{n} J_j(x) * \Delta_j(x, t, p). \qquad (5.2)$$

Acknowledgements

This work is supported by grant GA ČR 202/06/0041 and by project AV0Z10030501.

References

Bagnuolo, W.G., & Gies, D.R. 1991, *ApJ* 376, 266
Gonzáles, J. F., & Levato, H. 2006, *A&A* 44, 283
Hadrava, P. 1995, *A&AS* 114, 393
Hadrava, P. 1997, *A&AS* 122, 581
Hadrava, P. 2004a, in: *Spectroscopically and Spatially Resolving the Components of the Close Binary Stars*, R.W. Hilditch, H. Hensberge & K. Pavlovski eds., ASP Conf. Ser. 318, p. 86
Hadrava, P. 2004b, *Publ. Astron. Inst. ASCR* 92, 1
Hadrava, P. 2004c, *Publ. Astron. Inst. ASCR* 92, 15
Hadrava, P. 2005, *Ap&SS* 296, 239
Hadrava, P. 2006a, *A&A* 448, 1149
Hadrava, P. 2006b, *Ap&SS*, 304, 337
Hadrava, P. 2006c, ASP Conf. Ser., in press
Hadrava, P., & Kubát, J. 2003, in: *Stellar Atmosphere Modeling*, I. Hubeny, D. Mihalas & K. Werner eds., ASP Conf. Ser. 288, p. 149
Hill, G. 1993, in: *New Frontiers in Binary Stars Research*, K.-C. Leung & I.-S. Nhu eds., ASP Conf. Ser. 38, p. 127
Holmgren, D. 2004, in: *Spectroscopically and Spatially Resolving the Components of the Close Binary Stars*, R.W. Hilditch, H. Hensberge & K. Pavlovski eds., ASP Conf. Ser. 318, p. 95
Kallrath, J., & Milone, E.F. 1999, *Eclipsing Binary Stars: Modeling and Analysis*, Springer-Verlag, New York, Berlin
Rucinski, S. 1992, *AJ* 104, 1968
Simon, K.P., & Sturm, E. 1994, *A&A* 281, 286
Wilson, R.E. 1979, *ApJ* 234, 1054
Zucker, S., & Mazeh, T. 1994, *ApJ* 420, 806

Binary Stars as Critical Tools & Tests
in Contemporary Astrophysics
Proceedings IAU Symposium No. 240, 2006
W.I. Hartkopf, E.F. Guinan & P. Harmanec, eds.

Binary Stars in the Orion Nebula Cluster

Rainer Köhler[1]† **Monika G. Petr-Gotzens**[2], **Mark J. McCaughrean**[3],
Jerome Bouvier[4], **Gaspard Duchêne**[4], **Andreas Quirrenbach**[5]
and Hans Zinnecker[6]

[1]Sterrewacht Leiden, P.O. Box 9513, NL-2300 RA Leiden, The Netherlands
email: koehler@strw.leidenuniv.nl

[2]European Southern Observatory, Karl-Schwarzschild-Str. 2, 85748 Garching bei München,
Germany

[3]School of Physics, University of Exeter, Stocker Road, Exeter EX4 4QL, Devon, UK

[4]Laboratoire d'Astrophysique de Grenoble, Université Joseph Fourier,
BP 53, 38041 Grenoble Cedex 9, France

[5]ZAH Landessternwarte, Königstuhl, Heidelberg, Germany

[6]Astrophysikalisches Institut Potsdam, An der Sternwarte 16, 14482 Potsdam, Germany

Abstract. We report on a high-spatial-resolution survey for binary stars in the periphery of the
Orion Nebula Cluster, at 5–15 arcmin (0.65 – 2 pc) from the cluster center. We observed 228 stars
with adaptive optics systems, in order to find companions at separations of 0.13 – 1.12 arcsec
(60 – 500 AU), and detected 13 new binaries. Combined with the results of Petr (1998), we
have a sample of 275 objects, about half of which have masses from the literature and high
probabilities to be cluster members. We used an improved method to derive the completeness
limits of the observations, which takes into account the elongated point spread function of stars
at relatively large distances from the adaptive optics guide star. The multiplicity of stars with
masses $>2\,M_\odot$ is found to be significantly larger than that of low-mass stars. The companion
star frequency of low-mass stars is comparable to that of main-sequence M-dwarfs, less than
half that of solar-type main-sequence stars, and 3.5 to 5 times lower than in the Taurus-Auriga
and Scorpius-Centaurus star-forming regions. We find the binary frequency of low-mass stars
in the periphery of the cluster to be the same or only slightly higher than for stars in the
cluster core ($< 3'$ from θ^1C Ori). This is in contrast to the prediction of the theory that the
low binary frequency in the cluster is caused by the disruption of binaries due to dynamical
interactions. There are two ways out of this dilemma: Either the initial binary frequency in
the Orion Nebula Cluster was lower than in Taurus-Auriga, or the Orion Nebula Cluster was
originally much denser and dynamically more active. A detailed report of this work has been
published in *Astronomy & Astrophysics* (Köhler *et al.* 2006).

Keywords. techniques: high angular resolution, binaries: close, stars: formation, stars: pre–
main-sequence open clusters and associations: individual (Orion Nebula Cluster)

1. Introduction

Stellar multiplicity is very high among young low-mass stars, with companion star
frequencies close to 100 % for young stars in well-known nearby star-forming T associations like Taurus-Auriga (Leinert *et al.* 1993, Ghez *et al.* 1993, Ghez *et al.* 1997, Duchêne
1999). On the other hand, high binary frequencies are *not* observed among low-mass stars
in stellar clusters like the Orion Nebula Cluster (e.g., Prosser *et al.* 1994, Padgett *et al.*
1997, Petr *et al.* 1998, Petr 1998, Simon *et al.* 1999, Scally *et al.* 1999, McCaughrean
2001). The reason for this discrepancy is still unclear. Theoretical explanations include:

† Present address: ZAH Landessternwarte, Königstuhl, 69117 Heidelberg, Germany.

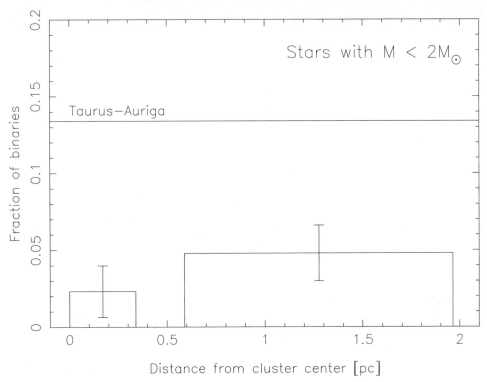

Figure 1. The binary frequency of low-mass stars (M<2M$_\odot$) in the center and the periphery of the Orion Nebula Cluster.

- Binaries are destroyed in close encounters with other stars (e.g., Kroupa 1995, Kroupa *et al.* 1999)
- The formation rate of binaries depends on the precollapse cloud conditions (e.g., temperature) (Durisen & Sterzik 1994, Sterzik *et al.* 2003)

The interaction time scale depends on the stellar density, so less binaries should be destroyed in the outer parts of the cluster.

2. Observations and Results

Our target list comprises some 230 stars in 52 fields, located at $5' - 15'$ $(0.7 - 2\,\mathrm{pc})$ from the cluster center. The adaptive optics systems at the 3.6-m telescope on La Silla and the Keck Telescope on Mauna Kea were used to observe them in the K-band. We found 13 companions in the separation range $0''.13 - 1''.12$ (60 – 500 AU). For comparison, we use the sample by Petr (1998), which contains 114 stars in the cluster core ($<3'$ or 0.4 pc from the center). We find the binary frequency of low-mass stars in the periphery of the cluster to be only slightly higher than in the core. In particular, the binary frequency in the periphery is significantly lower than in Taurus–Auriga.

3. Conclusions

We find no statistically significant difference of the binary frequency of low-mass stars between core and periphery. These results do not support the hypothesis that the binary frequency in Orion was initially as high as in Taurus and later reduced by dynamical interactions, unless the Orion Nebula Cluster was much denser in the past.

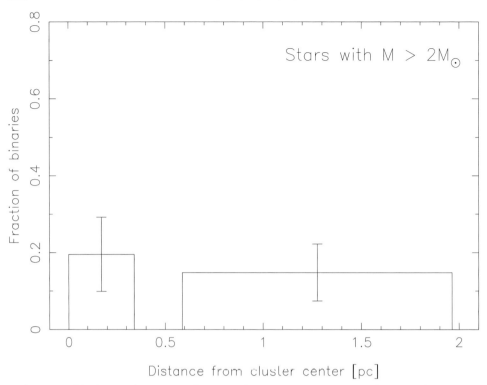

Figure 2. The binary frequency of intermediate- to high-mass stars (M>2M$_\odot$) in the center and the periphery of the Orion Nebula Cluster.

Acknowledgements

We thank the organizers for the opportunity to present our work.

References

Duchêne, G. 1999, *A&A* 341, 547

Durisen, R.H. & Sterzik, M.F. 1994, *A&A* 286, 84

Ghez, A.M., Neugebauer, G., & Matthews, K. 1993, *AJ* 106, 2005

Ghez, A.M., McCarthy, D.W., Patience, J., & Beck, T. 1997, *AJ* 481, 378

Köhler, R., Petr-Gotzens, M.G., McCaughrean, M.J., Bouvier, J., Duchêne, G., Quirrenbach, A., & Zinnecker, H. 2006, *A&A* 458, 461

Kroupa, P. 1995, *MNRAS* 277, 1491

Kroupa, P., Petr, M., & McCaughrean, M. 1999, *New Astronomy* 4, 495

Leinert, Ch., Zinnecker, H., Weitzel, N., *et al.* 1993, *A&A* 278, 129

McCaughrean, M.J. 2001, in: *The Formation of Binary Stars*, proceedings of IAU Symp. No. 200, eds. H. Zinnecker and R.D. Mathieu, ASP Conference Series, p. 169

Padgett, D.L., Strom, S.E., & Ghez, A. 1997, *ApJ* 477, 705

Petr, M.G., Coudé du Foresto, V., Beckwith, S.V.W., Richichi, A., & McCaughrean, M.J. 1998, *ApJ* 500, 825

Petr, M.G. 1998, *PhD Thesis*, University of Heidelberg

Prosser, C.F., Stauffer, J.R., Hartmann, L., *et al.* 1994, *ApJ* 421, 517

Scally, A., Clarke, C., & McCaughrean M.J. 1999, *MNRAS* 306, 253

Simon, M., Close, L.M., & Beck, T.L. 1999, *AJ* 117, 1375

Sterzik, M.F., Durisen, R.H., & Zinnecker, H. 2003, *A&A* 411, 91

Binary Stars as Critical Tools & Tests
in Contemporary Astrophysics
Proceedings IAU Symposium No. 240, 2006
W.I. Hartkopf, E.F. Guinan & P. Harmanec, eds.

© 2007 International Astronomical Union
doi:10.1017/S1743921307003924

Poster Abstracts (Session 2)

(Full Posters are available at http://www.journals.cambridge.org/jid_IAU)

Activities and Achievements of the Double Star Committee of the French Astronomical Society

J.-L. Agati, S. Caille, A. Debackère, P. Durand, F. Losse, Re. Manté, F. Mauroy, P. Mauroy, G. Morlet, C. Pinlou, M. Salaman, E. Soulié, Y. Thorel, & J.-C. Thorel

Societe Astronomique de France, Paris, France

Created in 1981 by Pierre DURAND with the support of Paul MULLER, the Double Star Committee constitutes ever since a forum of exchange of experiences and information in the field of double stars, particularly visual. The Committee relies on the advice of its scientific counsellors (in particular Pierre BACCHUS, Daniel BONNEAU, Paul COUTEAU and Jean DOMMANGET) to guide the work of its members. By fostering missions in observatories, it has stimulated the activities of observation and measurement of double stars. It has also encouraged the publication of measures (A&A and "Observations et Travaux") and raised up missions of verification of double star positions. Under its aegis, many series of measures of double stars made in particular with the 50-cm refractor at the Nice Observatory (either with a filar micrometer or with a CCD camera), were published. Uncertain positions of pairs have been checked and corrected. For the treatment of numerical images of double stars, software aiming in particular at the determination of position elements and the magnitude difference between components were tested and others created (e.g. REDUC and SURFACE). The spar plate double image micrometer of Lyot was developed and its fabrication raised up. Preliminary orbits of double stars have been calculated, as well. An amateur participates in the maintenance of the database of double star measures SiDoNie and pursues an historical research on the life and work of Robert JONCKHEERE. The Internet site of the Committee, created in 2005, informs laypersons as well as experienced amateurs (`http://saf.etoilesdoubles.free.fr`).

An Optical Search for Dwarf Novae in M22

A.P. Hourihane[1], P.J. Callanan[1], and A.M. Cool[2]

[1] *University College Cork, Cork, Ireland,*
[2] *San Francisco State University, San Francisco, CA, United States*

We report on the results of our 2004 monitoring program of M22, designed to search for CVs undergoing dwarf nova (DN) eruptions. We are interested to compare these globular cluster systems to similar ones in the field and determine if their different formation mechanisms result in different properties. We present a light curve for a CV candidate that went into outburst during May. Data were reduced using the ISIS image subtraction routine. Our ground-based results are consistent with previous HST measurements, and confirm the DN nature of the outburst. We also report on further attempts to apply the

ISIS software to look for additional outbursts of this object during 2004 as well as to identify other variable candidates in the core of M22.

An Investigation of the Small Eccentricity in the Spectroscopic Binary System ζ TrA

S. Komonjinda, J.B. Hearnshaw, and D.J. Ramm

Dept. of Physics and Astronomy, University of Canterbury, Christchurch, New Zealand

The orbital eccentricity of the SB1 system ζ TrA (S.T. F9V, $P \sim 13$ d) was found by Skuljan *et al.* (2004) to be $e = 0.01398 \pm 0.00019$. Lucy (2005) devised a statistical test of the significance of this result based on the amplitude and phase of the third harmonic in the Fourier analysis of the radial velocity data, and concluded that the non-zero eccentricity measured does not arise from a slightly eccentric Keplerian orbit, but from proximity effects in the binary. He therefore believes a circular orbit should be assigned to this system. In this paper, we investigate one possible proximity effect, namely the tidal distortion of the primary star, such that the measured Doppler shift does not accurately indicate the centre of mass radial velocity of the star as a whole.

The code of Wilson & Devinney (2003) was used to model the tidal distortion of the measured radial velocities, assuming a range of possible secondary masses, corresponding to M dwarf companions. The result is that even for the lowest possible mass secondary of 0.09 $M_\odot$ with $\sin i = 1$ (this gives the greatest tidal distortion, as it is closest to the primary) there is no significant effect on the radial velocities (the differences are of order 1 m/s as a result of the tidal effects). Similar negligible tidal effects arise using a white-dwarf companion.

We note that the difference between a circular orbit and the observations amounts to as much as 140 m/s at some phases, which is essentially the amplitude of the second harmonic in the data. Our conclusion is that this strong and highly significant second harmonic is most probably the result of a small orbital eccentricity as reported by Skuljan *et al.* (2004). We note that the observed third harmonic according to Lucy (2005) has an amplitude of only 5.2 ± 2.0 m/s, which is just over twice the error bar of its measurement, and that the predicted third harmonic for an eccentric orbit is only 1.6 m/s.

Discovery of a New Dwarf Nova, TSS J022216.4+412259.9: WZ Sge-type Dwarf Novae Breaking the Shortest Superhump Period Record

A. Imada, K. Kubota, T. Kato, D. Nogami, H. Maehara, K. Nakajima, M. Uemura, and R. Ishioka

Kyoto University, Kyoto, Japan

We report on the time-resolved CCD photometry of a newly discovered dwarf nova, TSS J022216.4+412259.9 during the outburst in the 2005 November-December. The best-estimated superhump period was 0.0554 days, which is the shortest superhump period among WZ Sge-type dwarf novae ever known. Double-peaked humps were also detected with a period of 0.05487 days in the early stage of the outburst. A rebrightening was exhibited after the end of the plateau phase. All of these observations indicate the WZ Sge nature of the system. We mainly discuss the rebrightening stage of the superoutburst, compared with other WZ Sge-type dwarf novae.

Periodic Perturbations of Relative Motion in the Multiple System ADS 15571.

E.A. Grosheva

Main Astronomical Observatory, St.Petersburg, Russia

For the multiple system ADS 15571, it was calculated a preliminary orbit of the visible secondary component (around the primary) and periodic perturbations of its relative motion were investigated. ADS 15571 has been observed at Pulkovo with a 26-in refractor since 1960; it was observed for the first time in 1832 by O. Strueve. Treatment of astronegatives was made with the scanner UMAX (resolution 1200 dpi, transparency adapter). The accuracy of annual average relative positions amounts to $0''.0048$ in angular separation ρ and $0°.018$ in positional angle θ.

As known, the secondary component of visual double star ADS 15571 is a spectroscopic binary with a period of 1.1522 days (Sanford,1927). Study of more then 40-yrs set of photographic observations with the 26-in refractor shows that the system has more one component with a period of 23 years. I obtained an astrometric orbit of the photocenter based on the apparent motion ellipse and concluded that the detected periodic perturbations of relative motion are caused by the existence of a low-magnitude companion. Its minimum mass was estimated at 0.62 $M_\odot$.

Determination of the Orbits and Estimation of the Masses of ADS 7251 and ADS 5983 (δ Gem)

N.A. Shakht, E.A. Grosheva, and D.L. Gorshanov

Main Astronomical Observatory of RAS (Pulkovo), Saint-Petersburg, Russia

The main purpose of our work was the investigation of the dynamics of two binaries; we also wish to define the masses of the components. After all the great interest was to search the existence a hidden mass or gravitational influence of probable dark satellites.

Our data have been founded on a homogeneous long-term series of photographic observations by means of Pulkovo observatory 65-cm refractor: 196 positions with m.e. $0''.004$ for ADS 7251 and 106 positions with m.e. $0''.020$ for δ Gem obtained in the years 1962-1996 and 1972-2006, respectively. All observations of δ Gem have been made with a diffraction grating because two its components have difference in magnitudes of about 5 mag.

For ADS 7251 the relative orbit, the total mass and the mass ratio have been obtained. Taking into account some contradictory data on masses and spectral classes of its components some discussion has been fulfilled. In the result of investigation of the individual motions of its components in the frame of reference stars the masses of A and B have been estimated. We were using different algorithms for the estimations of masses. As a result we have obtained the total mass and masses of components of ADS 7251. They are 0.56 and 0.54 $M_\odot$ for A and B respectively. The relative orbit for δ Gem has been obtained too, with a total mass of 2.5 $M_\odot$.

Any perturbations in the separation and position angles of ADS 7251 have not been discovered. The residuals in the motion of δ Gem are complicated character. In its separations a wave with a period of 6.1 year is revealed. The motion of a dark companion with mass about 0.2 $M_\odot$ is analysed, and some hypothesis on the other companion with a period about 1 year has been discussed.

Duplicity among Lambda Boo stars: The New Case of HD 204041

M. Gerbaldi[1] and R. Faraggiana[2]

[1] Institut d'Astrophysique, Paris, France,
[2] Dipartimento di Astronomia, Universita Trieste, Italy

The λ Boo stars are a fascinating class of stars: it is the only class of A-type stars with abundances lower than solar, in spite of the fact that they have the kinematical and photometric properties of Pop I stars. The most plausible theory, at present, to interpret this peculiar behaviour is based on the diffusion-accretion model proposed by Venn & Lambert (1990).

The key point of this model is the existence of a surrounding disk of dust and gas which would be the remnant of the star formation material and from which the — metal depleted — gas is accreted, the metals being incorporated into the dust.

The large variety of objects classified as λ Boo and the lack of any relation between their physical parameters and the measured abundance anomalies, coupled with the recent detection of bright companions near several λ Boo candidates (detected by ground-based speckle and our adaptive optics observations, as well as by the Hipparcos space experiment) prompted us to scrutinize all the high-resolution spectra we could obtain to search the objects that have composite spectra: a combination of two similar spectra which can be confused with that of a single metal poor star when analyzed by adopting their average T_{eff}, *log* g parameters.

The detection of duplicity is tricky because one of the characteristics of λ Boo stars is the mean-high v*sini* which produces broad and shallow lines. We have developed a strategy to detect such composite spectra by selecting the spectral features which are more sensitive to duplicity. We apply this analysis to one more object: HD 204041, a "well known" λ Boo star which is shown to be a further member of the "composite spectra" class.

Four Massive O-type Eclipsing Binaries

E. Fernández Lajús and V. Niemela

Facultad de Ciencias Astronómicas y Geofísicas, Universidad Nacional de La Plata, La Plata, Argentina

We have determined fundamental properties of four eclipsing massive O-type binaries. Three of these eclipsing binaries, namely FO15, CPD−59 2603 and CPD−59 2635 are members of the young open cluster Tr16 in the Carina Nebula. The fourth eclipsing binary, LS 1135, is located in the OB association Bochum 7. Light curves and radial velocities were analyzed with the Wilson-Devinney code. We find all four systems to be detached. The eclipsing binaries in the Carina Nebula have properties of stars which have recently arrived to ZAMS, as we find them to have smaller radii and luminosities than "normal" O-type stars. The properties of the components of LS 1135 correspond to "normal" massive stars.

LIADA's Double Star Section: Studies of Visual Double Stars by Amateurs

F.M. Rica[1], R. Benavides[2], and E. Masa[3]

[1] Astronomical Society of Mérida, Mida, Spain,
[2] Astronomical Society of Córdoba, Córdoba, Spain,
[3] Astronomical Society of Syrma, Valladolid, Spain

LIADA's Double Star Section has as its main goal to perform measures of relative astrometry of neglected and unconfirmed wide pairs, as well as to determine the astrophysical properties for their components and classify them, according to their nature, as physical, common origin, common proper motion or optical pairs.

$BVIJHK$ photometry, relative astrometry and kinematical data, in addition to other astrophysical parameters, were obtained from literature to characterize the components and the stellar systems. VizieR, Simbad, Aladin and the "services abstract" tools were used from the website of Centre de Données Stellaires de Strasbourg (CDS). USNO catalogs (USNO-B1.0 and UCAC-2) in addition to ESA catalogs (Tycho-2 and HIPPARCOS) were often used. Spectral types, luminosity classes, absolute magnitudes, and photometric distances were determined by using several tables, two colours and reduced proper motion diagrams. Astrophysical properties were corrected by reddening by using several maps.

CCD cameras, micrometric eyepieces, photographic plates from Digitalized Sky Survey (DSS) and other surveys were used to perform our astrometric measures.

According to their nature double stars are classified by using several professional criteria. Since 2001 LIADA has studied about 500 visual double stars, has discovered about 150 true binaries and several candidates to be white dwarfs, subdwarfs and nearby stars. Several orbits have been calculated.

Our results are published in national and international journals such as Journal of Double Star Observations (JDSO), in Information Circulars edited by Commision 26 of IAU and our measures were included in the WDS catalog. LIADA publishes a circular twice a year with our results.

Long-term Photometric Behaviour of the Symbiotic System AG Dra

R. Gális[1], L. Hric[2], L. Leedjärv[3], and R. Šuhada[1]

[1] Faculty of Sciences, P.J.Šafárik University, Košice, Slovak Republic,
[2] Astronomical Inst. of Slovak Academy of Sciences, Tatranská Lomnica, Slovak Republic,
[3] Tartu Observatory, Tõravere, Estonia

The general behaviour of the symbiotic system AG Dra was studied in the context of the long-term photometry and radial velocity analysis. The analysis of new and historical photometric data as well as radial velocities confirmed the continued presence of the second period, found by our previous analysis, which could be due to pulsation of the cool component of the AG Dra binary system. The discussion about very probable resonance of the orbital and pulsation periods as a general cause of the recurrence time of the active stages is also presented.

J-, H-, and K-band Spectra of Three SU UMa-type Dwarf Novae

R. Ishioka[1], K. Sekiguchi[1], and H. Maehara[2]

[1] Subaru Telescope, Hilo, HI, United States,
[2] Tokyo Univ., Tokyo, Japan

We present the results of J-, H-, and K-band spectroscopy on three SU UMa-type dwarf novae with orbital periods of 1.33 ~1.37 hr. We performed the SED fitting for the obtained spectra by assuming a power law distribution for the accretion disk and using template spectra of late-type dwarfs for secondary star.

ASAS 002511+1217.2 and EQ J183926+260409 are WZ Sge-type or WX Cet-type dwarf novae. We found strong water absorption features in their spectra, which are characteristic in late M- or L-type dwarfs. The SED fitting suggests that their secondary contributions to the overall SED are less than one third. We identified the secondary stars as M9 and L1 type dwarfs, which are rather less massive but still normal stars. The spectrum of SDSS J013701−091235 is dominated by the secondary component. Spectral features of this object are similar to those of an early M-type dwarf in spite of its short orbital period. The spectrum of SDSS0137 strongly suggests that the evolutionary path of this object is different from that of ordinary CVs, and this object is a candidate of EI Psc-like systems.

Long-Term Changes of the Supergiant in the X-Ray Binary Cyg X-1

E.A. Karitskaya[1], V.M. Lyuty[2], N.G. Bochkarev[2], V.V. Shimanskii[3], A.E. Tarasov[4], G.A. Galazutdinov[5], and B.-C. Lee[6]

[1] Institute of Astronomy, Rusian Academy of Sciences, Moscow, Russia,
[2] Sternberg Astronomical Institute, Moscow, Russia,
[3] Astronomy Department, Kazan' University, Kazan', Russia,
[4] Crimean Astrophysical Observatory, Nauchny, Crimea, Ukraine,
[5] Korean Astronomy Observatory, Daejon, Republic of Korea,
[6] Bohyunsan Optical Astronomy Observatory, KyungPook, Republic of Korea

Both photometrical and spectral variations point to supergiant parameter changes on the time scale of tens of years. A 35-year long homogeneous photometric series of UBV observations acquired at SAI Crimean Laboratory was used. The object brightness was slowly increasing from 1985 to 1995, and then decreasing to a minimum reached in 2003. The brightness minima were observed in 1971 and in 2003–2005. The largest amplitude was seen in U band (ΔU=0.1). During the transition from the maximal (1995–1999) to the minimal (2003–2005) brightness, the X-ray activity increased. High-resolution spectra were obtained in 1997 at the Crimean Astrophysical Observatory (the 2.6-m telescope, resolution R=35000), and in 2003-2004 at the Peak Terskol Observatory (the 2-m telescope, R=13000) and BOAO (the 1.8-m telescope, R=30000). Comparison of the observed and non-LTE simulated photometrical variations and He I λ4713Å line profiles leads to the conclusion that the star radius has increased by about 1-4% from 1997 to 2003-2004 and the supergiant effective temperature decreased by 1300-2400K.

The work is partially supported by RFBR grant 04-02-16924.

Light Curves of Two Eclipsing Binary Systems, BL Eridani and GW Cephei

W. Han[1], J.H. Youn[1], J.N. Yoon[1], H.S. Yim[1], C.U. Lee[1], J.W. Lee[2], H.I. Kim[1], Y.I. Byun[3], and S. Park[3]

[1] *Korea Astronomy and Space Science Institute, Taejon, South Korea,*
[2] *Sejong University, Seoul, South Korea,*
[3] *Yonsei University, Seoul, South Korea*

We present light curves of two short period binary systems, BL Eridani and GW Cephei. BL Eridani was observed with VRI filters on 5 nights at Siding Spring Observatory (SSO) by a 50-cm wide-field robotic telescope equipped with a 2K CCD camera, which was developed by the Korea Astronomy and Space Science Institute (KASI), and Yonsei University Observatory (YUO). Two nights were observed at SSO by automatic operation mode, and three nights were observed by automatic remote observation mode from KASI in Korea. With light curves covering a full phase, five new times of minimum lights were determined. The magnitude difference between primary minimum and secondary minimum with V filter appeared as 0.25 magnitude, similar to the observation by Yamasaki (1998). The light curves were analyzed by mode 3 of WD binary code, and q-search method, which shows the best fit at q = 0.48 and i = 88°.7.

The short-period contact binary system GW Cephei was observed on 5 nights by the 61-cm telescope with a 2K CCD camera at Sobaeksan Optical Astronomy Observatory (SOAO) of KASI with standard BVR filters. Full light curves were obtained, and five new times of minimum light were determined. We analyzed period variation of this system with 96 times of minimum light including our observations, which shows systematic variations that can be interpreted as the light time effect by a third body with a minimum mass of 0.17 $M_\odot$. We assumed a cool spot on the surface of the massive secondary component responsible for the asymmetry of the light curves, and calculated photometric parameters of GW Cephei by WD code, which shows q = 2.59, and i = 85°.4, with a 16% fill-out factor.

The Magnetic Cataclysmic Variable Star AM Her

B. Kalomeni[1], K. Yakut[2], and E.R. Pekunlu[3]

[1] *Dept. of Physics, Izmir Institute of Technology, Izmir, Turkey,*
[2] *Instituut voor Sterrenkunde, Katholieke Universiteit Leuven, Belgium,*
[3] *Department of Astronomy and Space Sciences, Ege University, Izmir, Turkey*

In this study the cyclotron spectrum from the accretion column of the magnetic cataclysmic variable star AM Her is deduced and compared with the observed one. In addition, R-band light curves of the system obtained with 1.5-m Russian-Turkish telescope during the observing seasons 2004-2005, when the system was in its low state, are presented. During 3 observing nights, brightness variations most probably due to the stellar activity of the companion star were detected. The system was also observed with ROTSE IIId during April – July 2005, when the system passes from low to high state. Significant brightness variations, more than 1 mag, were also observed on two observing nights.

Masses of Early-Type Contact Binary Systems

K.C. Leung

University of Nebraska, Lincoln, NE, United States

In general, the mass ratio derived from photometric analyses, q(pe), agrees quite well with the value derived from double-line spectroscopic binary, q(sp), for wide pairs as well as for close pairs. Therefore one wishes that masses could be determined for single-line systems with the help of photometric mass ratio.

It is believed that the large masses determined for early type contact systems may be not reliable or quite wrong due to misleading photometric solution mass ratios. Of course this includes some of my old published papers as well. The appearance of continuous light variation of their light curves resembles a system with contact configuration. In general, the temperature difference between the components is relatively well determined from the differential depths of the eclipses. Most of these systems have large temperature differences. In mode 3 (most popular) contact configuration of W-D method automatically ends up with a very large temperature discontinuity at the interface. Even though some of our (with D. Q. Zhou of Peking University) model calculations on circulation in contact atmosphere did arrive at stable flow. (The mathematics just becomes too difficult to handle.) One can easily argue that such temperature discontinuity in a system cannot be stable. If one utilizes the mode 1 configuration of the W-D method, there would not be a temperature discontinuity at the interface but then the temperature difference derived would not agree with the differential depths of eclipses. The key problem comes from the fact that there is only very slight difference in the shape of the light curves between a contact system and a very close semidetached system. Essentially there is no inflection point in the contact light curve while there is a slight inflection point for the latter. Since we are dealing with O and early B stars there are serious stellar winds and wind-wind interaction to be considered. These could well smooth out the slight inflection point of a semidetacted light curve so as to mimic a contact light curve. It is suspected that this misleading configuration led to the wrong mass ratio and in turn resulted in very large masses. Therefore one should not trust the automatically arrived contact solution for these systems. One should limit the solution to semidetacted configure (mode 3 or 4 of the W-D method) for those systems. Therefore it is recommended that the observations of these systems should be re-analyzed. Since most of these system are single-line binaries one might be able to obtain more reasonable masses with reliable photometric mass ratios.

Optical Spectra of Ultra-Compact X-ray Binaries

G. Nelemans

Radboud University Nijmegen, Nijmegen, Netherlands

We present optical spectroscopy of several (candidate) ultra-compact X-ray binaries (UCXBs) obtained with the ESO VLT and Gemini-North telescopes. In only one of five observed UCXB candidates did we we find evidence for H in its spectrum (4U 1556−60). We find some spectra consistent with C/O discs and one consistent with a He/N accretion disc. We discuss the implications of our findings for our understanding of the formation of UCXBs and the Galactic population of UCXBs. At the moment all studied systems are consistent with having white dwarf donors, the majority being C/O rich.

Eclipsing Binaries in Multiple Systems

E. Oblak[1], M. Kurpinska-Winiarska[2], and J.-M. Carquillat[3]

[1] *Observatoire de Besançon, France,*
[2] *Cracow Observatory, Jagiellonien University, Cracow, Poland,*
[3] *Observatoire Midi-Pyrénées, Toulouse, France*

To undertake dynamical studies on stellar triple systems and to test stellar models for subsolar mass stars, a photometric and spectroscopic observational campaign of newly discovered HIPPARCOS eclipsing binaries has been realized since 1997 through a collaborative international effort.

The radial velocity measurements were carried out, from 1997 to 2005, at the Haute-Provence Observatory, France, with CORAVEL and ELODIE spectrographs. The photometric observations started through a collaborative international effort at various observatories: Cracow, Poland; Krioneri, Greece; and more recently Lvov, Ukraine and AUG and TUG, Turkey.

Starting with a sample of 50 candidate objects, we retained at the end 36 systems, including 24 new double-lined eclipsing binaries.

We found seven objects to be new spectroscopic triple systems and confirmed the presence of the spectroscopically visible third body in the next three other systems.

Two triple systems, CU Cam and CN Lyn show evident long-period variations of the third body radial velocity as well as changes of the center-of-mass velocity of eclipsing system (Table 1).

Table 1 : Orbital Elements of the CU Cam and CN Lyn Systems.

Name Comp	P	T_0(JD)	e	ω	V_0	K	q/f(m)	asini	$M\sin^3 i$	σ
	days	+2450000		^{o}	km s^{-1}	km s^{-1}	M$_\odot$	Gm	M$_\odot$	km s^{-1}
CU Cam long-period orbit										
A	2478	1711	0.0	-	-19.62	6.33	0.45	216	1.52	0.89
	±39	±30				±.26	±.07	±14	±.35	
B						14.14		482	0.68	0.58
						±1.9		±64	±.14	
CU Cam short-period orbit										
Aa	3.363767	2820.770	0.0	-	var	59.08	0.0719	2.733		0.53
	-	±.002				±.17	±.0006	±.008		
CN Lyn long-period orbit										
A	1625	1908	0.59	318.8	-16.38	3.98	0.491	71.57	0.10	-
	±35	±13	±.04	±2.5	±.08	±.17	±0.10	±4.3	±.01	
B						8.10		146	0.051	0.29
						±.19		±7	±.005	
CN Lyn short-period orbit										
Aa	1.955505	2308.110	0.0	-	var	111.40	0.985	2.996	1.155	0.78
	±.000006	±.001				±.21	±.003	±.006	±.004	
Ab						113.12		3.042	1.138	0.90
						±.21		±.006	±.004	

We present and discuss the results from short and long-period orbital solutions for these two triple systems.

Orbit and Estimations of Masses of Components of ADS 14636 (61 Cygni) on the Basis of Photographic Observations at Pulkovo Observatory

D.L. Gorshanov and N.A. Shakht

Main Astronomical Observatory of RAS (Pulkovo), Saint-Petersburg, Russia

An investigation of the wide double star 61 Cyg was fulfilled by means of long-term series (40 years) of photographic observations of 26in refractor of Pulkovo observatory. The purposes of the investigation are

(*a*) determination of the pair's orbit and masses of its components, and
(*b*) detection of possible invisible satellites.

The two more long-term series of wide double stars – ADS 7251 (38 yr.) and ADS 14710 (23 yr.) – were used for control. Obtained photo-plates were measured by means of the automatic measurement machine "Fantazia" at Pulkovo Observatory.

Relative positions (distances between components and positional angles) were calculated for all pairs. The relative orbit of 61 Cyg was constructed and total mass of the pair was calculated by means of the Apparent Motion Parameters Method developed at Pulkovo Observatory.

These series were investigated by means of Scargle's and CLEAN methods to detect periodic deviations in orbital motion. The distant star ADS 14710 (π=0.0016) was used as a control star. Two peaks were detected in periodogram of X projection of components relative motion of 61 Cyg only. They correspond to periods of 6.4 and 11.2 years.

Investigation of the separate motion of each component of 61 Cyg relative to surrounded stars followed. The mass ratios and hence individual masses of its components were calculated. They are 0.74 ± 0.13 and $0.46 \pm 0.07 M_\odot$ for primary and secondary components respectively.

There is a small peak on periodogram of X projection of separate motion of 61 Cyg A. It corresponds to period of 6.2 years. We suppose that invisible satellites may be in 61 Cyg's system, but their mass does not exceed $0.01 M_\odot$.

Photometric Investigation of the Near-Contact Binary FR Ori

R. Gális[1], L. Hric[2], and E. Kundra[2]

[1]*Faculty of Sciences, P.J.Šafárik University, Košice, Slovak Republic,*
[2]*Astronomical Inst. of Slovak Academy of Sciences, Tatranská Lomnica, Slovak Republic*

An analysis of UBV photoelectric photometry for the eclipsing near-contact binary FR Ori using new observational data is presented. During four observational seasons 106.4 hours of observations were secured over 23 observational nights. The four new times of minima of the system FR Ori were determined from this photometric data. The analysis of the (O-C) diagram using method of weighted linear regression allowed us to state a new value of the orbital period of 0.883162859 day, and to construct new ephemerides for this eclipsing system. Detailed statistical analysis does not confirm the presence of intrinsic activity in this binary. The analysis of U, B and V light curves gave the geometrical and physical parameters of the system.

Radial-Velocity Analysis of the Post-AGB Star, HD 101584

F. Díaz[1], J. Hearnshaw[2], P. Rosenzweig[1], E. Guzman[1], T. Sivarani[3], and M. Parthasarathy[3]

[1] *Universidad de Los Andes, Merida, Venezuela,*
[2] *University of Canterbury, Christchurch, New Zealand,*
[3] *Indian Institute of Astrophysics, Bangalore, India*

This project concerns the analysis of the periodicity of the radial velocity of the peculiar emission-line supergiant star HD 101584 (F0 Ia), and also we propose a physical model to account for the observations.

From its peculiarities, HD 101584 is a star that is in the post-AGB phase. This study is considered as a key to clarify the multiple aspects related with the evolution of the circum-stellar layer associated with this star's last phase. The star shows many lines with P Cygni profiles, including Hα, Na D lines in the IR Ca triplet, indicating a mass outflow.

For HD 101584 we have performed a detailed study of its radial-velocity variations, using both emission and absorption lines over a wide range of wavelength. We have analyzed the variability and found a periodicity for all types of lines of 144 days, which must arise from the star's membership in a binary system. The data span a period of five consecutive years and were obtained using the 1-m telescope of Mt. John Observatory, in New Zealand, with the echelle and Hercules high resolution spectrographs and CCD camera.

HD 101584 is known to be an IRAS source, and our model suggests it is a proto-planetary nebula, probably with a bipolar outflow and surrounded by a dusty disk as part of a binary system. We have found no evidence for HD 101584 to contain a B9 star as found by Bakker *et al.* (1996). A low resolution IUE spectrum shows the absence of any strong UV continuum that would be expected for a B star to be in this system.

The BVRI Light Curves and Period Analysis of the Beta Lyrae System XX Leonis

P. Zasche[1], P. Svoboda[2], and M. Wolf[1]

Astronomical Institute of Charles University in Prague, Prague, Czech Republic

The contact eclipsing binary system XX Leonis (P = 0.97 days, sp A8) has been analysed using the PHOEBE programme, based on the Wilson-Devinney code. The *BVRI* light curves were obtained during spring 2006 using the 20-cm telescope and ST-7 CCD detector. The effective temperature of the primary component determined from the photometric analysis is T = (7889$\pm$61) K, the inclination of the orbit is i = (89.98$\pm$2.45) deg and the photometric mass ratio q = (0.41$\pm$0.01). Also the third body hypothesis was suggested, based on the period analysis using 57 minimum times and resulting the period of the third body p_3 = (52.96$\pm$0.01) yr, amplitude A = (0.057$\pm$0.029) d, and eccentricity e = (0.79$\pm$0.08), which gives the minimum mass $M_{3,min}$ = (3.6$\pm$0.8) $M_\odot$.

Photometric Study of the Short-Period W UMa System FZ Orionis

J.K. Pendharkar and P.V. Rao

Osmania University, Hyderabad, India

W Ursae Majoris (W UMa) variables are the most extreme examples of tidally distorted stars in contact binaries in which both components fills or overfills their inner Lagrangian zero-velocity surfaces, known as Roche lobes. FZ Orionis, one of the prototypes of W UMa variables, is studied. Photoelectric B and V observations of this eclipsing binary system obtained from the Japal-Rangapur Observatory show period changes indicating cyclic process of mass transfer and mass loss from the system. The light curves are analyzed using the Wilson-Devinney method and the system parameters are derived. The evolutionary status of the system is discussed.

Photometric Analysis of the Eclipsing Binary DE Canis Venatici (RX J1326.9+4532)

U.D. Goker and G. Tas

Ege University, Izmir, Turkey

White dwarfs and red dwarfs represent two different evolutionary stages of low-mass stars. In our Galaxy, the low-mass stars form the most numerous group of objects. For members of binary systems among them, one can derive their physical parameters like mass and radius. In addition, they include valuable information about the mass distribution of our galaxy. Different evolution phases of the binary stars consisting of white dwarfs and red dwarfs are very important for the astronomy because they allow us to test the theories of the stellar evolution. In this study, a literature survey about the structure and evolution of these systems is done and theoretical and observational results for DE CVn are presented. After obtaining new light curves, we derived the geometrical and physical parameters of the eclipsing binary DE CVn, consisting of a white dwarf and a red dwarf. We also discuss the problems of both DE CVn and related systems. DE CVn was observed with 3 different telescopes and 2 different receivers through the Johnson B, V, R filters in 2002-2003. Since the clearest variations were seen in the B filter, the B light curve was analysed using the Wilson-Devinney method with Mode 2 designed to solve detached binaries. The mass ratio q=1.1 was found. The visual magnitude of the white dwarf is 13.04 mag. in 0.0 phase and orbital period of the system is 0.364077 days. The DE CVn system consists of a DA-DB white dwarf (He-WDs) and a M1-M2 red dwarf according to our solution. The system conforms to the classical cataclysmic-variable definitions, but the P-M and P-R relation of cataclysmic variables which results from the light curve differs from that obtained from Patterson's P-T relation (1984). The latter indicates a different spectral class for the red dwarf. It is not well known whether the second companion of the system is in post-evolution phase or is not conformed to standard ZAMS M-R relation.

Parameters of the Two Helium-Rich Subdwarfs in the Short-Period Binary PG1544+488

A. Ahmad[1], C.S. Jeffery[1], R. Napiwotzki[2], and G. Pandey[3]

[1] *Armagh Observatory, Armagh, United Kingdom,*
[2] *Centre for Astrophysics Research, Univ. of Hertfordshire, Hatfield, United Kingdom,*
[3] *Indian Institute of Astrophysics, Bangalore, India*

Helium-rich subdwarf B (He-sdB) stars form a small group of chemically peculiar, early-type, low-mass stellar remnants. They are thought to be formed either as a result of mergers of white dwarfs or by convective mixing of a helium white dwarf envelope after a late helium flash. PG1544+488 is the prototype of the He-sdB stars. It was serendipitously found to be a short-period binary ($P \sim 0.5$ day) comprising two helium-rich subdwarfs. Here we report physical parameters and orbital solution for the two helium-rich subdwarfs in PG1544+488 from optical spectra obtained over a period of three years. The physical parameters – effective temperature ($T_{\rm eff}$), surface gravity ($\log g$) and helium abundances by number ($n_{\rm He}$) – for both subdwarfs were measured by fitting the observed spectra with LTE models using a χ^2 minimization procedure. The orbital solutions were obtained using radial velocities measured from the optical spectra. We also briefly discuss the implications of the discovery that PG1544+488 is a binary on our current understanding of the evolution of helium-rich subdwarf B stars and the possibility of a third formation channel for these stars involving a common-envelope in a close binary.

This research is supported by the UK Particle Physics and Astronomy Research Council through grants PPA/G/S/2002/00546 and PPA/G/O/2003/00044 and by the Northern Ireland Department of Culture, Arts and Leisure (DCAL).

The Mass Excess in the Systems of Wide Visual Double Stars on the Basis of Apparent Motion Parameters Method, Hipparcos Parallax and WDS Data

A.A. Kiselev, O.V. Kiyaeva, and I.S. Izmailov

Pulkovo Observatory, Saint Petersbourg, Russia

The formula for determination of the minimal value of the total mass of visual double star components is derived on the basis of the Apparent Motion Parameters method. To apply this formula, the trigonometric parallax p_t and the apparent motion parameters (distance ρ, position angle θ, apparent relative velocity μ and its direction ψ at a fixed epoch T_0, and also ρ_c – the radius of curvature of the observed short arc of the apparent motion near T_0) are to be known. We assume that Hipparcos parallax is determined with good precision. We selected 129 wide pairs ($\rho > 2''$, $p_t > 0''.01$, observed arc is enough for ρ_c determination) from the WDS catalog for investigation. We conclude that for 13 stars the value of minimal mass $M_{\rm min}$ is more than may be expected from mass luminosity relation $M_{\rm SP-L}$. Possible explanations include invisible satellites or some peculiarities.

(a) Component separation is more than $2''$.
(b) Parallax from Hipparcos catalogue is more than $0''.01$.
(c) Observations cover the arc of about $10°$ to $30°$ so that at least the orbit curvature can be determined, although the orbit elements and masses can not be determined.

We discovered several stars for which the minimum mass is essentially more than the value determined on the basis of the mass-luminosity relation.

Properties of the Cyg X-1 Optical Component

E.A. Karitskaya[1], V.V. Shimanskii[2], N.G. Bochkarev[3], N.A. Sakhibullin[2], G.A. Galazutdinov[4], and B.-C. Lee[5]

[1] *Institute of Astronomy RAS, Moscow, Russia,*
[2] *Astronomy Dept. of Kazan University, Kazan, Russia,*
[3] *Sternberg Astronomical Institute, Moscow, Russia,*
[4] *Korean Astronomy Obsevatory, Optical Astronomy Div., Daejon, Republic of Korea,*
[5] *Bohyunsan Optical Astronomy Observatory, KyungPook, Republic of Korea*

We report the results of V1357 Cyg optical spectral observations carried on through 2003-2004. The comparison of observed and non-LTE model HI, HeI and MgII profiles is given. Tidal distortion of the Cyg X-1 optical component and its illumination by X-ray radiation of secondary one are taken into account. We set restrictions on the optical component main characteristics: $T_{eff} = 30400 \pm 500$ K, $log\,g = 3.31 \pm 0.07$, and overabundance of He and Mg: [He/H] $= 0.43 \pm 0.06$ dex, [Mg/H] $= 0.4$–0.6 dex. We also found overabundances of C, N, O, Al, Si, S, Fe and Zn in respect to solar abundance.

The chemical composition points to:

- a metallicity ([Fe/H] $=0.34$ dex) typical for young stars;
- affected by matter transformation in the reactions of CNO cycle at the main sequence stage ([N/C] $=$ [N/O] $= 0.7$ dex);
- and by light elements burning ([Ne/H] $=$ [Si/H] $= 0.7$ dex).

The work is partially supported by RFBR grant 04-02-16924.

The Double-Lined Spectroscopic Binary θ^1 Ori E: An Intermediate-Mass, Pre-Main Sequence System

R. Costero, A. Poveda, and J. Echevarría

Instituto de Astronomía, UNAM, Mexico City, Mexico

Theta[1] Ori E $=$ ADS 4186 E $=$ NSV 2291, the fifth brightest star in the Orion Trapezium, was reported to be a double-lined spectroscopic binary by Costero *et al.* 2006 (IAUC 8669). In this paper we present the derived orbital elements of the binary system and physical parameters of its members. The velocity curve of each component was derived from 61 echelle spectra in which the absorption systems are not blended. The radial velocities were obtained by cross-correlating these spectra with those of two reference stars with well-measured radial velocities, in the $5120 - 5515$ Å spectral range.

The binary components are nearly identical, their composite spectral type being around G0IV. The Li I 6708Å absorption line is strong and the Ca II K line is in emission in both stars, indicative of their pre-main sequence evolutionary stage. The orbit is circular (e $<10^{-3}$). The orbital period and systemic velocity are 9.896 ± 0.001 d and 32.4 ± 1.0 km/s. The semi-amplitude of both components is 85.7 ± 3.0 km/s.

From the published K magnitude for the object and a suitable pre-main sequence stellar evolution model, we find the bolometric luminosity, radius and mass of each component to be, respectively, 89, 8.4 and 4.0 (in solar units), if the stars are identical to each other. Based on the latter values, the orbital inclination is about $59°$, while the minimum inclination for grazing eclipses to occur is $65°$. Hence, no observable eclipses in this binary are expected.

An Astrometric Study of the Triple System ADS 9173

Olga V. Kiyaeva

Pulkovo Observatory (GAO RAN), St. Petersbourg, Russia

The preliminary orbit of the wide pair ADS 9173 A-Bb is obtained by the Apparent Motion Parameters (AMP) method on the basis of Pulkovo 26-inch refractor uniform observations from 1982–2004, plus observations from the WDS catalog covering the period 1832–1980 and Hipparcos parallax. The spectroscopic orbit B-b is supplied by the elements i and Ω on the basis of deviations relative to the A-Bb orbit. The planes of the orbits are close to coplanarity. Pulkovo observations indicate perturbations with a period of more than 13 years. Probably the component A is also an astrometric binary.

The First Photometric Analysis of the Near-Contact Binary V370 Cygni

V.N. Manimanis and P.G. Niarchos

National and Kapodistrian University of Athens, Athens, Greece

The first complete CCD light curves of the near-contact eclipsing binary system V370 Cygni with an A0 primary have been obtained in the B, V, R and I filters during 5 consecutive nights in 2005 with the 122-cm telescope at the Kryoneri Station of the National Observatory of Athens. The light curves are analyzed with the PHOEBE version of the W-D program in order to determine the geometrical and physical parameters of the system. These parameters are used together with the available spectroscopic data to compute the absolute elements of the system in order to estimate its evolutionary status.

The Multiple System SZ Cam

Gabriela Michalska[1], Jiří Kubát[2], Daniela Korčáková[2], Adéla Kawka[2], Michal Ceniga[2,3], Blanka Kučerová[3], and Viktor Votruba[2,3]

[1] *Instytut Astronomiczny Uniwersytetu Wrocławskiego, Wrocław, Poland,*
[2] *Astronomický ústav AV ČR, Ondřejov, Czech Republic,*
[3] *Ústav teoretické fyziky a astrofyziky PřF MU, Brno, Czech Republic*

The multiple hierarchical system SZ Cam is one of the brightest members of the open cluster NGC 1502 and is the B component of visual double ADS 2984. It is composed of four components forming an SB2 eclipsing binary which is physically bound to an SB1 binary pair.

The system is solved using the method of spectra disentrangling. Hα line profiles of three components are obtained and spectroscopic orbit elements are redetermined.

The Orbit of the Visual Binary ADS 8630

M. Scardia[1], R.W. Argyle[2], J.-L. Prieur[3], L. Pansecchi[1], S. Basso[1], N. Law[2], and
C. Mackay[2]

[1] INAF – Osservatorio Astronomico di Brera, Merate, Italy,
[2] Institute of Astronomy, Cambridge, United Kingdom,
[3] Observatoire Midi-Pyrenees, Toulouse, France

We present a new orbit for the visual binary ADS 8630 = WDS12417-0127 = γ Vir. A series of new relative measurements, using the PISCO instrument on the 102-cm Zeiss telescope at Brera-Merate Observatory in speckle imaging mode, has been made. The PISCO observations cover an arc of 130 degrees and include the periastron passage of 2005. We discuss the possibility of a third body in the system.

The Remarkable Eclipsing Binary TW Draconis

M. Zejda[1], Z. Mikulasek[1], M. Wolf[2], and P. Svoboda[3]

[1] Institute of Theoretical Physics and Astrophysics, Faculty of Science, Masaryk
University, Brno, Czech Republic,
[2] Faculty of Mathematics and Physics, Charles University Prague, Czech Republic,
[3] Brno, Czech Republic

We analyzed new photometry of this well-known Algol-like eclipsing binary together with old photoelectric measurements, with the aim of better understanding of its orbital period changes and short-time light variations modulating the mean light curve. The analysis has been done by a new method based on the principal component analysis and robust regression. New spectroscopic observations and radial velocity curve are also presented, as well as the solution of the light curve of the system.

The Orbit of T Tauri South

R. Köhler and Th. Ratzka

[1] Sterrewacht, Leiden, Netherlands,
[2] MPI for Astronomy, Heidelberg, Germany

We report on previously unpublished diffraction-limited NIR observations of the young binary star T Tauri South. Orbital elements have been estimated by a least-squares fit to the relative positions. Although the parameters are not well constrained by the observations, we can derive a minimum system mass of about $3\,M_\odot$.

With the most recent astrometric measurements by Duchêne et al. (2006, A&A, astro-ph/0608018), hyperbolic (unbound) orbits can be excluded with high confidence. We conclude that T Tauri Sb is *not* in the process of being ejected from the system.

U.S. Naval Observatory Double Star CD 2006.5

B.D. Mason and W.I. Hartkopf

U.S. Naval Observatory, Washington, DC, United States

The U.S. Naval Observatory has produced its second CDROM of double star catalogs. This successor to the 2001.0 CDROM includes the latest versions (30 June 2006) of four major double star catalogs maintained at the USNO:

- *Washington Double Star Catalog* (WDS),
- *Second Photometric Magnitude Difference Catalog,*
- *Fourth Catalog of Interferometric Measurements of Binary Stars*, and
- *Sixth Catalog of Orbits of Visual Binary Stars.*

Each of these catalogs had seen significant changes during the past six years; for example, the WDS has grown by over 150,000 measures and the number of systems in the Interferometric Catalog has nearly doubled. Other improvements include precise coordinates for the vast majority of systems, as well as new observing lists for tens of thousands of "neglected" doubles.

Also included on this CDROM is a *Catalog of Linear Elements* for several hundred optical pairs. These elements should prove useful for improving the components' proper motions, as well as providing scale calibration out to several tens of arcseconds.

As was done with its predecessor, the new CDROM will automatically be distributed free of charge to members of the double star community and to astronomy libraries. Others may receive a complementary copy upon request.

The Orbit and Properties of the Massive X-Ray Binary BD+60 73 = IGR J00370+6122

C.T. Bolton and J.H. Grunhut

David Dunlap Observatory, University of Toronto, Richmond Hill, ON, Canada
e-mail: bolton@astro.utoronto.ca

Spectrograms of the blue and Hα regions of BD+60 73 obtained with the Cassegrain spectrograph on the David Dunlap Observatory 1.88-m telescope have been measured for radial velocities. These measures confirm that BD+60 73 is a single-line spectroscopic binary with the same period, 15.665 d, as the X-ray flux variations of IGR J00370+6122. The X-ray maxima occur at or just after the time of periastron passage. The orbital eccentricity, $e = 0.37$, and small $a \sin i$ are consistent with earlier suggestions that the X-ray flux variations are due to variations in the distance of the compact object from BD+60 73. Since $a \sin i$ is much less than the expected radius for a giant star, the orbital inclination must be low. Unless the optical companion has an unusually low mass for its spectral type, it seems more likely that the companion is a black hole rather than a neutron star, but this can only be resolved by further work on the properties of BD+60 73, since there are conflicting spectral classifications in the literature. The Hα line shows weak, variable emission, but we have insufficient data to test whether these variations are correlated with orbital phase. We note, as have other authors, that BD+60 73 is projected on the sky within the bounds of Cas OB5. However, the binary system has a radial velocity of approximately -40 km s^{-1} with respect to Cas OB5.

Session 3: New Observing Techniques and Reduction Methods:

Improved methods of data analysis

Figure 1. (top left to bottom right) Speakers Herman Hensberge, Geraldine Peters, Styliani Kafka, Mercedes Richards, Klaus Strassmeier, and Theo ten Brummelaar.

Figure 2. (top left to bottom right) Speakers Robert Wilson, Diego Falceta-Gonçalves, Robert Stencel, and Pavel Chadima. (Note: The meeting photo of Robert Stencel wasn't useable, so this one was provided by the author and is used with his permission.)

Figure 3. Statues of Brahe and Kepler at Pohořelec (erected 1984), in front of the Kepler Gymnasium. They are joined by rather more recent astronomer Bob Stencel.

Binary Stars as Critical Tools and Tests
in Contemporary Astrophysics
Proceedings IAU Symposium No. 240, 2006
W.I. Hartkopf, E.F. Guinan & P. Harmanec, eds.

Modern Analysis Techniques for Spectroscopic Binaries

Herman Hensberge[1] and Krešimir Pavlovski[2]

[1]Royal Observatory of Belgium, Ringlaan 3, B-1180 Brussels, Belgium
email: herman.hensberge@oma.be

[2]Department of Physics, University of Zagreb, Bijenička 32, HR-10000 Zagreb, Croatia
email: pavlovski@phy.hr

Abstract. Techniques to extract information from spectra of unresolved multi-component systems are revised, with emphasis on recent developments and practical aspects. We review the cross-correlation techniques developed to deal with such spectra, discuss the determination of the broadening function and compare techniques to reconstruct component spectra. The recent results obtained by separating or disentangling the component spectra is summarised. An evaluation is made of possible indeterminacies and random and systematic errors in the component spectra.

Keywords. methods: data analysis, techniques: spectroscopic, binaries: spectroscopic

1. Introduction

We summarize recent progress and discuss relations between analysis techniques to determine the orbital parameters and the intrinsic spectra of components in multiple systems. Progress in applying these techniques has been driven by very different astrophysical applications. The improvement of templates and the increase in sensitivity in cross-correlation techniques has been driven by planet search programmes. The emphasis is thus on the rich line spectra of cool stars. Broadening functions are of general interest, but at present applications are restricted to very-short period close binaries, where the rotational broadening hampers the detection and the analysis of the components. Techniques to separate and disentangle the spectra of the components from observed multi-component spectra have served mainly in the area of detached, hotter stars.

The emphasis in this contribution is put on practical issues, with the aim to guide potential users of these techniques to obtain in the most efficient way appropriate observation sets and to constrain the transfer of systematic errors to the output quantities. Excellent reviews on the different techniques used for reconstructing the component spectra are available in Gies (2004), Hadrava (2004) and Holmgren (2004). The use of broadening functions has been summarized in Rucinski (2002) and practical aspects are summarized in Rucinski (1998). An overview of 1D and 2D cross-correlation, with several examples, is found in Hilditch (2001, pp. 71-85).

This paper is structured as follows: Section 2 deals with improvements in measuring the orbital movement with cross-correlation techniques. Section 3 discusses the determination of broadening functions. Section 4 summarizes the progress in reconstructing component spectra since the Dubrovnik meeting in 2003 (Hilditch, Hensberge & Pavlovski 2004). Section 5 addresses the risk of introducing spurious patterns in the intrinsic component spectra.

2. Cross-correlation techniques

2.1. *Templates*

Cross-correlation of the stellar spectrum with a template spectrum supposed to represent well the intrinsic stellar one is already several decades a standard technique to measure Doppler shifts. Obviously, a single-lined template cannot represent well multiple component spectra with intrinsically different components. Therefore, the need arises to cross-correlate with a template involving two or more components, each with a time-dependent shift of its own,

$$t(x; s_1, \ldots, s_n) = t_1(x - s_1) + \sum_{i=2}^{n} \alpha_{i1} t_i(x - s_i) \tag{1}$$

This does not increase the computing time enormously, as it has been shown (Zucker & Mazeh 1994; Zucker, Torres & Mazeh 1995) that the n-dimensional surface of the cross-correlation function (CCF) can be reconstructed from a non-linear combination of $\frac{n(n+1)}{2}$ one-dimensional CCFs, namely the CCFs computed for all pairs (t_i, t_j), $i > j$, and $(t_i, \text{stellar spectrum})$. The relative strength parameters α_{i1} can be fixed by external conditions or optimized during the cross-correlation process. In the latter case, these parameters can be computed analytically as function of the one-dimensional CCFs. The technique has been applied successfully for $n = 2$ and $n = 3$. An example of a two-dimensional cross-correlation surface can be found in Zucker (2004).

2.2. CCF *sensitivity*

The need to detect faint components, or analyse low signal-to-noise data, often obtained with echelle spectrographs that impose on the data a strong modulation in each spectral order, has prompted investigations to find out how to improve the sensitivity level of the cross-correlation technique under the assumption that random noise is the dominating source of error.

Bouchy, Pepe & Queloz (2001) noted that bins at large spectral gradients and high signal-to-noise level contain most velocity information. They propose a weighting scheme, very similar to what is done in optimal extraction techniques in CCD spectroscopy, to obtain a CCF with minimum variance. Their paper includes also a discussion of the radial-velocity information content intrinsically present in the data as a function of the wavelength range, the spectral type (F to K), the rotational broadening and the spectrograph resolution.

Zucker (2003) used maximum likelihood principles to argue against a linear addition of CCFs obtained in different spectral regions. He gives a non-linear combination formula, which reduces in the limit of low signal-to-noise to a quadratic average of the CCFs.

Finally, Chelli (2000) argues for an alternative algorithm to determine Doppler shifts. It is based on a rigorous approach in the spectral Fourier domain that uses a weighted analysis of the cross spectrum phase between the high resolution spectra of the object and an appropriate template. In Galland *et al.* (2005) it is applied to stars of spectral type A and F.

In relation to earlier spectral types, Griffin, David & Verschueren (2000) investigated the impact of spectral mismatch in the B8–F7 spectral-type range on the accuracy of radial-velocity measurements and suggest a suitable window around $\lambda 4570$, though with fast rotation a very large window may be more appropriate to overcome the lower intrinsic radial velocity content.

2.3. *Single-lined spectroscopic binaries*

Very recently, Zucker & Mazeh (2006) presented a method to derive in a self-consistent way radial-velocity changes for a set of spectra of a single-lined binary, by considering the whole matrix of CCFs computed for all pairs of input spectra. In this case, an external template is not needed. It is shown to be equivalent to the use of a properly-weighted average of all input spectra (after alignment to compensate for the Doppler shifts) as a template. This technique goes strongly in the direction of the disentangling of spectra applied to a one-component system, the latter prefering to optimize for orbital parameters rather than radial velocity shifts in order to increase the robustness of the technique.

3. Broadening function

Despite its wide-spread use, the cross-correlation technique has few disadvantages which are particular relevant in the case of multiple systems. The shape of the CCF depends on the shape of the spectrum, because chance overlaps of different spectral lines in the stellar spectrum and the shifted template contribute to a fringing pattern in the CCF. In the case of the multiple-peaked CCF corresponding to a multiple system, this pattern may lead to biased radial velocities. In addition, the peaks in the CCF are wider than the spectral lines in the stellar spectrum, because the width of the spectral lines in the template adds to the resulting width of the CCF.

Both disadvantages are avoided when solving for a broadening function (BF) that, convolved with the template, represents the stellar spectrum. The BF then contains only the *additional* broadening mechanisms affecting the star (and not the template) and, possibly, reflects also time-dependent instrumental effects. As a consequence, different components separate easier in the BF than in the CCF. This method is developed by Rucinski (1992) and applied to close binaries with orbital periods less than one day in a series of papers (see Pribulla *et al.* (2006) and Rucinski & Duerbeck (2006) for recent papers on northern and southern stars).

The position of the BF reflects the radial velocity. In Pribulla *et al.* (2006) the position is measured fitting rotational profiles rather than Gaussian profiles. The integrated intensity of the different components in the BF is directly proportional to the light ratios, in the case of identical line blocking coefficients and on condition that the continuum is determined correctly. The latter is not trivial in view of the large rotational broadening in the rich line spectra. As noted various times by Rucinski, proper modelling of the BF extends its usefulness to studies involving stellar spots, limb darkening, stellar shape and other factors contributing to the broadening of spectral lines and its variation with orbital phase. The full exploitation of the technique lies clearly still in the future.

Technically, the determination of the BF reduces to a linear problem that is suitably overdetermined when the stretch of spectrum analyzed is much longer than the width over which the BF must be solved. The latter is of the order of the sum of the Doppler separations due to the orbital motion plus the width of the corresponding BFs.

4. Reconstruction of component spectra

4.1. *Separation and disentangling techniques*

When observed spectra are the sum of intrinsically time-invariant components that shift with respect to each other, depending on the time of observation, then the intrinsic component spectra can be reconstructed from a time-series of observed spectra by exploiting the relative Doppler shifts. The weight of a component may vary with time, but – at least in the original formulation – not its intrinsic spectrum. This excludes the use of spectra obtained in partial eclipses and systems with a component showing variations in spectral-line shape.

Early attempts to separate the spectra of the stars in composite spectra date back at least to Wright (1952). A series of papers was initiated by Griffin & Griffin (1986) for systems consisting of a cool giant and a hotter main-sequence star. They searched for a suitable template for the giant spectrum and reconstructed the hotter component by subtracting from the observed spectra the cooler template in the right amount, and properly shifted. As shown in Griffin (2002), the method fails when the cool giant is peculiar. A technique not based on assumptions about the shape of one of the component spectra is needed in such cases.

Several such methods were proposed in the last decennium of the 20th century. They require that the number of components is specified a priori. Bagnuolo & Gies (1991) introduced a tomographic technique to separate the component spectra once the mutual Doppler shifts are known. They propose an iterated least-squares technique (ILST) as solution scheme. Simon & Sturm (1994) formulated a solution for the more complex problem to separate the spectra of the components and to determine self-consistently the orbital parameters in an iteration scheme during which orbital parameters and component spectra are improved in turn. One refers to this more complex problem as the disentangling of the component spectra. Orbital parameters are optimised by a χ^2-type minimisation of the residuals between the observed spectra and their reconstructed model, while the problem is linear in the relative intensities of the component spectra. Solving for the latter unknowns involves a large set of overdetermined, but rank-deficient matrix equations (number of spectral bins times number of observed spectra) with a large number of unknowns (somewhat larger than the number of spectral bins times the number of components in the spectrum). This is performed by use of the singular value decomposition technique. The computational requirements were reduced significantly when Hadrava (1995) showed that in the space of the Fourier components of the spectra, the huge number of coupled equations reduces to many small sets of equations ($\frac{n_{bins}}{2} + 1$ independent sets of n_{comp} complex equations), each set corresponding to one Fourier mode.

Recently, González & Levato (2006) developed a method used earlier by Marchenko, Moffat & Eenens (1998). They use an iterative scheme, using alternately the spectrum of one component to predict the spectrum of the other one. In each step, the calculated spectrum of one star is used to remove its spectral features from the observed spectra and then the resulting single-lined spectra are used to measure the Doppler shifts for the remaining component and to compute its spectrum by an appropriately shifted combination of the single-lined spectra. This is a tomography-like method with iterations on the Doppler shifts.

In principle, solving the problem in velocity space or in Fourier space is equivalent, but there are some practical differences (see also Ilijić, Hensberge & Pavlovski 2001). One aspect relates to the edges of the considered spectral regions where, depending on the orbital phase, information on particular bins in the intrinsic spectra enters and leaves the selected spectral range in the observed spectra. Simon & Sturm (1994) solve for the component spectra over a spectral range slightly larger than in the observed spectra, although not all input spectra carry information on the outer bins. On the contrary, in Fourier space, the spectra are considered to be periodic and data are wrapped around. Another aspect relates to sampling non-integral bin velocity shifts. One can use interpolation schemes on the original grid or oversample the spectra in finer velocity grids. In pathological cases (strong lines at the edges of the selected interval, singular equations, ...) the result may be significantly different.

More important is the difference in weighting options: in velocity space, each bin can be weighted proportional to its precision, and blemished or useless data can be masked out (e.g. non-linear pixels, interstellar lines, telluric lines, . . .). Alternatively, the Fourier modes can be weighted allowing e.g. to diminish the impact of low-frequency Fourier components in the optimalisation process; it turns out to be easier in Fourier space to control and remedy the occurrence of spurious patterns in the component spectra due to numerical singularities or bias in the observed spectra. A combination of both techniques is an option: exploit the computational speed of the Fourier analysis to find the orbital solution, and separate the component spectra with known orbital parameters in velocity space to allow for the proper masking of the data. Both methods react also different on certain types of bias in the input data (Ilijić 2004; Torres, Hensberge & Vaz 2007).

4.2. *Input data*

The observed spectra must be sampled in velocity bins (logarithm of wavelength). In order to avoid resampling noise, and since the resolution of echelle spectra is often in good approximation proportional to velocity and not to wavelength, such sampling is best performed immediately when reducing the raw data. Ideally, any resampling during the iterative reconstruction process should start again from non-resampled data.

The techniques described here are differential in the sense that they rely on the time-variability of the Doppler shifts between pairs of components, and thus deliver differential velocities – the systemic velocity has to be determined separately and involves the identification of spectral lines. An ideal data set covers fairly homogeneously all relative Doppler shifts and does not concentrate on spectra observed near maximum line separation. In eccentric orbits, a fairly small range of orbital phases near periastron is suitable to cover all velocity shifts, although a good orbit determination may require a better phase coverage.

Spectra in mid-eclipse are extremely useful to stabilise the low-frequency components in the output spectra (see also Section 5). They also allow to circumvent the indeterminacy in the level of line blocking in the intrinsic component spectra (or, in other words, in their zero-point level) when the light ratio of the components is time-independent. These light ratios can be determined spectroscopically during the reconstruction process, or may be fixed by external conditions, as e.g. high-precision photometry, depending on which choice provides the most precise information. The relative light contributions or, equivalently, the line blocking in the component spectra, can also be estimated accurately from the observed spectra when the components have very deep absorption lines, since no spectral line in the intrinsic component spectra should cross the zero-intensity level. Hence, in absence of eclipses this calls for observation of spectral regions with deep absorption lines, which is especially feasible in slowly rotating cooler stars. In absence of this fortunate situation, light ratios can be bracketed in a more indirect way, e.g. by requiring that components have identical abundances (if realistic), or by requiring that faint and strong lines of the same ion should give the same abundance, or by bracketing the strength of specific absorption lines, etc. Fortunately, in the case of a constant light ratio between all components, the disentangling process can be separated from the decision which light ratio to apply (e.g. Ilijić *et al.* 2004).

The random noise in the output spectra is reduced by the combination of n_{obs} input spectra, but increases inversely proportional with the relative light contribution ℓ_j of each component j. A useful, but somewhat optimistic signal-to-noise estimate may be obtained from

$$(S/N)_j = (S/N)_{obs} n_{obs}^{\frac{1}{2}} \ell_j \tag{2}$$

Hence, the spectrum of the dominant component is often less noisy than the observed spectra, but a large set of input spectra is needed to obtain a high-quality spectrum of

Table 1. Recent applications of separating or disentangling of component spectra. Codes: FT Fourier analysis, IDD iterative Doppler differencing, ILST tomography, SVD velocity space analysis, NLLS non-linear least-squares

source	code	target(s)	comment
Budovičová *et al.* 2007	FT	o And	Be, 3 comp. disent. ; orbit
Harmanec *et al.* 2004	FT	κ Sco	NRP β Cep
Zwahlen *et al.* 2004	FT	Atlas	distance to Pleiades
Frémat *et al.* 2005	FT	DG Leo	(Am+Am)+A8 δ Sct; abund.
Hilditch *et al.* 2005	SVD/NLLS	SMC	40 OB-type EBs, fund. par.
Lehmann & Hadrava 2005	FT	55 UMa	triple, fund. p., 1300 sp.
Ribas *et al.* 2005	FT	EB in M31	fund. p. (TODCOR + separ.)
Pavlovski & Hensberge 2005	FT	V578 Mon	abund. early-B, NGC 2244
Saad *et al.* 2005	FT	κ Dra	Be, emiss.; sec. undetected
Uytterhoeven *et al.* 2005	FT	κ Sco	NRP β Cep, 700 sp.
Ausseloos *et al.* 2006	FT	β Cen	NRP β Cep, fund. p., 400 sp.
Bakış *et al.* 2006	FT	δ Lib	Algol-type
Boyajian *et al.* 2006	ILST	HD 1383	B0.5Ib+B0.5Ib, fund. p.
De Becker *et al.* 2006	IDD	HD 15558	IC 1805, detection sec. O7V
González & Levato 2006	IDD	HD 143511	fund. p., ecl. from sp., BpSi
González *et al.* 2006	IDD	AO Vel	quadruple, BpSi primary
Hensberge *et al.* 2007	FT	RV Crt	fund. p., pre-MS
Hillwig *et al.* 2006	ILST	Cas OB6	13 O-type stars, fund. p.
Hubrig *et al.* 2006	IDD	AR Aur	line shape var. B9(HgMn)
Koubský *et al.* 2007	FT	HD 208905	Cep OB2, triple
Linnell *et al.* 2006	FT	V360 Lac	crit. rot. Be, fund. p.
Martins *et al.* 2006	IDD	GCIRS16SW	Gal. Center, HeI 2.1μm
Pavlovski *et al.* 2006	FT/SVD	V453 Cyg	He abundance
Chadima *et al.* 2007	FT	β Lyr	distorted star + accr. disk
Lampens *et al..* 2007	FT	θ^2 Tau	δ Sct in Hyades; orbit
Pavlovski & Tamajo 2007	FT	CW Cep, V478 Cyg	He abundance

a faint component. With less random noise in the output spectra, systematic noise in the observed spectra may become the dominant source of uncertainty in the component spectra (Sect. 5).

4.3. *Application domain*

Recent applications, published after the reviews of Gies (2004) and Holmgren (2004), are given in Table 1. The columns give the names of the authors, the algorithm code used (acronyms as in Gies (2004) and IDD = iterative Doppler differencing for the González & Levato method), the target name and comments. The applications cover a wide range of spectral types (O to G) in binaries, spectroscopic triple systems and, in a single case, a spectroscopically quadruple system. Many of these systems proved intractable with classical techniques. Some of the components contribute less than 10% to the total light. In various applications, the data are combined with photometric and/or astrometric data. Some applications involve an impressive amount of several hundreds to more than one thousand spectra. Often, short spectral intervals are used, because they serve the purpose, but in other works large pieces of spectrum are successfully reconstructed.

These studies have led to the determination of flux ratios, the detection of eclipses, the spectroscopic detection of components, the analysis of the atmospheric parameters as for single stars, including (peculiar) abundances, the assignment of line profile variability to specific components and their study free of the diluting effects of other components, the detection of changes in a close-binary orbit caused by the tidal interaction of a third companion, and the determination of stellar masses and distances (the latter from the Pleiades to the Local Group galaxies).

Among the high S/N and high-resolution applications, several scientific programmes aim to study the chemical composition of the atmosphere: an observational study of rotational mixing during the main-sequence life-time of high-mass stars is performed by means of helium abundances. Abundances of several light elements were also obtained for a zero-age main-sequence eclipsing binary in NGC 2244, profiting from a precise determination of the gravity and the temperature ratio between the components, and thus leaving less ambiguity in the chemical composition. Note that several systems mentioned in Table 1 have components with a peculiar atmospheric composition, some of them revealing their peculiarity only after the spectra were disentangled. The first direct determination of the mass of a BpSi-type star was performed in AO Vel, and this quadruple system deserves better than the limited data set studied at present. Among the multiple systems studied to provide clues to the inter-relations between pulsation, rotation, chemical peculiarities and binarity in the domain of intermediate-mass stars (around late-A spectral types), DG Leo consists of a close binary with two metallic-line stars and an equal-mass wide companion that is pulsating.

Although pulsating stars, and especially line-profile variables, violate the basic assumptions, the technique has proven its usefulness. Several applications deal with β Cep-type stars. The disturbance of the companion on the line-profile variations of the pulsating component can be removed in order to facilitate the identification of the pulsation modes and the assignment to a particular component (see also Aerts 2007 in this symposium). The success of these studies is for part due to the large number of input spectra which de-correlated effectively the line-shape variability from the orbital phase, such that the procedure used to disentangle the spectra sees the variability merely as an extra "random noise" relative to orbital phase. This is not guaranteed, as pulsation periods and orbital periods, although very different, may by chance be aliases of each other. It is e.g. also untrue for line-profile changes in semi-detached systems, where the changes are phase-locked to the orbital cycle, which may lead to the detection of spurious components (Bakış et al. 2006). Hadrava (2004) has described how to generalise the technique to disentangle spectra in order to include certain types of intrinsic stellar variability and how to probe the stellar atmosphere by analyzing spectra obtained in partial eclipses, especially in the presence of the Schlesinger-Rossiter effect. The development of such generalised algorithms would significantly broaden the range of systems to which the reconstruction techniques can be applied with high confidence.

Several papers deal with the fundamental parameters of high-mass stars, some of them highly evolved, in the Cas OB6 region (incl. IC 1805), in the Galactic Center (an extremely high-mass binary), and beyond our Galaxy. An important aim of studies of eclipsing binaries in other galaxies is to contribute to the calibration of the distance scale. The most extensive application since previous reviews was performed by Hilditch, Howarth & Harries (2005) in the Small Magellanic Cloud. Together with their previous work (Harries, Hilditch, & Howarth 2003) they alltogether disentangled 50 eclipsing binaries. Their sample comprises detached, semi-detached, and contact binaries. Ribas et al. (2005) and Bonanos et al. (2006) studied eclipsing binaries in M31 and M33, respectively. The low S/N spectra, even while secured at the worlds largest telescopes, apparently hamper disentangling efforts (although Ribas et al. 2005 succeeded to separate the component spectra with fixed orbital parameters), but it might be worthwhile to investigate whether the limitation is due to too low S/N or to bias in the input data (normalisation, blemishes at low light level, interstellar bands, etc).

In several of the analyses of the Ondřejov group, the telluric lines are separated from the stellar components in Fourier space. The approximation is good as long as the telluric line does not move across a large stellar spectral gradient. Hadrava (2006) showed that variability in the intensity of spectral lines can be used to disentangle one component from another, even in absence of Doppler shifts. He separated in this way telluric from stellar lines in a set of spectra obtained in a short time-interval. The same paper discusses an extension of the disentangling of spectra to include components with a known spectrum (*constrained disentangling*). Avoiding in this way the introduction of a large amount of parameters, by exploiting prior knowledge e.g. on the telluric line spectrum or on the interstellar spectrum, increases the robustness of the analysis. The first application of this concept is shown in Hadrava (2007).

5. Bias in the reconstructed spectra

5.1. *Nearly-singular equations*

Depending on the data set, some of the equations may become (nearly)-singular. Insight can be gained from studying the case of a binary star in the algorithm using the Fourier components of the input spectra, as the singularity can be coupled directly to specific Fourier modes. The determinant D of the set of equations for Fourier mode m is, for N bins in the observed spectra,

$$D^{\frac{1}{2}} = \sum_{k'=1}^{k-1}\sum_{k=2}^{K} (\ell_1(\phi_k) - \ell_1(\phi_{k'}))^2 + 2\sum_{k'=1}^{k-1}\sum_{k=2}^{K} \ell_1(\phi_k)\ell_1(\phi_{k'})\ell_2(\phi_k)\ell_2(\phi_{k'})\left(1 - \cos x\right) \quad (3)$$

with $x = 2\pi\frac{m}{N}\left(v_2(\phi_k) - v_1(\phi_k) - v_2(\phi_{k'}) + v_1(\phi_{k'})\right)$ and $-\frac{N}{2} + 1 < m < \frac{N}{2}$. The square-root of D is expressed as a sum of non-negative terms. Each term refers to a pair of orbital phases $\phi_k, \phi_{k'}$ and involves the corresponding light contributions ℓ_1 and $\ell_2 = 1 - \ell_1$ and the relative Doppler shifts $v_2 - v_1$.

In the case of significant light variability, the first sum of terms guarantees that no singularities occur. The continuum level in the component spectra is then well-determined. In absence of light variability, the determinant is strictly 0 for $m = 0$ corresponding to the intrinsic uncertainty how to distribute the observed line blocking over the two components, as mentioned earlier (Sect. 4.2). Near-singularities then exist likely for other low-frequency modes ($m \ll N$), responsible for the undulations in the component spectra mentioned in various papers (e.g. Hensberge *et al.* 2000; Fitzpatrick *et al.* 2003; Pavlovski & Hensberge 2005; González & Levato 2006). Often no attention is paid to the fact that the bias introduced in one component is strictly correlated with the bias in the other component (in antiphase and amplitude proportional to ℓ_j^{-1}). Continuum windows in one component suffice to remove the bias in both component spectra.

Numerical singularities appear in high-frequency modes when the argument of the cosine function is a multiple of 2π for all (most) pairs of observed spectra, which occurs for integer values of $\frac{N}{m}\left(v_2(\phi_k) - v_1(\phi_k) - v_2(\phi_{k'}) + v_1(\phi_{k'})\right)$ (Fig. 1). The equations for (nearly)-singular modes should be solved using the singular value decomposition technique. On this condition, the singularity in high-frequency modes can be shown to be of practical concern only when N is small i.e. when applying the method on single spectral lines, since the amplitude of the noise pattern is inversely proportional to N.

The key point is that the occurrence of singularities depends in a predictable way on the distribution of the observations over the orbit, on the level of time-variability in the relative light contributions, and on the chosen log-wavelength sampling.

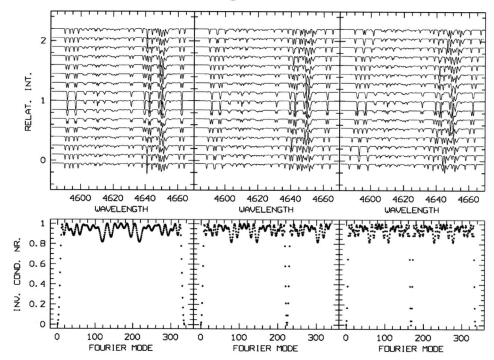

Figure 1. Series of 16 artificial two-component spectra and corresponding inverse condition numbers for non-negative Fourier modes. The relative Doppler velocities, from left to right: $\frac{K2}{K1} = 1, 2, 3$ respectively, and the orbital phases were chosen in order to reproduce cases with singular low- and high-frequency Fourier modes (inverse condition number equal to zero).

5.2. *Biased input data*

Multi-component spectra are often quite complex, because of the twice higher line density and the dilution of spectral lines. Especially in late-type spectra of close binaries, the synchronisation of the orbital motion and the stellar rotation may cause lines to be broader and shallower than in single stars. All these elements conspire to obscure the position of the continuum and the time-dependent Doppler shifts may lead to trace an observed (pseudo)-continuum that is biased with a dependence on orbital phase. How will the process of separation of spectra react on such type of bias?

Experiments with artificial data to which different types of phase-dependent bias was added show that the amplitude of the bias in the component spectra may be significantly larger than in the input spectra (Fig. 2). The amplification is proportional to the ratio of the length of the spectral interval to the sum of the maximum Doppler shifts and inversely proportional to the relative light contribution ℓ_j. However, mid-eclipse spectra reduce such low-frequency bias to a fraction of the bias in the input spectra when weights are applied in the low m modes (Fig. 3). While the shape of the line profiles during eclipse might cast doubt about the usefulness of mid-eclipse spectra in the high-frequency modes, the advantages of their inclusion in the solution for low m must be emphasized.

Other types of bias encountered in observed spectra include shallow features, e.g. weak interstellar bands, detector blemishes or unidentified faint stellar components. Static features will either be included in a static stellar component and be amplified by ℓ^{-1}, or, in the absence of such a component, they are at least slightly deformed and enter partially in the different components, inversely proportional as well to the stellar velocity amplitude K_j as to ℓ_j. Such features not belonging to any of the components and undetected in the observed spectra are sometimes clearly recognized in one of the output spectra.

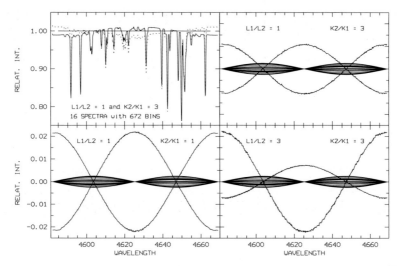

Figure 2. Separation of component spectra applied the data set shown in the rightmost panel of Fig. 1 (upper panels) and similar ones (same component spectra, but different light and velocity amplitude ratios, as indicated in the panels. All spectra have $K1 = 12$ bins). An orbital-phase dependent bias was added to the observed spectra $S_{obs}(\phi, \ln\lambda)$, i.e. $S_{obs} \rightarrow \frac{S_{obs}}{1-0.0025\sin 2\pi\phi \sin 2\pi\psi}$ with $\psi = \frac{\ln\lambda - \ln\lambda_{start}}{\ln\lambda_{end} - \ln\lambda_{start}}$. Resulting separated component spectra (left panel) show an amplified sinusoidal bias, of which a detailed view is shown in the other panels for three different cases. The set of 16 thin-line sine curves in these panels show the bias in the input spectra (never and nowhere larger than 0.25%), the two thick-line sine curves indicate the amplified bias in the output spectra.

The previous comments apply to separation of the spectra with known orbital parameters. The disentangling process is more complex, since any bias in the input will also influence the orbital parameters. Hynes & Maxted (1998) discuss, based on numerical simulations, the relation between random noise in the input data and the uncertainty of the velocity amplitudes of the components. Also Ilijić showed, in the aforementioned meeting in Dubrovnik, that the uncertainty on the velocity estimates may be significantly too optimist when it is derived from a cross-correlation of the observed spectra with the disentangled component spectra *without* taking into account that the intensities in the component spectra were also parameters. There is indeed feed-back between residuals in the velocities and residuals in the component spectra. Realistic χ^2–surfaces taking into account all parameters do not have the symmetry expected when the velocity amplitude estimates were independent of the errors in the reconstructed component spectra. The matter is relevant for the precision on the stellar masses. It relates also to the question in which conditions the reconstructed spectra do a better job, in terms of velocity amplitudes, than methods using "independent" templates. Are there limits depending on S/N, orbital coverage, richness of the line spectrum, ...?

Acknowledgements

HH acknowledges the project "IUAP P5/36" financed by the Belgian Science Policy. KP acknowledges funding by the Croatian Ministry of Science under the project #0119254. We thank Saša Ilijić and Kelly Torres for contributions to and discussions on Section 5.

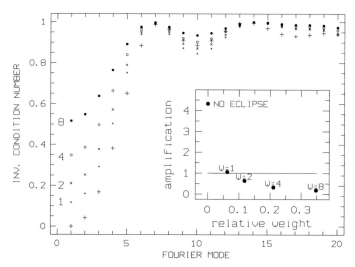

Figure 3. Removal of the (near)-singularity of the equations in low Fourier modes, and reduction of the amplitude of the bias in the component spectra (small frame, "amplification" is the ratio of the amplitude of the bias in the output spectra relative to the input spectra) in case one of the spectra is taken in a total eclipse is shown for different weights W given to the eclipse spectrum.

References

Aerts, C. 2007, these proceedings, 432

Ausseloos, M., Aerts, C., Lefever, K., Davis, J., & Harmanec, P. 2006, *A&A* 455, 259

Bagnuolo, W.G. Jr. & Gies, D.R. 1991, *ApJ* 376, 266

Bakış, V., Budding, E., Erdem, A., Bakış, H., Demircan, O., & Hadrava, P. 2006, *MNRAS* 370, 1935

Bonanos, A.Z., Stanek, K.Z., Kudritzki, R.P., Macri, L.M., Sasselov, D.D., Kaluzny, J., Stetson, P.B., Bersier, D., Bresolin, F., Matheson, T., Mochejska, B.J., Przybilla, N., Szentgyorgyi, A.H., Tonry, J., & Torres, G. 2006, *ApJ* (in press)

Bouchy, F., Pepe, F. & Queloz, D. 2001, *A&A* 374, 733

Boyajian, T.S., Gies, D.R., Helsel, M.E., Kaye, A.B., McSwain, M.V., Riddle, R.L., & Wingert, D.W. 2006 *ApJ* 646, 1209

Budovičová, A., Štefl, S., Hadrava, P., Rivinius, Th., & Stahl, O. 2005, *ApSS* 296, 169

Chadima, P., Harmanec, P., Ak, H., Demircan, O., Yang, S., Koubský, P., Škoda, P., Šlechta, M., Wolf, M., Božić, H., Ruždjak, D., & Sudar, D. 2007, these proceedings, 205

Chelli, A. 2000, *A&A* 358, L59

De Becker, M., Rauw, G., Manfroid, J., & Eenens, P. 2006, *A&A* 456, 1121

Fitzpatrick, E.L., Ribas, I., Guinan, E.F., *et al.* 2003, *ApJ* 587, 685

Frémat, Y., Lampens, P., & Hensberge, H. 2005, *MNRAS* 365, 545

Galland, F., Lagrange, A.–M., Udry, S., *et al.* Galland, F., Lagrange, A.–M., Udry, S., Chelli, A., Pepe, F., Queloz, D., Beuzit, J.–L., & Mayor, M. 2005, *A&A* 443, 337

Gies, D.R. 2004, *ASP Conf. Ser.* 318, 61

González, J.F. & Levato, H. 2006, *A&A* 448, 283

González, J.F. & Hubrig, S., Nesvacil, N., & North, P. 2006, *A&A* 449, 327

Griffin, R.E. 2002, *AJ* 123, 988

Griffin, R. & Griffin, R. 1986, *JApA* 7, 195

Griffin, R.E.M., David, M., & Verschueren, W. 2000, *A&AS* 147, 299

Hadrava, P. 1995, *A&AS* 114, 393

Hadrava, P. 2004, *ASP Conf. Ser.* 318, 86

Hadrava, P. 2006, *A&A* 448, 1149

Hadrava, P. 2007, *poster,* these proceedings, 111

Harmanec, P., Uytterhoeven, K., & Aerts, C. 2004, *A&A* 422, 1013

Harries, T.J., Hilditch, R.W., & Howarth, I.D. 2003, *MNRAS* 339, 157

Hensberge, H., Pavlovski, K., & Verschueren, W. 2000, *A&A* 358, 553

Hensberge, H., Vaz, L.P.R., Torres, K.B.V., & Armond, T. 2006, in: *Multiple Stars across the HR Diagram*, ESO Astrophysics Symposia (in press)

Hilditch, R.W. 2001, *An Introduction to Close Binary Stars*, CUP, Cambridge, pp. 71-85

Hilditch, R.W., Hensberge, H., & Pavlovski, K. (Eds.) 2004, Proc. Worskop on *Spectroscopically and Spatially Resolving the Components of Close Binary Stars*, ASP Conf. Ser., vol. 318

Hilditch, R.W., Howarth, I.D., & Harries, T.J. 2005, *MNRAS* 357, 304

Hillwig, T.S., Gies, D.R., Bagnuolo, W.G.Jr., Huang, W., McSwain, M.V., & Wingert, D.W. 2006, *ApJ* 639, 1069

Holmgren, D.E. 2004, *ASP Conf. Ser.* 318, 95

Hubrig, S., González, J.F., Savanov, I., Schöller, M., Ageorges, N., Cowley, C.R., & Wolff, B. 2006, *MNRAS* 372, 286

Hynes, R.I. & Maxted, P.F.L. 1998, *A&A* 331, 167

Ilijić, S. 2004, *ASP Conf. Ser.* 318, 107

Ilijić, S., Hensberge, H., & Pavlovski, K. 2001, *Fizika B* 10, 357

Ilijić, S., Hensberge, H., Pavlovski, K., & Freyhammer, L.M. 2004, *ASP Conf. Ser.* 318, 111

Koubský, P., Daflon, S., Hadrava, P, Cunka, K., Kubát, J., Korčáková, D., Škoda, P., Šlechta, M., Votruba, V., Smith, V.V., & Bizyaev, D. 2006, in: *Multiple Stars across the HR Diagram*, ESO Astrophysics Symposia (in press)

Lampens, P., Frémat, Y., De Cat, P., & Hensberge, H. 2007, these proceedings, 213

Lehmann, H., & Hadrava, P. 2005, *ASP Conf. Ser.* 333, 211

Linnell, A.P., Harmanec, P., Koubský, P., Božić, H., Yang, S., Ruždjak, D., Sudar, D., Libich, J., Eenens, P., Krpata, J., Wolf, M., Škoda, P., & Šlechta, M. 2006, *A&A* 455, 1037

Marchenko, S.V., Moffat, A.F.J., & Eenens, P.R.J. 1998, *PASP* 110, 1416

Martins, F., Trippe, S., Paumard, T., Ott, T., Genzel, R., Rauw, G., Eisenhauer, F., Gillessen, S., Maness, H., & Abuter, R. 2006, *ApJL* 649, L103

Pavlovski, K. & Hensberge, H. 2005, *A&A* 439, 309

Pavlovski, K., & Tamajo, E. 2007, these proceedings, 209

Pavlovski, K., Holmgren, D.E., Koubský, P., Southworth, J., & Yang, S. 2006, *ApSS* (in press)

Pribulla, T., Rucinski, S.M., Lu, W., Mochnacki, S.W., Conidis, G., Blake, R.M., DeBond, H., Thomson, R.J., Pych, W., Ogloza, W., & Siwak, M. 2006, *AJ* 132, 769

Ribas, I., Jordi, C., Vilardell, F., Fitzpatrick, E.L., Hilditch, R.W., & Guinan, E.F. 2005, *ApJ* 635, L37

Rucinski, S. 1998, *Tr. J. of Physics* 1, 1

Rucinski, S. 2002, *AJ* 124, 1746

Rucinski, S.M. 1992, *AJ* 104, 1968

Rucinski, S.M. & Duerbeck, H.W. 2006, *AJ* 132, 1539

Saad, S.M., Kubát, J., Hadrava, P., Harmanec, P., Koubský, P., Škoda, P., Šlechta, M., Korčáková, D., & Yang, S. 2005, *ApSS* 296, 173

Simon, K.P. & Sturm, E. 1994, *A&A* 281, 286

Torres, K.B., Vaz, L.P., & Hensberge, H. 2007, these proceedings, 210

Uytterhoeven, K., Briquet, M., Aertc, C., Telting, J.H., Harmanec, P., Lefever, K., & Cuypers, J. 2005, *A&A* 432, 955

Wright, K.O. 1952, *Publ. DAO* 9, 189

Zucker, S. 2003, *ApJ* 342, 1291

Zucker, S. 2004, *ASP Conf. Ser.* 318, 77

Zucker, S. & Mazeh, T. 1994, *ApJ* 420, 806

Zucker, S. & Mazeh, T. 2006, *MNRAS* 371, 1513

Zucker, S., Torres, G., & Mazeh, T. 1995, *ApJ* 452, 863

Zwahlen, N., North, P., Debernardi, Y., Eyer, L., Galland, F., Groenewegen, M.A.T., & Hummel, C.A. 2004, *A&A* 425, L45

Binary Stars as Critical Tools & Tests
in Contemporary Astrophysics
Proceedings IAU Symposium No. 240, 2006
W.I. Hartkopf, E.F. Guinan & P. Harmanec, eds.

Bipolar Jets, Hot Interaction Regions, and Colliding Winds in OB Interacting Binaries

Geraldine J. Peters†

Space Sciences Center & Department of Physics & Astronomy, University of Southern
California, Los Angeles, CA 90089-1341.
email: gjpeters@mucen.usc.edu

Abstract. Contemporary information on the nature of the circumstellar environment in early-type interacting binaries derived from observation is discussed. Emphasis is placed on results from FUV spectroscopy. New spectra from the *FUSE* spacecraft and earlier FUV observations with *IUE* have revealed the presence of hot interaction regions, bipolar outflows, splash outflows, and accretion hot spots in Algol-type systems with B-type mass gainers. In addition, close O-type pairs and systems containing O + W stars show a shock-heated region generated from their colliding winds. Recent *FUSE* observations of an apparent hot accretion spot and associated *splash* plasma in the direct-impact system U Cephei are presented.

Keywords. binaries: close, circumstellar matter, stars: early-type, ultraviolet: stars

1. Introduction

It has been more than 30 years since the publication of the landmark papers on mass flow in semidetached binary stars by Lubow and Shu (1975, 1976). FUV spectroscopy from the *IUE, ORFEUS-SPAS II, HST*, and *FUSE* spacecrafts and Doppler tomography using optical data have provided observational confirmation of many structures discussed in the Lubow and Shu papers and revealed other components of the circumstellar (CS) material that were not predicted from their gas dynamical calculations. In this paper we discuss some of the key results from FUV observations. Information derived from tomography is presented by Richards in another article contained in these proceedings.

The behavior of the mass accretion in Algol binaries can be understood with the aid of the r–q diagram in which the fractional radius of the mass gainer (R_p/a) is plotted versus the mass ratio, $q = M_{loser}/M_{gainer}$, and compared with the computations of gas stream hydrodynamics by Lubow & Shu (1975). A representative r–q is shown in Figure 1. The upper *dashed* curve, ϖ_d, delineates the fractional radius of the dense accretion disk for systems of different mass ratios. The lower *dashed* curve, ϖ_{min}, shows the minimum distance that a stream particle will achieve relative to the center of the gainer. If a system falls above the upper curve, the gas stream will strike the photosphere of the mass gainer and if any Hα emission from CS material is seen, it will be transient. If the system falls below the lower curve, one will see a prominent accretion disk that emits strongly in the Balmer lines and other species. The most variable accretion disks are found in systems that fall between the two curves, since the inner part of the *wide* gas stream will strike or graze the photosphere of the primary while the outer edge will feed an accretion disk. These concepts underlie the discussion below.

† Based on observations made with the NASA-CNES-CSA Far Ultraviolet Spectroscopic Explorer. FUSE is operated for NASA by the Johns Hopkins University under NASA contract NAS5-32985.

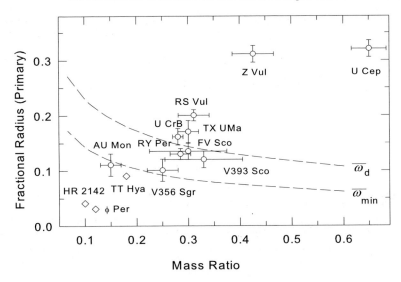

Figure 1. The locations of several well-known Algol systems in the r–q diagram. The *diamonds* show systems with prominent disks.

2. Hot Interaction Regions

Evidence for a high temperature plasma above the *trailing* hemisphere in an Algol system was first reported by Plavec (1983). *IUE* observations of the direct-impact system U Cep during the total phase of its primary eclipse revealed the presence of prominent emission lines from highly-ionized species such as N V, C IV, and Si IV that imply the presence of a CS plasma with a temperature of $\sim$100,000 K. The strongest component of the emission was mapped to CS gas on the side of the binary that sustained the gas stream impact. About the same time Peters & Polidan (1984, PP84) found absorption lines from the latter species in the phase interval 0.5-0.9 in *IUE* spectra of AU Mon, CX Dra, and U CrB. The latter authors called the CS plasma in which these absorption lines are formed the *High Temperature Accretion Region* (HTAR). Both Plavec and PP84 suggested that the plasma is heated from a shock, or a series of shocks, produced by the impacting gas stream and PP84 concluded that the HTAR features are formed by resonance scattering in a plasma of high temperature ($T_{ion} \sim 10^5$ K), intermediate density ($N_e \sim 10^9$ cm^{-3}), and moderate-high carbon depletion. *IUE* data acquired later in the mission revealed that the HTAR prevails in the systems in the upper two-thirds of the r–q diagram where stable disks cannot form and the azimuthal extent of the HTAR depends on the period of the binary (Peters 2001).

3. Bipolar Flows

Although *IUE* spectra provided considerable information on the nature of the CS material in the orbital plane in Algol binaries (Peters 2001), it has been the *FUSE* spacecraft (Moos *et al.* 2000) that has revealed the prevalence of a highly-ionized plasma flow above/below the plane. Prior to *FUSE*, research teams found evidence for jet-like structures in the disk system β Lyr. Combining optical interferometry, spectroscopy, and photometry a consortium of researchers in the mid-1990s concluded that most of the Hα and He II emission in β Lyr is formed in bipolar jets (Harmanec *et al.* 1996). This conclusion was reinforced by Hoffman *et al.* (1998) who combined optical and FUV

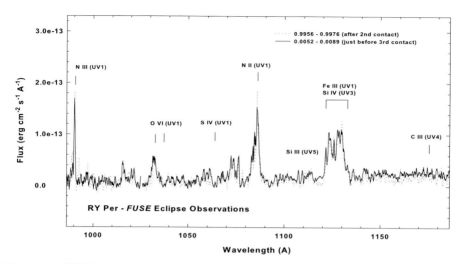

Figure 2. *FUSE* observations of RY Per taken during totality on 2002 December 8 reveal the presence of emission lines from highly-ionized species and evidence for extreme CNO processing in the material that is currently being transferred.

spectropolarimetry (from the WUPPE instrument flown on the *ASTRO 2* shuttle mission) to produce a model where the line formation region for the Balmer emission, UV emission lines, and NUV continuum is oriented perpendicular to the orbital plane where the visible continuum originates.

FUV spectra from *FUSE* have recently been used to map the location of the hot CS plasma in three Algol binaries (cf. Peters & Polidan 2004). Phase-resolved spectra in the region $\lambda\lambda950$–1190 Å of V356 Sgr (B3V + A2II, P=8.90^d), TT Hya (B9.5Ve + K1III-IV, 6.95^d), and RY Per (B4V + F7III-IV, 6.86^d) obtained during the total eclipse† have revealed the presence of optically-thin, broad (FWHM $\sim$600–1200 km s^{-1}) emission lines from O VI in all three systems. O VI $\lambda1032$ Å is the dominant feature in the *FUSE* spectrum of V356 Sgr. These features suggest the presence of a 300,000 K component to the CS plasma. Since only slight variations are observed in the O VI lines throughout totality, we have concluded that the emission originates in a turbulent outflow perpendicular to the orbital plane. In contrast, the emission from moderate-ionization species such at C III, N II, and Si III tends to be double-peaked and displays a type of V/R variability that suggests the features are formed in a disk (cf. Peters 2001, Peters & Polidan 2004).

The *FUSE* totality observations of RY Per are especially noteworthy (Figure 2). The level of ionization appears to be slightly higher in this system than in the other two program objects. O VI and the lines from the moderately-ionized species display a FWHM of about 800 km s^{-1} and very little change between phase 0.997 and 0.007. All of the emission appears to be originating in a bipolar outflow. There is evidence for extreme CNO-processing in the CS plasma that forms the emission features. Emission from the resonance lines of N II, III is especially strong and there is *no* emission from C III $\lambda977$ and 1176 Å that *is* observed in V356 Sgr and TT Hya. From an analysis of *IUE* LORES spectra of RY Per taken during totality, Olson & Plavec (1997) also concluded that the

---

† The emission lines can be seen only during totality in Algol systems in the upper two-thirds portion in the r–q diagram. Emission is visible at all phases in the prominent disk systems.

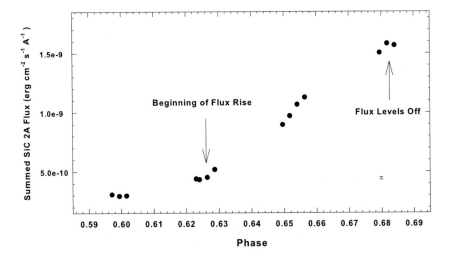

Figure 3. Summed flux from the *FUSE* SiC 2A detector for each 8^m exposure obtained during the long 2004 February 09 observation plotted versus phase. Note that an apparent accretion hot spot rotates into our line-of-sight at phase 0.625 and is fully visible at phase 0.682. A typical error bar (lower right) is about the width of the plotted points.

CS plasma in which the emission lines are formed is N-rich and C-poor. But the spectral resolution of the *IUE* data was too low to obtain information on the velocity behavior of the lines. Since photospheric absorption from C III 1176 Å is clearly seen in RY Per, we conclude that the material in the bipolar outflow originates in the gas steam from the mass-losing star and has had only minimal interaction with the photospheric material.

4. Hot Accretion Spots and Splash Regions

In Peters (2001) I identified a *splash plasma* as one of the components of the CS material in Algol-type interacting binaries. This region is detected from the presence of violet-shifted absorption lines from moderate to high-ionization species and is located downstream from the impact site. The material appears to have been reflected from the mass gainer's photosphere and heated by the shock from the encounter. The *splash plasma* is mostly seen in systems in the middle portion of the r–q diagram (e.g., RY Per, FV Sco, V393 Sco) but as we will see from the *FUSE* observations of U Cep presented below, it can also reside above the impact site of a short-period system.

U Cep (B7V + G8III-IV; P=2.49^d) has been observed with the *FUSE* spacecraft at random phases since 2002 as part of the *FUSE* Survey and Supplemental Program. During the course of a five hour observation on 2004 February 9, we observed a rather spectacular rise in the level of the FUV flux that is shown in Figure 3. We have plotted the sum of the flux recorded by the SiC 2A detector for each of the 14 exposures versus phase. The flux begins to rise at phase 0.625 and levels off by phase 0.682. We interpret this photometric behavior as evidence for a "hot" accretion spot that rotated into our line-of-sight. The spot would be centered at about phase 0.90, very close to the azimuth where we would expect the gas stream to impact photosphere of the B star. From the time it took for the spot to fully come into view, we can place a lower limit of 20° on its size. The spot would occupy ∼2% of the facing hemisphere of the B star. The FUV flux level observed in a *FUSE* observation taken at a phase of 0.28 on the following day, 2004 February 10,

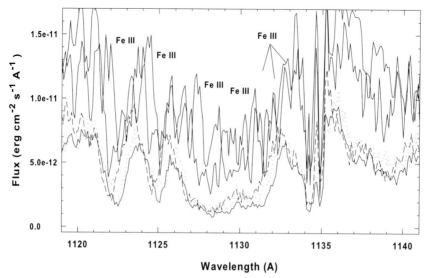

Figure 4. Selected exposures from the *FUSE* LiF 2a/Lif 1b observations of U Cep of 2004 February 9 & 10 compared with a spectrum taken on 2002 September 1 at a phase of 0.83 (*bold* line). Exposure 16 on February 9 obtained at a phase of 0.68 at flux maximum is plotted with a *thin* line. The spectra showing a lower flux level are Exposures 2 (phase 0.60, *medium* solid line) and 7 (phase 0.63, *dotted* line, just before flux rise) on February 9, and Exposure 1 on February 10 (phase 0.28, *dashed* line). The locations of Fe III shell-type lines seen in the 2002 spectrum are indicated.

was comparable to the pre-rise value. An elevated flux was observed on two other dates (2002 September 1, phase 0.83, and 2003 February 13, phase 0.12). At these phases the hot spot would be located to the left (later phase) of our line-of-sight, and on the receding limb, respectively. A lower flux level was observed in six other observations taken between phases 0.22–0.58 (spot behind the facing hemisphere). Obvious "shell-type" features of Fe III and similar species, that are redshifted by $\sim$150 km s^{-1} relative to the B star, were observed at phase 0.83 (cf. Figure 4). Weaker, blueshifted ($\sim$-200 km s^{-1}) sharp lines were seen on 2004 February 9 (post-rise), but there was no discernible shell structure at phase 0.12. The brightest FUV flux was observed at phase 0.83. The plasma in which these features are formed appears to be limited in angular extent and has a temperature in the range 20,000–40,000 K (based upon the collisional ionization calculations of Shull & Van Steenberg 1982) that is 2–3 times higher than the photosphere of the mass gainer. The Fe III plasma is most certainly associated with the impact site. The component observed above the impact region is clearly associated with the mass infall, but the material downstream appears to have been splashed from the impact site.

5. Colliding Winds

In close O star pairs and O + W binaries, a collision is inevitable between their massive, fast-moving (V$_\infty$ $\sim$-1000–4000 km s^{-1}) winds. Plasma heating occurs along a bow shock that is curved around the star with the weaker wind (St-Louis *et al.* 2006). A shock cone produces P Cygni features of C III, IV, O VI, Si IV, S IV, and P V. St-Louis *et al.* (2005) have recently modeled the bow shock region in Sanduleak 1 (W04+O4) in the SMC from *FUSE* data. Iping *et al.* (2006, poster presentation at this meeting) are currently analyzing new *FUSE* observations of five other massive O star pairs with short periods (2.2-15.1^{d}) in the our Galaxy and the Magellanic Clouds that will ultimately reveal further information on the interaction zone.

6. Concluding Remarks

During the past two decades FUV spectroscopy has provided a wealth of information on OB interacting binaries. Collisions between a gas stream and the mass gainer's photosphere or stellar winds produce a shock-heated plasma that radiates in spectral lines from highly-ionized species that can only be observed in the FUV. Bipolar jets may be commonplace in Algol-type systems. Evidence for a hot bipolar outflow has been found in all three Algol binaries that have been observed during totality with the *FUSE* spacecraft. From polarimetry Piirola *et al.* (2005) concluded that a high-latitude spot or flow exists in the disk system W Ser. Evidence for a high velocity polar stream has recently been found in the direct-impact system U CrB from a 3-D Doppler tomography analysis of Hα data (Agafonov *et al.* 2006). The relationship between the bipolar flow, *splash* plasma, and the HTAR is presently unknown but all of these regions appear be formed by the impacting gas stream. The plasmas reside around or downstream from the impact site. The bipolar flow could simply be splashed material that is shock-heated then injected into the CS environment. The observations of U Cep described above support this idea. The possible role of a magnetic field that originates in the late-type mass loser and becomes "frozen-in" to the gas stream has yet to be investigated. Information on the CS environment in OB interacting binaries will continue to improve with future FUV observations and 3-D magnetohydrodynamic simulations of mass flow.

Acknowledgements

The analysis of the *FUSE* observations on U Cep is a collaborative effort with B-G Andersson, T. Ake, and R. Sankrit, whom I would like to thank for fruitful discussions. The author appreciates support from NASA grants NAG5-12253 and NNG04GL17G.

References

Agafonov, M., Richards, M.T., & Sharova, O. 2006, *ApJ* 652, 1547

Harmanec, P. *et al.* 1996, *A&A* 312, 879

Hoffman, J.L., Nordsieck, K.H., & Fox, G.K. 1998, *AJ* 115, 1576

Lubow, S.H. & Shu, F.H. 1975, *ApJ* 198, 383

Lubow, S.H. & Shu, F.H. 1976, *ApJ* 207, L53

Moos, H.W. *et al.* 2000, *ApJ* 538, L1

Olson, E.C. & Plavec, M.J. 1997, *AJ* 113, 425

Peters, G.J. 2001 in: D. Vanbeveren (ed.), *The Influence of Binaries on Stellar Population Studies* (Kluwer), p. 79

Peters, G.J. & Polidan, R.S. 1984, *ApJ* 283, 745 (PP84)

Peters, G.J. & Polidan, R.S. 2004, *Astron. Nachr.* 325, 225

Piirola, V., Berdyugin, A., Mikkola, S., & Coyne, G.V. 2005, *ApJ* 632, 576

Plavec, M.J. 1983, *ApJ* 275, 251

Shull, J.M. & Van Steenberg, M. 1982, *ApJS* 48, 95

St-Louis, N., Moffat, A.F.J., Marchenko, S., & Pittard, J.M. 2005, *ApJ* 628, 953

St-Louis, N., Moffat, A.F.J., Marchenko, S., Pittard, J.M., & Boisvert, P. 2006, in: G. Sonneborn, Moos, H. W., & B.-G. Andersson (eds.), *Astrophysics in the Far Ultraviolet, Five Years of Discovery with FUSE* (ASP Conference Series, Vol. 348), p. 121

Binary Stars as Critical Tools & Tests
in Contemporary Astrophysics
Proceedings IAU Symposium No. 240, 2006
W.I. Hartkopf, E.F. Guinan & P. Harmanec, eds.

Detecting Chromospheric Activity on the Secondary Star of Cataclysmic Variables

Styliani Kafka

NOAO/CTIO, 603 Casilla, La Serena, Chile
email:skafka@noao.edu

Abstract. Chromospheric activity on the secondary stars of cataclysmic variables (CVs) is a key ingredient for angular momentum loss from the system via magnetic braking. This effect is thought to drive the evolution of the system and is invoked to explain a number of observed properties of CV light curves, such as long-term modulations and high/low states. However, obtaining observational support for magnetic activity has proven difficult. We present a new method of studying chromospheric activity on the secondary stars of CVs, using near-IR spectral features. We discuss in particular the magnetic CV AM Herculis, in which satellites to the H-alpha emission line are interpreted as arising from magnetically confined gas streams (prominences). This phenomenon provides a new technique for mapping magnetic structures on CV secondaries, and advances our understanding of the nature of magnetic structures and activity on CV secondaries.

Keywords. cataclysmic variables, stars: individual (AM Her, ST LMi, VV Pup), stars: activity stars: flare

1. Introduction

Cataclysmic variables (CVs) are semi-detached binaries in which a white dwarf (primary star) is accreting from its lower main sequence (K/M dwarf) Roche-lobe filling companion. Depending on the strength of the magnetic field of the white dwarf, CVs can be classified in two major categories. In disk systems (DCVs) the accreted material forms an accretion disk around the white dwarf primary. In magnetic systems (hereafter MCVs) the strong magnetic field of the WD ($B \geqslant 10^7 G$) disrupts the formation of a disk and leads the accretion stream through the magnetic field lines of the white dwarf towards its magnetic poles. A detailed review of all the subcategories of CVs can be found in Warner (1995) and will not be repeated here.

The secular evolution of CVs is controlled by angular momentum loss via magnetic braking through a wind originating from the secondary star (Ritter 1984). Magnetic activity on the secondary star is also considered responsible for a number of observed phenomena in CVs. Semi-periodic long-term modulations of the optical brightness of several CVs are attributed to magnetic cycles on the secondary (Bianchini 1987). Warner (1988) proposed that variations of the mean luminosity and outburst intervals of dwarf novae are in agreement with the presence of magnetic cycles on the secondary star. Erratic but prominent (up to 4 magnitudes) drops of brightness in both DCVs and MCVs are interpreted as temporary interruption of the mass transfer process by the presence of starspots in front of the L1 point (Livio & Pringle 1994; Kafka & Honeycutt 2005). In disk systems (IP Peg and SS Cyg), low velocity components revealed the presence of a stationary "compact source" located outside the orbital plane of the systems, but near the center of mass of the two stars and is attributed to slingshot prominences, similar to those observed in fast rotating single stars (Steeghs *et al.* 1996). Finally, the detection of intense radio outbursts from a MCV (AM Her) is interpreted as originating from electron-cyclotron masers near the surface of the magnetic (∼1000G) secondary (Chanmugam & Dulk 1982; Dulk, Bastian & Chanmugam 1983). The above are just a few examples on the variety of observed phenomena in CVs pointing to activity on their secondary star for their interpretation. However, direct observations of such activity appear to be difficult to gather, since accretion masks or, even worse, mimics signatures of activity such as X-rays and Hα emission.

Perhaps the only time where accretion is absent, or significantly reduced, is during VY Scl low states. Such low states, which are defined photometrically as intense (up to 5 mag) drops of the brightness of the system, occur in both DCVs and MCVs. They are erratic and unpredictable in occurence and duration; however, they allow the two components to be revealed, offering a unique opportunity to explore the presence of activity on the secondary star.

2. AM Her leading the way

We were fortunate to observe one of the MCVs, AM Her, during its extended 2003-2005 low state both photometrically and spectroscopically. AM Her is the prototype for the MCV (or polar) category; as such, it has been the subject of a large number of studies. Therefore, we now know that it consists of a magnetic white dwarf (B∼12.5±0.5 MG; Bonnet-Bidaud *et al.* 2000) which is accreting material from its ∼M4–M5 donor star (Bailey *et al.* 1988; Gänsicke *et al.* 1995; Kafka *et al.* 2005a) through one of its magnetic poles. Its long-term optical light curve (Figure 1, top) reveals numerous low states, in which the magnitude of the system drops to 15.6 in V (from 13-14 mag in the high state). When in the low state, it exhibits events, which have been attributed to either accretion bursts or activity (flares) on the secondary star. One of the better resolved ones it presented by Shakhovskoy *et al.* (1993): it appeared as a 2-mag, 20-min brightening during the 1992 low state, similar to UV Cet-type (M dwarf) flares. During a photometric monitoring campaign of the system, Kafka *et al.* (2005a) revealed a series of events having durations of ∼15-90 min, amplitudes of ∼0.2-0.6 mag and duty cycles of ∼35%; these were tentatively attributed to stellar activity (flares) on the secondary star. Furthermore, low state X-ray flux has been assigned to coronal emission from an active secondary star (de Martino *et al.* 1998). Therefore, there were previous indications that the secondary star in AM Her could be active.

Armed with this, we monitored the system spectroscopically during three observing runs in 2003-2005, using the WIYN 3.5-m and the KPNO 2-m telescopes. Typical exposure times range between 450 and 750 sec. We chose to observe the system in the near IR region (5700-8500 Å) with the hope to distinguish components that do not originate from the white dwarf or any residual accretion. A sample spectrum of the system is presented in Figure 1 (bottom, left). The continuum is modulated by pronounced TiO bands, which are characteristics of an M-type secondary star. Features from the secondary's photosphere, such as the KI and NaI doublet allowed us to determine a new spectroscopic ephemeris for the system, defining phase zero at inferior conjunction of the secondary star:

$$T_0 = \text{HJD } 2,446,603,403(5) + 0.12892704(1) \text{ E}$$

The Hα line is in emission; occasionally, we see the HeI lines (at 5876, 6678 and 7065 Å) in emission too.

The interesting finding in our work was that the Hα line has structure: the central peak appears to have blue/red satellites, whose visibility changes with orbit. These satellites

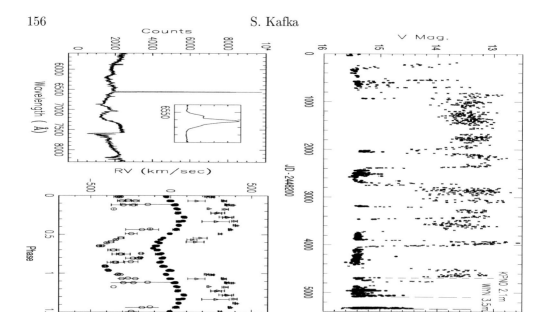

Figure 1. Top: 1990-2005 light curve of AM Her (from Kafka & Honeycutt 2005); Bottom left: low-state spectrum of AM Her, with the main features labeled (inset: zoom in on Hα), right: Hα radial velocity of the central peak (filled circles) and the satellites (open circles) for one night in 2005 (from Kafka *et al.* 2006).

appeared immediatelly after the system dropped in the low state in 2003 indicating that they were probably present even when the system was in the high state, but were masked by accretion (Kafka *et al.* 2005b). The presence of the satellites in 2005 imply that the underlying structures are persistent and long-lived. To decompose the lines, we used Gaussian fits to the satellites and the central peak. The number of Gaussians are, generally speaking, unambiguous for most of the components, reproducing the cumulative profile of the line. Using the spectroscopic ephemeris from above, we constructed phased RV curves for the central peak and its components; an example is presented in Figure 1 (bottom, right).

A striking first characteristic of all the RVs is that they follow the motion of the secondary star. The velocity of the central peak advocates for an origin to the inner hemisphere of the secondary star; we tentatively attribute it to irradiation. The RVs of the satellites are more intricate. They appear for half a cycle each, whereas both of them are present around phases 0.0 and 0.5, and they have structure at phase 0.8. Their phasing, amplitude and curvature advocate for an origin on the secondary star; however their γ velocity of ∼300 km sec^{-1} does not agree with the γ velocity of the secondary star. Also, the satellites can not originate from a jet-like structure, otherwise their RV would cross the path of the RV of the central peak (instead of transitioning from one side to the other). They rather have the appearance of a partial ring about the secondary. However, a circumbinary ring would not follow the orbital motion of the secondary alone, and there are no stable orbits around the secondary star alone, given that the it fills its Roche lobe. Ruling out the white dwarf, irradiation and system dynamics, the only logical choice for the origin of the satellites is that they are produced by the magnetic field of the secondary star, likely in the form of prominences or prominence-like structures that extend half way around the star, similar to slingshot prominences in fast rotating stars (Collier-Cameron & Robinson 1989). Such loops can be transient and extended, resembling a partial ring, with material flowing along the magnetic field lines of the secondary. We cannot completely rule out an alternative explanation of infall velocites

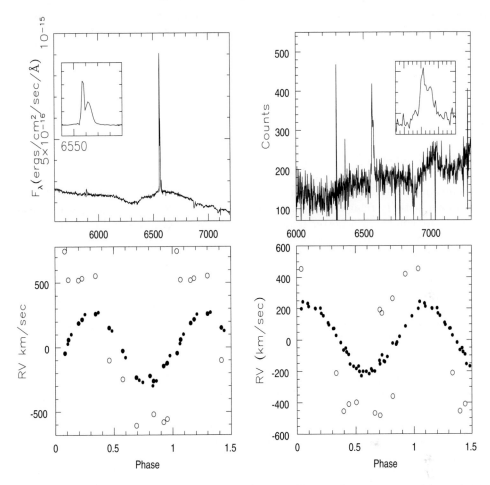

Figure 2. VV Pup (left; from Mason *et al.* 2007) and ST LMi (right; from Kafka *et al.* 2006). In both cases, the top panel displays a low state spectrum of the system (with the Hα line in the inset) and the bottom panel the radial velocity of the central Hα peak (filled circles) and the satellites (open circles).

from residual accretion being responsible for the satellites (or a part of the observed emission). However we have concluded that gas motions in loops about the secondary star are strongly favored mainly because 1) accretion is very low or absent at the times of our observations, and 2) the radial velocities of the satellites appear to follow the motion of the secondary star.

Our 3Å resolution allows us to resolve only three components in the profile of the Hα line; however, it is likely that this line has a more complicated structure. Considering that the satellites are present in the system for more than one year, irradiation from the L1 point of the system should not be the major component of the observed emission from the L1 point. Future higher resolution data could distinguish the various components and determine the presence of active regions at L1.

3. Other CVs?

The question that arises here is, whether this signature of activity is common in CV secondary stars, or if AM Her is just an exception. Recent observations came to test our discovery: Mason *et al.* (2007) in their low state data of the MCV VV Pup, present three epochs of spectroscopic observations of the system. During one of these epochs, the Hα line has satellites similar to those in AM Her (Figure 2, left). The RVs of all three peaks of the lines follow the secondary star, and reveal structures with velocities reaching $500\,\mathrm{km\,sec^{-1}}$. In addition, Kafka *et al.* (2006) present new data of the recent low state of ST LMi, in which the Hα line is also triple-peaked, pointing to the secondary star for the origin of all three components (Figure 2, right). In VV Pup, the satellites appear during only one of the epochs of observing, indicating that the underlying structures are variable on timescales of weeks. In ST LMi, the EW variations of the Hα line indicate the presence of an active region on the side of the secondary (Kafka *et al.* 2006).

With VV Pup and ST LMi in the growing group of systems in which the Hα line reveals magnetic activity on the secondary stars, we have now a new technique of tracing and mapping activity-related magnetic structures in CV secondaries. Using the observed structures, we can reach a better understanding of the formation and evolution of such structures allowing us to test our theories of angular momentum loss via magnetic braking on CV evolution and the effect of fast rotation and tidal/magnetic interactions on activity. Differences in the character of magnetic activity in various systems is expected, since the secondary stars are of different spectral types. For CVs above the period gap (such as AM Her), the secondary star is expected to have a radiative core; activity is then induced via an α-Ω dynamo, in a manner similar to activity in the sun. In CVs such as ST LMi and VV Pup, the secondary star is fully convective; however the presence of activity and loop prominences advocate towards the presence of a global magnetic field in their secondary. All three systems in our sample are magnetic CVs, which raises the question of the effect of the white dwarf's magnetic field on the activity levels of its companion. Therefore, a critical test for this technique is to look for similar structures during the low state of a DCV, which is among our future plans.

Acknowledgements

I would like to thank Kent Honeycutt and Steve Howell for hours of exciting conversations. I would also like to acknowledge an AAS travel grant which allowed my participation to this IAU symposium.

References

Bearman, P.W. & Graham, J.M.R. 1980, *J. Fluid Mech.* 99, 225
Bailey, J., Hough, J. H., & Wickramasinghe, D. T. 1988, *MNRAS*, 233, 395
Bianchini, A. 1987, Memorie della Societa Astronomica Italiana, 58, 245
Bonnet-Bidaud, J. M., *et al.* 2000, *A&A*, 354, 1003
Chanmugam, G., & Dulk, G. A. 1982, *ApJL*, 255, L107
Collier Cameron, A., & Robinson, R. D. 1989, *MNRAS*, 238, 657
Dulk, G. A., Bastian, T. S., & Chanmugam, G. 1983, *ApJ*, 273, 249
Gänsicke, B. T., Beuermann, K., & de Martino, D. 1995, *A&A*, 303, 127
Kafka, S., & Honeycutt, R. K. 2005, *AJ*, 130, 742
Kafka, S., Robertson, J., Honeycutt, R. K., & Howell, S. B. 2005a, *AJ*, 129, 2411
Kafka, S., Honeycutt, R. K., Howell, S. B., & Harrison, T. E. 2005b, *AJ*, 130, 2852
Kafka, S., Honeycutt, R. K., & Howell, S. B. 2006, *AJ*, 131, 2673 —c1995,
Kafka, S. Robertson, J., Howell, S.B, & Honeycutt, R.K. 2006, *AJ*, submitted
Livio, M., & Pringle, J. E. 1994, *ApJ*, 427, 956
Mason E., Wickramasinghe, D., Howell, S.B., & Szkody, P. 2007, *A&A*, in press
Ritter, H. 1985, *A&A*, 145, 227

Shakhovskoy, N. M., Alexeev, I. Y., Andronov, I. L., & Kolesnikov, S. V. 1993, Cataclysmic Variables and Related Physics, 10, 237

Warner, B. 1995, Cambridge Astrophysics Series, Cambridge, New York: Cambridge University Press,

Warner, B. 1988, *Nature*, 336, 129

Discussion

JUAN MANUEL ECHEVARRIA: Have you tried to combine Roche Tomography with your analysis and can you rule out Zeeman splitting in the Hα line?

KAFKA: We can securely rule out Zeeman splitting as a possible source for the lines. I am not aware of a case in which Zeeman splitting results in line emission. Furthermore, the separation of the "peaks" of the satellites (or "dips" in the line) does not coincide with the expected position of the $\sigma^{\pm}$ components of a 12MG magnetic field (which is the white dwarf field for AM Her). We have not attempted Roche Tomography due to the low resolution of our data.

GEORGE SONNEBORN: Magnetic activity and prominences are usually hotter than the K/M V secondary. Are there other spectral signatures of chromospheric activity, especially higher ionization lines?

KAFKA: In bluer wavelengths, one would expect the CaII H&K lines to be in emission, in a manner similar to chromospherically active stars. In the UV, the CIV and SiIV lines are good indicators of chromospheric activity. For AM Her, signatures of activity in the UV lines are reported in poster presented in this meeting by Steve Saar *et al.*("The Ultraviolet Universe: Stars from Birth to Death", 26th meeting of the IAU, Joint Discussion 4, 16-17 August 2006, Prague, Czech Republic, JD04, #30, 4) Our observations are concentrated around H-alpha; there are no higher ionization species there.

Binary Stars as Critical Tools & Tests
in Contemporary Astrophysics
Proceedings IAU Symposium No. 240, 2006
W.I. Hartkopf, E.F. Guinan & P. Harmanec, eds.

Doppler Tomography of Accretion Disks and Streams in Close Binaries

Mercedes T. Richards

Department of Astronomy & Astrophysics, Pennsylvania State University, 525 Davey
Laboratory, University Park, PA, 16802, USA
email: mrichards@astro.psu.edu

Abstract. The application of tomography to the study of gas flows in interacting binaries has led to fascinating images of the Cataclysmic Variables and Algol-type binaries. Such detailed images are currently unachievable using direct-imaging techniques. Numerous images of accretion flows have now been derived from optical and ultraviolet spectra and they have been used to identify multiple emission sources including the gas stream, accretion disk, accretion annulus, shock regions, and the chromosphere of the mass loser. It was difficult to distinguish between the separate sources of emission since these sources have overlapping velocities in the Doppler tomogram. However, with the aid of a new spectral synthesis code, we can now systematically extract the individual emission sources to sequentially isolate the images of the disk and gas stream. With these new tools, we have begun to extract the critical properties of the disk and gas stream more accurately than previously possible.

Keywords. accretion, accretion disks, stars: activity, (stars:) binaries: eclipsing, hydrodynamics, radiative transfer, (stars:) novae, cataclysmic variables

1. Introduction

Since the technique of Doppler tomography (Marsh & Horne 1988) was introduced 18 years ago, it has been used to provide indirect images of accretion structures in close binaries which cannot be resolved spatially with the largest telescopes. The only direct images of distinct gas flows between stars in an interacting binary were produced in 2005 by Karovska *et al.* (2005) from soft X-ray *Chandra* and HST images of Mira. In all other cases, we have to resort to indirect techniques to study the emission sources and gas flows in the binary. The more general technique of tomography has been used successfully in medicine, geophysics, archaeology, and oceanography to construct three-dimensional (3D) images from two-dimensional (2D) pictures or "slices" through the object collected at many positions around the object. These slices or projections are represented by the Radon transform (Radon 1917); the 3D image is recovered through a summation process called *back projection*; and the overall image reconstruction procedure is known as *tomography*. In astronomy, the technique can be readily applied to eclipsing binaries and rotating stars, which provide changing views of the system, and the process is called *Doppler Tomography* because the gas motions detected through Doppler shifts provide an image of the accretion flows in velocity coordinates.

Doppler tomography has revitalized the study of accretion structures in interacting binaries which contain both compact and normal mass gainers. This procedure has been described in several papers: Marsh & Horne (1988), Robinson, Marsh, & Smak (1993), Kaitchuck *et al.* (1994), Staude *et al.* (2001), Richards (2001), and Richards (2004). The technique has now been extended to the third dimension by Agafonov *et al.* (2006) to reveal the gas flows beyond the orbital plane. To date, there have been two conferences on the application of tomography to astronomy: "Astrotomography: An International

Workshop on Indirect Imaging," held in Brussels in 2000 July, published in Lecture Notes in Physics, Vol. 573 (edited by H.M.J. Boffin, D. Steeghs, & J. Cuypers), and Joint Discussion No. 9 on "Astrotomography" at the IAU General Assembly in Sydney, in 2003 July, which was published in *Astronomische Nachrichten*, Vol. 325, No. 3 (2004). This latter volume provides a very good summary of the various astrotomography techniques.

The basic technique of tomography introduced by Johann Radon (Radon 1917; Herman 1980; Shepp 1983) can be used to create an image of an object from projections. In the most common application, if a 2D object is described by the function $f(x, y)$, then the Radon transform of that 2D object is a set of 1D projections along the line $s = x \cos \theta + y \sin \theta$ for the full range of angles θ:

$$p(s, \theta) = \int_{-\infty}^{\infty} \int_{-\infty}^{\infty} f(x, y)\, \delta(x \cos \theta + y \sin \theta - s)\, dx\, dy$$

where, δ is the Dirac delta function such that $\delta(a) = \infty$ if $a = 0$, and $\delta(a) = 0$, otherwise. The technique of *back projection* is the process in which this 1D projection is inverted to reveal the 2D function $f(x, y)$. This is fairly easy to do since the 1D Fourier transform of the projection function $p(s, \theta)$ is equal to the 2D Fourier transform of the 2D object, $f(x, y)$, calculated along the projection line. The result is

$$f(x, y) = \int_{0}^{2\pi} \int_{-\infty}^{\infty} \int_{-\infty}^{\infty} p(s, \theta)\, |\omega|\, e^{2\pi i \omega (x \cos \theta + y \cos \theta - s)}\, ds\, d\omega\, d\theta.$$

Similarly, in the standard two-dimensional case of Doppler tomography, the Radon transform of the function, $I = f(v_x, v_y)$, is a set of 1D projections, $p(v_r, \phi)$, which is the *line profile* at each orbital phase, ϕ. The 2D function is obtained by integrating over the line: $v_r = v_x \cos \phi + v_y \sin \phi$ and with a change of variables from the Cartesian to the velocity frame: $x \to v_x$, $y \to v_y$, $s \to v_r$, $\theta \to \phi$.

$$f(v_x, v_y) = \int_{0}^{2\pi} \int_{-\infty}^{\infty} \int_{-\infty}^{\infty} p(v_r, \phi)\, |\omega|\, e^{2\pi i \omega (v_x \cos \phi + v_y \sin \phi - v_r)}\, dv_r\, d\omega\, d\phi.$$

This change of variables produces the same formulation as the filtered back projection method derived for Doppler tomography (see Marsh & Horne 1988, Robinson *et al.* 1993, and Kaitchuck *et al.* 1994). This process can be readily extended to the case of the 3D tomogram, since the Radon transform of $f(x, y, z)$, is a set of 2D projections, $p(s, \theta, \psi)$.

2. Application to Compact Binaries and Algol-type Binaries

Accretion disks are found in binaries which contain white dwarfs and neutron stars (e.g., cataclysmic variables (CVs) and soft X-ray binaries), and non-compact main sequence stars (e.g., Algol-type binaries). The main differences between these systems are: (1) mass transfer occurs onto a compact object in the CVs and X-ray binaries, while the mass gainer in the Algols is a main sequence star; (2) the accretion structures in the compact systems are bright relative to the stars, while the structures in the Algols are faint relative to the luminous main sequence primary star; (3) the large size of the mass gainer in the Algols leads to the direct impact of the gas stream onto the stellar photosphere in the short-period Algols ($P_{\rm orb} \leqslant 5$ days), while this type of impact does not occur in the long-period Algols ($P_{\rm orb} > 5 - 6$ days) or the compact systems. The result is that a complex set of accretion structures is formed in the short-period Algols compared to the classical accretion disks in the other systems.

2.1. *Images of Cataclysmic Variables and X-ray Binaries*

The non-magnetic CVs (e.g., dwarf novae) contain a white dwarf with no field or a weak magnetic field, the magnetic CVs (or polars) have very strong magnetic fields which inhibit the formation of an accretion disk, and the intermediate group of intermediate polars contains a white dwarf with a moderate magnetic field. Images of over 33 CVs in outburst and quiescent states have now been produced (see reviews by Marsh 2001, Morales-Rueda 2004, Steeghs 2004, Schwope *et al.* 2004). These include 18 dwarf novae (e.g., U Gem, WZ Sge, IP Peg), 2 intermediate polars (EX Hya, V1025 Cen), 9 polars (e.g., AM Her, UZ For, V1309 Ori), 2 old novae (V841 Oph, RR Pic), and 2 helium CVs (AM CVn, GP Com). The complete list is given in Morales-Rueda (2004). Tomograms have been produced of spectra at the following wavelengths: Hα, Hβ, Hγ, Hδ, He I λ4472 Å, He I λ5876 Å, He I λ6678 Å, He I λ8236 Å, He I λ1.083μ, Na I, Ca II, He II λ4686 Å, and He II λ1.163μ.

The tomograms of non-magnetic CVs are usually dominated by the image of the accretion disk, with less prominent gas streams. It was in these images that spiral structure in disks was first discovered (e.g., Steeghs *et al.* 1997, Harlaftis *et al.* 1999, Morales-Rueda 2000, Groot 2001). A movie showing the changes in the disk structure from quiescent to outburst states in WZ Sge was made by Steeghs (2004) using multiple He II and Hβ tomograms. Hydrodynamic simulations of these systems by Sawada *et al.* (1986) had earlier proposed that the spiral structure was produced by tidally–induced spiral density wave. These simulations have been refined by Bisikalo & Matsuda (2007).

The magnetic CVs have provided the most impressive images of accretion streams (e.g., HU Aqr: Schwope *et al.* 1997; V1309 Ori: Staude *et al.* 2001). In these systems, the strong magnetic field inhibits the formation of an accretion disk, so the gas stream dominates the image. Ultraviolet and X-ray images of two X-ray binaries, Her X-1 and SMC X-1, have been produced (see review by Vrtilek *et al.* 2004) and suggest that there are still too few studies to make a coherent picture of the accretion structures in these systems. Her X-1 displays a prominent accretion disk, while most of the emission in the massive wind-fed system SMC X-1 is associated with the mass losing star.

2.2. *Images of the Algols*

Doppler tomograms of 12 eclipsing Algols, 1 non-eclipsing Algol (CX Dra), and 1 RS CVn binary have been produced (see reviews by Richards 2001 and Richards 2004). The RS CVn binary V711 Tau (HR 1099) was included to illustrate the location and appearance of chromospheric Hα emission in the tomogram since this binery was not predicted to be in the process of Roche lobe overflow. The images of the Algols were made from over 2500 spectra collected systematically from 1992 - 2004: RZ Cas, δ Lib, RW Tau, β Per (Algol), TX UMa, U Sge, S Equ, U CrB, RS Vul, SW Cyg, CX Dra, TT Hya, AU Mon, in order of increasing orbital period, $P_{orb} = 1.2 - 11.1$ days (Richards 2004). Most of the tomograms are based on optical spectra at the following wavelengths: Hα, Hβ, He I (6678 Å) and Si II (6371 Å). In the ultraviolet, a tomogram of U Sge was produced from the Si IV (1394 Å) line (Kempner & Richards 1999), while MgII h & k lines were used to image the chromosphere in V711 Tau (Richards & Rosolowsky 1998).

The greatest challenge in the study of accretion structure in the Algols is that these structures are faint relative to the luminous primary star. So, in order to extract information about these structures, it is necessary to compute the stellar contribution to the line profile and remove it from the observed spectra (Richards 1993). This procedure assumes that the accretion structure is optically thin; an assumption that has been tested successfully in the case of long-period systems where tomograms can be derived from both the observed and difference spectra. The resulting difference profiles display enhanced emission profiles which can readily be processed into Doppler images.

The $H\alpha$ tomograms show that the Algols display several sources of emission in the orbital plane (Richards 2001, Richards 2004): (1) the gas stream along the predicted gravitational trajectory (Richards *et al.* 1995); (2) transient or permanent accretion disks (Richards *et al.* 1995, Albright & Richards 1996); (3) a star-stream impact region where the gas stream strikes the stellar photosphere; (4) a disk-stream impact region where the disk strikes the incoming gas stream; (5) a circumprimary bulge produced by the impact of the high velocity gas stream onto the slowly rotating photosphere; (6) the chromosphere of the secondary (Richards & Albright 1996); and (7) an absorption zone that overlaps with the locus of hotter gas seen in the UV tomogram (Kempner & Richards 1999).

The gas stream emission detected in the Algols represented the first ever images of these structures for the entire class of interacting binaries. Later, even more impressive gas streams were found in polars by Schwope *et al.* (1997). The comprehensive study of the multiwavelength tomograms of the non-eclipsing system of CX Dra (Richards *et al.* 2000) demonstrated that multiwavelength tomograms could be used effectively to derive the properties of the accretion structures in the binary; including a well-formed disk (derived from He I and Si II spectra) and a truncated gas stream (derived from the $H\alpha$ and $H\beta$ lines). Moreover, multi-epoch tomograms have shown that the emission sources in systems like β Per and TX UMa are quite stable; while other systems like U Sge and U CrB display structures which alternate between stream-like and disk-like states. The hydrodynamic simulations of Richards & Ratliff (1998) suggested that this variability was caused by changes in the mass transfer rate over intervals within a few orbital periods.

3. Separation of the Emission Sources Using Synthetic Spectra

The complex emission sources in the Algols cannot readily be separated into the individual components because the various sources overlap in the velocity domain. In order to study these components, Budaj & Richards (2004) developed a spectrum synthesis code called SHELLSPEC. Synthetic spectra were calculated to sequentially model (1) the stars, (2) the stars and disk, and (3) the stars, disk, and gas stream combined. Doppler tomograms were produced by incorporating these three sets of synthetic spectra. The technique was illustrated in the case of the long-period Algol, TT Hya ($P_{orb} = 6.96$ days) by Budaj *et al.* (2005) and Miller *et al.* (2007).

3.1. *The Synthetic Spectra*

The SHELLSPEC code solves in LTE the simple radiative transfer along the line of sight in an optional optically thin 3D moving medium (Budaj & Richards 2004). Transparent (or non-transparent) objects such as a spot, disk, stream, jet, shell, or stars may be defined (or embedded) in three dimensions and their composite synthetic spectrum calculated (Budaj & Richards 2004, Budaj *et al.* 2005). The stars may have Roche geometry and known intrinsic spectra. The scattered light from a central object can be taken into account assuming an optically thin environment. Intrinsic spectra (intensity or flux from a unit surface area not broadened by the rotation) of the primary and secondary star can be precalculated using the codes TLUSTY and SYNSPEC, (Hubeny 1998, Hubeny & Lanz 1992, Hubeny & Lanz 1995, Hubeny *et al.* 1994) as well as with the Kurucz (1993) model atmospheres code. These intrinsic spectra can be assigned to the primary and secondary star and a complex spectrum of both stars and the circumstellar matter can

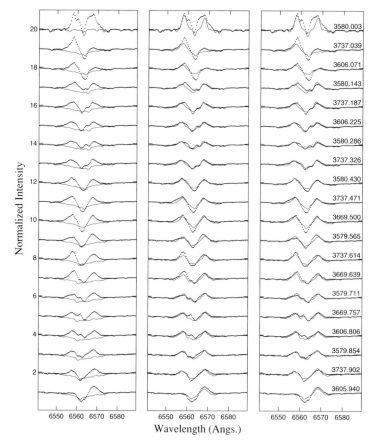

Figure 1. Comparison of the observed (dotted line) and synthetic (solid line) Hα spectra of TT Hya. The epochs and orbital phases are listed to the right of the figure. The synthetic spectra include calculations of (a) stars only (left frame); (b) stars and disk (middle frame); and (c) stars, disk, and gas stream (right frame). There is a significant improvement in the fit between the observed and synthetic spectra once the accretion disk is included, but only a minor improvement once the gas stream is included in the model (courtesy: Miller *et al.* 2007).

be calculated using SHELLSPEC. Synthetic light curves or trailing spectrograms can be produced by changing the view points on the 3D object.

The SHELLSPEC code was first applied to the long-period Algol, TT Hya. Budaj *et al.* (2005) compared observed and synthetic Hα spectra and found that the disk extends to about 10 solar radii in the orbital plane and has a vertical thickness comparable to the diameter of the primary star. They also found that the overall strength of the emission is regulated mainly by the density and temperature of the disk, while the position and separation of the emission peaks are influenced by the outer disk radius, the inclination, and the radial density profile. In addition, the depth of the central absorption is very sensitive to the temperature, inclination, geometry, and dynamics of the disk. Further analyses of the optical and ultraviolet spectra of TT Hya were used by Miller *et al.* (2007) to refine the properties of the disk. These refined disk and system properties were used to generate synthetic spectra of the disk for input in the tomography code.

For TT Hya, the assumed temperature of the disk was $T = 7000$K and the disk density at the inner radius was 3.3×10^{-14}g cm^{-3} (Miller *et al.* 2007). The initial gas stream temperature was set to $T = 8000$K to correspond to the peak in Hα emissivity.

The estimated density of the stream at the starting point ($2 \times 10^{-14} \mathrm{g\,cm^{-3}}$) was chosen to improve the fit between the observed and synthetic emission at the quadratures of the orbit and during the eclipses. For these calculations, the density was allowed to vary along the stream to satisfy the continuity equation. The electron number densities in the disk and stream were calculated from the temperature, density, and chemical composition using the Newton-Raphson linearization method. Miller *et al.* (2007) obtained a mass transfer rate of $\sim 2 \times 10^{-10} M_\odot yr^{-1}$, based on the velocity, density, and cross-section of the gas stream above and below the disk. This rate is two orders of magnitude higher than the lower limit of $10^{-12} M_\odot yr^{-1}$ obtained by Peters & Polidan (1998).

3.2. *Extraction of the Disk and Stream and Discovery of an Eccentric Disk*

The observed Hα spectra of TT Hya were compared with synthetic spectra of (1) the stars only, (2) stars and disk, and (3) stars, disk, and stream. Figure 1 illustrates that the stellar spectrum is not sufficient to explain the observed profile, but the inclusion of the disk emission made a remarkable improvement to the fit between observed and synthetic spectra. The inclusion of the gas stream only slightly improves the fit, especially near the quadratures.

To illustrate the quality of these fits, Miller *et al.* (2007) generated tomograms from (a) the observed spectra, (b) the observed spectra minus the synthetic stellar spectra, (c) the observed spectra minus the synthetic spectra of the stars and accretion disk, and (d) the observed spectra minus the synthetic spectra of the stars, disk, and gas stream. Hα spectra of TT Hya spanning 16 years from 1985–2001 were used to make the images. The resulting tomograms are shown in Figure 2. This figure shows the tomograms in velocity coordinates, the locations of the stars, and the predicted gravitational trajectory of the gas stream from the L_1 point. In addition, the space between the large solid circle and smaller dashed circle represents the locus of a Keplerian disk which extends from the surface of the mass gainer to its Roche lobe in Cartesian space.

The most apparent structure in the tomogram derived from the observed spectra corresponds to the accretion disk surrounding the primary star (Figure 2, top left frame). The emission seen in the observed spectrum is heavily distorted by the underlying spectrum of both stars. The subtraction of the stellar contribution produced a much more pronounced image of the accretion disk (Figure 2, bottom left frame). When the spectra of the stars and disk were removed from the observed spectrum, the tomogram was nearly clear, suggesting that the disk had been successfully modeled. However, when the tomogram was enhanced (Figure 2, top middle frame), it revealed the presence of two weaker features: emission from a gas stream along the predicted gravitational trajectory, and a smaller arc-like structure within the locus of the accretion disk. Finally, when the combined synthetic spectrum of the stars, disk, and gas stream was removed from the observed spectrum, the tomogram was once again clear, but on enhancement (Figure 2, bottom middle frame), the tomogram revealed that the gas stream emission had been adequately modeled. The only remaining structure in the tomogram was the arc-like feature within the accretion disk. This arc was finally understood with the help of hydrodynamic simulations of Richards & Ratliff (1998) (Figure 2, right frames); it represents the portion of an asymmetric elliptical disk that was not modeled because SHELLSPEC assumes a symmetric circular Keplerian accretion disk (Miller *et al.* 2007).

The detection of the arc-shaped structure demonstrated for the first time that the accretion disks in Algol systems are asymmetric and elliptical in the velocity domain. Moreover, the isolation of the low-luminosity gas stream in TT Hya also marked the first time that the gas stream had been imaged in this binary. The successful removal of the gas stream emission from the tomogram by Miller *et al.* (2007) also suggests that the

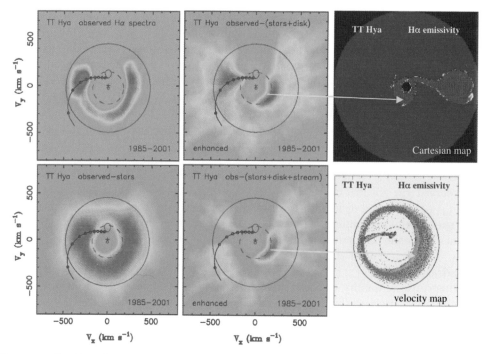

Figure 2. Hα Doppler tomograms of TT Hya: (a) the observed spectra (intensity, I = 0.98 – 1.12), (b) observed spectra minus stars (I = 0.98 – 1.6), (c) enhanced version of observed spectra minus stars and disk (I = 0.99 – 1.2), (d) enhanced version of observed spectra minus stars, disk, and gas stream (I = 0.99 – 1.2); Miller *et al.* (2007). The eccentric disk in TT Hya was predicted by the hydrodynamic simulations of Richards & Ratliff (1998).

properties of the gas stream, including the mass accretion rate, can be derived using this technique.

These results demonstrate that the SHELLSPEC code can be used to sequentially model the stars, disk, and gas stream. Only a very faint portion of the disk could not be modeled. Comparisons with simulated Hα tomograms of TT Hya based on hydrodynamic simulations by Richards & Ratliff (1998) suggested that the unmodeled region was indeed part of an asymmetric accretion disk. This region was not modeled by SHELLSPEC because the code assumes a circular Keplerian disk structure. This result provides the first observational confirmation that the disk in TT Hya is elliptical and asymmetric. Moreover, the similarity between the tomogram based on the observed Hα spectra only and that based on the observed minus stellar contribution suggests that the gas in the accretion disk is optically thin at Hα, as assumed in the differencing procedure (see Richards 1993, Richards 2004).

4. The First 3D Tomograms

It is well-known that the accretion structures should be fully three dimensional, especially near the hot spot where gas stream comes in contact with the outer edge of the disk (in CVs and long-period Algols) or the actual surface of the mass gainer (in the short-period Algols). Evidence of these 3D gas flows is based on studies of ultraviolet spectra and eclipse mapping (see, e.g., Peters 2007, in these proceedings). Nevertheless normal back-projection process translates the images onto the 2D frame in the orbital plane of the binary. The first 3D tomograms have now been produced by Agafonov *et al.* (2006) for the short-period Algol system, U CrB. They used a "radioastronomical approach"

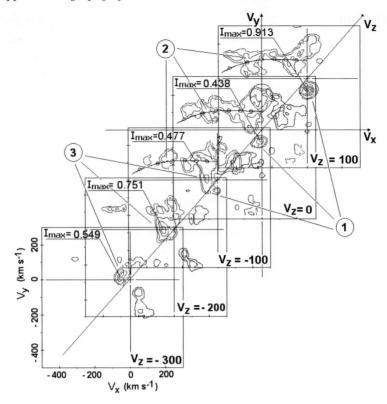

Figure 3. The three-dimensional Doppler tomogram of U CrB from spectra taken in 1994. The three strongest Hα emission features are (1) the accretion annulus, (2) the gas stream moving along the ballistic trajectory from the L1 point, and (3) the high velocity jet moving in the V_z direction. (Courtesy: Agafonov *et al.* 2006).

(RA) which is more effective than the filtered back projection technique, especially under conditions of incomplete orbital coverage resulting in insufficient sampling of the spectra.

The radioastronomical approach is equivalent to the convolution of the unknown 3D Doppler tomogram $f(x, y, z)$ with the summarized point spread function, $h(x, y, z)$: $g(x, y, z) = f(x, y, z) * * * h(x, y, z) + n(x, y, z)$. Here, the summarized image or "dirty map" is

$$g(x, y, z) = \sum_{i=1}^{N} u_i(x, y, z) \quad \text{and} \quad h(x, y, z) = \sum_{i=1}^{N} h_i(x, y, z),$$

and $n(x, y, z)$ is the noise (Agafonov *et al.* 2006). In Doppler tomography, the summarized image is constructed by means of back-projecting from the set of line profiles. The function $h(x, y, z)$ is constructed from the set of transfer functions h_i by taking into consideration the angles corresponding to the orbital phases and system inclination. These are the same angles used to construct the summarized image.

The 3D tomography of U CrB using the radioastronomical approach found that the gas flow was typically confined to within ± 30 km s^{-1} of the orbital plane. The tomograms also display the first distinct images of the gas stream moving with high velocity ($\sim$200 km s^{-1}) across the orbital plane of the binary. This is probably a jet of material being ejected from the orbital plane near the splash zone at the star-stream impact site.

5. Conclusions

Doppler tomography has been used successfully to produce images of accretion flows in a variety of interacting binaries including the CVs, X-ray binaries, and Algols. These images would not have been otherwise possible since these systems and too distant to be directly imaged. The accretion structures identified in the compact systems and the long-period Algols display many similarities, while the short-period Algols present a variety of accretion structures. Miller *et al.* (2007) have demonstrated that synthetic spectra can be used effectively to extract the properties of the individual contributors to the observed spectra. Their procedure can be used to study other Algol disk systems (e.g., AU Mon). With minor modifications, this approach can also be used to study the disks in novae and cataclysmic variables.

The major highlights of Doppler tomography have been the discovery of spiral structure in the accretion disk (e.g., Steeghs *et al.* 1997, Harlaftis *et al.* 1999), the images of gas streams along the predicted gravitational trajectory in the polars and Algols (e.g., Schwope *et al.* 1997, Richards *et al.* 1995), and the discovery that at least one Algol disk in not symmetric (Miller *et al.* 2007). The 3D tomogram produced by Agafonov *et al.* (2006) using their radioastronomical approach has also revealed, for the first time, an image of gas jets beyond the orbital plane. The application of this technique opens new possibilities for the study of the gas motions beyond the orbital plane in binaries because this provides a clearer understanding of the evolution of close binaries and the physical conditions governing mass transfer.

The new 3D hydrodynamic simulations of Bisikalo & Matsuda (2007) in these proceedings extends the work of Richards & Ratliff (1998) to much more exciting directions. There is now the opportunity to compare the hydrodynamic velocity maps with the Doppler tomogram derived from observations to study the flow structure and stream disk impact site more deeply than before. So, this is just the beginning.

Acknowledgements

I thank my many collaborators on several recent projects related to tomography, especially, Michail Agafonov, Geary Albright, Ján Budaj, Milind Cholkar, Ryuko Hirata, Ilian Iliev, Pavel Koubský, Seiji Masuda, Brendan Miller, Geraldine Peters, Mark Ratliff, Olga Sharova, Vojtěch Simon, and Petr Škoda for their contributions to this review. This research was partially supported by NSF-NATO grant DGE-0312144 and NASA ADP grant NNG04GC48G. I am also grateful for HET observing time from Penn State University. Participation in this symposium was supported by NSF/AAS and IAU travel grants. I dedicate this work to Emilios Harlaftis (1965-2005) who was passionate about Doppler tomography and was a co-discoverer of spiral structure in CV disks.

References

Agafonov, M.I., Richards, M.T., & Sharova O.I. 2006, *ApJ*, 652, 1547
Albright, G.E. & Richards, M.T. 1996, *ApJ*, 459, L99
Bisikalo, D.V. & Matsuda, T. 2007, these proceedings, 356
Budaj, J. & Richards, M.T. 2004, *Contrib. Astron. Obs. Skalnate Pleso*, 34, 167
Budaj, J., Richards, M.T., & Miller B. 2005, *ApJ*, 623, 411
Groot, P.J. 2001, *ApJ*, 551, 89
Harlaftis, E., Steeghs, D., Horne, K., Martín, E., & Magazzú 1999, *MNRAS*, 306, 348
Herman, G.T. 1980, *Image Reconstruction from Projections: The Fundamentals of Computerized Tomography* (New York: Academic Press)
Hubeny, I. 1988, *Comput. Phys. Comm.*, 52, 103
Hubeny, I. & Lanz, T. 1992, *A&A*, 262, 501
Hubeny, I. & Lanz, T. 1995, *ApJ*, 439, 875

Hubeny, I., Lanz, T., & Jeffery, C. S. 1994, *Newsletter on Analysis of Astronomical Spectra* No.20, C.S. Jeffery (CCP7; St. Andrews: St. Andrews Univ.), 30

Kaitchuck, R.H., Schlegel, E.M., Honeycutt, R.K., Horne, K., Marsh, T.R., White, J.C., & Mansperger, C.S. 1994, *ApJS*, 93, 519

Karovska, M. Schlegel, E., Hack, W., Raymond, J. C., & Wood, B. E. 2005, *ApJ*, 623, L137-L140

Kempner, J.C. & Richards, M.T. 1999, *ApJ*, 512, 345

Kurucz, R.L. 1993, *SYNTHE Spectrum Synthesis Programs and Line Data* (CD-ROM 18)

Marsh, T. R. 2001, *Lecture Notes in Physics*, 573, 1

Marsh, T.R. & Horne, K. 1988, *MNRAS*, 235, 269

Miller, B., Budaj, J., Richards, M.T., Koubský, P., & Peters, G.J. 2007, *ApJ*, 656, 1075

Morales-Rueda, L. 2004, *Astron. Nach.*, 325, 193

Morales-Rueda, L., Marsh, T. R., & Billington, I. 2000, *MNRAS*, 313, 454

Peters, G.J. 2007, IAU Symp. 240, W. Hartkopf, E. Guinan, & P. Harmanec (eds.), in press

Peters, G.J. & Polidan, R.S. 1998, *ApJ*, 500, L17

Radon, J. 1917, *Berichte Sächsische Akademie der Wissenschaften Leipzig Math. Phys. Kl.*, 69, 262 (reprinted in 1983: Proc. Symposia Appl. Math, 27, 71)

Richards, M.T. 1993, *ApJS*, 86, 255

Richards, M.T. 2001, *Lecture Notes in Physics*, 573, 276

Richards, M.T. 2004, *Astron. Nach.*, 325, 229

Richards, M.T. & Albright, G.E. 1996, in *Stellar Surface Structure*, ed. K. Strassmeier & J. Linsky (Dordrecht: Kluwer), 493

Richards, M.T., Albright, G.E., & Bowles, L. M. 1995, *ApJ*, 438, L103

Richards, M.T., Koubský, P., Šimon, V., Peters, G.J., Hirata, R., Škoda, P., & Masuda, S. 2000, *ApJ*, 531, 1003

Richards, M.T. & Ratliff, M.A. 1998, *ApJ*, 493, 326

Richards, M.T. & Rosolowsky, E.W. 1998, *ASP Conf. Ser.*, 154, 2038

Robinson, E.L., Marsh, T.R., & Smak, J.I. 1993, in *Accretion Disks in Compact Stellar Systems*, ed. J.C. Wheeler (Singapore: World Scientific), 75

Schwope, A.D. 2001, *Lecture Notes in Physics*, 573, 127

Schwope, A.D., Mantel, K.-H., & Horne, K. 1997, *A&A*, 319, 894

Schwope, A.D., Staude, A., Vogel, J., & Schwarz, R. 2004, *Astron. Nach.*, 325, 197

Shepp, L.A. 1983, *Proc. Symposia in Appl. Math*, 27, 1

Sawada, E., Matsuda, T., & Hachisu, I. 1986, *MNRAS*, 219, 75

Staude, A., Schwope, A. D., & Schwarz, R. 2001, *A&A*, 374, 588

Steeghs, D. 2004, *Astron. Nach.*, 325, 185

Steeghs, D., Harlaftis, E. T., & Horne, K. 1997, *MNRAS*, 290, L28

Vrtilek, S.D., Quaintrell, H., Boroson, B., & Shields, M. 2004, *Astron. Nach.*, 325, 209

Discussion

JUAN ECHEVARRIA: A basic assumption of Doppler Tomography is that the accretion material is in the orbital plane. Have you taken into account magnetic fields to explain the partially broken disks you show in some of the systems?

RICHARDS: No, we have not included any treatment of magnetic fields. However, we have checked to see if the magnetic field could have deflected the gas streams from the gravitational path and we plan to add a treatment of magnetic effects in SHELLSPEC, our synthesis code. I should also note, however, that coronal mass ejections (flares) may be the cause of changes in the structure of accretion flows, by changing the mass transfer rate on a time scale of several months.

NICK ELIAS: How many phases are required to reconstruct an image via Tomography?

RICHARDS: Hundreds would be ideal; 25 is OK if they're evenly spaced around the orbit.

Binary Stars as Critical Tools & Tests
in Contemporary Astrophysics
Proceedings IAU Symposium No. 240, 2006
W.I. Hartkopf, E.F. Guinan & P. Harmanec, eds.

Doppler Imaging of Close Binaries

Klaus G. Strassmeier

Astrophysical Institute Potsdam (AIP), Potsdam, Germany
email: kstrassmeier@aip.de

Abstract. Many of the interesting spotted stars are in close binaries where one can find almost any rotational period due to the rotational synchronization with the orbital motion. Binaries are thus good laboratories to study the impact of particular astrophysical parameters that nature usually does not make easily observable. On rapidly-rotating stars, we can indirectly resolve the surface by tomographic imaging techniques and map the surface temperature distribution as a proxy of the (predominantly radial) magnetic field. Binaries are not as straightforward to map as single stars and I will show some examples where it was successful and some where it failed. Eclipses may give some clues on the amount of unresolved features in the images. I present one case of a bright giant of $100L_\odot$ in a close binary with even a deformed surface geometry but otherwise solar-type behavior. One of the basic goals is to learn about the impact of inter-binary magnetic fields on the evolution of its components in general and to eventually provide conclusive constraints for numerical MHD models on the other hand.

Keywords. stars: activity, stars: binaries: close, stars: spots, stars: rotation, techniques: spectroscopic, techniques: polarimetric

1. Introduction and motivation

Stellar activity is a collective term for all phenomena inside and outside of a star that are related to its magnetic field, e.g., the rise of flux tubes in a stellar convection zone and their appearance as spots and plages on the stellar surface, or magnetically induced particle acceleration and its braking effect on the stellar rotation, or simply the interplay between a magnetic field and its surrounding plasma. The magnetic field plays the key role in many astrophysical processes but, partly because its mathematical treatment can be somewhat complicated, it is usually the first process that is neglected in evolutionary and other models.

A most important ingredient for our understanding of the solar-stellar connection is the interaction of the magnetic field with the astrophysical plasma. The magnetic field produces the pressure $B^2/8\pi$ (in cgs units) perpendicular to the field lines and a tension $B^2/4\pi$ along the field lines but since charged particles move freely parallel to the field, the plasma is not affected by the magnetic tension. For example, in a young sunspot horizontal equilibrium is reached when the difference of the gas pressure outside and inside the spot equals its magnetic pressure while the vertical equilibrium is basically given by the hydrostatic equilibrium. The plasma in a sunspot reaches a stable equilibrium by lowering its vertical position (the so-called Wilson depression) because the pressure in its surrounding increases with depth. Of course, spots can sink because they are cooler than their surrounding. If the magnetic pressure exceeds the gas pressure a current must flow along the field lines.

Therefore, a stellar magnetic field is intimately related to flows, most notably due to stellar rotation and differential rotation. The quest to understand stellar magnetic activity is thus also always a quest to understand stellar rotation and stellar differential rotation. In Doppler imaging, one uses the fact that stars rotate in order to map their surface inhomogeneities and to detect, e.g., differential surface rotation.

Several reviews appeared on the subject of "magnetic activity of binary stars", e.g., Guinan & Giménez (1993) or Strassmeier (2002). A more recent review on the subject of Doppler imaging of binaries is Strassmeier (2005) but see also the many papers in the "Bruxelles" proceedings (Boffin *et al.* 2001). In the present paper I will emphasize the technical requirements for Doppler and Zeeman-Doppler imaging and update on the literature of binary-star results.

2. The technique of Doppler imaging

2.1. *Principle*

Doppler imaging (DI) is a computational technique similar to medical tomography that inverts a series of high-resolution spectral line profiles into an "image" of the stellar surface (Deutsch 1958, Vogt *et al.* 1987, Rice *et al.* 1989, Strassmeier 1990, Collier Cameron 1992, Piskunov & Rice 1993, Berdyugina *et al.* 1998, Rice & Strassmeier 2000, Rice 2002). Cool starspots produce distortions in the spectral line profiles that systematically change during a star's rotation. It is the way in which these distortions change with time which allows to reconstruct the stellar surface temperature distribution. The technique is very similar to that described by Richards (2007) in these proceedings except that the velocity range in the line profile is restricted to the (rotating) stellar surface and the intensity range is given by the temperature contrast between spotted and unspotted photosphere.

2.2. *Observational requirements*

Two stellar parameters — brightness and rotational broadening — dictate the instrumental requirements for Doppler imaging. The former sets a limit for the achievable S/N ratio and the latter determines the size of a resolution element across the stellar surface. So far, current telescope and spectrograph combinations limited Doppler imaging to stars with $v \sin i$ as low as 17.5 km s^{-1} (Strassmeier & Rice 1998; using the CFHT and its Gecko spectrograph on EK Dra, G1.5V, at $R \approx$120,000) and 18 km s^{-1} (Washuettl & Strassmeier 2001; using ESO's CAT and CES for AG Dor, K0V+K4V, at $R \approx$50,000) and V as faint as $\approx$14.3mag (Strassmeier *et al.* 2005; using ESO's VLT and UVES on MN Lupi, M0, at $R \approx$60,000). Least-squares deconvolution techniques (Donati *et al.* 1997) may eventually push the brightness limit several magnitude further.

2.3. *Computational requirements*

As an example, our DI code TEMPMAP (e.g. Rice 2002) performs either single-line, multi-line or full-spectrum inversion in integral light. A comparison of these modes was presented in a poster at the *Cool Star 14* workshop (Rice & Strassmeier 2007). In any mode the code solves the equation of transfer at 72 depth points on 2592 surface elements through a set of 10 Kurucz ATLAS-9 or 12 model atmospheres. Either the default mixing-length description of convective flux or the non-parameterized convective-flux description of Canuto & Mazzitelli (1992) may be used. We found generally good agreement with local line profiles computed from either of them (see Strassmeier & Rice 1998). Our code uses either Maximum Entropy or a Tikhonov regularization and can also include two-bandpass photometry in its solution simultaneously. Atomic line data must be known accurately and we usually adopt transition probabilities from VALD (Kupka *et al.* 1999), or use our own values. The computing demand increases with the number of spectral lines, its sampling — in phase as well as in wavelength — its number of surface elements and the number of model atmospheres and their depth points. Generally speaking, a high-end

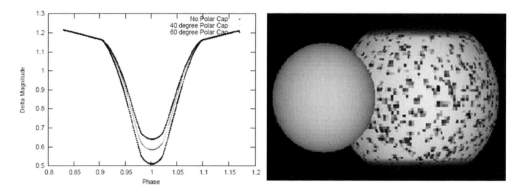

Figure 1. Left: The effect of a polar spot on the primary eclipse light curve of SV Cam. Right: Visualization of SV Cam from eclipse mapping. The primary star (the larger star) has two large cap-like polar spots and is peppered with unresolved spots amounting to a filling factor of alone 28%. Adapted from Jeffers *et al.* (2006).

PC can handle a single-line inversion with, say, ten blends in 10 minutes. A small 4-8 CPU Beowulf-type cluster is needed to run the full-spectrum mode.

Atomic-line contamination by molecular lines (mostly TiO, OH, CO, FeH, etc.) becomes an issue even at effective temperatures as high as the Sun's. New codes that take into account molecular bands now exist (Berdyugina 2002, Savanov & Strassmeier 2005) but did not yet lead to published results. The computing demands become significant but can still be handled on a single high-end PC.

The inversion process for Zeeman Doppler imaging (ZDI) requires particularly extensive calculations of local Stokes profiles over the entire stellar disk. Usually these data are pre-tabulated but in case of ZDI become unrealistically large due to an arbitrarily complex magnetic-field structure. Also, the weak-field approximation does not provide the needed accuracy, besides that it is only applicable for Stokes V. Forward numerical tabulation of the integration of the polarized radiative transfer equation becomes therefore a numerically not manageable task. Kopf *et al.* (2007) have developed an approximation method based on Multi Layer Perceptrons (MLP), a common type of artificial neural networks, that uses the decomposition of local Stokes profiles into their eigenspectra via a Principal Component Analysis. The adaptation (training) of the MLPs is based on a conventional numerical integration of the polarized radiative transfer with the quadratic DELO method. After training with tens of thousands of line profiles, the MLP method yields a speed-up of a factor 1,000 compared to DELO (e.g., Kochukhov & Piskunov 2007). Calculating the Stokes V profile with the MLP method yields an error of 0.18%, while the weak-field approximation gives an error of 2.3% (Kopf *et al.* 2007). Computations currently require a medium-size PC cluster like our own 320-PC 700 Gflops/s machine.

3. Some applications to close binary stars

In the following, I will discuss some of the results on a few selected binary systems.

3.1. *SV Cam: an eclipsing SB2 binary with a hot active component*

Jeffers *et al.* (2006) obtained spectrophotometric eclipse light curves with HST/STIS and mapped the primary F9 component of this F9V+K4V-IV binary (P_{orb}=0.593 days). They found that the observed surface flux from the eclipsed low-latitude regions of the

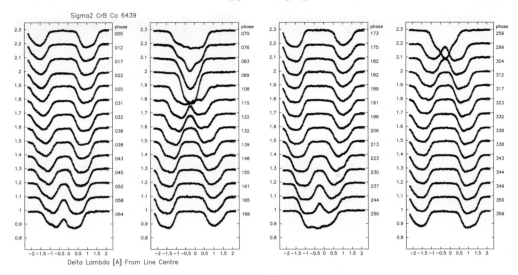

Figure 2. CFHT data for Doppler imaging of σ^2 CrB. Observations and fits for both stellar components are shown from the inversion of the Ca I 643.9-nm region. The inversion is done simultaneously for both components with orbital parameters fxed. Adapted from Strassmeier & Rice (2003).

F9 primary was 30% lower than what is predicted from PHOENIX model atmospheres and the *Hipparcos* parallax, despite that the eclipse light-curve shape did not reveal spatial surface inhomogeneities. This could be explained only if there is about a third (28%) of the eclipsed region covered by unresolved cool starspots. Even if this "dark" spot component is assumed to also exist over the rest of the surface, a huge cap-like polar spot down to a latitude of ($\pm$)48$\pm$6° was required in order to fit the eclipse light curves (Figure 1, Jeffers *et al.* 2006). This is to be seen in contrast to the earlier DI results of Lehmann *et al.* (2002) who found no polar-cap like spot but only high-latitude spots. On the other hand, spot coverage and temperatures from differential TiO-band modelling had indicated spot filling factors of the order of 30–50% (O'Neil *et al.* 1998), usually more than what DI from optical spectral line profiles had been given. However, DI of eclipsing binaries is a particularly ill-posed situation due to the north-south mirroring of the Doppler signal and, moreover, the presence of two components in the combined spectrum. Nevertheless, we conclude that SV Cam had shown us a problem that we need to address in future applications.

3.2. *The F/G-type ZAMS binary σ^2 CrB*

This particular target is a nice example of a close binary of two solar-type overactive stars (F9V+G0V) that were spun up due to spin-orbit coupling ($P_{\rm orb}$=1.14 days). Single stars with such properties usually do not exist. Both stars resemble the primary of SV Cam (see above). A Doppler image was presented by Strassmeier & Rice (2003). In this application we have extended TEMPMAP to invert the combined spectrum from both stars simultaneously over a large wavelength range. The full-spectrum solution considered a total of 171 spectral lines and two continuum bandpasses. As for the single-line maps, a total of 56 spectra from all phases was used. Figure 2 shows the entire input data for *one* spectral-line region.

The σ^2 CrB surface structure exhibits a complex symmetry within the two stellar components. Both component's high-latitude spots appear asymmetric with respect to their rotational poles and tend to anti-face each other with respect to the apsidal line.

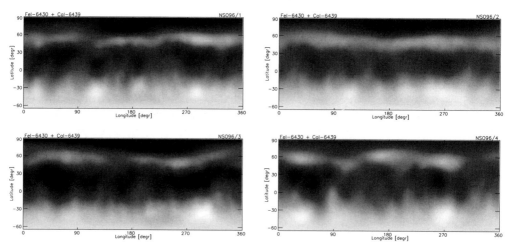

Figure 3. Doppler images of ζ And for four consecutive stellar rotations in 1996. The maps show temperature between 4800K (white) and 3400K (black) with spots at high-latitude and even polar spots with temperatures of about 800–1200 K below the effective photospheric temperature. Data from Kövari *et al.* (2007b).

A large equatorial warm belt seems to exist on the trailing hemisphere of each of the two stars with respect to the orbital motion, i.e., they appear centered at the central meridian during the respective quadrature phases, and thus are not facing each other as would be the case for a hypothetical irradiation effect. The leading hemispheres on both components appear with an effectively lower temperature than the trailing hemispheres.

We believe that a meaningful interpretation must involve a strong coupling of the individual stellar magnetic fields and its associated electron acceleration and braking along particular field lines. Whether such magnetic coupling in tidally locked stars has an impact on the dynamo and the formation and rise of flux tubes remains to be determined but is suggested from the current results.

3.3. *The ellipsoidal K-giant binary* ζ *And*

The star is a magnetically active ≈100 $L_\odot$ K1 giant (with an unseen companion) with its rotation synchronized to the 17.8-day orbital period. Its lithium abundance of $\log n$ =1.2 places it in the vicinity of Li-rich RGB stars but it is nevertheless a Li-normal chromospherically active binary star. The primary seems to undergo its first standard dredge-up dilution. The unprojected equatorial rotational velocity, and the fact that the rotational period appears synchronized to the orbital period, suggests a likely radius of the primary of 16.0 ± 0.2 $R_\odot$. The mass function a.o. parameters together with the most likely inclination from DI suggests a 0.75-$M_\odot$ K-dwarf secondary circling the primary at a distance of just 2.7 (primary) stellar radii, i.e., 2.2 stellar radii above its surface, giving rise to a tidally deformed primary-star surface.

Kövari *et al.* (2007b) presented an extensive DI study of this system. Our DI code TEMPMAP had to be modified to work with gravitationally deformed surfaces. Kövari *et al.* (2007a) (in this proceeding) discusses various tests that were performed during the line-profile inversions of ζ And and I refer to this paper for more details.

Our Doppler maps revealed cool high-latitude and even polar spots with temperatures of about 800–1200 K below the effective photospheric temperature (Figure 3). The polar spot faded by several hundred degrees within one to two stellar rotations during at least one occasion in 1996/97. It was recovered with the previous contrast (1200 K) one year later though. The low-latitude spots tended to group on the two hemispheres visible during quadrature, i.e., $\pm 90°$ from the apsidal line following and preceding the location of the secondary star. This seemed to be the case for both observing seasons we had data for. At the same time the cool polar spot had a large appendage near the phase of conjunction with the secondary behind in 1996/97, but less determinable in 1997/98. A cross-correlation analysis of consecutive maps and a fit with a solar-type quadratic law revealed a more rapidly rotating equator with a surface shear with respect to higher latitudes of four times lower than for the Sun and with a lap time of 360 days.

4. Summary

The sum of the activity phenomena detected up to now is staggering and significantly more complex than what just scaled-up solar analogy would imply. Various groups found

(a) large spots on or near the rotational pole (e.g., Vogt *et al.* 1999, Strassmeier 1999, Berdyugina *et al.* 1998, Collier-Cameron 1995),

(b) differential rotation both in the same sense and in the opposite sense than on the Sun (e.g., Vogt *et al.* 1999, Donati & Collier-Cameron 1997, Hatzes 1998, Rice & Strassmeier 1996, Strassmeier *et al.* 2003, Donati *et al.* 2003, Marsden *et al.* 2006), and

(c) seasonal variations thereof (e.g., Donati *et al.* 2003, Marsden *et al.* 2007),

(d) possibly detected (meridional) flows toward the pole (Weber & Strassmeier 2001, Strassmeier & Bartus 2000, Kövari *et al.* 2007b, Weber & Strassmeier 2007),

(e) detected and mapped transiting prominences (Dunstone *et al.* 2007, Jardine *et al.* 1998, Collier-Cameron *et al.* 1996),

(f) warm spots (e.g., Unruh *et al.* 1998, Strassmeier 1994,1999),

(g) and possibly detected the impact regions from accretion streams in very young pre-main-sequence photospheres (Strassmeier *et al.* 2005),

(h) observed active longitudes and related activity flip-flops (e.g., Korhonen *et al.* 2004, Berdyugina 2003, Oláh *et al.* 2002),

(i) and found good evidence for uniformly distributed and therefore unresolved spots (Jeffers *et al.* 2006, Hussain *et al.* 2007),

(j) and found spot evolution on very short- and on very long timescales (e.g. Washuettl *et al.* 2007, Petit *et al.* 2004, Barnes *et al.* 2001) and, finally,

(k) directly detected complex magnetic surface fields (e.g. Donati *et al.* 2006, Petit *et al.* 2004, Piskunov & Kochukhov 2002, Wade *et al.* 2000). The latter is still a rather virgin field because of the lack of appropriate spectro-polarimetric data.

Finally, after 20 years of Doppler Imaging,

Yes, the stars, they always make me laugh.

(A. Saint Exupéry, The Little Prince).

Acknowledgements

I'd like to thank my colleagues John Rice from Brandon University, and Michi Weber, Thorsten Carroll, Heidi Korhonen and Igor Savanov from AIP for many fruitful discussions on various related software issues.

References

Afram, N., Berdyugina S.V., Fluri D.M., Suwald, F., Kuhn, J.R., Harrington, & D.M. 2007, in J. Stauffer *et al.* (eds.), *14th Cool Stars, Stellar Systems, and the Sun*, Pasadena, ASPC, CD-ROM (in press)

Barnes, J.R. & Collier Cameron, A. 2001, *MNRAS*, 326, 950

Berdyugina, S.V. 2002, in Brown A., Harper G., & Ayres, T. (eds.), 12th Workshop, *Cool Stars, Stellar Systems, and the Sun*, www-publication, p.210

Berdyugina, S.V. 2003, in Benz, A. & Dupree, A. (eds.), IAU Symp. 219, *Stars as Suns: Activity, Evolution and Planets*, Sydney, p.181

Berdyugina, S.V., Berdyugin, A.V., Ilyin, I., & Tuominen, I. 1998, *A&A*, 340, 437

Boffin, H.M.J., Steeghs, D., & Cuypers, J. (eds.) 2001, *Astrotomography*, Lecture Notes in Physics, Springer, Vol. 573

Canuto V.M. & Mazzitelli I. 1992, *ApJ*, 389, 724

Collier Cameron, A. 1992, in Byrne P.B., & D.J. Mullan (eds.), *Surface Inhomogeneities on Late-Type Stars*, Lecture Notes in Physics, Vol. 397, Springer-Verlag, Berlin, p. 33

Collier Cameron, A. 1995, *MNRAS*, 275, 534

Collier Cameron, A. 1996, in Linsky, J.L., & Strassmeier, K.G. (eds.), IAU Symp. 176, *Stellar Surface Structure*, Kluwer, p.449

Deutsch, A. 1958, in Lehnert B. (ed.), *Electromagnetic Phenomena in Cosmological Physics*, IAU Symp. 6, Cambridge Univ. Press, Cambridge, p. 209

Donati, J.-F. & Collier Cameron, A. 1997, *MNRAS*, 291, 1

Donati, J.-F., Forveille, T., Collier Cameron, A., *et al.* 2006, *Science* 311, 633

Donati, J.-F., Semel, M. Carter, B.D., Rees, D.E., & Collier Cameron, A. 1997, *MNRAS*, 291, 658

Donati, J.-F., Collier Cameron, A., & Petit, P. 2003, *MNRAS*, 345, 1187

Dunstone, N., Collier Cameron, A., Barnes, J., & Jardine, M. 2007, in J. Stauffer *et al.* (eds.), 14th *Cool Stars, Stellar Systems, and the Sun*, Pasadena, ASPC, CD-ROM (in press)

Guinan, E.F. & Giménez, A. 1993, in J. Sahade *et al.* (eds.), *The Realm of Interacting Binary Stars*, Kluwer, Dordrecht, p. 51

Hatzes, A.P. 1998, *A&A*, 330, 541

Hussain, G.A.J., Saar, S.H., Aufdenberg, J., Ringwald, F., & Johns-Krull, C. 2007, in J. Stauffer *et al.* (eds.), 14th *Cool Stars, Stellar Systems, and the Sun*, Pasadena, ASPC, CD-ROM (in press)

Jardine, M., Barnes, J., Unruh, Y.C., & Collier-Cameron, A. 1998, in D. Webb *et al.* (eds.), *IAU Colloq. 167*, PASPC

Jeffers, S.V., Aufdenberg, J.P., Hussain, G.A.J., Collier Cameron, A., & Holzwarth, V. 2006, *MNRAS*, 367, 1308

Kochukhov, O. & Piskunov, N.E. 2007, these proceedings

Kopf, M., Carroll, T., & Strassmeier, K.G. 2007, in J. Stauffer *et al.* (eds.), 14th *Cool Stars, Stellar Systems, and the Sun*, Pasadena, ASPC, CD-ROM (in press)

Korhonen, H., Berdyugina S.V., & Tuominen I. 2004, *AN* 325, 402

Kövari, Zs., Bartus, J., Oláh, K., Strassmeier, K.G., Rice, J.B., Weber, M., & Forgacs-Dajka, E. 2007a, these proceedings, 212

Kövari, Zs., Bartus, J., Strassmeier, K.G., Oláh, K., Washuettl, A., Weber, M., & Rice, J.B. 2007b, *A&A*, 463, 1071

Kupka, F., Piskunov, N.E., Ryabchikova, T.A., Stempels, H.C., & Weiss, W.W. 1999, *A&AS*, 138, 119

Lehmann, H., Hempelmann, A., & Wolter, U. 2002, *A&A*, 392, 963

Marsden, S.C., Donati, J.-F., Semel, M., Petit, P., & Carter, B.D. 2006, *MNRAS*, 370, 468

Marsden, S.C., Berdyugina, S.V., Carter, B.D., *et al.* 2007, in J. Stauffer *et al.* (eds.), 14th *Cool Stars, Stellar Systems, and the Sun*, Pasadena, ASPC, CD-ROM (in press)

Oláh, K., Strassmeier, K.G., & Weber, M. 2002, *A&A*, 389, 202

O'Neil, D., Neff, J., & Saar, S.H. 1998, *ApJ*, 507, 919

Petit, P., Donati, J.-F., Wade, G.A., *et al.* 2004, *MNRAS*, 348, 1175

Piskunov, N.E. & Kochukhov, O. 2002, *A&A*, 381, 736

Piskunov, N.E. & Rice, J.B. 1993, *PASP* 105, 1415

Rice, J.B. 2002, *AN* 323, 220

Rice, J.B. & Strassmeier, K.G. 1996, *A&A*, 316, 164

Rice, J.B. & Strassmeier, K.G. 2000, *A&AS*, 147, 151

Rice, J.B. & Strassmeier, K.G. 2007, in J. Stauffer *et al.* (eds.), 14th *Cool Stars, Stellar Systems, and the Sun*, Pasadena, ASPC, CD-ROM (in press)

Rice, J.B., Wehlau, W.H., & Khokhlova, V.L. 1989, *A&A*, 208, 179

Richards, M. 2007, these proceedings, 160

Savanov, I. & Strassmeier, K.G. 2005, *A&A*, 444, 931

Strassmeier, K.G. 1990, *ApJ*, 348, 628

Strassmeier, K.G. 1994, *A&A*, 281, 395

Strassmeier, K.G. 1999, *A&A*, 347, 225

Strassmeier, K.G. 2002, in F.C. Lazaro & M.J. Arevalo (eds.), *Selected Topics on Binary Stars: Observations and Physical Processes*, Lecture Notes in Physics, Springer Verlag, p.48

Strassmeier, K.G. 2005,, in Pavlovski *et al.* (eds.), *Physics of Close Binary Stars*, Dubrovník, Croatia, ASPC 318, p.69

Strassmeier, K.G. & Bartus, J. 2000, *A&A*, 354, 537

Strassmeier, K.G., Kratzwald, L., & Weber, M. 2003, *A&A*, 408, 1103

Strassmeier, K.G. & Rice, J.B. 1998, *A&A*, 330, 685

Strassmeier, K.G. & Rice, J.B. 2003, *A&A*, 399, 315

Strassmeier, K.G., Rice, J.B., Ritter, A., Küker, M., Hussain, G.A.J., Hubrig, S., Shobbrook, R. 2005, *A&A*, 440, 1105

Unruh, Y.C., Collier-Cameron, A., & Guenther, E. 1998, *MNRAS*, 295, 781

Vogt, S.S., Hatzes, A.P., Misch, A.A., & Kürster, M. 1999, *ApJS*, 121, 547

Vogt, S.S., Penrod, G.D., & Hatzes, A.P. 1987, *ApJ*, 321, 496

Wade, G.A., Donati, J.-F., Landstreet, J.D., & Shorlin, S.L.S. 2000, *MNRAS*, 313, 823

Washuettl, A. & Strassmeier, K.G. 2001, *A&A*,, 370, 218

Washuettl, A., Strassmeier, K.G., & Weber, M. 2007, *AN*, submitted

Weber, M. & Strassmeier, K.G. 2001, *A&A*, 373, 974

Weber, M. & Strassmeier, K.G. 2007, in J. Stauffer *et al.* (eds.), 14th *Cool Stars, Stellar Systems, and the Sun*, Pasadena, ASPC, CD-ROM (in press)

Binary Stars as Critical Tools & Tests
in Contemporary Astrophysics
Proceedings IAU Symposium No. 240, 2006
W.I. Hartkopf, E.F. Guinan & P. Harmanec, eds.

© 2007 International Astronomical Union
doi:10.1017/S1743921307003997

Reducing Binary Star Data from Long-Baseline Interferometers

Theo A. ten Brummelaar

Center for High Angular Resolution Astronomy, Georgia State University,
The CHARA-Array, Mount Wilson Observatory, CA 91023, USA
email: theo@chara-array.org

Abstract. Processing long-baseline interferometry data presents a unique set of complications: How does one derive relative astrometry from interferometric data? How can 1-D interferometric results be used to solve a 2-D orbit? How can baseline-only solutions be combined with historical data and how should interferometric data be published so that they can be combined with archived data? What new techniques for interferometers are coming on line? This paper contains a brief review of interferometric data analysis in the context of binary star astrometry.

Keywords. Binary Stars, Interferometry, Astrometry, Closure Phase, Imaging.

1. Introduction

Ground based optical and infrared long baseline interferometry (OLBI) has in recent years come of age and is now producing some outstanding results in many areas of stellar astrophysics. With more than 60 publications in 2006 at the time of writing this paper, and many more on the way, the comparatively small number of instruments available to the community are very scientifically productive and with new instrumentation now being commissioned this productivity will continue to expand.

The techniques of OLBI are particularly well suited to the study of binary stars because the high resolution of these instruments makes it possible to resolve astrometric orbits with unprecedented precision and accuracy. In the next few years the gap between astrometric and radial velocity measurements will rapidly close and calculating masses, stellar radii, and other fundamental characteristics of stars at the 1% level or better will become common place.

In this paper, I will attempt to give a broad overview of how OLBI data is collected and interpreted in the context of binary star astrometry. A more detailed explanation of the techniques of OLBI (Lawson (2000)), in the form of lecture notes for the 1999 Michelson Summer School run by the Michelson Science Center†, is available for free from the Jet Propulsion Laboratory‡ (JPL). A good source of historical information on the subject can be found in Lawson (1997) and more up to date information, including a list of publications, can be found on the OLBIN web page¶ maintained by JPL. Note that while many of the examples in this text come from the CHARA Array (ten Brummelaar *et al.* (2005)), principally because I have easy access to these data, they are generally applicable to any existing or planned long baseline interferometer.

† msc.caltech.edu
‡ email: lawson@huey.jpl.nasa.gov
¶ olbin.jpl.nasa.gov

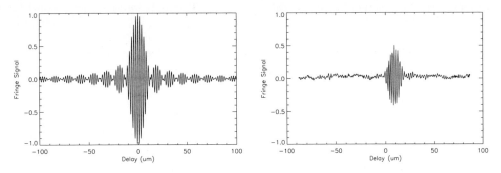

Figure 1. Examples of raw fringe data: *Left*: Modeled data *right*: Real data.

2. Basic Theory

An interferometer takes the light from two, or more, telescopes and combines this light together in a single position known as the beam combiner. Each pair of telescopes, whose separation is known as the baseline, will create a fringe packet, that is, an interference pattern, whose width depends on the optical bandwidth of the system and whose magnitude and phase depend on the object, the baseline, the central wavelength of the optical filter, and numerous instrumental and atmospheric effects. The mathematical form of a fringe packet is given by

$$I(t,\nu) = 1 + V(t,\nu) \times \frac{\sin\left(\pi\, x(t,\nu)\, \Delta\nu\right)}{\pi\, x(t,\nu)\, \Delta\nu} \times \cos\left(2\pi\, x(t,\nu)\, \nu + \Phi_{\mathrm{obj}} + \Phi_{\mathrm{atm}}(t)\right) \quad (2.1)$$

where I is the measured intensity, t is time, V is the fringe amplitude, x is the path length difference between the two beams, $\nu = \frac{1}{\lambda}$ and $\Delta\nu$ are the central wavenumber and width of the optical filter, Φ_{obj} is the visibility phase of the object and Φ_{atm} is the phase error introduced by the atmosphere. An example of a theoretically perfect fringe is given in the left plot of Figure 1.

The Van Citter-Zernike theorem (See for example Thompson *et al.* 2001 or Born & Wolf 1999) states that the complex quantity called the visibility, that is the amplitude and phase of the fringe packet, represents one Fourier component of the intensity distribution of the object on the sky. In principle, one need only measure the visibility of the fringes at many baselines and perform an inverse Fourier transform to derive an image of the object in question. The power of interferometry derives from the fact that the angular resolution of this image is proportional to the central wavelength divided by the largest baseline. Thus for a baseline of 100 meters and a wavelength of 1 μm the best resolution attainable will be 1.0^{-8} radians or about 2 mas. For binary star observations the resolution can be much higher.

Of course there are many factors that make the real situation much more complex. In order to see any fringes at all one must overcome many engineering challenges, and like any other ground based technique the data are distorted by the atmosphere and instrumental effects. For example, one must have tip/tilt servo systems on each telescope, or even a full blown adaptive optics system on the larger apertures. Furthermore, the optical path length of the light from the star, through the two telescopes and within the instrument all the way to the beam combiner, must be controlled to a precision of much better than 1 μm. For example, at the CHARA Array, the delay lines are kept stable with an RMS error of less than 10 nm, and this figure will be similar at other facilities. These path lengths are always changing as the earth rotates, and the atmosphere is constantly adding wavefront and phase errors that change on the time scale of milliseconds which can severely restrict sample times and data rates. Thus, an interferometer consists of many servo systems, each with time constants faster than the atmospheric modulations. An example of real fringe data is given in the right hand plot in Figure 1.

All of these instrumental and atmospheric effects cause changes in the fringe amplitude and phase, almost always reducing the amplitude and totally destroying the phase information on a single baseline. Worse yet, the amplitude modulations are constantly changing as the turbulent atmosphere blows by the telescopes. A comparison of the modeled and real fringe data shown in Figure 1 reveals many of these problems. First of all, despite the fact that this is unusually high signal to noise data†, there is still a great deal of noise present in the signal, so much so that the first side lobes of the fringe packet are barely visible and the rest not visible at all. Some of this noise is due to scintillation, but most of it is camera and photon noise. Since the output signal is a subtraction of the two outputs of a beam splitter scintillation noise cancels out to a large extent. Secondly, the fringe is not centered within the delay scan indicating that the phase information has been lost. Thirdly, the shape of the fringe envelope has been distorted due to the fact that the atmospherically induced phase and wavefront errors are constantly changing. Finally, despite the fact that the object is unresolved, the fringe amplitude is much less than 1.0.

In order to get around these difficulties the standard practice is to measure many, often several hundred, fringe scans and take the mean of the fringe amplitude. In this way the random fluctuations of amplitude are averaged out. One then repeats this measurement on a nearby unresolved source and takes the ratio of the, hopefully resolved, science target and the, hopefully unresolved and spherical, calibrator object. This yields a good estimate of the fringe amplitude of the science target for the baseline in question.

In a so called "open air" beam combiner with no spatial filtering this calibration procedure typically yields precisions of $\approx 5\%$, while a single mode fiber based instrument will filter out much of the atmospheric distortion and achieve results of $\approx 1\%$. Spatial filtering does, however, come at the cost of a lower magnitude limit. In excellent seeing conditions one can do better, but these values are fairly representative of the two techniques.

This process is repeated for many baselines in order to collect a data set of fringe amplitudes for many spatial frequencies. One can then fit a model to these data in order to extract the scientific parameters of interest. Three examples of such models, for a resolved symmetric uniform disk and two binary stars of different differential magnitude, are given in Figure 2.

Unfortunately, with single baseline measurements the phase information can not be calibrated and is normally ignored. Techniques for recovering phase information for multiple baseline measurements will be discussed in section 6.

3. Separated fringe packet astrometry

It is of course, not all bad news. There are numerous techniques for getting around the difficulty in calibrating visibility amplitudes. Consider, for example, the case when the stars in a binary system are far enough apart so that the fringe packets from each star do not overlap in delay space. For example, taking once again a filter centered at 1 μm with a 10% bandpass, the fringe packet width will be $\frac{1^2}{0.1} = 10$ μm, so on a 100m baseline this represents an angular separation of $10^{-5}/100 = 10^{-7}$ radians or about 20 mas. These numbers scale directly with wavelength and baseline. By measuring the separation of the two fringe packets in a scan one can derive a one-dimensional measurement of the

† One might say "typical" data.

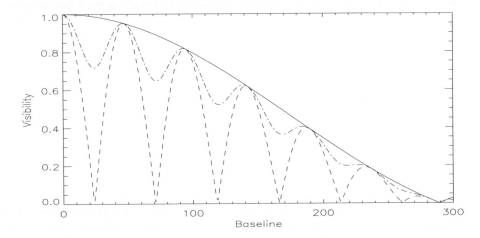

Figure 2. Examples of visibility functions in K band (2.3 μm): *Solid line*: Uniform disk of diameter 2 mas. *Dashed line*: Binary star, each star of diameter 2 mas with a separation of 10 mas and a differential magnitude of 0. *Dash-dot line*: Same binary star with a differential magnitude of 2.0.

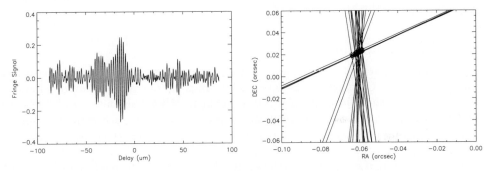

Figure 3. Examples of separated fringe packet data of HD 19332 taken at the Chara Array. *Left*: Single scan showing two fringe packets, one from the primary and one from the secondary. Note that since the primary is resolved, it is in fact the fringe packet on the left, the one with a lower fringe amplitude. Each scan of this type gives a single 1-D separation of the binary star on the sky. *Right*: Resulting astrometry from many scans on two baselines. Each line in this shows the possible position of the secondary with respect to the primary, and the intersection of these lines gives the full astrometric solution of $\rho = 64.0 \pm 1.8$mas and $\theta = 289 \pm 1.6°$ for this epoch.

separation of the binary system on a line parallel to the projected baseline. With more than one baseline one can get a full astrometric data point including ρ and θ, for that epoch. This technique was pioneered by Dyck *et al.* (1995) and an example data set from the CHARA Array for the star HD 19332 is given in Figure 3.

The precision of this technique depends, for the most part, on the measurement of the fringe packet separation. This in turn depends on the calibration of the system used to scan through the fringes, knowledge of the projected baseline, and the algorithm used to identify the center of the fringe envelope. In the case of the data shown in Figure 3 the fringe envelope center was found by fitting a Gaussian to the demodulated fringe amplitude and the scanning mirror calibration was good to about 1%. Since the position of the baseline is known to within 100 μm, or one part in 10^6 it has little effect on the final precision. Furthermore, since it takes a finite amount of time to scan from one fringe packet to the next, the atmosphere can move the fringes around further blurring the results.

There are several ways to avoid these difficulties. If you are lucky, the fringe packet of the secondary lands on one of the side lobes of the primary and in these cases the central wavelength of the filter itself becomes an excellent reference for phase (Bagnuolo, *et al.* 2006). In this case, the precision is dependent on the knowledge of this wavelength, so using the baseline and wavelength of our previous example, and assuming we know this wavelength to 5% the precision can be as high as $\frac{0.05^{-6}}{100} = 5^{-10}$ radians or about 100 μas, and on larger baselines much better than that. If you have more than one delay and beam combiner, the two stars in the system can be measured simultaneously, totally removing the effects of the atmosphere and even greater precision can be achieved (Muterspaugh, *et al.* 2006).

There is another way in which these separated fringe packet objects can be used to study binary stars. In a surprising number of cases, the "primary" is in fact itself a spectroscopic binary star. For objects of this type, you have the lucky coincidence of having a calibrator within the same scan as the object of scientific interest. So, rather than slewing between object and calibrator, it is possible to collect both calibration and science data in each single scan. This means that the calibration object is spatially and temporally very close to the science target and this not only more than doubles the data throughput but improves the calibration process considerably. Work is now underway to take advantage of this special class of object.

4. Visibility amplitude astrometry

If the fringe packets of the two stars in the system do overlap you must measure the fringe amplitude on a range of baselines and a range of epochs. For example, Figure 4, shows data from the Michigan Infra Red Combiner (MIRC, Monnier, *et al.* (2006)) at the CHARA Array using four telescopes, yielding six baselines, each with eight spectral channels. These data took about 20 minutes to collect and the binary star signature is clear yielding an instant ρ and θ measurement, as well as the diameters of both components. Using the phase closure techniques outlined in section 6, these data can also be used to create an image of the system.

Data of this type can be directly fitted to models of the binary star yielding astrometric results. Furthermore it is possible, indeed it is now standard practice, to produce a combined fit of the interferometric and radial velocity data resulting in a full three dimensional solution of the system. This has been the aim of interferometric binary stars studies for many years, and will certainly become a very powerful method for measuring stellar masses, and other fundamental parameters, in the years to come. A great deal of exciting work in this area has been done at the Palomar Testbed Interferometer (PTI Colavita *et al.* 1999) for example in Torres *et al.* 2002, a study of HD 195897, where they measured the mass of the primary to 2%, the secondary to 1% and have a factor of two better precision on the parallax than Hipparcos. This sort of precision is an upper limit of what can now be achieved and we should expect this to improve in the near future.

5. How should we present our results?

There are many in the community who would prefer us to report all our results in the form of ρ, θ and epoch, as has been traditionally done in binary star science. This simplifies cataloging of these data and makes it easier to combine them with other astronomical measurements. In many cases this is possible — separated fringe packets, dual

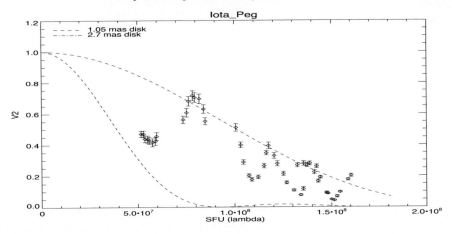

Figure 4. Sample data of the binary star ι Peg taken with the MIRC beam combiner at the CHARA Array in H band (1.6 μm). *Dashed lines*: Models of uniform disks of 1.05 and 2.7 mas. *Data points*: Measured visibilities on six baselines each with eight spectral channels.

star systems, slow moving systems, objects or system for which we can obtain a lot of data quickly, phase closure data (see section 6) — but in many cases it is not possible, nor even appropriate.

Since it is the binary stars with full double lined spectroscopic orbits we are most interested in, and these tend to have short periods, it isn't always possible to collect data quickly enough to solve for ρ and θ for a single epoch. By the time you have moved to a calibrator and returned to the object enough times to get an astrometric solution, the system can move significantly and a single astrometric data point no longer makes any sense. The same is true for an object for which you have a very dispersed data set and objects close to the resolution limit of the interferometer. In these cases you might as well combine the spectroscopic and interferometric data and solve for the complete orbit and stellar parameters all at once.

So how then do we report our results so that they can, at some future time, be combined with other measurements of the same object? The same way we have traditionally combined our data with radial velocity measurements. First of all, we could report our visibility amplitudes and baseline projects as raw data and an IAU standard format already exists for this kind of data exchange (Pauls *et al.* 2005). If it is possible to derive ρ and θ that should be done and reported. The same goes for the 1 dimensional data of separated fringe packet astrometry.

In the not-too-distant future we are all going to have to learn to deal with complex data sets that cross many boundaries between experimental methods. What is really needed is software that will combine radial velocity, ρ/θ, visibility amplitude, phase closure, lunar occultation and Ouija board data into a single orbital solution. This may be, as they say, an excellent exercise for a student.

6. Phase Closure and Imaging

So far we have only dealt with fringe amplitude data, having dismissed fringe phase as being totally washed out by the atmosphere. Fortunately, if we have three or more telescopes, we can borrow a technique developed for radio interferometry called phase closure (see again Thompson *et al.* 2001). If you add the phases of the fringe packets in

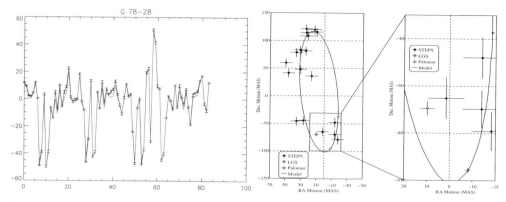

Figure 5. *Left*: Closure phase data on the Binary star G78-28 of data taken with an aperture mask at Palomar. *Right*: Orbit for G78-28 with data from STEPS (Pravdo *et al.* 2006), LGS data and two new points from aperture masking. Note the size of the error bars for the aperture masking data.

a closed circle of baselines you have the closure phase, something very similar to a bi-spectrum, and it turns out that the atmospheric phase errors are related to the telescopes and not the baselines and cancel out. While you are still throwing away some phase information, this yields a good measurable which is extremely sensitive to asymmetries in the object being studied. Since binaries stars are inherently asymmetric this technique is ideal for binary star astrometry.

The most impressive results to date using the technique of closure phase have been obtained by placing a mask on a large aperture telescope. These masks have many small holes of approximately $3r_0$ diameter in a two dimensional non redundant pattern. In this way, the telescope becomes an interferometer and is forced to have an extremely well defined point spread function. For example, Figure 5 shows on the left the raw and fitted closure phase signals measured at one epoch using a mask at the Palomar telescope (Pravdo *et al.* 2006). It is not possible to differentiate the raw and fitted signal as the fit is so good. On the right of Figure 5 is shown the resulting new orbit for this object combined with some data obtain with other techniques. The interferometric data are clearly the most precise and help to confine the orbit. This represents a λ/D result.

Phase closure data can, as it does in radio interferometry, also be used to create images. Because imaging requires many baselines and closure phases, this has been most clearly demonstrated with aperture masking. For example, Figure 6 shows images of the prototype pinwheel nebula WR 104 (Tuthill *et al.* 1999). This is a binary star with a period of 243.5 days and an internal separation of 1 mas with a Roche Lobe overflow precipitating the WR stage in this system.

Despite the relatively small number of apertures, ground base long baseline interferom-eters have been making images of binary stars for some time. The first of these was done by the COAST group in Cambridge England (Baldwin *et al.* 1996), followed closely by the NPOI group (Benson *et al.* 1997) and later the IOTA group (Monnier *et al.* (2004)) in the United States. More recently, interferometers with more than three apertures and multi-way beam combiners have, or will soon, come on line like Amber/VLTI (Malbet *et al.* (2006)) and MIRC/CHARA Array (Monnier, *et al.* (2006)), and not far off in the future the MROI (Creech-Eakman *et al.* (2006)). These new generation instruments combine many telescopes at once and can collect more data more quickly than single baseline instruments. For example, the MIRC/CHARA system currently combines four telescopes, with eight spectral channels for each baseline, resulting in 48 amplitude and

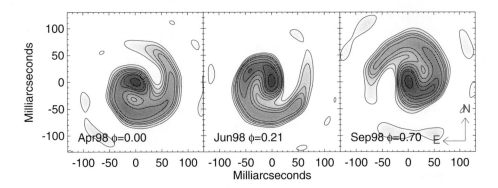

Figure 6. Images of WR 104 at three separate epochs taken using an aperture mask on one of the Keck telescopes. This is a binary star system and the motion of the spiral is clearly visible with time.

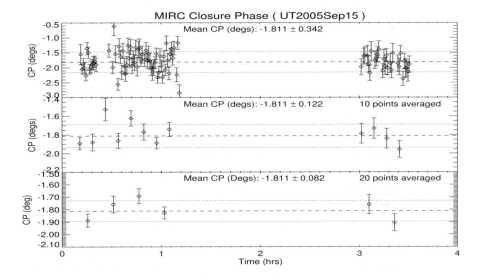

Figure 7. Closure phase measurements taken using the MIRC beam combiner at the CHARA Array. Re-binning and averaging of these data shows that the errors in these data are close to the limit required for planet detection.

32 closure phase measurements in a single data cycle. MIRC will, in 2007, be expanded to combine all six of the CHARA Array telescopes. High resolution imaging of binary stars, and more complex structures, will soon be common place.

As a final example, we can consider the use of an interferometer in the study of extra-solar planets. In order to directly detect the planet known to exist in the 51 Peg system and using the largest baselines at the CHARA Array, a closure phase precision of 0.05 degrees or better is needed. Figure 7 shows the first closure phase results using MIRC at CHARA in late 2005, and even with this early data set a precision of 0.082 degrees was achieved. It is not unreasonable to expect that direct planet detection from the ground using this sort of technique will one day be possible, even routine.

7. Conclusions

I have attempted in this short review, to cover the basics of the reduction of binary star data using a long baseline interferometer. This technique is not new, indeed it has been around for almost 100 years, but is only now beginning to live up to its long touted potential. It is the highest resolution technique available and will make visual binaries out of many double lined spectroscopic binaries, furthermore, it has an accuracy that is competitive with eclipsing systems and can provide more, and complimentary, information for these objects. There is also great potential for studies of tidal interactions. Brown dwarfs are now being measured with aperture masks and planet detection is not far away. It is my belief that long baseline interferometry will soon do for binary star astrometry what Speckle interferometry did in the seventies and eighties. It's obvious: if you have access to an interferometer you should be studying binary stars.

Acknowledgements

I would like to thank a number of people who provided essential material for this talk and paper: Dr. Andy Boden (Michelson Science Center), Dr. John Monnier (University of Michigan), Dr. Peter Tuthill (University of Sydney), Dr. Bill Bagnuolo (Georgia State University) and Dr. Hal McAlister (Georgia State University). This would have been an extremely dull talk without their help.

References

Bagnuolo, W.G., Jr., Taylor, S.F., McAlister, H.A. ten Brummelaar, T.A., Gies, D.R., Ridgway, S.T., Sturmann, J., Sturmann, L., Turner, N.H., Berger, D.H., & Gudehus, D. 2006, *ApJ* 131, 2695

Baldwin, J. E., Beckett, M. G., Boysen, R. C., Burns, D., Buscher, D. F., Cox, G. C., Haniff, C. A., Mackay, C. D., Nightingale, N. S., Rogers, J., Scheuer, P. A. G., Scott, T. R., Tuthill, P. G., Warner, P. J., Wilson, D. M. A., & Wilson, R. W. 1996, *A&A* 306, L13

van Belle, G.T., Ciardi, D.R., ten Brummelaar, T.A., McAlister, H.A., Ridgway, S.T., Berger, D.H., Goldfinger, P.J., Sturmann, J., Sturmann, L., Turner, N.H., Boden, A.F., Thompson, R.R., & Coyne, J 2005, *Apj* 637, 494

Benson, J.A., Hutter, D.J., Elias, N.M., II, Bowers, P.F., Johnston, K.J., Hajian, A.R., Armstrong, J.T., Mozurkewich, D., Pauls, T.A., Rickard, L.J., Hummel, C.A., White, N.M., Black, D., & Denison, C.S. 1997, *AJ* 114, 1221

Born, M & Wolf, E. 1999, *Cambridge University Press*, United Kingdom

ten Brummelaar, T.A, McAlister, H.A., Ridgway, S.T., Bagnuolo, Jr, W.G., Turner, N.H., Sturmann, L., Sturmann, J., Berger, D.H., Ogden, C.E., Cadman, R., Hartkopf, W.I., Hopper, C.H., & Shure, M.A. 2005, *ApJ* 628, 453.

Colavita, M.M., Wallace, J.K., Hines, B.E., Gursel,Y., Malbet, F., Palmer, D.L., Pan, X.P., Shao, M., Yu, J.W., Boden, A.F., Dumont, P.J., Gubler, J., Koresko, C.D., Kulkarni, S.R., Lane, B.F., Mobley, D.W., & van Belle, G. T. 1999, *ApJ* 510, 505.

Creech-Eakman, M.J., Bakker, E.J., Buscher, D.F., Coleman, T.A., Haniff, C.A., Jurgenson, C.A., Klinglesmith, D.A., III, Parameswariah, C.B., Romero, V.D., Shtromberg, A.V., & Young, J.S. 2006, *SPIE* 6268

Dyck, H.M., Benson, J.A., Schloerb & F.P. 1995, *AJ* 110, 1433

Lawson, P.R. (Ed) 1997, *Selected Papers on Long Baseline Stellar Interferometry* SPIE Milestone volume of papers covering stellar interferometry from 1868 to 1996.

Lawson, P.R. (Ed) 2000, *Principles of Long Baseline Stellar Interferometry* Course notes from the 1999 Michelson Interferometry Summer School

Malbet, F., Petrov, R.G., Weigelt, G., Stee, P., Tatulli, E., Domiciano de Souza, A., & Millour, F. 2006, *SPIE* 6268

McAlister, H.A., ten Brummelaar, T.A., Gies, D.R., Huang, W., Bagnuolo, W.G., Jr., Shure, M.A., Sturmann, J., Sturmann, L., Turner, N.H., & Taylor, S. F. 2005, *ApJ* 628, 439

Monnier, J.D., Traub, W.A., Schloerb, F.P., Millan-Gabet, R., Berger, J.P., Pedretti, E., Carleton, N.P., Kraus, S., Lacasse, M.G., Brewer, M., Ragland, S., Ahearn, A., Coldwell,

C., Haguenauer, P., Kern, P., Labeye, P., Lagny, L., Malbet, F., Malin, D., Maymounkov, P., Morel, S., Papaliolios, C., Perraut, K., Pearlman, M., Porro, I. L., Schanen, I., Souccar, K., Torres, G., & Wallace, G. 2004, *ApJ* 602, L57

Monnier, J.D., Pedretti, E., Thureau, N., Berger, J.P., Millan-Gabet, R., ten Brummelaar, T.A., McAlister, H.A., Sturmann, J., Sturmann, L., Muirhead, P., Tannirkulam, A., Webster, S., & Zhao, M. 2006, *SPIE* 6268

Muterspaugh, M.W., Lane, B.F., Konacki, M., Burke, B.F., Colavita, M.M., & Kulkarni, S.R., Shao, M. 2006, *A&A* 446, 723

Pauls, T.A., Young, J.S., Cotton, W.D., & Monnier, J.D. 2005, *PASP* 117, 1255

Pravdo, S.H., Shaklan, S.B., Wiktorowicz, S.J., Kulkarni, S., Lloyd, J.P., Martinache, F., Tuthill, P.G., & Ireland, M.J. 2006, *ApJ* 649, 389

Thompson, A.R., Moran, J.M., & Swenson, G.W., Jr. 2001, *John Wiley and Sons*, New York

Torres, G., Boden, A.F., Latham, D.W., Pan, M., & Stefanik, R.P. 2002, *AJ* 127, 1716

Tuthill, P. G., Monnier, J. D., & Danchi, W. C. 1999, *Nature* 398, 487

Discussion

NANCY EVANS: Can you say anything about (say) the fraction of triples as a function of mass? Let me challenge you. For well studied Cepheid binaries (massive stars) we find as many as half the binaries are triple.

TEN BRUMMELAAR: I apologize, I was obviously not clear enough. While we do have surveys underway for double and triple systems they are far from complete. My intention here was to point out that once you have found a triple system, you can take advantage of the fact that it can contain a calibrator and binary star in a single scan. I'm afraid I can not, as yet, give you an answer to your question.

TED GULL: Re: the assumption that the stars are spherical. Massive stars $> M_\circ$, with rotation, are predicted to be non-spherical. Have you looked at the these massive star systems for evidence of oblate geometry?

TEN BRUMMELAAR: Indeed we have, and checking that calibrator stars are not likely to be non-spherical is extremely important. Two of the first three CHARA Array scientific publications dealt with exactly this sort of object (McAlister *et al.* (2005) & van Belle *et al.* 2005), and it is likely to be a very hot topic in the field of interferometry for some time to come.

Binary Stars as Critical Tools & Tests
in Contemporary Astrophysics
Proceedings IAU Symposium No. 240, 2006
W.I. Hartkopf, E.F. Guinan & P. Harmanec, eds.

© 2007 International Astronomical Union
doi:10.1017/S1743921307004000

Close Binary Star Observables: Modeling Innovations 2003-06

R. E. Wilson

[1]Astronomy Department, University of Florida, Gainesville, FL, USA
email:wilson@astro.ufl.edu

Abstract. Innovative work on close binary models in 2003-06 improved upon synthesized line spectra, line profiles, and polarimetry; developed new ways of parameter estimation; and increased solution effectiveness and efficiency. Recent applications demonstrate the analytic power of binary system line spectrum models that pre-date the triennium. X-ray binary line profiles and radial velocity curves were refined by solution of the radiative transfer problem with specific inclusion of X-irradiation. Model polarization curves were generated by Monte Carlo experiments with multiple Thomson scattering in thin and thick binary system disks. In the parameter estimation area, independent developments by two groups now allow measurement of ephemerides, apsidal motion, and third body parameters from whole light and velocity curves, to supplement the traditional way of eclipse timings. Although the new route to those parameters is not well known within the ephemeris community, there are accuracy advantages and the number of applications is increasing. Numerical solution experiments on photometric mass ratios have checked two views of their intuitive basis, and show that mass ratios are well determined where star radii and limiting lobe radii are both well determined, which is for semi-detached or over-contact binaries with total-annular eclipses. Solution efficiency and automatic operation is needed for processing of light curves from large surveys, and will also be valuable for preliminary solutions of individually observed binaries. Neural networks have mainly been used for classification, and now a neural network program reliably finds preliminary solutions for W UMa binaries. Archived model light curves and Fourier fitting also are being pursued for classification and for preliminary solutions. Light curves in physical units such as $erg \cdot sec^{-1} \cdot cm^{-3}$ now allow direct distance estimation by combining the absolute accuracy of model stellar atmospheres with the astrophysical detail of a physical close binary model, by means of rigorous scaling between surface emission and observable flux. A Temperature-distance (T-d) theorem specifies conditions under which temperatures of both stars and distance can be found from light and velocity curves.

Keywords. radiative transfer, polarization, scattering, methods: analytical, stars: atmospheres, stars: distances, (stars:) binaries: eclipsing, (stars:) binaries (including multiple): close, (stars:) binaries: spectroscopic

---

1. Innovations in Overview

The three years following the Sydney General Assembly saw renewed progress in modeling of close binary star observables, partly spurred by large scale space and ground based observational programs. This review covers essential ideas for some of the more innovative work in the window 2003 to 2006, with mention of applications that demonstrate usefulness of earlier conceptual advances. The number of innovations precludes detailed examination of all areas, so one development, solution of light curves in standard physical units, is explained conceptually while others are described in terms of overall aims and motivations.

Disks are interesting in terms of their formation, structure and stability, and they occur in several binary star contexts as a challenge to modelers of light curves, line spectra, and time-dependent polarization. The variety of synthesized observables is increasing for stars as well as for disks, with notable attention to X-ray binaries. Examples of work on solutions include parameter estimation for third stars and for ephemerides, while numerical experiments have clarified the conceptual basis for photometric mass ratios. Solution efficiency and automatic operation are indispensible for surveys such as OGLE, ASAS, and Gaia, with neural networks and approximate fitting schemes now being adapted to several kinds of classification and to initial parameter estimates. Absolute flux solutions explore whether working in standard physical units might be advantageous, with the obvious connection being to distances.

2. Synthesized Observables

Models of 2003-06 demonstrated the power of earlier line spectrum syntheses, refined line profile and radial velocity (hereafter *RV*) curve computations, and produced signatures of disk polarization by Monte Carlo experiment.

2.1. *Line Spectra*

Whether for stars or disks, analysis of binary system line spectra is a difficult problem that stretches machine resources. However model line spectra can be powerful probes and, for non-eclipsing binaries, essentially the only probes of system astrophysics. The basic programming requirement is to attach model stellar atmosphere output to local star or disk surface elements, then integrate observable flux into the momentary line of sight throughout the spectrum, allowing for the same phenomena (eclipses, gravity effect, etc.) as for ordinary light curves. Computational and memory efficiencies and perhaps parallel processing will be needed to deal with the slowness of model atmosphere programs, with having another dimension (wavelength), with having sufficient resolution in that dimension, with smooth superposition of local Doppler shifts, and with having additional parameters. Chemical abundances can be a lure toward parameter proliferation, so hard decisions on parameterization must be made. A line spectrum generator based on Kurucz (1998) atmospheres is embedded within Prša & Zwitter's (2005a) user-friendly interface (called *PHOEBE*). Several applications (e.g., Hoard *et al.* 2004 and Linnell *et al.* 2005) on cataclysmic variable MV Lyrae; Hoard *et al.* (2005) on magnetic white dwarf YY Draconis; Linnell *et al.* (2006) on the non-eclipsing double-contact† binary V360 Lacertae) demonstrate effectiveness of the binary spectrum and light curve program by Linnell & Hubeny (1994, 1996) that is based on Hubeny (1990, 1991) atmospheres. Earlier spectral computations for disks and stars are cited by Linnell & Hubeny. Parameter extraction has so far been by trial and error. Trials on single stars and single star models can guide reasonable parameter choices in chemical abundances, damping, microturbulence, and perhaps even differential rotation.

2.2. *X-ray Binary Line Profiles and Radial Velocity Curves*

Problems of binary line profile distortions due to tides and irradiation and their effect on *RV*'s go back at least to Sterne (1941), with effects due to eclipses extending back to Schlesinger (1909, 1916) and Rossiter (1924), and with newer references in Wilson & Sofia (1976) and the Wilson (1994) review. The essential *RV* phenomena are included in several public *EB* observables models. Antokhina, Cherepashchuk & Shimanskii (2005) have now refined X-ray binary line profiles and *RV* shifts by solving the radiative transfer problem with specific treatment of X-radiation. Line distortions caused by stellar winds

---

† The double-contact morphological type is defined in Wilson (1979).

were modeled by Abubekerov, Antokhina & Cherepashchuk (2004), with applications to the High Mass X-ray Binaries LMC X4, Cen X3, SMC X1, Vela X1, and 4U 1538−52, particularly including consequences for mass estimates.

2.3. *Polarization*

By far the main present impediment to progress on time-dependent binary star polarization is lack of observations, as modelers are unlikely to work in an area devoid of data. Among the few observational papers over the years, a particularly illogical practice has been tabulation in terms of orbital phase† (with the whole cycles removed) instead of time. As some polarization phenomena are episodic rather than periodic, time records are essential – and require no more journal space than phases. A good start on the observational side was made with polarization curves of Algol by Kemp *et al.* (1983), but to date there have been no extensions of time-dependent polarization to fainter stars, either by refinement of polarimeters or by use of large telescopes. However Hoffman, Whitney & Nordsieck (2003) made exploratory radiative transfer computations of polarization due to multiple Thomson scattering in illuminated thin and thick circumstellar binary system disks. They give extensive tabular and graphical results of Monte Carlo simulations that follow radiative transfer in internal and external radiation fields. The disks are defined geometrically by a central opening angle and a radius. The parameter definition problem will need development when there are observations for applications.

3. Parameter Estimation

3.1. *Whole Light and Velocity Curve Ephemerides and Extension to Third Bodies*

The idea of finding *EB ephemerides from whole light and RV curves* is more than a decade old, yet remains nearly unknown in the "timing diagram" community, although good demonstrations of improvement over eclipse timing ephemerides (for comparable time spans) continue to appear. The same can be said for the extension to third bodies. Indeed, recent otherwise excellent accounts and reviews of *EB* ephemeris work (*e.g.* Kreiner *et al.* 2001; Rovithis-Livaniou 2005) do not mention the idea. Accordingly, inclusion within this review may foster awareness, whose present lack is likely due to the concept's rather quiet introduction in two independent developments – so quiet that the early conceptual work can be traced only through applications, with the two camps becoming aware of each others work only recently. Ephemeris parameters are a heliocentric reference time (HJD_0), the period at the reference time (P_0), the rate of period change (dP/dt), and the rate of advance ($d\omega/dt$) of the argument of periastron. The reference time may refer to periastron or to conjunction. For third bodies we have another example of the "Astronomy of the Invisible" that began with Neptune's discovery and continues through extra-solar planets, supermassive black holes in galactic centers, and Universal dark matter. Third body information is in Doppler shifts of the *EB* and in phase excursions of light curves and *RV* curves. Third body (subscript 3b) parameters include $HJD_{0,3b}$, $P_{0,3b}$, e_{3b}, ω_{3b}, and orbital semi-major axis, a_{3b}. Essential to the procedure is to fit multiple curves in *time* rather than phase. Some applications are to light curves, some to *RV* curves, and others to light and *RV* curves combined. The main difficulty for a third body is identification of the correct orbit period among many aliases that follow from the typical large data gaps in astronomical data. Power spectral analyses

---

† Publication of phases in place of time also has become distressingly common for light curves. Although light variation is typically more nearly periodic than polarization, phased data cannot be used for ephemeris work and have greatly reduced value for investigation of cycle to cycle and epoch to epoch changes. Referees and editors should be aware of this problem.

of several kinds may find correct periods in reasonably favorable circumstances and the situation is helped by *simultaneous* solution of light and *RV* curves, as the combined data types partly fill gaps. The numerous successful applications demonstrate the power and adaptability of the overall process.

Applications on one side include Mayer *et al.* (1991) [HJD_0, P_0, and $d\omega/dt$ from *RV*'s and light curves of the *EB* V1765 Cygni]; Harmanec & Scholz (1993) [HJD_0, P, and dP/dt from a century of *RV*'s for β Lyrae†]; Tarasov *et al.* (1995) [HJD_0 and P from *RV*'s of the non-eclipsing triple system ϵ Persei]; Harmanec *et al.* (2004) [P and dP/dt for ER Vulpeculae, a solar-type *EB*]; Janik *et al.* (2003) [$d\omega/dt$ for the well-detached *EB* V436 Persei]; and Horn *et al.* (1996) [$HJD_0, dP/dt, d\omega/dt, HJD_{0,3b}$, and P_{3b} from *RV*'s of the non-eclipsing triple system 55 Ursae Majoris]. The Horn *et al.* third body results for 55 UMa agree accurately with speckle (*i.e.* visual binary) parameters by McAlister, Hartkopf & Franz (1990). All of the above-mentioned applications are by means of the combination light-*RV* program by P. Hadrava, with the basic idea having been briefly mentioned in Hadrava (1990), and are by the Simplex algorithm, as formulated by Kallrath & Linnell (1987). A guide to specifics is in Hadrava (2004).

Ephemeris solutions for the parallel development are by the Differential Corrections (*DC*) algorithm, with the basic logic and mathematics and many references on applications (starting with Elias *et al.* 1997) in Wilson (2005). The extension to third bodies is in Van Hamme & Wilson (2005, 2007). Basic to the *DC* version is that ephemeris parameters are found along with all other parameters in a general solution, and with standard errors, rather than separately in an initial step. The [HJD_0, P, dP/dt, $d\omega/dt$] solution facility has been in the public *W-D* program since 2003.

3.2. *Photometric Mass Ratios*

Incorrect published remarks about photometric mass ratios (q_{ptm}) probably outnumber correct ones. A common misconception is that q_{ptm}'s mainly derive from ellipsoidal (*i.e.* tidal) variation, whereas they mainly derive from relative radii ($r_{1,2} = R_{1,2}/a$) in relation to limiting lobe radii. The origins of photometric mass ratios are recounted in Wilson (1994) for the logically distinct cases of semi-detached (*SD*) and over-contact (*OC*) binaries. For *SD*'s, q_{ptm} follows from the condition of the contact star's mean radius matching its mean lobe radius, which is a definite function of q. A solution constraint that one star accurately fills its limiting lobe is required to exploit the *SD* condition properly. The situation is slightly more complicated for *OC*'s, where *two* radii are involved. There the essential q_{ptm}-related quantity is R_2/R_1, the ratio of mean star radii, which depends strongly on q and relatively weakly on over-contact level (f), with the relation being inverted to find $q(R_2/R_1, f)$. Since R_2/R_1 and f are measurable from light curves, q_{ptm} naturally follows. Many *OC*'s are only slightly over-contact so their q relation approximately reduces to $q(R_2/R_1)$. With radii the links to q_{ptm} and total-annular eclipses the link to strong measures of radii, *completely eclipsing OC* 's and *SD*'s should have the strongest q_{ptm}'s. Indeed, experiments (Terrell & Wilson 2005) find correct and fast-converging q's within standard error expectations for *OC*'s and *SD*'s from solutions of noisy synthetic light curves with total-annular eclipses (*viz.* Fig. 1).

The q_{ptm} concept basically does not apply to detached binaries (*DB*), except that a weak q estimate can sometimes be found by graphing variance against q.

---

† Harmanec & Scholz conclude that β Lyr's ephemeris can be derived more accurately by whole curve *RV* fitting than by times of eclipse minima.

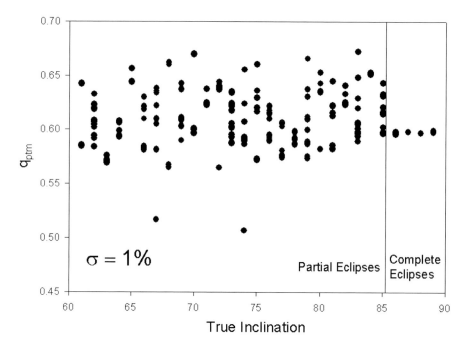

Figure 1. Dramatic q_{ptm} accuracy improvement in passing from partial to total-annular eclipses as the orbital inclination is stepped in 1° increments (Terrell & Wilson 2005). Results are from *DC* solutions of noisy synthetic *OC* light curves for a true mass ratio of 0.60. The standard deviation of the Gaussian noise is 1%.

4. Solution Efficiency and Automatic Operation

Although schemes for automatic light curve classification and for starting parameter generation have been stimulated by existing and anticipated large databases, applications to individually observed binaries also can profit from artificial intelligence and other automation, thereby saving computational as well as human time and eliminating incorrect solutions before publication. Competition should identify the leading programs for preliminary solutions and for classification, and also suggest ways to extract their best features for building later generations of programs.

4.1. Neural Networks

Neural networks have been utilized for automatic classification by Sarro, Sanchez-Fernandez & Gimenez (2006), and for starting parameters in the *Eclipsing Binary Artificial Intelligence* (*EBAI*) project by Devinney *et al.* (2006). Sarro *et al.* briefly summarize previous neural network usage in astronomy. Aims of their program are to separate pulsating stars from *EB*'s and to classify *EB*'s into four categories according to characteristics such as eclipse depths and widths. Sarro *et al.* then statistically relate their four categories to the morphological types of *DB*, *SD*, and *OC* via numerical experiments on 81 systems from the literature. *EBAI* has so far been tried on synthetic W UMa light curves, with added noise, and seems impressively reliable.

4.2. Archived Model Light Curves

The idea of Archived Model Light Curves (*AMLC*) for starting parameter generation is to store large numbers of theoretical light curves over parameter ranges for comparison with observed curves. The central point is that even with millions of archived synthetic light curves, computation of variance on modern computers is simple and goes very fast. Of course, the theoretical and observational data are definite numbers and require no model computations, once an archive has been generated. *AMLC* goes back at least to Wyithe & Wilson (2001, 2002), where it was tried on the full dataset of *OGLE EB*'s in the Small Magellanic Cloud with good results, based on some thousands of archived curves. Now Kallrath & Wilson (2007) are developing archives with millions of curves. Efficient data packing algorithms and substantial computing times in the generation step are needed for such large archives.

4.3. *Fourier Fitting*

The long history of Fourier series applications to light curves has continued in the recognition of *EB*'s among other variables in work by Groenewegen (2004, 2005) and in morphological *EB* classification (*DB, SD, OC*) by Pojmanski (2002), who applied ideas on cosine series fitting by Rucinski (1993). Although slightly before the 2003-2006 triennium, Pojmanski's work is incorporated in analysis of *ASAS*, the All Sky Automated Survey (Paczynski *et al.* 2006) that has observed light curves of more than 10,000 *EB*'s, many being discoveries. Although the scheme's true uncertainties remain to be established, its potential for efficient impersonal classification appears promising.

4.4. *Words of Caution*

All procedures for starting parameter estimation and for classification should be tested on *noisy synthetic data*, as that is the only way to know true values and morphological types for comparison with procedure results. Most schemes have not been so tested and can give false impressions of reliability (*EBAI* is an exception, as it has been so tested). Tests on real binaries have considerably reduced value as their characteristics are known only approximately, with some published solutions even being in local minima of parameter space and therefore wrong. The common practice of testing against real data can likely be traced to recognition that models can and do have shortcomings. However one must remember that a model is involved either way, whether the test data are real or synthetic, as adopted parameters for real binaries are based on a model.

4.5. *A User-friendly Interface*

Among several programs constructed over recent decades to serve as user-friendly interfaces to *EB* light curve and solution programs, that by Prša & Zwitter (2005a) is especially well designed and multi-faceted. It offers a variety of solution algorithms (*DC*, Simplex, others), generates line spectra with broadened profiles, and conveniently makes pictures and graphs. It also introduces a conceptual innovation (*viz.* Prša & Zwitter 2005a) by utilizing the color index information in *standardized* light curves to allow solution for a second component temperature. Of course, tradition is to set the temperature of one star from spectra or other evidence and solve light curves for the other. Another innovation is to compensate the effect of interstellar extinction on light curves (a problem of finite bandwidth) so as to reduce or eliminate corresponding solution errors (Prša & Zwitter 2005b).

5. Absolute Light Curves and Distance

That light curves need not be in standard physical flux units to yield much of their astrophysical information has been recognized from the earliest work on *EB* analysis, but the question arises: might there be advantages in working with "absolute" light curves? A recently suggested answer (Wilson 2006, 2007) invokes the following 3-part idea:

> 1: The absolute emission accuracy of modern model stellar atmospheres can now be combined with the structural detail of *EB* models.
> 2: Conventional synthetic light curves can be made absolute by simple but rigorous scaling (*not* global scaling for spherical stars).
> 3: Parts 1 and 2 together allow distance to be an ordinary solution parameter by *DC* or another algorithm (Direct Distance Estimation = *DDE*), and with a standard error. A side benefit is that bandpass luminosities can directly be produced in standard physical units or in solar luminosities. Potential benefits are uniform and accurate distance statistics, and improved thruput (reduced human work). Experience shows that the procedure works as least as well (actually better) for *SD*'s and *OC*'s as for *DB*'s, so there no longer is any reason to limit *EB* distance targets to *DB*'s.

Model stellar atmospheres accurately give absolute emission, so they allow computation of *EB* light curves for comparison with observed light curves, which can be made absolute via the Johnson (1965) or Bessell (1979) calibrations. These calibrations differ by only 4% in *U, B,* and *V*, with the Johnson/Bessell ratio essentially the same in the three bands, so corresponding derived distances (*i.e.* Johnson *vs.* Bessell) differ by only 2%. Neither active model atmosphere computation nor storage of theoretical spectra nor integration over photometric bands is needed because those steps have already been taken via the Van Hamme & Wilson (2003) Legendre polynomial representation of model atmosphere emission. Observable flux over bandwidth in cm can be in erg $\cdot$ sec$^{-1}\cdot$ cm^{-3}. The bolometric luminosity parameter, L_{bol}, of traditional distance scaling is in erg$\cdot$sec^{-1}, but L_{bol} – being emission over all directions and all wavelengths – is not observable in practice. An impediment to accuracy in traditional L_{bol}-distance scaling is that author-dependent bolometric corrections are needed. The *DDE* idea is to work from the observational side with the directly observable quantity $F(t)$ (time dependent flux in a given band), and from the model side with local intensity I in erg $\cdot$ sec$^{-1}\cdot$ sr$^{-1}\cdot$ cm^{-3} that becomes integrated into theoretical $F(t)$. Thus Part 1 (above) can be realized, but let us ask whether *local* introduction of *cgs* units – an annoying programming problem – is really necessary. Part 2 says that it is not, provided that one starts from a model such as *W-D* that is fully consistent in regard to radiative units. Even the oldest *W-D* versions produce observable fluxes (output) and local intensities (internal quantities) that correspond rigorously, unit-wise, to bandpass luminosity (input). So the user's choice of luminosity unit fixes the flux unit, the program operates in user-defined flux units, and a readily computable scaling factor converts *W-D* flux from user-defined to *cgs* units. Computation of that factor might be done in several ways, with a straightforward way being to scale from normal emergent intensity at a surface reference point (Wilson 2003, 2005). A convenient reference point is one of the poles. The scaling relation naturally involves the star-observer distance, d, and is

$$F_d^{\text{abs}} = 10^{-0.4\text{A}} \left[F_{a,1}^{\text{prog}} \left(\frac{I_1^{\text{abs}}}{I_1^{\text{prog}}} \right) + F_{a,2}^{\text{prog}} \left(\frac{I_2^{\text{abs}}}{I_2^{\text{prog}}} \right) \right] \left[\frac{a}{d} \right]^2 , \qquad (5.1)$$

where F_d is flux at the observer's location, F_a is flux at distance a (the orbital semi-major axis length), I is polar normal emergent intensity, and A is bandpass interstellar extinction in stellar magnitudes. Of course the units of a and d need only be the same. Superscripts *abs* and *prog* mean "absolute" and "program" and subscripts $1, 2$ denote the binary components. The computation is thereby easy, yet brings the sophistication of stellar atmosphere and *EB* models (tides, irradiation, gravity effect, etc.) to bear on the problem.

5.1. *A Temperature-distance Theorem*

In traditional *EB* solutions, where the flux unit is comparison star flux, the idea is to set one temperature from "external" information (*e.g.* spectra) and find the other from light curves (essentially from relative eclipse depths). Distance can be estimated separately from a standard magnitude measured outside eclipse, corrected for interstellar extinction. The *DDE* way is to make full use of the absolute flux and color information in *standardized* light curves to find distance (d) and the second temperature, and with standard errors.† A Temperature-distance (*T-d*) theorem (Wilson 2006) specifies conditions under which various combinations of $[T_1, T_2, d]$ can be measured: *Temperatures of both stars and distance can be found objectively from standard light curves in 2 or more bands* (*e.g. U, B, V*, etc., not differential). A solution of only one standard light curve must sacrifice one of the three parameters, thus finding $[T_1, T_2]$, $[T_1, d]$, or $[T_2, d]$ while assuming the third parameter. A solution of three or more curves will be over-determined in $[T_1, T_2, d]$, with consequent biases that may or may not be significant, depending on calibration consistencies in particular bands, on assumed chemical composition, and on accuracy of the adopted model stellar atmospheres. The logical basis is explained in Wilson (2006) and numerical experiments that agree with the theorem's predictions will be described later.

Acknowledgements

I am pleased to thank Prof. P. Harmanec for alerting me to the work by P. Hadrava on ephemerides by whole light and *RV* curve fitting, and for supplying references to several applications papers. Thanks are also due to W. Van Hamme for discussions and other input and to D. Terrell for his contribution of Figure 1, which originally appeared as Figure 4 of Terrell & Wilson (2005), and is reprinted with the kind permission of Springer Science and Business Media. The work was supported by U.S. National Science Foundation grant 0307561.

References

Abubekerov, M.K., Antokhina, E.A., & Cherepashchuk, A.M. 2004, *Astr. Reports* 48, 89

Antokhina, E.A., Cherepashchuk, A.M., & Shimanskii, V.V. 2005, *Astr. Reports* 49, 109

Bessell, M.S. 1979, *PASP* 91, 589

Devinney, E.J., Guinan, E.F., Bradstreet, D., DeGeorge, M., Giammarco, J., Alcock, C., & Engle, S. 2006, *BAAS* 37, 1212

Elias, N.M., Wilson, R.E., Olson, E.C., Aufdenberg, J.P., Guinan, E.F., Guedel, M., Van Hamme, W.V., & Stevens, H.L. 1997, *ApJ* 484, 394

Groenewegen, M.A.T. 2004, *A&A* 439, 559

Hadrava, P. 2004, WWW site http://www.asu.cas.cz/%7Ehad/fotel.html

Harmanec, P., Božić, H., Thanjavur, K., Robb, R.M., Ruždjak, D., & Sudar, D. 2004, *A&A* 415, 289

Harmanec, P., Scholz, G. 1993, *A&A* 279, 131

Hoffman, J.L., Whitney, B.A., & Nordsieck, K.H. 2003, *ApJ* 598, 572

Horn, J., Kubát, P., Harmanec, P., Koubský, P., Hadrava, P., Šimon, V., Štefl, S., & Škoda, P. 1996, *A&A* 309, 521

Hubeny, I. 1990, *ApJ* 351, 632

† In §4.5 we have already seen the Prša & Zwitter idea for finding two temperatures from standardized light curves.

Hubeny, I. 1991, in: C. Bertout, S. Collin-Souffrin & J.P. Lasota (eds.), *Proc. IAU Colloq. 129* (Gif-sur-Yvette: Editions Frontières, Singapore: Fong & Sons), p. 227

Janík, J., Harmanec, P., Lehmann, H., Yang, S., Božić, H, Ak, H., Hadrava, P., Eenens, P., Ruždjak, D., Sudar, D., Hubeny, I., & Linnell, A.P. 2003, *A&A* 408, 611

Johnson, H.L. 1965, *Comm. Lunar & Planetary Lab.* 3, 73

Kallrath, J. & Linnell, A.P. 1987, *ApJ* 313, 346

Kemp, J.C., Henson, G.D., Barbour, M.S., Kraus, D.J., & Collins, G.W. 1983, *ApJ* 273, L85

Kreiner, J.M., Kim, C.H., & Nha, I.S. 2001, "An Atlas of O-C Diagrams of Eclipsing Binary Stars", (Krakow: Wydawnictwo Naukowe Akademii Pedagogicznej)

Kurucz, R.L. 1998, in Proc. IAU Symp. 189, ed. T.R. Bedding, A.J. Booth, & J. Davis (Dordrecht: Kluwer), p. 217

Linnell, A.P., Szkody, P., Gansicke, B., Long, K., Sion, E.M., Hoard, D.W., & Hubeny, I. 2005, *ApJ* 624, 923

Linnell, A.P., Harmanec, P., Koubský, P., Božić, H., Yang, S., Ruždjak, D., Sudar, D., Libich, J., Eenens, P., Krpata, J., Wolf, M., Škoda, P., & Šlechta, M. 2006, *A&A* 455, 1037

Linnell, A.P. & Hubeny, I. 1994, *ApJ* 434, 738

Linnell, A.P. & Hubeny, I. 1996, *ApJ* 471, 958

Mayer, P., Hadrava, P., & Harmanec, P. 1991, *Bull. Astr. Inst. Czech.* 42, 230

McAlister, H.A., Hartkopf, W.I., & Franz, O.G. 1990, *AJ* 99, 965

Paczynski, B., Szczygiel, D.M., Pilecki, B., & Pojmanski, G. 2006, *MNRAS* 368, 1311

Pojmanski, G. 2002, *Acta Astr.* 52, 397

Prša, A. & Zwitter, T. 2005a, *ApJ* 628, 426

Prša, A. & Zwitter, T. 2005b, *Ap&SS* 296, 315

Rossiter, R.A. 1924, *ApJ* 60, 15

Rovithis-Livaniou, H. 2005, *Ap&SS* 296, 91

Rucinski, S. 1993, *AJ* 105, 1433

Sarro, L.M., Sanchez-Fernandez, C., & Gimenez, A. 2006, *A&A* 446, 395

Schlesinger, F. 1909, *Publ. Allegheny Obs.* 1, 123

Schlesinger, F. 1916, *Publ. Allegheny Obs.* 3, 23

Sterne, T. 1941, *Proc. Nat. Acad. Sci. (U.S.)* 27, 168

Tarasov, A.E., Harmanec, P., Horn, J., Lyubimkov, L.S., Rostopchin, S.I., Koubský, P., Blake, C., Kostunin, V.V., Walker, G.A.H., & Yang, S. 1995, *A&AS* 110, 59

Terrell, D. & Wilson, R.E. 2005, *Ap&SS* 296, 221

Van Hamme, W. & Wilson, R.E. 2003, in: U. Munari (ed.), *Gaia Spectroscopy, Science and Technology* (San Francisco: ASP), vol. 298, p. 323

Van Hamme, W. & Wilson, R.E. 2007, *ApJ*, 662, in press

Van Hamme, W. & Wilson, R.E. 2005, *Ap&SS*, 296, 121

Wilson, R.E. 1979, *ApJ* 234, 1054

Wilson, R.E. 1990, *ApJ* 356, 613

Wilson, R.E. 1994, *PASP* 106, 921

Wilson, R.E. 2005, *Ap&SS*, 296, 197

Wilson, R.E. 2006, *Proc. Seventh Pacific Rim Conference on Stellar Astrophysics*, ASP Conf. Ser. vol. 362, 3

Wilson, R.E. & Devinney, E.J. 1971, *ApJ* 166, 605

Wilson, R.E. & Sofia, S. 1976, *ApJ* 203, 182

Wyithe, J.S.B. & Wilson, R.E. 2001, *ApJ* 559, 260

Wyithe, J.S.B. & Wilson, R.E. 2002, *ApJ* 571, 293

Discussion

PETR HARMANEC: Bob, I find it appropriate to mention that many things you were describing in your talk with reference to studies dated 2005–2007, such as including time derivatives, including apsidal motion into solutions, modeling triple star motion or dereddening, were successfully realized in Petr Hadrava's program FOTEL already early in the nineties (Hadrava 1990, 2004).

WILSON: Thanks for the information. Foundations of the ephemeris work that I described go back to 1997, but I was not aware that Petr Hadrava's program had those capabilities since 1990. I will reference his work in the review's printed version. It seems that we have two independent developments.

EDGARD SOULIE: In order to calculate third order corrections, did you resort to the technique branded <<Automatic differentiation of algorithms>>, which is implemented efficiently? Proceedings of conferences were published notably by Corliss & Griewank (1991) and Faure, Griewank, & Hascoet (2000).

WILSON: Do you mean "third *body* corrections"? Anyway, the derivations were found by hand, just differentiating the functions.

CARLSON CHAMBLISS: $O - C$ data on eclipsing binary minima were often used to "discover" 3rd components. Some of these are spurious. Can new procedures be used to determine which are real and which are spurious?

WILSON: Yes, and our work (Van Hamme & Wilson 2007) finds that in one "classical" triple system (very well known), the third body doesn't seem to be there.

Binary Stars as Critical Tools & Tests
in Contemporary Astrophysics
Proceedings IAU Symposium No. 240, 2006
W.I. Hartkopf, E.F. Guinan & P. Harmanec, eds.

Line-Profile Variations on Massive Binary Systems: Determining η Carinae Orbital Parameters

D. Falceta-Gonçalves[1], Z. Abraham[1] and V. Jatenco-Pereira[1]

[1]Instituto de Astronomia, Universidade de São Paulo, Rua do Matão 1226, CEP 05508-900, São Paulo, Brazil
email: diego@astro.iag.usp.br

Abstract. When the winds of two massive stars orbiting each other collide, an interaction zone is created consisting of two shock fronts at both sides of a contact surface. During the cooling process, elements may recombine generating spectral lines. These lines may be Doppler shifted, as the gas stream flows over the interaction zone. To calculate the stream velocity projected into the line of sight we use a simplified conical geometry for the shock fronts and, to determine the synthetic line profile, we have to sum the amount of emitting gas elements with the same Doppler shifted velocity. We show that the stellar mass loss rates and wind velocities, and the orbital inclination and eccentricity, are the main parameters on this physical process. By comparing observational data to the synthetic line profiles it is possible to determine these parameters. We tested this process to Brey 22 WR binary system, and applied to the enigmatic object of η Carinae.

Keywords. binaries: general, Wolf-Rayet, winds, line: profiles

1. Introduction

Wolf-Rayet (WR) stars are believed to be at the end of the evolutionary history of massive stars. For this reason, they play a key role in the stellar evolution theories. Typically, they present fast winds (~ 3000 km s^{-1}) with extreme mass-loss rates, ranging from 10^{-6} to 10^{-4} M$_\odot$yr^{-1} (Lamers 2001). WR stars are also known as variable stars, due to strong variability in their spectral lines. This variability is believed to be originated by wind clumpiness. However, in some objects, the spectral lines present periodic variability, possibly revealing duplicity (Bartzakos, Moffat & Niemela 2001). In some objects, these lines present anomalous behaviour, varying not only their intensity, but their velocity and profile and, eventually, presenting a 2-peaked profile.

In Falceta-Gonçalves, Abraham & Jatenco-Pereira (2006), we presented a model to explain the periodic variability of the line profiles for WR binary systems. We assumed that these lines are not generated in the stellar wind, but in the shock region between the winds of both stars. In a WR+O system, both stars present high velocity winds, leading to a strong shock region, which can be understood as a high temperature and density gas. In Figure 1 (left panel) we show the contact surface (S$_C$), calculated as the momentum equilibrium region for the ratio $\eta = \dot{M}_2 v_2/\dot{M}_1 v_1 = 0.1$. Neglecting the curvature due to the orbital motion of the secondary star, the shock becomes asymptotically conical with an opening angle $\beta \simeq 44°$. The dashed line represent the line of sight intercepting two fluid elements of the cone, resulting in two different Doppler-shifted velocities. This explains the double-peaked profiles in some WR stars.

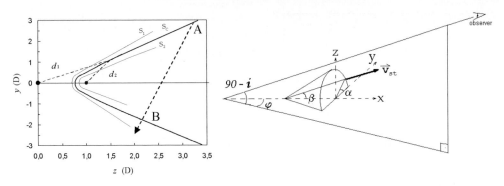

Figure 1. Schematic view of the shock surfaces and the assumed geometry.

2. The Model

To calculate the stream velocity projected into the line of sight we use the schematic geometry shown in the right panel of Figure 1. The stream velocity projected over the line of sight, as shown in Figure 1, is given by:

$$\mathrm{v}_{\mathrm{obs}} = \mathrm{v}_{\mathrm{flow}}\left(-\cos\beta\cos\varphi\sin i + \sin\beta\cos\alpha\sin\varphi\sin i - \sin\beta\sin\alpha\cos i\right), \qquad (2.1)$$

where $\mathrm{v}_{\mathrm{flow}}$ is the stream velocity, i is the inclination of the orbital plane with respect to the line of sight, α is the cone azimuthal angle and φ is the orbital phase, defined here as the angle between x and the projection of the line of sight on the orbital plane. As these shocks are strongly radiative, they present high turbulence amplitudes. To account for the line broadening we used Equation 2.2:

$$I(\mathrm{v}) = \mathcal{C}\int_0^\pi \exp\left[-\frac{(\mathrm{v}-\mathrm{v}_{\mathrm{obs}})^2}{2\sigma^2}\right]d\alpha, \qquad (2.2)$$

where $\mathcal{C}$ is the normalization constant and $\sigma = \delta\mathrm{v}_{\mathrm{turb}}/\mathrm{v}_{\mathrm{flow}}$. Substituting the mean value $\mathrm{v}_{\mathrm{obs}}$, given by Equation 2.1, and integrating the right-hand side of the equation over α, we obtain the relative line intensity for each projected velocity into the line of sight v. In the computation of the syntetic line profiles we also take into account the optical depth of the shock region. If $\tau \gg 1$, the redshifted peak is absorbed, resulting in a single peaked profile.

3. The case of η Carinae

η Carinae is one of the most interesting objects in our Galaxy. Famous by the eruptions in the 19th Century, it presents complex light curves at all frequencies. Damineli (1996), using the HeI 10830 Å line light curve, first determined the periodic behaviour of this object. He found the system to have a period of 5.52yr and, using the HeII 4686 Å (Steiner & Damineli 2004) an eccentricity of ~ 0.8, assuming that the line is produced by the wind of the primary at 500 km s^{-1}, and that the periastron occurs in opposition.

However, Falceta-Gonçalves, Abraham & Jatenco-Pereira (2005) analysing the X-ray data, and Abraham *et al.* (2005a,b) using radio light curves, determined that the orbit should have a higher eccentricity ($e = 0.9 - 0.95$) and that the periastron should occur near conjuntion. In order to verify which orbital model is correct, we decided to apply the last to reproduce the HeII 4686 Å used by Steiner & Damineli (2004). Also, instead of assuming the line to be generated by the wind of the primary star, we assume that the

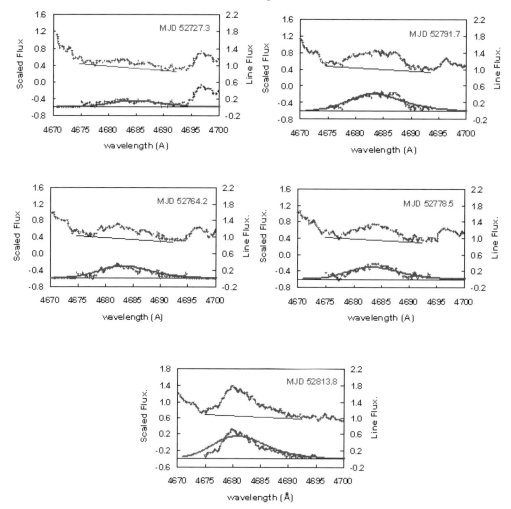

Figure 2. For different epochs, the observed HeII 4686 line profiles (up) and (bottom) the subtracted data and the adjust from the present model (red line).

line is generated at the shock cone. We calculated the synthetic line profiles for different orbital phase angles, and identified the best fitting of each of them to the more complete set of data from Martin & Davison (2006).

In Figure 2 we show the observational data from Martin & Davison (2006), and the best fitting results for $i = 90°$, $\beta = 56°$, $\sigma = 0.2$, $v_{flow} = 450$km s^{-1} and $\tau \gg 1$. To determine the orbit shape, we have to determine the phase angle at each orbital phase. From the fitting of the observed epochs, we found an eccentricity $e = 0.95$, and we determined that the periastron, supposed to occur in 2003 June, 28th, occurs with an angle of $\sim 30°$ regarding the conjunction. These results are in full agreement with the previous works on X-rays and radio frequencies.

Acknowledgements

D.F.G thanks FAPESP (No. 04/12053-2) for financial support. Z.A. and V.J.P. thank FAPESP, CNPq and FINEP for support.

References

Abraham, Z., Falceta-Gonalves, D., Dominici, T.,Caproni, A., & Jatenco-Pereira, V. 2005a, *MNRAS*, 364, 922.

Abraham, Z., Falceta-Gonalves, D., Dominici, T., Nyman, L.-A., Durouchoux, P., McAuliffe, F., Caproni, A., & Jatenco-Pereira, V. 2005b, *A&A*, 437, 977.

Bartzakos, P., Moffat, A. F. J., & Niemela, V. S. 2001, *MNRAS*, 324, 33.

Damineli, A. 1996, *ApJ*, 460, 49.

Falceta-Gonçalves, D., Abraham, Z., & Jatenco-Pereira, V. 2005, *MNRAS*, 357, 895.

Falceta-Gonçalves, D., Abraham, Z., & Jatenco-Pereira, V. 2006, *MNRAS*, 371, 1295.

Lamers, H. J. G. L. M. 2001, *PASP*, 113, 263.

Martin, J. C., Davidson, K., Humphreys, R. M., Hillier, D. J., & Ishibashi, K. 2006, *ApJ*, 640, 474

Binary Stars as Critical Tools & Tests
in Contemporary Astrophysics
Proceedings IAU Symposium No. 240, 2006
W.I. Hartkopf, E.F. Guinan & P. Harmanec, eds.

© 2007 International Astronomical Union
doi:10.1017/S1743921307004024

Evidence for a Precessing Disk in the Extreme Binary ε Aurigae

Robert E. Stencel[1]

[1]Department of Physics and Astronomy, University of Denver,
2112 E. Wesley Ave., Denver CO 80208, USA
email: rstencel@du.edu

Abstract. Among the longest known eclipse durations and binary periods is that of the star ε Aurigae, which exhibits 2-year long eclipses every 27.1 years. Oddly, the nature of the secondary in the system continues to elude ready identification. In 1965, Huang proposed a massive disk as the eclipsing body, and study of the 1984 eclipse led Lissauer and Backman to suggest an embedded B star binary in the disk to maintain it. A collaboration of observers allows me to present recent optical photometry and spectroscopy, near-IR spectroscopy and Spitzer space telescope IRS and MIPS observations of ε Aurigae as it approaches its next eclipse. These data argue for current detectability of the embedded binary, and precession of the disk axis, suggesting a radical change is possible for the next mid-eclipse brightening. An international monitoring campaign for the 2009-2011 is being organized, and participation invited via website http://www.du.edu/~rstencel/epsaur.htm .

Keywords. stars: epsilon Aur, binaries, eclipsing, supergiant, circumstellar matter, accretion disks, infrared, photometry, polarimetry, interferometry

1. Introduction

The prevailing hypothesis concerning the nature of the long period eclipsing binary HD 31964 (ε Aurigae) features a (contact?) B star binary embedded in a massive, 20 AU diameter cold disk [475K], all of which orbits an F-type supergiant star (Carroll *et al.* 1991). Total system mass is approximately 29 solar masses, with an orbital separation of 27.6 AU and period of 27.1 years. Flat-bottomed eclipses of two years duration and 0.7 mag depth optically suggest that the cold disk covers half the surface area of the F star (Huang, 1965). The next eclipse is predicted to start in 2009. Kemp *et al.* (1986) analyzed polarimetry of the 1984 eclipse and argued that the disk is inclined 2 to 5 degrees from its orbital plane. Together with a central eclipse brightening that has varied over the past 3 eclipse events, disk tilt could signal precession of its orientation. Disk precession could also affect view-ability of the disk-centered B-star binary. New observations are being sought to test this possibility.

Key to testing this model is assembly of an observational spectral energy distribution (SED). In addition to UBV photometry described below, recent non-eclipse near and mid-IR spectra have been obtained with MIMIR and Spitzer instruments, respectively. These fluxes seem at odds with IRAS photometry reported by Backman and Gillett (1985) during the last eclipse. The prominent 475K IR excess appears to have been replaced by a hydrogen emission line source and power law extension of the mid-IR spectrum.

2. Recent Observations

Optical photometry in the UBV system has been resumed by J. Hopkins at Hopkins Phoenix Observatory as of 2003 (www.hposoft.com/Astro/PEP.html), which supplements

Table 1. SPITZER Space Telescope observations of ε Aur

Date	MJD	Instrument	Mode
2005 Sep 25	53639	MIPS	PHOT and SED
2005 Oct 19	53663	IRS	Stare/HiRes
2006 Feb 23	53790	MIPS	PHOT and SED
2006 Mar 17	53812	IRS	Stare/HiRes

his 1982-88 coverage of the past eclipse. From these recent UBV photometry, he found the mean V magnitude was 3.05 with a cyclic variation range of 0.132 magnitude, having a basic period of 66.2 days (based on Peranso/ANOVA software analysis). V band also shows a brightening trend over the 2003-2006 seasons. Low amplitude variability of ε Aur out of eclipse has long been known (Shapley 1928), and its detectability may be orbital phase dependent. Mean B magnitude is 3.625 with a variation of 0.175 magnitude and similar period. Mean U band is 3.725 with a variation of 0.35 magnitude and similar period. The B-V and U-B colors both show a redder trend, ranging from 0.54 to 0.58 magnitude and 0.10 to 0.14 magnitude, respectively, corresponding to F8 supergiant colors, comparable to SED fitting results. The B-V colour is significantly redder than the canonical F0 I spectral type assigned ε Aur, and may represent spectral evolution during the 20th century.

A minor revolution in CCD cameras and small spectrometers allows accomplished amateurs to obtain spectra with modest telescopes. Both Mais and Schanne have reported variations in Hα profiles, including transient double peaked emission wings, in spectra obtained during 2004-2006 (cf. Hopkins 2006). Mais utilizes an SBIG spectrometer/ST10XME on a C-14/Paramount, yielding 0.4 Åper pixel resolution. Schanne's home-built spectrometer on a 5 inch Maksutov-Newtonian achieves somewhat higher resolution. Schanne reports variation of the intensities of the wings. The red wing is increasing, while the blue one is decreasing with time, during spring 2005.

New JHKLM region spectra were obtained with the 1-5 micron MIMIR instrument on 20 January 2006 [MJD 53756] by D. Clemens and A. Pinnick with the 1.83 m Perkins telescope near Flagstaff, Arizona, with NSF PREST program sponsorship. The spectra were obtained during an interval when V band magnitude was approaching a local minimum in brightness. The spectra show many lines of neutral hydrogen, including Brα [5-4, weakly in emission], Pfγ [5-8, in absorption] and many of the Hu series [6-n, strongly in absorption] between 3.35 and 4.15 microns. Unidentified emission features are seen at 3.87, 3.89 and 3.99 microns.

Clemens noted that hydrogen lines dominate the spectra, with helium and metals absent. Bracketα is in emission, all others are in absorption – including the rest of the Brackett series, the Humphreys series, and the maybe an occasional Pfund line. The only stars in a paper by Hanson, Conti and Rieke (1996) that show Brα emission without He I 2.05 micron emission are X-ray binary stars (mass transfer systems). Clemens reports that Paschenβ is absent, which is normally a very strong line in other sources (in absorption in stars and emission in PN, both from prior Mimir observations). The Paβ line could be optically thick (perhaps self-absorbed, like Lyα).

Under Spitzer cycle 2 observing time, I obtained IRS high resolution mid-IR spectra and MIPS photometry of ε Aur on the following dates (Table 1): The spectra show a continuum with weak emission lines, consistent with the MIMIR spectra. The MIPS photometry reveals some variation between observational epochs. These data are being combined to establish an overall spectral energy distribution. A complete report is being prepared for publication elsewhere.

Table 2. Predicted contact times for the 2009-2011 eclipse of ϵ Aur

Contact	MJD	Date	Duration
First	55050	2009 Aug 06	137 d
Second	55187	2009 Dec 21	223 d
Mid	55410	2010 Jul 09	...
Third	55633	2011 Mar 11	223 d
Fourth	55897	2011 May 15	64 d

While the opaque cold disk remains the working hypothesis to explain the eclipse behavior and apparent infrared excess during the last eclipse, Dana Backman (private communication) has been advocating a bright spot model to account for behavior between eclipses – namely, a heated sub-stellar point on the disk, facing the F supergiant. Visibility of this bright spot will vary with orbital phase, and possibly other factors. Details regarding these sets of observations, and updates, can be found at the webpage mentioned below.

3. An international campaign for the 2009-11 eclipse

Given the remarkable suite of astronomical facilities on earth and in orbit, it should be feasible to obtain excellent multi-wavelength coverage of the upcoming eclipse phenomena. Table 2 shows the predicted times of eclipse, based on analysis of previous eclipsing timings, by J. Hopkins (private communication). We have set up a campaign website to enlist collaborators and collate results, much as we did for the prior eclipse cycle (Stencel 1985). The website is: http://www.du.edu/~rstencel/epsaur.htm, and we invite interested persons to share in this multi-generational adventure.

Acknowledgements

I am grateful to my collaborators in this effort, who deserve credit but not blame for theories advanced here: Tom Ake, Dana Backman, Dan Clemens, Ed Guinan, Jeff Hopkins, Dale Mais and Lothar Schanne. Some of this work is based on data obtained with the Spitzer Space Telescope and was supported in part by NASA through contract agreement 1275955 issued by the Jet Propulsion Laboratory, California Institute of Technology. I am also grateful to the estate of William Herschel Womble for support of astronomy at the University of Denver, and to my colleague, Toshiya Ueta, for helpful comments on this maunscript.

References

Carroll, S., Guinan, E., McCook, G., & Donahue, R. 1991, *Ap. J.* 367, 278

Hanson, M., Conti, P., & Rieke, M. 1996, *Ap.J.Sup.* 107, 281

Hopkins, J. 2006, http://www.hposoft.com/Astro/PEP/EAUR/EAurSpect.html

Kemp, J., Henson, G., Kraus, D., Beardsley, I., Carroll, L., & Collins, G. 1986, *Ap. J.* 300, L11

Shapley, H. 1928, *Harvard Obs. Bull.* 858, 5

Stencel, R.E. 1985 (ed.), *The 1982-1984 Eclipse of Epsilon Aurigae (Washington DC: NASA Conf. Publ. 2384)*.

Binary Stars as Critical Tools & Tests
in Contemporary Astrophysics
Proceedings IAU Symposium No. 240, 2006
W.I. Hartkopf, E.F. Guinan & P. Harmanec, eds.

© 2007 International Astronomical Union
doi:10.1017/S1743921307004036

New Findings Supporting the Presence of Several Distinct Structures of Circumstellar Matter in β Lyræ

P. Chadima[1], P. Harmanec[1,2], H. Ak[3], O. Demirçan[4], S. Yang[5], P. Koubský[2], P. Škoda[2], M. Šlechta[2], M. Wolf[1], H. Božić[6], D. Ruždjak[6] and D. Sudar[6]

[1]Astronomical Institute of the Charles University, Faculty of Mathematics and Physics,
V Holešovičkách 2, CZ-180 00 Praha 8, Czech Republic
email: pavel.chadima@gmail.com, hec@sunstel.asu.cas.cz, wolf@cesnet.cz
[2]Astronomical Institute of the Academy of Sciences, CZ-251 65 Ondřejov, Czech Republic
email: koubsky(skoda,slechta)@sunstel.asu.cas.cz
[3]Department of Astronomy & Space Sciences, Faculty of Arts & Sciences, Erciyes University,
38039 Kayseri, Turkey, email: hasan@physics.comu.edu.tr
[4]Department of Physics, Faculty of Sciences and Arts, Çanakkale Onsekiz Mart University,
17100 Çanakkale, Turkey, email: demircan@comu.edu.tr
[5]Department of Physics and Astronomy, University of Victoria, P.O. Box 3055 STN CSC,
Victoria, B.C., Canada V8W 3P6, email: yang@uvastro.phys.uvic.ca
[6]Hvar Observatory, Faculty of Geodesy, Kačićeva 26, 10000 Zagreb, Croatia
email: hbozic(dsudar,rdomagoj)@geof.hr

Abstract. (1) 52 photographic and 651 electronic spectra were disentangled using the program KOREL. This led to the detection of a number of weak absorption lines originating in the atmosphere of the accretion disk. So far, the detection of this disk spectrum was only reported for the Si 6347 and 6371 Å doublet. (2) The basic spectrophotometric quantities of 15 absorption lines of the primary were measured and corrected for the orbital light changes in order to eliminate the contribution of the secondary light to the observed continuum level. After the correction, a significant phase dependency of the spectral-line characteristics near the primary eclipse was detected. In all probability, another so far unknown absorption-line spectrum was thus found. This spectrum may be due to additional absorption of light of the primary in one of the jet-like structures and in the spherical gas envelope surrounding the accretion disk. Both these structures are seen projected against the primary during its eclipse.

Keywords. stars: individual (β Lyr), stars: variables: eclipsing, binaries: close, circumstellar matter, accretion disks

1. Introduction

β Lyr is eclipsing and interacting binary which is at the stage of high mass-transfer rate between the components. Circumstellar matter most probably consists of a geometrically and optically thick accretion disk which entirely hides the light from the secondary, jet-like structures perpendicular to the orbital plane and a scattering gas envelope surrounding the disk around the secondary. (See Harmanec 2002 and references therein for the full history of investigation of β Lyr).

In the present study, we used two sets of spectra – 52 "blue" digitized photographic spectra secured at the Ondřejov Observatory, Czech Republic and 651 "red" electronic spectra, secured at Ondřejov and at the Dominion Astronomical Observatory, Canada.

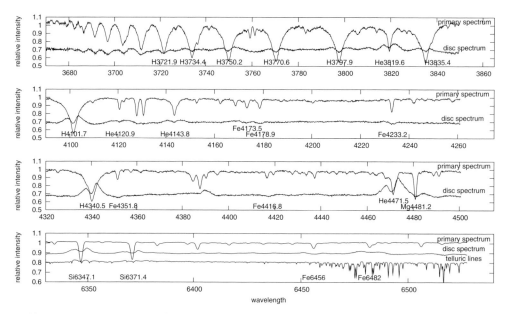

Figure 1. Disentangling of four spectral regions. Emission lines, seen in the disk spectra, originate in fact in the jet-like structures but our attempt to disentangle them separately failed.

2. Disentangling of spectra

Four selected spectral regions were disentangled using the KOREL program (Hadrava 2004). The results of the disentangling are displayed in Figure 1. In each panel, the primary spectrum is shown on the top and the disk spectrum below it. For electronic spectra longward of about 6300 Å, it was also necessary to include the telluric lines into the solution.

Note that the disentangling procedure led to the discovery of a number of weak absorption lines originating in the pseudophotosphere of the accretion disk. This way, we obtained a rich line spectrum of the disk, not limited to the previously known Si II doublet. Moreover, the radial velocities from the accretion-disk spectra define well a sinusoidal curve in antiphase to that of the primary.

3. Spectrophotometry of absorption lines

Central intensity (CI) and equivalent width (EW) of 15 stronger absorption lines were measured and further proceeded. These lines belong to the primary but their measurements refer to the common continuum of both, the primary and the accretion disk. The measured quantities must therefore be corrected to the continuum of the primary star only. The correction equations are as follows

$$CI_{corr} = 1 - \frac{F_p(\lambda, f) + F_d(\lambda, f)}{F_p(\lambda, f)}(1 - CI_{obs}), \quad EW_{corr} = \frac{F_p(\lambda, f) + F_d(\lambda, f)}{F_p(\lambda, f)} EW_{obs}.$$

Subscript *obs* means observational and subscript *corr* corrected values. F_p and F_d denote the relative monochromatic fluxes of the primary and of the accretion disk at given wavelength λ and orbital phase f. To evaluate this correction factor, it is necessary to use a light-curve solution which takes into account the presence of the thick accretion disk. To this end, we used the latest version of the BINSYN program (Linnell 2000).

In Figure 2, there are the phase plots of CI and EW of Si II 6371 line. Note a very pronounced line strengthening in the phases around the primary eclipse. The same phase

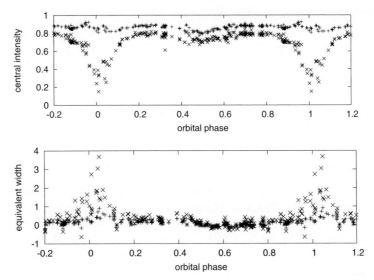

Figure 2. Phase dependency of central intensity and equivalent width of Si II 6371 line. Observed values are shown as +, corrected values as ×.

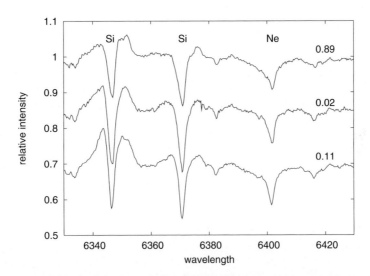

Figure 3. Three spectra around the primary eclipse. Orbital phases are given on the right side.

dependence was found also for all other investigated lines. These changes are too large to be caused by physical or geometrical properties of the primary. This effect can be also demonstrated in Figure 3. Note that the absorption lines in phases near the primary mid-eclipse are deeper which contradicts the fact that most of the primary surface is hidden from the view in this interval of phases.

It seems that the only plausible explanation is the following one: During the eclipse, there is always a certain part of the primary which remains uneclipsed by the disk and the light from it passes through both, the gaseous envelope and one of the jets which leads to formation of additional absorption lines. It means that there is another absorption-line spectrum observable only in the phases of the primary eclipse.

4. Conclusion

Both analyses led to the discovery of a new set of spectral lines – the accretion disk spectrum and the "shell" spectrum originating at the jet and/or the scattering envelope above the disk, providing additional support to the current model of β Lyr.

Acknowledgements

We acknowledge the use of the programs SPEFO, BINSYN and KOREL, made available by their authors J. Horn & J. Krpata, A.P. Linnell & I. Hubeny and P. Hadrava. Our special thanks go to A.P. Linnell for his patient advice concerning the proper use of BINSYN. Some of the Ondřejov spectra were obtained by P. Hadrava, D. Holmgren, J. Horn, J. Kubát, V. Šimon and S. Štefl. We profitted from the use of electronic bibliography maintained by the NASA/ADS system. Czech authors were supported from the research plans J13/98: 113200004 of Ministry of Education, Youth and Sports and AV 0Z1 003909, from project K2043105 of the Academy of Sciences of the Czech Republic and from the grants GA ČR 205/02/0788 and 205/06/0304 of the Czech Science Foundation. Turkish authors were partly supported by the Research Council of Turkey and the Research Foundation of Çanakkale University. Collaboration between Turkish and Czech authors was facilitated thanks to a NATO collaborative linkage grant PST.CLG.979343.

References

Hadrava, P. 2004, *Publications of Astron. Inst. of Academy of Sciences of the Czech Rep.* 92, 15
Harmanec, P. 2002, *Astronomische Nachrichten* 323, 87
Linnell, A.P. 2000, *MNRAS* 319, 255

Binary Stars as Critical Tools & Tests
in Contemporary Astrophysics
Proceedings IAU Symposium No. 240, 2006
W.I. Hartkopf, E.F. Guinan & P. Harmanec, eds.

© 2007 International Astronomical Union
doi:10.1017/S1743921307004048

Poster Abstracts (Session 3)

(Full Posters are available at http://www.journals.cambridge.org/jid_IAU)

Abundances from Disentangled Component Spectra: High-mass Stars in the Eclipsing Binaries CW Cep and V478 Cyg

K. Pavlovski and E. Tamajo
Department of Physics, University of Zagreb, Zagreb, Croatia

Techniques for the spectral disentangling of the composite spectra of close binaries, in combination with light-curve analysis, make it possible to reconstruct individual component spectra. In an on-going project we are making photospheric elemental abundance determinations for high-mass stars in OB binaries. The principal goal of the project is an observational test of rotational evolutionary stellar models of high-mass stars, and calibration of the photospheric abundance changes in terms of the stellar mass, the fractional life-time on the Main Sequence, and the rotational velocity.

In the present work two more OB binaries, CW Cep and V478 Cyg, with masses of the components in the range of 12–$15\,M_\odot$ were analyzed. Results would be discussed in the context of our previous studies in this series, and compared to the theoretical predictions.

Accretion Disks of Binary Stars as a Probe of Accretion Disks in Other Astrophysical Objects

R.N. Eze
University of Nigeria, Nsukka, Enugu, Nigeria

The understandings of the processes that lead to accretion in astrophysical objects have been a central problem for some decades now. However, we have a more detailed knowledge of binary disks accretion process because, first there are large number of such systems within 100 pc; and such proximity allows fainter spectral features to be analyzed. Secondly, the primary star in a binary can sometimes be used as probe, eclipsing portions of the disks at different times, thereby allowing disks properties to be mapped. Finally, accretion disks occur over a time scale convenient for observational monitoring and much can be leaned by studying the evolving spectra. In this work an attempt has been made to use the results from spectral analyses of the accretion disks of cataclysmic variables to make generalization on accretion disks of other astrophysical objects. For the 15 cataclysmic variables studied there is sufficient evidence to show that their accretion disks are highly ionized. This result supports the fact that the angular momentum transport which lead to accretion is purely by magneto rotational instability if the accretion disks is adequately ionized.

A Potential Stabilizing Magnetic Effect for Astrophysical Jets

M.V. Viallet and H. Baty
Astronomical Observatory, Strasbourg, France

Observations show that astrophysical jets are able to propagate over huge distances in comparison with their radial extents. This challenges theoretical and numerical studies, which predict a fast disruption of the jet by the development of internal magnetohydrodynamic instabilities. Particularly important is the Kelvin-Helmoltz instability, known to be the most dangerous for the jet's collimation. Here we use the modern finite-volume based MHD code, VAC, to reinvestigate the development of the Kelvin-Helmholtz instabilities by using high-resolution spatial simulations. We show that an initial large-scale disruption occurs after a propagation distance of a few tens of jet radius, in agreement with previous studies. In addition, we establish the existence of a new mechanism driven by the initial disruption of the jet. This mechanism gives rise to a local reinforcement of the magnetic field, which in turn leads to a new more stable configuration of the jet.

Comparison of Different Spectral Disentangling Techniques Applied to a Triple System

K.B.V. Torres[1,2], L.P.R. Vaz[1], and H. Hensberge[2]
[1] *Dept. de Fisica, ICEx, Universidade Federal de Minas Gerais, Belo Horizonte, Brazil,*
[2] *Royal Obervatory of Belgium, Brussels, Belgium*

Spectral disentangling allows us to reconstruct the contribution of multiple stars to a composite spectrum. Different algorithms have been proposed, which fall in two categories: algorithms operating on the Fourier components of the spectra and algorithms operating on the velocity bins (logarithm of wavelength). Almost all applications use the former algorithms because of public availability of software code and the computing time involved. Nevertheless, the simultaneous use of both techniques would give a better insight in the error budget, since each algorithms is sensitive in a different way to uncertainties in the input data and the options available to assign weights to the input data are different.

We will discuss the application of both techniques to a triple system, consisting of a dominant stationary third component dominating the composite spectrum and a close binary with one component much fainter than the other. Several spectral regions are selected, some of which contain telluric lines. When necessary for a better understanding, simpler artificial spectra are also analysed by both techniques.

An Expanded Bandpass List for Atmospheric Emission in Eclipsing Binary Models

W. Van Hamme[1] and R. E. Wilson[2]

[1] Department of Physics, Florida International University, Miami, FL, USA,
[2] Astronomy Department, University of Florida, Gainesville, FL, USA

Programs for modeling binary star observables compute emergent intensity for a given composition as it varies with local effective temperature, local gravity, and direction. With the arrival of huge data sets from Gaia and other surveys, the benefits of fast, compact, and accurate computation of atmospheric radiation is likely to remain critical for the foreseeable future. Experience has shown that accurate radiative modeling is important for good parameter estimation. Here we augment the radiative treatment by Van Hamme & Wilson (2003, ASP Conf. Ser. 298, *Gaia Spectroscopy, Science and Technology*, ed. U. Munari (San Francisco: ASP), 323) with a procedure by which individuals can generate the needed Legendre coefficients for arbitrary photometric bands. Resulting files can be inserted directly into the Wilson-Devinney program without sacrifice of portability or program unity, and should easily be adaptable to other binary star programs. We expect the new bandpass options to become part of the public W-D program. Limb-darkening tables will be placed at `http://www.fiu.edu/~vanhamme/limdark.htm`.

HD 61273: a New Semi-Detached Binary with an Accretion Disk

F. Royer[1], P. North[2], D. Briot[1], G. Burki[3], and F. Carrier[4]

[1] Observatoire de Paris, Meudon, France,
[2] Laboratoire d'Astrophysique - EPFL, Sauverny, Switzerland,
[3] Observatoire de Genève, Sauverny, Switzerland,
[4] Instituut voor Sterrenkunde, Leuven, Belgium

A detailed spectroscopic and photometric analysis shows for the first time that HD 61273 is a close binary system, composed of a dwarf early-type star and a K0 giant.

The system underwent and perhaps still undergoes a mass transfer. The most evolved star fills its Roche lobe and transfers mass to the hotter companion which is now the most massive component of the system. The Hα line displays a variable emission: signature of an accretion disk. This new semi-detached system is not eclipsing, but only shows photometric ellipsoidal variability due to the elongated shape of the giant component filling its Roche lobe, with a 12.919 day period.

Doppler Imaging of Stars with Roche-Geometry

Zs. Kövári[1], J. Bartus[2], K. Oláh[1], K.G. Strassmeier[2], J.B. Rice[3], M. Weber[2], and E. Forgács-Dajka[4]

[1]*Konkoly Observatory, Budapest, Hungary,*
[2]*Astrophysical Institute Potsdam, Potsdam, Germany,*
[3]*Brandon University, Brandon, Canada,*
[4]*Eötvös University, Dept. of Astronomy, Budapest, Hungary*

Tests are carried out on retrieving Doppler maps from distorted stars in close binaries to estimate how Doppler imaging may be aliased by the ellipticity. Maps obtained for the distorted shape are compared with the results of the simple spherical approximation, using real data of the RS Cvn-type close binary star ζ Andromedae.

Hydrodynamic Simulations of Illuminated Secondary Atmospheres in Dwarf Novae

M. Viallet and J.M. Hameury

Astronomical Observatory, Strasbourg, France

Dwarf novae (DN) are a subfamily of cataclysmic variable, i.e., binary stellar systems where a white dwarf accretes matter through an accretion disc from a stellar companion. They undergo regular outbursts lasting a few days with a recurrence time of a few tens of days. The outburst phenomenon is best interpreted as being due to a thermal/viscous limit cycle of the accretion disc. The general framework of the disc instability is now well understood, but some points are still unclear. In particular, the possibility that the illumination of the secondary atmosphere leads to an enhancement of the mass transfer rate is still under debate. We present here the first 2D hydrodynamic simulations of the surface flows in the secondary that result from the strong inhomogenous heating of the atmosphere during outbursts. We also discuss the possibility that the L1 point, from which matter leaves the secondary, is directly heated by the disc rim. During an outburst this could in turn contribute to the mass transfer enhancement.

Properties of Circumstellar Dust in Symbiotic Miras

D. Kotnik-Karuza[1], T. Jurkic[1], and M. Friedjung[2]

[1]*Department of Physics, University of Rijeka, Croatia,*
[2]*Institut D'Astrophysique de Paris, Université Pierre & Marie Curie, Paris, France*

We present a study of the properties of circumstellar dust in symbiotic Miras during sufficiently long time intervals of minimal obscuration. The published $JHKL$ magnitudes of o Ceti, RX Pup, KM Vel, V366 Car, V835 Cen, RR Tel and R Aqr have been collected. In order to investigate their long-term variations, we removed the Mira pulsations to correct their light curves. Assuming spherical temperature distribution of the dust in the close neighbourhood of the Mira, the DUSTY code was used to solve the radiative transfer in order to determine the dust temperature and its properties in each particular case. The preliminary results of this systematic study of dust envelopes in symbiotic stars with Miras as cool components provide information on nature of dust in these objects.

MIDI Observations of IRCs : Constraining the Geometry of the Warm Circumstellar Environment

S. Correia[1], Th. Ratzka[2], G. Duchene[3], and H. Zinnecker[1]
[1] *Astrophysikalisches Institut Potsdam (AIP), Potsdam, Germany,*
[2] *MPIA, Heidelberg, Germany,*
[3] *LAOG, Grenoble, France*

Despite more than a decade of investigations, the nature of Infrared Companions (IRCs) is still a matter of debate. While the hypothesis that IRCs could be in an earlier evolutionary stage than their primaries implies that they are embedded in an optically thick envelope, recent high spectral resolution near-infrared spectroscopy would rather favor the scenario of IRCs being normal T Tauri stars seen through an almost edge-on disk. We will report on recent high-spatial resolution interferometric observations of the IRCs Glass-I, Haro 6-10 and VV CrA obtained in the Mid-IR with MIDI/VLTI which provide further insights into the geometry of their dusty environment and will in general contribute to a better understanding of this intriguing class of objects.

Window to the Stars

R.G. Izzard and E. Glebbeek
University of Utrecht, Utrecht, Netherlands

We present a graphical user interface to the popular TWIN stellar evolution code. It removes the drudgery associated with the traditional approach to running the code, while maintaining the power, output quality and flexibility a modern stellar evolutionist requires.

Spectral Disentangling and Combined Orbital Solution for the Hyades Binary θ^2 Tau

P. Lampens, Y. Frémat, P. De Cat, and H. Hensberge
Koninklijke Sterrenwacht van Belgie, Brussel, Belgium

Theta^2 Tau is a detached, "single-lined" binary and the most massive resolved spectroscopic binary of the Hyades cluster. It also shows a complex pattern of pulsations of type δ Scuti. Its eccentric orbit (e=0.7) has a period of 140.7 days. Component B is very difficult to detect spectroscopically. For this reason, the few radial velocities published in the literature are inaccurate. From recent high-resolution spectroscopic data obtained with the ELODIE spectrograph at the OHP (France), we derived accurate radial velocities for *both* components, applying a spectral disentangling algorithm (by P. Hadrava, 1999). We then combined these measurements with available very-large baseline interferometric data in order to improve the knowledge of the orbital parameters and derived fundamental properties in a self-consistent way. Such determination is also very pertinent to revisit the evolutionary status of both components.

RR Lyrae in M15: Fourier Decomposition and Physical Parameters

G. Garcia Lugo[1], A. Arellano Ferro[2], and P. Rosenzweig[1]
[1] *Universidad de Los Andes, Mérida, Venezuela,*
[2] *Instituto de Astronomía, Universidad Nacional Autónoma de México, D F, Mexico*

In the present study, V and R images of M15 were obtained in two seasons during the year 2000 and 2001, respectively. These images were taken using the 1.5-m telescope of San Pedro Mártir Observatory, in Baja California, México. The telescope was equipped with a CCD Tektronix of 1024×1024 pixels with a size of $24\mu^2$.

In this work, results of CCD photometry are reported for 33 known RR Lyrae stars in M15. The periodicities of some variables have been recalculated and new ephemeredes are given. The Blazhko effect, reported previously for the V12 star, has not been detected. Through the use of the technique of Fourier decomposition of the light curves, the physical parameters of the type RRab and RRc variables have been estimated. The cluster is Oosterhoff type II and the values determined for the iron content and the distance are [Fe/H] = -1.98 ± 0.24 y and d = 8.67 ± 0.41 kpc, respectively. The mean values of the physical parameters determined for the RR Lyrae stars, place the cluster correctly in the sequences Oosterhoff type – metallicity and metallicity – effective temperature, valid for globular clusters. These sequences suggest that the origin of the Oosterhoff dichotomy is of evolutive nature.

Session 4: New Observing Techniques and Reduction Methods:

Observing in the Era of Large-Scale Surveys

Figure 1. (top left to bottom right) Speakers Dimitri Pourbaix, Andrej Prša, Tsevi Mazeh, David Koch, Panos Niarchos, and Tomaž Zwitter.

Figure 2. (top left to bottom right) Speakers Sandrine Thomas, Deepak Raghavan, Carla Maceroni, and Ralph Neuhäuser.

Figure 3. (left to right) Kosmas Gazeas, Andrej Prša, and Alceste Bonanos.

Binary Stars as Critical Tools & Tests
in Contemporary Astrophysics
Proceedings IAU Symposium No. 240, 2006
W.I. Hartkopf, E.F. Guinan & P. Harmanec, eds.

Pipeline Reduction of Binary Light Curves from Large–Scale Surveys

Andrej Prša[1,2] and Tomaž Zwitter[2]

[1]Villanova University, Dept. of Astronomy, 800 Lancaster Ave, Villanova, PA 19085, USA

[2]University of Ljubljana, Dept. of Physics, Jadranska 19, SI-1000 Ljubljana, EU
emails: andrej.prsa@fmf.uni-lj.si, tomaz.zwitter@fmf.uni-lj.si

Abstract. One of the most important changes in observational astronomy of the 21st Century is a rapid shift from classical object-by-object observations to extensive automatic surveys. As CCD detectors are getting better and their prices are getting lower, more and more small and medium-size observatories are refocusing their attention to detection of stellar variability through systematic sky-scanning missions. This trend is aditionally powered by the success of pioneering surveys such as ASAS, DENIS, OGLE, TASS, their space counterpart Hipparcos and others. Such surveys produce massive amounts of data and it is not at all clear how these data are to be reduced and analysed. This is especially striking in the eclipsing binary (EB) field, where most frequently used tools are optimized for object-by-object analysis. A clear need for thorough, reliable and fully automated approaches to modeling and analysis of EB data is thus obvious. This task is very difficult because of limited data quality, non-uniform phase coverage and solution degeneracy. This paper reviews recent advancements in putting together semi-automatic and fully automatic pipelines for EB data processing. Automatic procedures have already been used to process Hipparcos data, LMC/SMC observations, OGLE and ASAS catalogs etc. We discuss the advantages and shortcomings of these procedures.

Keywords. methods: data analysis, numerical; catalogues, surveys; binaries: close, eclipsing, fundamental parameters; techniques: photometric, spectroscopic

1. Introduction

Doing astronomy today is simply unimaginable without computers. To facilitate observing preparations, we use databases; to observe, we use control software; to reduce the acquired data, we use reduction programs. Just how far the computer autonomy of the data acquisition process goes is best described by the increasing trend of refurbishing small and medium-size telescopes into fully automatic, robotic instruments†. Surveys such as OGLE (Udalski *et al.* 1997), EROS (Palanque-Delabrouille *et al.* 1998), ASAS (Pojmanski 2002), space mission Hipparcos' epoch photometry (Perryman & ESA 1997), and others, have changed observational astronomy: streams of data produced by automatic telescopes around the world and in space are overwhelming for currently existing tools and astronomers cannot cope anymore.

Take eclipsing binaries, for example. So far there have been about 500 published papers with physical and geometrical parameters determined to better than 3% accuracy. For a skilled eclipsing binary guru it takes 1–2 weeks to reduce and analyse a single eclipsing binary by hand. To date, there are about 10 000 photometric/RV data-sets that in principle allow modeling to a 3% accuracy. By 2020, the upcoming missions such as Pan-Starrs (Kaiser *et al.* 2002) and Gaia (Perryman *et al.* 2001) will have pushed this

---

† A comprehensive list of more than a hundred such facilities may be found, e.g., at
http://www.astro.physik.uni-goettingen.de/~hessman/MONET/links.html.

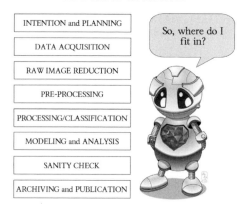

Figure 1. Schematic view of a typical EB reduction and analysis pipeline.

number to ~10 000 000. Even if all observational facilities collapsed at that point so that no further data got collected, it would take 12 500 astronomers to analyse these data in the next 100 years! Since there are currently about 13 000 members of the IAU, the only way to achieve this in the next 100 years by traditional methods is to have *every* astronomer in the world doing eclipsing binaries. And of course, do not forget to shut down all robotic telescopes out there!

With the change in observational astronomy, traditional analysis methods and tools need to change too. This paper overviews most important aspects of automatic procedures, tiers that form a pipeline reduction of eclipsing binary light curves. Next Section deals with basic principles of the reduction and analysis pipeline; Section 3 reviews most important applications of automatic pipelines on large-scale survey data. Section 4 stresses the everlasting importance of dedicated observations. Finally, Section 5 concludes and gives some prospects for the future.

2. Tiers of the reduction and analysis pipeline

A full-fledged pipeline for reduction and analysis of photometric data of eclipsing binary stars would ideally consist of 8 distinct tiers depicted in Figure 1.

2.1. *Intention and planning*

For as long as we discuss stellar objects in general, and eclipsing binaries in particular, there are two apparently frightening facts that need to be considered: **1)** a target star has already been observed and **2)** a target star has already been observed many times. There are literally hundreds of photometric survey missions that have been swiping the sky across and over in a very wide magnitude range, and chances are indeed slim that a given star has not been observed yet.

According to Hipparcos results, there are about 0.8% of eclipsing binaries in the overall stellar population (917 out of 118 218 stars, Perryman & ESA 1997). Projecting these statistics to other large surveys gives an estimate of how many eclipsing binaries are expected to be present in survey databases: ~136 000 in ASAS (11 076 detected by Paczyński *et al.* 2006), ~ 56 000 in the OGLE LMC field (2 580 detected by Wyrzykowski *et al.* 2003), ~ 16 000 in OGLE SMC field (1 350 detected by Wyrzykowski *et al.* 2004), ~80 000 in TASS (Droege *et al.* 2006) etc. Gaia will make a revolution in these numbers since the aimed census of the overall stellar population is ~ 1 billion up to $V = 20$ (Perryman *et al.* 2001). Admittedly, magnitude levels and variability detection threshold change from survey to survey, but a shortage of eclipsing binaries in the databases is

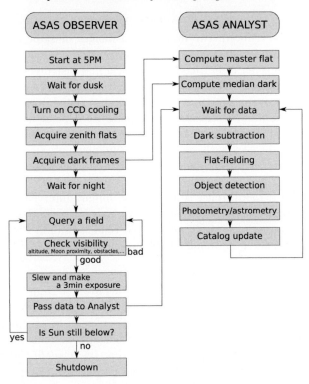

Figure 2. Automatic pipeline of the ASAS project. The pipeline consists of two separate (yet connected) engines: Observer and Analyst. The Observer takes care of the data acquisition, and the Analyst takes care of data reduction and analysis. The only human intervention needed is closing the observatory in case of bad weather and changing the DAT-2 storage tapes. The schematic view was adapted from the description of the ASAS project (Pojmanski 1997).

more than obvious. In other words, there are many eclipsing binaries out there that are either undetected, unconfirmed or misclassified. Stressing a well-known fact that eclipsing binaries are unique in their potential to yield accurate masses, radii, temperatures and distances, and realizing that many of them are reachable by small-size ground instruments, eclipsing binaries should definitely hold one of the top positions on observational candidates list.

2.2. *Data acquisition*

Most automated of all pipeline tiers, data acquisition has become a truly reliable run-of-the-mill. An example of a fully automatic data acquisition and analysis pipeline is that of the All-Sky Automated Survey (ASAS, Pojmanski 1997), depicted on Figure 2. The level of sophistication is already such that it assures accurate and reliable data both from ground-based and space surveys – and in plenty. A more serious problem for space surveys seems to be telemetry: how do we get the data down to Earth? For instance, Hipparcos' downlink rate was only 24 kbit/s, Gaia's will be 5 Mbit/s (Lammers 2005) – significantly less than the bandwidth we are used to from everyday life. To avoid using lossy compression algorithms, surveys must use optimized telemetry for the given field and/or data pre-processing (e.g. binning, filtering, selective downloads). Reliable and lossless I/O pipelines and finding ways to store all the acquired data are definitely two of the greatest challenges for data acquisition of the future.

2.3. *Raw image reduction*

Acquired data must be reduced: two-dimensional images must be converted to the observed quantity (magnitudes, fluxes, ...). To fully appreciate the need for accurate image reduction, one must consider a multitude of physical and instrumental effects that influence the observed data. Some of them – e.g., telescope optics, CCD quantum efficiency and non-linearity, filter response – may be adequately treated during the reduction process. Others – sky variability, instrumental temperature dependence, cosmic rays, interstellar and atmospheric extinction – usually demand more involvement because of their dependence on time and wavelength, or because of unknown physical conditions. Raw image reduction consists of taking the acquired image, extracting the data and removing all instrumental artifacts contained in that data. This procedure, along with the developed tools (e.g., IRAF, Tody 1986), relies somewhat on human intervention, but in principle it could be automated to meet the accuracy of today's surveys. One of the steps in the ASAS pipeline, for example, is a fully automated reduction (c.f., Figure 2): subtracting dark current and flat-fielding (Pojmanski 1997).

2.4. *Pre-processing*

Once the images have been reduced, the data are ideally free from instrumental systematics, but imprints of other effects (most notably atmospheric extinction and variable seeing) in phased data are still present. These effects may be significant and, as such, they should be removed from the data. To this procedure we refer to as pre-processing.

There are two approaches to pre-processing: *parametric modeling* and *detrending*. The former uses modeling functions and seeks optimal parameters to reproduce the effect at hand; since it relies on physical insights, its application is more-or-less transparent. Detrending, on the other hand, is based on statistical properties of the observed time series and uses mathematical tools to achieve the same goal. Treating atmospheric extinction with parametric models is given e.g., by Prša & Zwitter (2005b), while detrending is presented e.g. by Tamuz *et al.* (2005). Since the application of the latter is not limited to just a given physical effect, it is well worth stressing its major strengths.

Strictly speaking, a *trend* in a time series is a slow, gradual change in observables that obscures parameter relationships under investigation. *Detrending* is a statistical operation of removing stochastical dependence in consecutive observations, thus making the pre-processed data distributed according to the normal (Gaussian) probability distribution function. Tamuz *et al.* (2005) proposed a generalized Principal component analysis (PCA) method that accounts for variable observation uncertainties. The method is able to remove systematics from the data without any prior knowledge of the effect. Figure 3 shows an example of how the algorithm is able to process noisy planetary transit data (top row) by consecutively detrending four distinct systematic effects, yielding the detrended data (bottom row). Strengths of the method are its universality and little importance of the starting values of trend parameters, and the reduction to ordinary PCA in case of constant observation uncertainties. Its deficiencies are non-orthogonal eigenvectors (and thus deteriorated statistical properties in cases of a highly variable S/N ratio), a danger of filtering out intrinsic long-term variability and no relation to physical background of the trend. That said, generalized PCA method has proven to be one of the most successful methods for detrending that has been applied so far.

2.5. *Processing/classification*

By the time the observed data is ready for scientific munching, most of the non-intrinsic artifacts should have been removed. By *processing* we refer to seeking broad scientific properties of the observed object: analysis of variance, period determination, phased curve

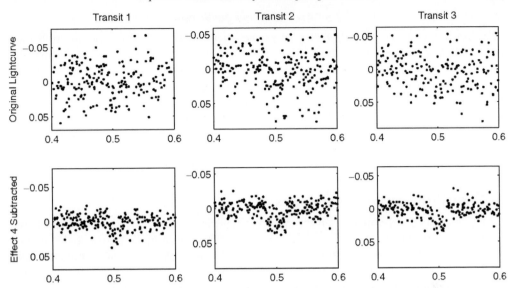

Figure 3. Detrending results for 3 planetary transit examples. The plotted diagrams depict relative magnitude vs. phase. The top row shows the original data, and the bottom row shows the detrended data, after 4 consecutive detrending iterations. Adapted from Tamuz *et al.* (2005).

folding etc. Closely related is *classification*: based either on the processing results or on statistical pattern analysis, the observed objects are classified into their respective groups. While manual approaches usually rely on the former principle (we recognize the shape of the light curve and evaluate it critically in a broader physical context – do parameters make sense, is the period plausible for a given type of object, ...), automatic approaches will prefer the latter principle, e.g., through the use of Fourier fitting, inversions, neural networks etc. Either way, processing and classification aim to discriminate gems from ordinary rocks in terms of our primary interest.

2.6. *Modeling and analysis*

Computationally most demanding task, at least with respect to eclipsing binaries, is their modeling and analysis. Seeking and interpreting a set of physical and geometrical parameters involves solving the inverse problem. There are many dedicated codes that enable accurate modeling – WD (Wilson & Devinney 1971), WINK (Wood 1971), NDE (Nelson & Davis 1972), EBOP (Etzel 1981), FOTEL (Hadrava 1990) and many others. We discuss their usage and application to survey data in detail in the following section.

2.7. *Sanity check*

A famous statement by R.E. Wilson, "There is more to modeling eclipsing binaries than parameter fitting," pretty much encapsulates the idea of sanity check. Solving the inverse problem does not only mean finding physical and geometrical parameters that best reproduce the data, it also means seeking parameter inter-dependencies, understanding hyperspace non-linearity and, above all, being aware of the limitations of the data-set at hand and the used modeling engine. Since eclipsing binaries are used for "calibrating the calibrations", mis- and over-interpreting the data may have tragic consequences on solution reliability. Getting a solution from a model is only a fraction of the work; the majority is assessing its uniqueness and physical feasibility of that solution.

2.8. *Archiving and publication*

More important than the publication of papers themselves is the question on publishing data. What to do with the immense data flow that is expected from large-scale surveys? How to set standards and specifications for publishing and storing data? How to coordinate efforts and how to distribute the results? Finally, what is our next step in terms of model enhancements? Let us face it – missions such as CoRoT (Baglin *et al.* 2002) and Kepler (Koch *et al.* 2004) will deliver milli-magnitude accuracies in just a few years – do we honestly believe that our models can support such accuracies? All of these are still open questions that demand our immediate attention.

3. First bites on large databases

One of the first attempts to survey eclipsing binaries in the LMC goes back to Payne-Gaposchkin (1971), who visually examined about 2000 photographic plates, and classified and listed the main characteristics of 78 eclipsing binaries. At that time computers only started infiltrating modern astronomy and automatic handling was not possible. Yet at the same time, the first EB modeling codes were emerging, most notably those of Horák (1966, 1970), Wilson & Devinney (1971), Wood (1971), Nelson & Davis (1972), Mochnacki & Doughty (1972) and somewhat later Hill (1979), Etzel (1981), Hadrava (1990) and Linnell & Hubeny (1994), that would eventually form the base of automatic pipelines.

In the early nineties, surveys began to yield first databases that were used for EB detection and analysis. Grison *et al.* (1995) assembled a list of 79 EBs in the bar of the LMC from the EROS survey data. Of those, only one system was previously identified as an EB, so this work effectively doubled the number of known EBs in the LMC. In the year that followed, Friedemann *et al.* (1996) used IRAS data (Neugebauer *et al.* 1984) to look for coincidences in the positions of EBs taken from the 4th edition of the GCVS (Kholopov *et al.* 1992) and about 250 000 IRAS sources. They found 233 candidates, of those 63% Algol-type binaries where accretion disks could be responsible for the IR imprint.

Attacks on LMC continued by Alcock *et al.* (1997), who used the MACHO database (Cook *et al.* 1995) to analyse 611 bright EBs. The selection was based on visual identification by examining phase plots. They pointed out two physical quantities that, besides inclination, account for most variance in light curves: the sum of relative radii and the surface brightness ratio. For preliminary analysis the authors used the Nelson & Davis (1972) code and, following the GCVS designation types, they proposed a new decimal classification scheme depicted in Figure 4.

The next survey to provide results for 933 EBs was OGLE (Szymanski *et al.* 1996). Series of systematic analyses were conducted by Rucinski (1997b,a, 1998) and later Maceroni & Rucinski (1999); Rucinski & Maceroni (2001) that stressed the success and importance of the Fourier decomposition technique (FDT) for classification of variable stars. The technique itself – fitting a 4th order Fourier series to phased data curves and mapping different types of variables in Fourier coefficient space (c.f., Figure 5, left) – was first proposed for EBs already by Rucinski (1973) and has been used ever since, most notably for classifying ASAS data (Pojmanski 2002; Paczyński *et al.* 2006).

Somewhat ironically, the first one to implement a fully automatic analysis pipeline for obtaining physical parameters of EBs was the most vocal advocate against any automated approaches: R.E. Wilson. In their two papers, Wyithe & Wilson (2001, 2002) carried out an automatic search from 1459 EBs in the SMC detected by OGLE to find ideal distance

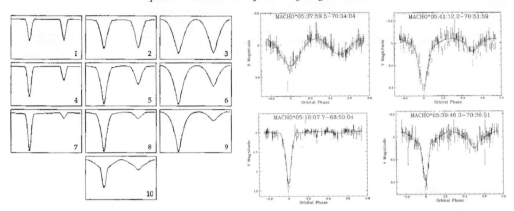

Figure 4. A decimal classification scheme proposed by Alcock *et al.*. The scheme relies on two physical parameters: the sum of relative radii and the surface brightness ratio. Four plots on the right are classified data from the MACHO survey. Adopted from Alcock *et al.* (1997).

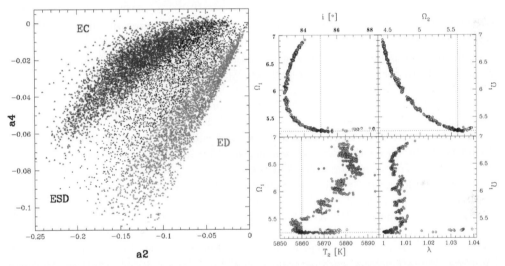

Figure 5. Left: three types of eclipsing binaries (detached, semi-detached, and contact) mapped in the a_2–a_4 Fourier composition space; adopted from Paczyński *et al.* (2006). Right: heuristic scanning with Powell's direction set method. Converged results are shown for different parameter cross-sections; cross-hairs denote the right solution, and the symbol's shade of gray corresponds to the reached χ^2 value: the darker the tone, the lower the χ^2. Taken from Prša & Zwitter (2006).

estimators. WD was run in an automatic mode for the first time, although on a stripped level of complexity: the model assumed canonical values for physical parameters poorly defined by a single-passband photometric data: mass ratio $q = 1$, argument of periastron $\omega = 0$ or π, the temperature of the secondary $T_2 = 15\,000\mathrm{K}$, no spots, simple reflection, synchronous rotation etc. Yet for the first time, an automatic, decision-making pipeline was tested against synthetic data and then applied to observations. Despite several deficiencies (systematics introduced through assertions, DC-based method without heuristical search for solution uniqueness, no account of reddening) the authors succeeded to come up with two groups of candidates for ideal distance indicators: widely detached EBs and EBs with total eclipses. A manual follow-up analysis of 19 bright, large-amplitude

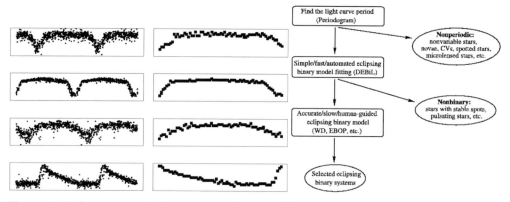

Figure 6. Left: an example of conversion of phased light curves to 70×15 pixel images, which are fed to the neural network image recognition algorithm. Taken from Wyrzykowski *et al.* (2003). Right: a tier-based pipeline proposed by Devor (2005): observed light curves are passed sequentially through filters and only the ones that fulfil all criteria make it to the next tier.

candidates in their list was done e.g., by Graczyk (2003), deriving the distance modulus to the SMC to be ∼18.9±0.1.

Meanwhile, a number of reliable solutions of individual EB solutions was steadily growing by a dedicated series of manual analyses, e.g., by Andersen *et al.*, Munari *et al.* and others. Instead of immediately going for survey data, our group decided to test fully automatic pipelines on these high-quality data. In our early work (Prša 2003) we obtained encouraging results for 5 morphologically different EBs, stressing importance of data diversity – photometric data without RVs does not suffice for accurate modeling results. Trying to follow up on our devised scheme, we soon identified main deficiencies of the DC algorithm: since it is based on numerical derivatives, it may frequently diverge, and it gets stuck in local minima. To overcome this, we proposed two types of derivative-less methods: Nelder & Mead's downhill Simplex method (Prša & Zwitter 2005c) and Powell's Direction set method (Prša & Zwitter 2006). To understand and explore parameter degeneracy, heuristic scanning and parameter kicking were introduced (Prša & Zwitter 2005a, c.f., Figure 5, right) – the problem does *not* lie in the DC, but in the inverse problem itself: its non-linearity, parameter degeneracy and data quality limitations. With this in mind we created a new modeling environment called PHOEBE† (PHysics Of Eclipsing BinariEs; Prša & Zwitter 2005a) that features a flexible scripting language. This language is developed specifically with modeling and analysis of large surveys in mind.

Continuing with the OGLE data harvest, Wyrzykowski *et al.* (2003, 2004) identified 2580 EBs in the LMC and 1351 EBs in the SMC. The novelty of their classification approach is using Artificial neural networks (ANN) as an image recognition algorithm, based on phased data curves that have been converted to low-resolution images as depicted on Figure 6. Their classification pipeline was backed up by visual examinations of results. Although there were no physical analyses in their pipeline, observational properties of the sample, as well as 36 distance estimator candidates for the LMC, have been derived.

In 2005, Devor implemented a tier-based elimination pipeline: observed light curves are sequentially passed through filters in the order of increasing computational time cost. Each tier filters out light curves that do not conform to the given criteria. Once a clean sample of light curves is available, it is submitted to a central part of the pipeline, a

† More information on PHOEBE may be found at `http://phoebe.fiz.uni-lj.si`.

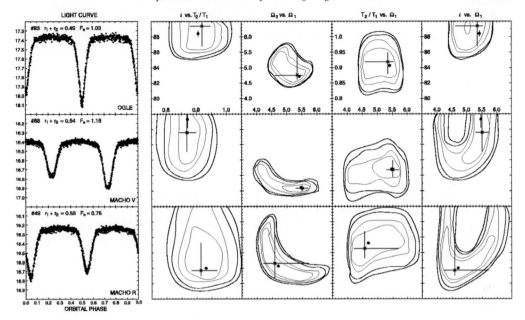

Figure 7. Examples of Monte-Carlo simulations for three EB light curves. Most importantly, the authors stress the importance of examining phase space cross-sections depicted on the right. Please refer to the original paper for further details. Taken from Michalska & Pigulski (2005).

dedicated program DEBiL (Detached eclipsing binary light curve solver; c.f., Figure 6, right), fitting a simplified EB model (spherical, limb-darkened stars on a classical Keplerian orbit) to observations. The pipeline yielded 10 861 eclipsing binaries out of 218 699 bulge field variables from OGLE II data (Udalski *et al.* 1997). Its main advantage is speed: $\sim$ 1 minute per light curve on a 333MHz Sun UltraSparc 5 workstation. Its main deficiencies are lack of the 3rd proposed tier (accurate analysis) and an over-simplified model that may produce false positives among other variables.

One of the best papers, in our opinion, that dealt with eclipsing binaries from OGLE II data, was the one by Michalska & Pigulski (2005). Its thorough analysis and deep insight into caveats of the EB field make it exemplary for all similar undertakings in the future. The authors limited their analysis to bright ($V < 17.5$, $V - I < 0.5$), high S/N, EA type binaries that exhibit small proximity effects. After proving by example that the original differential image analysis (DIA) calibration is flawed due to uncertainty of reference flux in the flux-to-magnitude calibration, they proposed a novel method of calibrating DIA data and demonstrated its significantly better results. Once the OGLE II data has been re-calibrated, the authors added MACHO, OGLE I and EROS data (when available). The data have been submitted to a WD-based pipeline: the first step was to find initial parameter estimates by the Monte-Carlo method (c.f., Figure 7), and the second step was to converge to the final solution by DC. A result is a list of 98 proposed candidates for distance estimates to the LMC, along with accurately determined parameters in relative units. Out of the sample, 58 stars are found to have eccentric orbits, and 14 systems are exhibiting apsidal motion.

Out of the crowd emerges yet another program to tackle the problem: EBAI (Eclipsing Binaries with Artificial Intelligence; Devinney *et al.* 2005). This project does not only classify the data, it does more: blindingly fast, it determines coarse parameters of eclipsing binaries in a large data set. Study is underway for these parameters to be fed to a WD-based solver within PHOEBE. This solver maps the hyperspace around the solution, verifying its uniqueness and heuristically determining error estimates.

Another recent work that we wish to draw specific attention to has been done by Tamuz *et al.* (2006). The authors devised a new algorithm called EBAS (Eclipsing Binary Automatic Solver), aimed specifically to large datasets and thus based on the faster, yet less accurate EBOP code (Etzel 1981). Similarly to the discussed predecessors, EBAS also uses the sum of relative radii as a principal parameter. Yet there are two important novelties of their approach: instead of inclination the authors introduced the impact parameter – the projected distance between the centers of the two stars during the primary eclipse, measured in terms of the sum of radii – and they introduced a new "alarm" statistics, the goal of which is to automatically discriminate best-fit χ^2 values from still apparently acceptable values, but corresponding to distinctively wrong solutions. A follow-up application of EBAS on 938 OGLE LMC binaries with B-type main-sequence primary stars (Mazeh *et al.* 2006a) yielded the distributions of the fractional radii of the two components and their sum, the brightness ratios and the periods of the short-period binaries. Intriguingly, they observed that the distribution in $\log P$ is *flat* on the 2-10 days interval and that the detected frequency of their target stars is significantly smaller than the frequency deduced by dedicated RV surveys. The details on these findings are also given by Mazeh *et al.* (2006b).

Our attempt to preserve paper readability, and struggling against page limits at the same time, regrettably prohibits us to summarize all the work done so far. That is why we wish to at least acknowledge other important developments of this field – and to apologize for any unintentional omissions in this brief review. Reader interested in pipeline reduction of binary light curves from large-scale surveys will surely benefit from the work of Lastennet & Valls Gabaud (2002), Brett *et al.* (2004), Ribas *et al.* (2004), Wilson (2004), Hilditch *et al.* (2004, 2005), Eyer & Blake (2005), Groenewegen (2005), Naficy *et al.* (2005), Sarro *et al.* (2006) and many others.

4. Traditional observations are *not* obsolete

After so much stress on surveys, missions and sophistication in fully automatic approaches it is tempting to conclude that traditional object-by-object observations have become obsolete. This is one of most dangerous misconceptions, apparently powered even by our own statement in the introduction that most (if not all) of the candidates have already been observed a number of times. Although these hot topics are appealing because of shear numbers of observed objects, there are several deficiencies in the context of eclipsing binaries that we should be aware of:

- Surveys and missions have a limited life-time that is generally not governed by the eclipsing binary harvest. Rather, limitations arise on account of funding, technology and reaching primary scientific objectives. A direct consequence is the selection effect in observed EBs: only the ones with suitable periods will have been detected.
- The main driving idea of surveys is to acquire as much data as quickly as possible. Due to adopted sky scanning laws, the sky coverage is typically non-uniform and the observations are thus clustered in time. Although this might not seem too important for close binaries, it is critical in case of well detached binaries where there is practically no surface deformation and where eclipses occur only on a narrow phase interval. Having a point or two within the eclipse is hardly any different than having no point at all.
- In order to reach survey completeness in terms of object counts during the mission life-time, the number of data points per object is usually poor. This means that the phase coverage for eclipsing binaries is often not sufficient for recognition and classification purposes, because of the strong sensitivity of period detection algorithms to phase completeness.

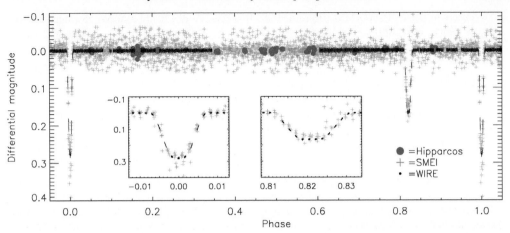

Figure 8. Phased light curve of ψ Cen. Data points from WIRE (black dots) may be compared to those from SMEI (gray plus symbols) and from Hipparcos (grey circles). Taken from Bruntt *et al.* (2006).

• Unprecedented in numbers does not mean unprecedented in accuracy. At least so far, survey data has been more challenging to reduce and analyse than a typical dedicated observation because of the significant data scatter. However, WIRE (the Wide-field Infra-Red Explorer, Hacking *et al.* 1999), despite its failure to achieve primary scientific goals, had already given us a snapshot of the milli-magnitude photometric precision of the future (c.f., Figure 8).

• Most importantly: surveys usually lack data diversity. In order to get absolute temperatures of both stars and interstellar extinction, multi-passband photometry is needed; to get reliable estimates of absolute sizes of an eclipsing binary system, radial velocities are needed. To break inter–parameter correlations and solution degeneracy, as many diverse data-sets as possible are needed: astrometry and parallaxes, photometry, polarimetry, spectroscopy — the more the better. Different physical and geometrical parameters, and their inter-dependencies, are revealed by different types of data-sets.

If we take all of the above into account, we may only conclude that follow-up observations are still badly needed.

5. Conclusions and prospects for the future

Overwhelming data quantities are upon us and changing traditional ways of modeling and analysis of eclipsing binaries is thus inevitable. There are many fine studies that bring us closer to this goal. One of the greatest properties of astronomy, when compared to other sciences, is a strong sense of collaboration, and absence of blind competition, between astronomers; our questions, therefore, on how to facilitate and how to propagate the idea of joint development of these new approaches, and how to handle huge data-sets that are pouring in, are aimed at every single individual interested in contributing its own piece to this fascinating puzzle. These are indeed scientifically challenging times and it would be too bad if we missed them.

References

Alcock, C., Allsman, R.A., Alves, D., Axelrod, T.S., Becker, A.C., Bennett, D.P., Cook, K.H., Freeman, K.C., Griest, K., Lacy, C.H.S., Lehner, M.J., Marshall, S.L., Minniti, D.,

Peterson, B.A., Pratt, M.R., Quinn, P.J., Rodgers, A.W., Stubbs, C.W., Sutherland, W., & Welch, D.L. 1997, *AJ*, 114, 326

Andersen, J., Clausen, J.V., Nordstroem, B., & Reipurth, B. 1983, *A&A*, 121, 271

Baglin, A., Auvergne, M., Barge, P., Buey, J.-T., Catala, C., Michel, E., Weiss, W., & COROT Team. 2002, in *Stellar Structure and Habitable Planet Finding*, ESA SP-485, ed. B. Battrick, F. Favata, I.W. Roxburgh, & D. Galadi, 17–24

Brett, D.R., West, R.G., & Wheatley, P.J. 2004, *MNRAS*, 353, 369

Bruntt, H., Southworth, J., Torres, G., Penny, A.J., Clausen, J.V., & Buzasi, D.L. 2006, *A&A*, 456, 651

Cook, K.H., Alcock, C., Allsman, H.A., Axelrod, T.S., Freeman, K.C., Peterson, B.A., Quinn, P.J., Rodgers, A.W., Bennett, D.P., Reimann, J., Griest, K., Marshall, S.L., Pratt, M.R., Stubbs, C.W., Sutherland, W., & Welch, D.L. 1995, in IAU Colloq. 155, *Astrophysical Applications of Stellar Pulsation*, ASP Conf. Ser. 83, ed. R. Stobie & P. Whitelock, 221

Devinney, E., Guinan, E., Bradstreet, D., DeGeorge, M., Giammarco, J., Alcock, C., & Engle, S. 2005, *BAAS*, 1212

Devor, J. 2005, *ApJ*, 628, 411

Droege, T.F., Richmond, M.W., Sallman, M.P., & Creager, R.P. 2006, *ArXiv Astrophysics e-prints*

Etzel, P.B. 1981, in *Photometric and Spectroscopic Binary Systems*, ed. E.B. Carling & Z. Kopal, 111

Eyer, L. & Blake, C. 2005, *MNRAS*, 358, 30

Friedemann, C., Guertler, J., & Loewe, M. 1996, *A&AS*, 117, 205

Graczyk, D. 2003, *MNRAS*, 342, 1334

Grison, P., Beaulieu, J.-P., Pritchard, J.D., Tobin, W., Ferlet, R., Vidal-Madjar, A., Guibert, J., Alard, C., Moreau, O., Tajahmady, F., Maurice, E., Prevot, L., Gry, C., Aubourg, E., Bareyre, P., Brehin, S., Gros, M., Lachieze-Rey, M., Laurent, B., Lesquoy, E., Magneville, C., Milsztajn, A., Moscoso, L., Queinnec, F., Renault, C., Rich, J., Spiro, M., Vigroux, L., Zylberajch, S., Ansari, R., Cavalier, F., & Moniez, M.1995, *A&AS*, 109, 447

Groenewegen, M.A.T. 2005, *A&A*, 439, 559

Hacking, P., Lonsdale, C., Gautier, T., Herter, T., Shupe, D., Stacey, G., Fang, F., Xu, C., Graf, P., Werner, M., Soifer, B., Moseley, H., & Houck, J. 1999, in *Astrophysics with Infrared Surveys: A Prelude to SIRTF*, ASP Conf. Ser. 177, ed. M.D. Bicay, R.M. Cutri, & B.F. Madore, 409

Hadrava, P. 1990, *Contributions of the Astronomical Observatory Skalnate Pleso*, 20, 23

Hilditch, R.W., Harries, T.J., & Howarth, I.D. 2004, *New Astronomy Review*, 48, 687

Hilditch, R.W., Howarth, I.D., & Harries, T.J. 2005, *MNRAS*, 357, 304

Hill, G. 1979, *Publications of the Dominion Astrophysical Observatory Victoria*, 15, 297

Horák, T. 1966, *Bulletin of the Astronomical Institutes of Czechoslovakia*, 17, 27

Horák, T.B. 1970, *AJ*, 75, 1116

Kaiser, N., Aussel, H., Burke, B.E., Boesgaard, H., Chambers, K., Chun, M.R., Heasley, J.N., Hodapp, K.-W., Hunt, B., Jedicke, R., Jewitt, D., Kudritzki, R., Luppino, G.A., Maberry, M., Magnier, E., Monet, D.G., Onaka, P.M., Pickles, A.J., Rhoads, P.H.H., Simon, T., Szalay, A., Szapudi, I., Tholen, D.J., Tonry, J.L., Waterson, M., & Wick, J. 2002, in *Survey and Other Telescope Technologies and Discoveries*, Proceedings of the SPIE, Volume 4836, ed. J.A. Tyson & S. Wolff, 154–164

Kholopov, P.N., Samus, N.N., Durlevich, O.V., Kazarovets, E.V., Kireeva, N.N., & Tsvetkova, T.M. 1992, *Bulletin d'Information du Centre de Donnees Stellaires*, 40, 15

Koch, D.G., Borucki, W., Dunham, E., Geary, J., Gilliland, R., Jenkins, J., Latham, D., Bachtell, E., Berry, D., Deininger, W., Duren, R., Gautier, T.N., Gillis, L., Mayer, D., Miller, C.D., Shafer, D., Sobeck, C.K., Stewart, C., & Weiss, M. 2004, in *Optical, Infrared, and Millimeter Space Telescopes*. Proceedings of the SPIE, Volume 5487, ed. J.C. Mather, 1491–1500

Lammers, U. 2005, in ESA SP-576: *The Three-Dimensional Universe with Gaia*, ed. C. Turon, K.S. O'Flaherty, & M.A.C. Perryman, 445

Lastennet, E. & Valls-Gabaud, D. 2002, *A&A*, 396, 551

Linnell, A.P. & Hubeny, I. 1994, *ApJ*, 434, 738

Maceroni, C., & Rucinski, S.M. 1999, *AJ*, 118, 1819

Mazeh, T., Tamuz, O., & North, P. 2006a, *MNRAS*, 367, 1531

—. 2006b, *Ap&SS*, 44

Michalska, G. & Pigulski, A. 2005, *A&A*, 434, 89

Mochnacki, S.W. & Doughty, N.A. 1972, *MNRAS*, 156, 51

Munari, U., Tomov, T., Zwitter, T., Milone, E.F., Kallrath, J., Marrese, P.M., Boschi, F., Prša, A., Tomasella, L., & Moro, D. 2001, *A&A*, 378, 477

Naficy, K., Riazi, N., & Kiasatpour, A. 2005, *AJ*, 130, 1862

Nelson, B. & Davis, W.D. 1972, *ApJ*, 174, 617

Neugebauer, G., Habing, H.J., van Duinen, R., Aumann, H.H., Baud, B., Beichman, C.A., Beintema, D.A., Boggess, N., Clegg, P.E., de Jong, T., Emerson, J.P., Gautier, T.N., Gillett, F.C., Harris, S., Hauser, M.G., Houck, J.R., Jennings, R.E., Low, F.J., Marsden, P.L., Miley, G., Olnon, F.M., Pottasch, S.R., Raimond, E., Rowan-Robinson, M., Soifer, B.T., Walker, R.G., Wesselius, P.R., & Young, E. 1984, *ApJL*, 278, L1

Paczyński, B., Szczygieł, D.M., Pilecki, B., & Pojmański, G. 2006, *MNRAS*, 368, 1311

Palanque-Delabrouille, N., Afonso, C., Albert, J.N., Andersen, J., Ansari, R., Aubourg, E., Bareyre, P., Bauer, F. *et al.* & the EROS Collaboration. 1998, *A&A*, 332, 1

Payne-Gaposchkin, C.H. 1971, *The variable stars of the Large Magellanic Cloud*, (Smithsonian Contributions to Astrophysics, Washington: Smithsonian Institution Press, —c1971)

Perryman, M.A.C., de Boer, K.S., Gilmore, G., Høg, E., Lattanzi, M.G., Lindegren, L., Luri, X., Mignard, F., Pace, O., & de Zeeuw, P.T. 2001, *A&A*, 369, 339

Perryman, M.A.C. & ESA. 1997, *The Hipparcos and Tycho catalogues. Astrometric and photometric star catalogues derived from the ESA Hipparcos Space Astrometry Mission*, Publisher: Noordwijk, Netherlands: ESA Publications Division, Series: ESA SP Series vol no: 1200, ISBN: 9290923997 (set)

Pojmanski, G. 1997, *Acta Astronomica*, 47, 467

—. 2002, *Acta Astronomica*, 52, 397

Prša, A. 2003, in *GAIA Spectroscopy: Science and Technology*, ASP Conf. Ser. 298, ed. U. Munari, 457

Prša, A. & Zwitter, T. 2005a, *ApJ*, 628, 426

—. 2005b, *Ap&SS*, 296, 315

Prša, A., & Zwitter, T. 2005c, in *The Three-Dimensional Universe with Gaia*, ESA SP-576, ed. C. Turon, K.S. O'Flaherty, & M.A.C. Perryman, 611

—. 2006, *ArXiv Astrophysics* e-prints

Ribas, I., Jordi, C., Vilardell, F., Giménez, Á., & Guinan, E.F. 2004, *New Astronomy Review*, 48, 755

Rucinski, S.M. 1973, *Acta Astronomica*, 23, 79

Rucinski, S.M. 1997a, *AJ*, 113, 1112

—. 1997b, *AJ*, 113, 407

—. 1998, *AJ*, 115, 1135

Rucinski, S.M. & Maceroni, C. 2001, *AJ*, 121, 254

Sarro, L.M., Sánchez-Fernández, C., & Giménez, Á. 2006, *A&A*, 446, 395

Szymanski, M., Udalski, A., Kubiak, M., Kaluzny, J., Mateo, M., & Krzeminski, W. 1996, *Acta Astronomica*, 46, 1

Tamuz, O., Mazeh, T., & North, P. 2006, *MNRAS*, 367, 1521

Tamuz, O., Mazeh, T., & Zucker, S. 2005, *MNRAS*, 356, 1466

Tody, D. 1986, in Proceedings of *Instrumentation in astronomy VI*; Part 2 (A87-36376 15-35). Bellingham, WA, Society of Photo-Optical Instrumentation Engineers, ed. D.L. Crawford, 733

Udalski, A., Kubiak, M., & Szymanski, M. 1997, *Acta Astronomica*, 47, 319

Wilson, R.E. 2004, *New Astronomy Review*, 48, 695

Wilson, R.E. & Devinney, E.J. 1971, *ApJ*, 166, 605

Wood, D.B. 1971, *AJ*, 76, 701

Wyithe, J.S.B. & Wilson, R.E. 2001, *ApJ*, 559, 260

—. 2002, *ApJ*, 571, 293

Wyrzykowski, L., Udalski, A., Kubiak, M., Szymanski, M., Zebrun, K., Soszynski, I., Wozniak, P.R., Pietrzynski, G., & Szewczyk, O. 2003, *Acta Astronomica*, 53, 1

Wyrzykowski, L., Udalski, A., Kubiak, M., Szymanski, M.K., Zebrun, K., Soszynski, I., Wozniak, P.R., Pietrzynski, G., & Szewczyk, O. 2004, *Acta Astronomica*, 54, 1

Binary Stars as Critical Tools & Tests
in Contemporary Astrophysics
Proceedings IAU Symposium No. 240, 2006
W.I. Hartkopf, E.F. Guinan & P. Harmanec, eds.

Analysis of the Eclipsing Binaries in the LMC Discovered by OGLE: Period Distribution and Frequency of the Short–Period Binaries

Tsevi Mazeh[1], Omer Tamuz[1] and Pierre North[2]

[1]School of Physics and Astronomy, Raymond and Beverly Sackler Faculty of Exact Sciences,
Tel Aviv University, Tel Aviv, Israel
email: mazeh@wise.tau.ac.il

[2]Laboratoire d'Astrophysique, Ecole Polytechnique Fédérale de Lausanne (EPFL),
Observatoire, CH-1290 Sauverny, Switzerland

Abstract.
We review the results of our analysis of the OGLE LMC eclipsing binaries (Mazeh, Tamuz & North 2006), using EBAS — Eclipsing Binary Automated Solver, an automated algorithm to fit lightcurves of eclipsing binaries (Tamuz, Mazeh & North 2006). After being corrected for observational selection effects, the set of detected eclipsing binaries yielded the period distribution and the frequency of all LMC short-period binaries, and not just the eclipsing systems. Somewhat surprisingly, the period distribution is consistent with a flat distribution in $\log P$ between 2 and 10 days. The total number of binaries with periods shorter than 10 days in the LMC was estimated to be about 5000. This figure led us to suggest that $(0.7 \pm 0.4)\%$ of the main-sequence A- and B-type stars are found in binaries with periods shorter than 10 days. This frequency is substantially smaller than the fraction of binaries found by smaller radial-velocity surveys of Galactic B stars.

Keywords. methods: data analysis, binaries: eclipsing, Magellanic Clouds

1. Introduction

The OGLE project yielded a huge photometric dataset of the LMC (Udalski *et al.* 2000), which includes a few thousand eclipsing binary lightcurves (Wyrzykowski *et al.* 2003). This dataset allows for the first time a statistical analysis of the population of short-period binaries of an entire galaxy. To analyse this set of lightcurves we constructed EBAS (Eclipsing Binary Automated Solver), an automated algorithm (Tamuz, Mazeh & North 2006) to fit eclipsing lightcurves. Having solved the lightcurves with EBAS, we proceeded to derive the statistical features of the eclipsing binaries of the LMC (Mazeh, Tamuz & North 2006). In this short paper we present the EBAS algorithm and show two results of interest: a flat log-period distribution and a lower-than-expected binary frequency of the short-period early-type binaries.

2. The EBAS Algorithm

EBAS is based on the EBOP code (Popper & Etzel 1981, Etzel 1981), which consists of two main components. The first component generates a lightcurve for a given set of orbital elements and stellar parameters, while the second finds the parameters that best fit the observational data. We only used the lightcurve generator, and wrote our own code to search for the best-fit set of elements that minimize the χ^2 statistic. The search for the global χ^2 minimum is performed in two stages. We first find a good initial guess, and then use a simulated annealing algorithm to find the global minimum. While the first stage is merely aimed at finding an initial guess for the next stage, in most cases it already converges to a very good solution. A full description of the algorithm is given in Tamuz, Mazeh & North (2006).

During the development of EBAS we found that some solutions with low χ^2 might be unsatisfactory. While, for such solutions, the value of χ^2 is reasonable, a visual inspection of the residuals, plotted as a function of phase, revealed that the fit is sub-optimal. For such cases, human interaction was needed to improve the fit, or to otherwise decree the solution unsatisfactory. In order to allow an automated approach, an automatic algorithm must replace human evaluation.

We therefore defined a new estimator which is sensitive to the correlation between adjacent residuals of the measurements relative to the model. This feature is in contrast to the behaviour of the χ^2 function, which measures the sum of the squares of the residuals, but is not sensitive to the signs of the different residuals and their order.

Denoting by k_i the number of residuals in the i-th run (a series of consecutive residuals with the same sign), we defined the 'alarm' $\mathcal{A}$ as:

$$\mathcal{A} = \frac{1}{\chi^2} \sum_{i=1}^{M} \left(\frac{r_{i,1}}{\sigma_{i,1}} + \frac{r_{i,2}}{\sigma_{i,2}} + \cdots + \frac{r_{i,k_i}}{\sigma_{i,k_i}} \right)^2 - \left(1 + \frac{4}{\pi}\right) , \tag{2.1}$$

where $r_{i,j}$ is the residual of the j-th measurement of the i-th run and $\sigma_{i,j}$ is its uncertainty. The sum is over all the measurements in a run and then over the M runs. Dividing by χ^2 assures that, in contrast to χ^2 itself, $\mathcal{A}$ is not sensitive to a systematic overestimation or underestimation of the uncertainties. It is easy to see that $\mathcal{A}$ is minimal when the residuals alternate between positive and negative values, and that long runs with large residuals increase its value.

When a solution found by EBAS showed high $\mathcal{A}$, EBAS automatically classified the solution as unsatisfactory and started a modified search of the parameter space in order to find a better solution. Systems for which a low enough $\mathcal{A}$ solution could not be found were marked as such. Visual inspection of these lightcurves and their fitted models showed that most of them are close contact systems which EBOP did not model well.

3. Analysis of the OGLE LMC Eclipsing Binaries

The LMC OGLE-II photometric campaign (Udalski *et al.* 2000) was carried out from 1997 to 2000, during which between 260 and 512 measurements in the I-band were taken for 21 fields (Zebrun *et al.* 2001). Wyrzykowski *et al.* (2003) searched the photometric data base and identified 2580 binaries. We analysed those systems with EBAS and found 1931 acceptable solutions. We excluded all binaries with sum of radii larger than 0.6 of the binary separation. Furthermore, we excluded also binaries with high $\mathcal{A}$ values, which indicated that they are probably contact systems that EBOP can not model properly (see Mazeh, Tamuz & North (2006) for details).

The very large sample of 1931 short-period binaries enables us to derive some statistical features of the population of short-period binaries in the LMC. However, the sample suffers from serious observational selection effects, which affected the discovery of the eclipsing binaries. To be able to correct for the selection effects we needed a well-defined homogeneous sample. We therefore trimmed the sample before deriving the period distribution. Our trimmed sample consisted of all systems of magnitude between 17 and 19 in the I band with periods shorter than 10 days and having main-sequence color (see

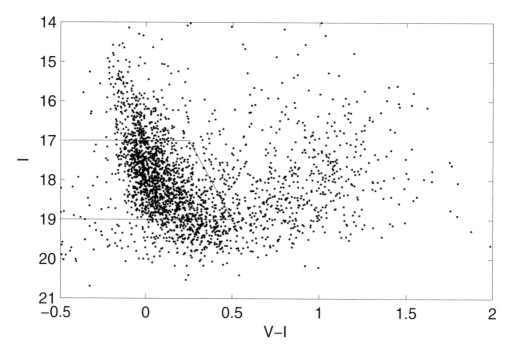

Figure 1. The *I*-magnitude as a function of the averaged $V - I$ colour of all eclipsing binaries of the sample. Only binaries that appear inside the trapezoid were included in the trimmed sample.

Figure 1). In this way we chose only systems with A and B main-sequence primaries. See Mazeh, Tamuz & North (2006) for a full description of the trimmed sample.

3.1. *The period distribution*

In order to explore the period distribution of the LMC short-period binaries, and not just the eclipsing binaries, we needed to correct for the selection effects which affect the likelihood of a binary system to be eclipsing, and to be detected by the OGLE survey. We did this by calculating for each eclipsing binary in the trimmed sample a weight w, which was the reciprocal of the probability of being detected as an eclipsing binary, assuming random orientation and phase (see Tamuz, Mazeh & North 2006). When calculating the period distribution of the binary population, we then considered that eclipsing binary as w systems.

We plotted in Figure 2 two period histograms of the trimmed sample, before and after the correction for the observational effects was applied. We emphasize that if the correction was applied properly, the right panel represents the period distribution of all binaries in the LMC with I magnitude between 17 and 19, and with sum of radii smaller than 0.6, and not only the eclipsing binaries.

The corrected period histogram shows clearly a distribution that rises up to about $\log P = 0.3$, and then flattens off, suggesting that the period distribution of the short-period binaries with early-type binaries is consistent with *a flat log distribution between 2 and 10 days*.

3.2. *The frequency of the short-period binaries*

As stated above, the corrections applied by Mazeh, Tamuz & North 2006 allowed us to study the distributions of all short-period binaries, up to 10 days, and not only the

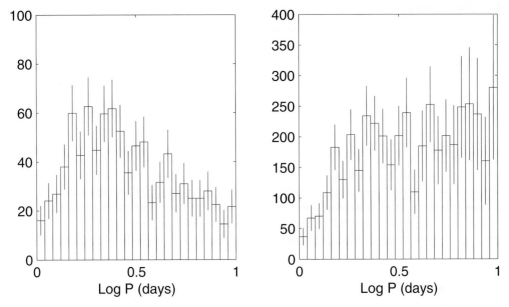

Figure 2. The period distribution of the binaries in the LMC. The left panel shows the period histogram of the trimmed sample, while the right one shows the corrected histogram

eclipsing binaries. We wished to take advantage of this feature of our analysis and estimate, within the limits of the sample, the total number of binaries in the LMC and their fractional frequency.

The sum of weights of the 938 binaries in the trimmed sample was 4585. This means that to our best estimate there are 4585 short-period binaries in the LMC that fulfill the constraints we had on the trimmed sample. To get the total number of binaries with periods shorter than 10 days we had to add the high-alarm binaries and the ones with large radii; since the added systems are close binaries, their detection probability reaches almost unity so we did not correct for undetected binaries in their case. When we added those, we ended up with 5004 binaries.

We therefore suggest that the number of binaries in the LMC, with period shorter than 10 days, with I between 17 and 19, and for which the $V - I$ colour of the system indicates a main-sequence primary, is about 5000. Assuming Wyrzykowski *et al.* (2003) might not have detected all the eclipsing binaries in the dataset and classified them as such, we arbitrarily assigned an error of 30% to the number of binaries in the sample.

In order to estimate the fractional frequency of A- and B-type main-sequence stars which reside in binaries that could have been detected by OGLE, we had to estimate how many main-sequence *single* stars were found by OGLE in the same range of magnitudes and colours. To do that, we applied exactly the same trimming procedure we performed above to the whole OGLE dataset of LMC stars and found 332,297 main-sequence stars in the I-range of 17–19 mag. However, the luminosity of a binary is brighter than the luminosity of its primary by a magnitude that depends on the light ratio of the two stars of the system. Therefore, we considered stellar range of $I = 17.5 - 19.5$, and found 705,535 stars. We therefore adopted 700,000 as a representative number of single stars in the LMC similar to the population of our binary sample. We arbitrarily assign an error of 50% to this figure.

We therefore concluded that $(0.7 \pm 0.4)\%$ of the main-sequence A- and B-type stars in the LMC are found in binaries with periods shorter than 10 days.

4. Discussion

4.1. *The period distribution*

The flat log-period distribution derived here is probably not consistent with the period distribution of Duquennoy & Mayor (1991), who adopted a Gaussian with $\overline{\log P} = 4.8$ and $\sigma = 2.3$, P being measured in days. Their distribution would provide within the same log P interval almost twice as many systems at $\log P = 1.0$ than at $\log P = 0.3$ (more precisely, the factor would be 1.7), while the new derived distribution is probably flat.

It is interesting to note that Heacox (1998) reanalysed the data of Duquennoy & Mayor (1991) and claimed that $f(a)$, the distribution of the G-dwarf orbital semi-major axis, a, is $f(a) \propto a^{-1} da$, which implies a flat log orbital separation distribution. This is equivalent to the present probable result, although the latter refers to LMC binaries with A- and B-type primaries, and is limited only to a very small range of orbital separation.

The flat log-period distribution implies a flat log orbital-separation distribution for a given total binary mass. Such a distribution indicates that there is no preferred length scale for the formation of short-period binaries (Heacox 1998), at least in the range between 0.05 and 0.16 AU, for a total binary mass of 5 $M_\odot$. Alternatively, the results indicate that the specific angular momentum distribution is flat on a log scale at the 10^{19} cm^{2}s^{-1} range.

4.2. *The binary fraction*

The fractional frequency found here is surprisingly small — $(0.7 \pm 0.4)\%$. It would be interesting to compare the frequency of B-star binaries found here for the LMC with a similar frequency study of B stars in our Galaxy. However, such a large systematic study of eclipsing binaries is not available. Instead, a few radial-velocity and photometric searches for binaries in relatively small Galactic samples were performed. Wolff (1978) studied the frequency of binaries in sharp-lined B7–B9 stars. Her Table 1 presents 73 such stars with $V \sin i < 100$ km s^{-1}, among which 17 are members of binary systems with $P_{orb} < 10$ days. Her sample yields a percentage of 23%, one and a half orders of magnitude larger than what we found in the LMC.

Similar and even higher values of binary frequency were derived for B stars in Galactic clusters and associations. Morrell & Levato (1991) have determined the frequency of binaries among B stars in the Orion OB1 association, and obtained an overall frequency of binaries with $P_{orb} < 10$ days of $26 \pm 6\%$. Garcia & Mermilliod (2001) found a rate of binaries as high as 82% among stars hotter than B1.5 in the very young open cluster NGC 6231, though in a sample of 34 stars only. Most orbits have periods shorter than 10 days. Raboud (1996) estimated a rate of binaries of at least 52% for 36 B1–B9 stars in the same cluster.

These studies show that binaries hosting B-type stars tend to be more frequent in young Galactic clusters than in the field. However, even though our LMC binaries are representative of the LMC field rather than of LMC clusters, the frequency we derived appears much lower than that of the binaries in the Galactic field.

A few photometric studies were devoted to the search of eclipsing binaries in Galactic globular clusters (e.g., Yan & Reid 1996, Kaluzny 1997). One such study was the HST 8.3-day observations of 47 Tuc (Albrow *et al.* 2001), which monitored 46,422 stars and discovered 5 eclipsing binaries with periods longer than about 4 days. From their conclusion we derive $(2 \pm 1)\%$ for periods between 2 and 10 days. This value, derived for late-type stars, is much smaller than the frequency derived by the radial-velocity surveys of B stars, and is close to the $(0.7 \pm 0.4)\%$ frequency we derived for the LMC.

We conclude, therefore, that the frequency of binaries we found in the LMC is substantially smaller than the frequency found by Galactic radial-velocity surveys. The detected frequency of B-type binaries is larger by a factor of 30 or more than the frequency we found, while the binary frequency of K- and G-type stars is probably larger by a factor of four (Halbwachs *et al.* (2003) found 2.7 ± 0.8 per cent binaries with $P < 10$ days in the solar neighbourhood). On the other hand, the binary frequency found by photometric searches in 47 Tuc is only slightly higher and still consistent with the frequency we deduced for the LMC. It seems that the frequency derived from photometric searches is consistently smaller than the one found by radial-velocity observations. We are not aware of any observational effect that could cause such a large difference between the radial velocity and the photometric studies. Therefore, we suggest that the large difference in the binary frequencies is probably real and remains a mystery. Obviously, it would be extremely useful and interesting to have studies of eclipsing binaries similar to the present one in our Galaxy, as well as in other nearby galaxies.

Acknowledgments

We are grateful to the OGLE team, and to L. Wyrzykowski in particular, for the photometric data set and for the eclipsing binary analysis. This work was supported by the Israeli Science Foundation through grant no. 03/233.

References

Albrow, M.D., Gilliland, R.L., Brown, T.M., Edmonds, P.D., Guhathakurta, P., & Sarajedini, A. 2001, *ApJ* 559, 1060

Duquennoy, A. & Mayor, M. 1991, *A&A* 248, 485

Etzel, P.B. 1981, in *Photometric and Spectroscopic Binary Systems*, p. 111-120

Garcia, B. & Mermilliod, J.-C. 2001, *A&A* 368, 122

Halbwachs, J.L., Mayor, M., Udry, S., & Arenou, F. 2003, *A&A* 397, 159

Heacox, W.D. 1998, *AJ* 115, 325

Kaluzny, J., Krzeminski, W., Mazur, B., Wysocka, A., & Stepien, K. 1997, *Acta Astronomica* 47, 249

Mazeh, T., Tamuz, O., & North, P. 2006, *MNRAS* 367, 1531

Morrell, N. & Levato, H. 1991, *ApJS* 75, 965

Popper, D.M. & Etzel, P.B. 1981, *AJ* 86, 102

Raboud, D. 1996, *A&A* 315, 384

Tamuz, O., Mazeh, T., & North, P. 2006, *MNRAS* 367, 1521

Udalski, A., Szymanski, M., Kubiak, M., Pietrzynski, G., Soszynski, I.,Wozniak, P., & Zebrun K. 2000, *Acta Astronomica* 50, 307

Wolff, S.C. 1978, *ApJ* 222, 556

Wyrzykowski, L., Udalski, A., Kubiak, M., Szymanski, M., Zebrun, K., Soszynski, I., Wozniak, P.R., Pietrzynski, G., & Szewczyk, O. 2003, *Acta Astronomica* 53, 1

Yan, L. & Reid, I.N. 1996, *MNRAS* 279, 751

Zebrun, K., Soszynski, I., Wozniak, P.R., Udalski, A., Kubiak, M., Szymanski, M., Pietrzynski, G., Szewczyk, O., & Wyrzykowski, L. 2001, *Acta Astronomica* 51, 317

Binary Stars as Critical Tools & Tests
in Contemporary Astrophysics
Proceedings IAU Symposium No. 240, 2006
W.I. Hartkopf, E.F. Guinan & P. Harmanec, eds.

The *Kepler Mission* and Eclipsing Binaries

David Koch[1], William Borucki[1], Gibor Basri[2], Timothy Brown[3],
Douglas Caldwell[4], Jorgen Christensen-Dalsgaard[5], William
Cochran[6], Edna DeVore[4], Edward Dunham[7], Thomas N. Gautier[8],
John Geary[9], Ronald Gilliland[10], Alan Gould[11], Jon Jenkins[4], Yoji
Kondo[12], David Latham[9], Jack Lissauer[1], and David Monet[13]

[1]NASA Ames Research Center, Moffett Field, CA, USA
[2]University of California-Berkeley, Berkeley, CA, USA
[3]Las Cumbres Observatory Global Telescope, Golenta, CA USA
[4]SETI Institute, Mountain View, CA, USA
[5]Aarhus University, Denmark
[6]University of Texas at Austin, Austin, TX, USA
[7]Lowell Observatory, Flagstaff, AZ, USA
[8]Jet Propulsion Laboratory, Pasadena, CA, USA
[9]Smithsonian Astrophysical Observatory, Cambridge, MA, USA
[10]Space Telescope Science Institute, Baltimore, MD, USA
[11]Lawrence Hall of Science, UC-Berkeley, Berkeley, CA, USA
[12]NASA Goddard Space Flight Center, Greenbelt, MD, USA
[13]United States Naval Observatory, Flagstaff, AZ, USA

Abstract. The *Kepler Mission* is a space-based photometric mission with a differential photometric precision of 14 ppm (at $V = 12$ for a 6.5 hour transit). It is designed to continuously observe a single field of view (FOV) of greater then 100 square degrees in the Cygnus-Lyra region for four or more years. The primary goal of the mission is to monitor more than one-hundred thousand stars for transits of Earth-size and smaller planets in the habitable zone of solar-like stars. In the process, many eclipsing binaries (EB) will also be detected and light curves produced. To enhance and optimize the mission results, the stellar characteristics for all the stars in the *Kepler* FOV with $V < 16$ will have been determined prior to launch. As part of the verification process, stars with transit candidates will have radial-velocity follow-up observations performed to determine the component masses and thereby separate eclipses caused by stellar companions from transits caused by planets. The result will be a rich database on EBs. The community will have access to the archive for further analysis, such as, for EB modeling of the high-precision light curves. A guest observer program is also planned to allow for photometric observations of objects not on the target list but within the FOV.

Keywords. Planet detection, exoplanets, differential photometry, eclipsing binaries

1. Introduction

The *Kepler Mission* is NASA's first mission capable of detecting Earth-size and smaller planets in orbit around solar-like stars. It is a photometric space-based mission designed specifically to search for habitable planets orbiting in or near the habitable zone (HZ) of solar-like stars by detecting sequences of planetary transits, a method first described by Borucki & Summers (1984). A habitable planet is taken to be from 0.8 to 2.2 $R_\oplus$ or, if one assumes an Earth-like density, from about 0.5 to 10 $M_\oplus$. For planets less than about 0.5 $M_\oplus$ the surface gravity is too small to retain a life sustaining atmosphere. If $M_\oplus$ is greater than about 10, then there is sufficient gravity for a planet to hold onto the lightest and most abundant gases, H and He, and become a gas-giant. The HZ is taken to be the range of distances from a star where liquid water can exist on the surface of a planet (Kasting *et al.* 1993). Specifically the *Kepler Mission* is designed to be capable of detecting a 1.0 Earth-size planet at 1 AU from a solar-like (G2V) star in four years (four transits) with a single transit signal-to-noise ratio of four for a grazing transit of just 6.5 hours duration. An Earth-size transit of a solar-like star produces a relative change in brightness of 84 parts per million and lasts for 13 hours when it crosses the center of the star. Given this detection sensitivity, *Kepler* has a broad detection capability from a Mars-size planet in the HZ of a $V = 9$ G8-dwarf or $V = 13$ M2-dwarf to a two-Earth radii planet in the HZ of a $V = 13$ F0-dwarf to planets significantly smaller than the Earth for orbital periods on the order of a few days. The latter will produce hundreds of transits during the four year mission lifetime.

The mission will produce an unprecedented photometric database in terms of the continuity, duration, photometric precision and number of stars observed. This unique database should provide a wealth of information for other astrophysical uses. Examples are given below.

2. Mission Overview

The *Kepler Mission* is based upon a classical Schmidt telescope design with a 95-cm aperture and more than one-hundred square degree field of view (FOV). The FOV is equivalent to about six Palomar Schmidt plates. The FOV is located in the Cygnus-Lyra region centered on RA = 19^h 22^m 40^s, Declination = $44°$ $30'$, just above the galactic plane and looking down the Orion arm of the Galaxy. This provides a rich star field that is continuously viewable throughout the year as the spacecraft orbits the Sun and a sample of stars similar to our local neighborhood. The typical stellar distances for most of the usable stars are from a few hundred parsecs to about 1 kpc. Shot noise limits the distance of usable stars. The apparent magnitude range is from about $V = 9$ to 15.

The FOV is oriented such that all but twelve stars brighter than V = 6 are placed in the gaps between the CCDs. The focal plane consists of 42 CCDs, which are 2200 columns by 1024 rows of 27 micron pixels, thinned-back illuminated and anti-reflection coated. Each CCD has two output amplifiers and a pair of CCDs is mounted together to form a square module. The modules are mounted in a 5×5 array with each of the four corner modules having instead a 512×512 pixel CCD. These are used as fine guidance sensors.

The single FOV will be viewed for the entire mission. However, every three months the photometer-spacecraft must be rotated 90° about the optical axis to keep the fixed-solar array pointed to the Sun and the focal-plane radiator pointed to deep space. The positions and orientations of the CCD modules have been chosen so that the focal plane is four-fold symmetric. Thus after a 90° rotation all of the selected target stars remain on active pixels. The photometer-spacecraft will be launched into an Earth-trailing heliocentric orbit, similar to Spitzer. It is expected to drift away from the Earth at the rate of about 0.1 AU per year. The mission is scheduled for launch on a Delta II in November 2008.

The mission has been described in a number of papers (Borucki *et al.* 2005, Koch *et al.* 2004). The basic photometric capability and mission design have remained nearly the same since selection by NASA in December 2001 as the tenth Discovery mission. The detailed design has matured, and the three fundamental design requirements have not been descoped:

1. Detection of an Earth-size transit at 4σ for a $V = 12$ solar-like star in 6.5 hours;

2. The capacity to observe 170,000 stars for at least the first year and 100,000 stars at the end of the mission; and

3. A mission duration of at least four years.

A recent significant change that was made to the mission design was to go from an articulated high-gain antenna (HGA), which allowed for downlinking of the science data every four days without interrupting the observing and having a fifteen-minute integration time, to a body-mounted HGA, which requires a one-day interruption in observing every 31 days for data downlink and required the integration time to double to thirty minutes in order to halve the amount of downlink data.

3. Photometric Characteristics

To understand both how planet detection is performed and how the data might be used for other astrophysical purposes we describe here the basic characteristics of the *Kepler* photometer and the data processing.

We describe the photometric precision of *Kepler* in terms of all the possible sources of noise that have to be taken into account when looking for transits in the light curves. We refer to this as the Combined Differential Photometric Precision (CDPP) and we use the units of parts per million (ppm). The CDPP includes the photon shot noise, variability of the source and the measurement noise. The latter includes not only the CCD read noise and electronics noise, but also the shot noise from the sky due to background stars, zodiacal light, stray light in the photometer, ghosting in the optics, etc. The design requirement is a CDPP $\leq$ 20 ppm for a $V = 12$ G2V star and 6.5-hour integration. For design purposes we have assumed stellar variability, which is red noise, to be 10 ppm for a solar-like star for a 6.5-hour integration. For a 6.5 hour integration, a $V = 12$ G2V star produces 5×10^9 electrons in the *Kepler* photometer, yielding a shot noise of 14 ppm. Thus the measurement noise by design has to be $\leq$10 ppm at $V = 12$ for the CDPP to meet the requirement. Design and testing indicate that this will be achieved. The dynamical range of the photometer for planet detection is roughly from $V = 9$ to $V = 15$. This depends on the stellar type and apparent brightness, the planetary size and orbit and the mission duration. At $V = 9$ the required CDPP $\leq$ 12 ppm and at $V = 14$ the required CDPP $\leq$ 41 ppm for a 6.5-hour integration.

Unlike most conventional telescopes, the *Kepler* photometer has a rather broad point spread function (psf) of about six arcsec and large pixels. One pixel is $3''.98$. This is necessary due to the very large FOV, the need for large well depth to accommodate the shot noise requirement and the desire to minimize the photometric sensitivity to intrapixel quantum efficiency variations.

The system has a capacity at the beginning of the mission for 170,000 stars. Sometime prior to the fourth year this needs to be cut back to 100,000 stars, simply being limited by the telecom system usage. All of these stars will have a sampling resolution of thirty minutes. A subset of 512 stars will have a sampling resolution of one minute. Initially this will be used to measure p-mode oscillations in the brighter stars. Once transits are detected, some portion of these short-cadence targets will be used to observe high signal-to-noise ratio (SNR) transits.

To minimize the shot noise, the photometer has a single broad bandpass from 430 to 890 nm FWHM. The short wavelength cutoff was selected to avoid the Ca II H&K lines. For the Sun, 60% of the irradiance variation is at wavelengths less than 400 nm, but only accounts for about 12% of the total flux (Krivova *et al.* 2006). Hence the 430 nm cutoff helps to improve the system SNR. The red cutoff was chosen to avoid fringing within the CCDs and to minimize the potential for false positives produced by faint reddened background EB.

4. Target Selection

The primary goal of *Kepler* is to detect terrestrial planets around solar-like stars, that is, late-dwarf (F, G, K and M) stars. The magnitude range is roughly $V = 9$ to 15. Since there did not exist a catalog to this depth near the galactic plane with the necessary information on which to base the selection, the project undertook the process of creating the *Kepler Input Catalog* (KIC) led by a team from SAO. The KIC is based on new multi-band photometric observations using the same $g - r - i - z$ filter sequence as the Sloan survey (Abazajian *et al.* 2003) plus an additional filter for the Mg b lines at 516.7, 517.3 and 518.4 nm. The net result will be a catalog with classification information for the two million stars in the *Kepler* FOV to $K < 14$. Using model fitting and a newly expanded Kurucz library of spectra (Castelli & Kurucz 2003), the catalog will contain the effective temperature, $log(g)$, metallicity [Fe/H], reddening-extinction, mass and radius. The catalog will be federated at the USNO Flagstaff Station with other catalogs, including 2MASS and USNO-B for cross reference. Each star will be ranked for its potential for terrestrial planet detection using the KIC and a merit function to determine the minimum detectable planet size in or near the HZ of each star. This process will be reapplied post-launch incorporating the measured CDPP for each star to re-rank the stars on the target list as part of a necessary on-going down-selection process.

5. Data Processing

On-board the integration time for the CCDs may be set from 2.5 to 8.0 sec. The longer integration time helps to improve the CDPP for the fainter objects, but also results in saturation of more stars. With an integration time of 5 sec, stars brighter than about $V = 12$ saturate. However, we have demonstrated in laboratory testing that precision photometry can still be achieved with bloomed pixels provided the rail voltages are properly set on the CCD and that the full scale on the analog-to-digital converter (ADC) is greater than full well. The CCDs are read out in half a second without a shutter. Individual integrations are co-added on-board for thirty minutes, although for a subset of 512 target stars, one-minute co-additions are preserved. Once a thirty-minute co-add is accumulated in computer memory for all the 95 megapixels in the focal plane, the pixels of interest for the target stars are read out, about 3% of the pixels. There are additional collateral pixels from over-clocking and from masked regions. These are used for removing the bias, smear and determining the dark level. The values are re-quantized to account for the larger value of noise on the high end of the scale. The data are then Huffman encoded (compressed) and stored on the solid-state recorder for later transmission to the ground.

Smear is a result of clocking out the data without a shutter. Every pixel in a column passes under every piece of sky in a column during a read out. This produces a column unique constant offset. And it also produces an optical fat-zero that helps to keep the traps full.

On the ground, the raw data are unpacked, decompressed and archived. The bias and smear are removed. Cosmic ray hits are removed and the background flux is estimated and removed. Then two parallel paths are followed to obtain stellar flux time series, one using aperture photometry and the other using difference image analysis. Ensemble normalization is applied to the raw flux time series to remove common-mode noise at each cadence in time. These data are then used to produce de-trended relative flux time series for each target. These light curves are then archived and will be made available for others to use, for example for modeling eclipsing binary systems. To each time series a wavelet transform is applied and conditioned with a whitening matched filter. The time series are then folded modulo all possible periods and searched for a multiple-event-detection statistic above a threshold of seven sigma for planetary transits, yielding an 84% detection rate for an eight sigma folded transit signal. A similar process using a Fourier transform rather than a wavelet transform is used to conduct the reflected light search for short period non-transiting giant planets.

6. Eclipsing Binary Detection

From the processing described above, we expect to detect all the eclipsing binaries from within our target set with periods shorter than the mission duration as well as a significant fraction of those with even longer periods that exhibit only a single eclipse. Note that a single grazing Earth-like transit of a $V = 12$ solar-like star has an SNR $= 4$ and a gas giant like Jupiter produces a 1% transit depth which for *Kepler* will have an SNR ≈ 400. So even a single grazing eclipse of a long period binary will have a recognizable character in the data with a significant SNR. Recall that an eclipsing binary signature of a V-shape signal will be significantly different from the U-shape character of a planetary transit. The eclipsing binary events will be cataloged for the astronomical community to further analyze as part of the Data Analysis Program (see below), perhaps using automated processes such as eBAI (Guinan *et al.*, 2007).

For a typical terrestrial planetary transit, the SNR will be too low to distinguish the event from a grazing eclipsing binary (a transiting white dwarf will actually cause a brightening (Sahu & Gilliland 2003)). Stellar companions will induce radial velocity variations that are typically $\gg 1$ $\mathrm{km\,s^{-1}}$. The *Kepler* project will conduct moderate resolution spectroscopy to identify these situations using, for example, the SAO Digital Speedometer (Latham 1992). The net result is that the *Kepler Mission* will produce high precision light curves for more than a thousand eclipsing binaries that are expected to be detected by the mission and available for use by the astrophysical community.

What about non-eclipsing binaries that are on the *Kepler* target list? Many of these can also be identified in a rather circuitous fashion. The most accurate way (to a few percent) to obtain the stellar size is to use the distance to the star, the measured flux, the extinction-reddening, the effective temperature from the KIC and to apply the Stefan-Boltzmann law to derive the stellar area. If the area appears to be something like twice what one would expect for a dwarf with the given effective temperature, based on the modeling that went into deriving $\log(g)$ in the KIC, then the target must be either a binary or perhaps a chance optical double star. Spectroscopic observations will in most cases lead to recognition of its true nature. With regard to the distance, we are expecting that this can be derived from astrometric parallaxes out to 1 kpc using the *Kepler* data. Based upon laboratory measurements (Koch, *et al.* 2000), simulations and the high SNR of the data, this does appear to be plausible. The GAIA mission (Niarchos, Munari and Zwitter 2006) results will also be very useful in refining and confirming some of these cases. Getting the stellar areas correct, especially eliminating a factor of two ambiguity, is necessary for correctly determining the planetary areas.

7. Additional Astrophysical Uses for the Data

Given the uniqueness in precision, completeness, duration and number of stars in the archived data there are potentially a host of other astrophysical uses for the *Kepler* data. These include such things as; measuring p-mode oscillations (Brown & Gilliland 1994), which yield the density and age of stars; analysis of stellar activity, which can yield star spot cycles (especially if the mission is extended beyond about half of a solar cycle) and white light flaring; the frequency of Maunder minimums for solar-like stars, which has implications for paleoclimatology and perhaps the future of our Earth's climate; stellar rotation rates; cataclysmic variables, providing pre-outburst activity and mass transfer rates; and active galactic nuclei, providing a measure of the "engine" size in BL Lacs, quasars and blazers. Performing p-mode observations of all of the usable stars in the FOV (about 5000 are bright enough) has already been planned as part of the *Kepler* science team effort.

8. Community Participation and Data Access

The community can participate in the *Kepler* results in several ways: a guest observer (GO) program, a data analysis program (DAP) and a participating scientist program (PSP), all of which will be competed by NASA Headquarters through the annual Research Opportunities in Space and Earth Sciences (ROSES). Although the program is open to scientists worldwide, only United States proposals may receive funding.

8.1. *Guest Observer Program*

Within the GO program, scientists may propose to view objects within the *Kepler* FOV which are not already on the planet detection target list, whether galactic or extra-galactic. These objects may be intrinsically variable stars, such as, pulsating (Cepheids, RR Lyrae, Mira, etc.), rotating (ellipsoidal, etc.), eruptive (T Tauri, Wolf-Rayet, etc.) and explosive (novae, super-novae, cataclysmic variables). In general one may assume that all F-, G-, and K-dwarf stars to $V = 14$–15 and M-dwarf stars to $K = 14.5$ are already on the list and any other object is not. Proposed objects will typically be observed for a minimum of three months and to as long as the mission duration based on justification. We will try to schedule the observations to coincide with any other coordinated observing. Capacity for three thousand objects has been set aside for this program. These will be at thirty-minute cadence except that twenty-five can be allocated at one-minute cadence. The data will be processed using the standard *Kepler* pipeline to produce de-trended light curves. The raw data will also be archived and available if the investigators choose to perform their own processing, for example, if one wants to carry out psf fitting to the pixel level data.

8.2. *Data Analysis Program*

All the *Kepler* data will be placed in the Multi-mission Archive at the STScI (MAST) and open to the public once the data have been validated and release approved by the principal investigator and NASA Headquarters, based on a pre-arranged schedule. There will be a latency of about one year due to the one-month intervals between data downlinks, transmission gaps which may not be recovered for another month, the need to de-trend the data in blocks of three months (each roll orientation), etc. Keep in mind that the nature and usefulness of the data has to do with the greatest extent of the time series for each object, not what is contained in a single snapshot of the FOV. This is not an imaging mission. Thus there will probably be more value and emphasis on DAP investigations that make use of the full mission length rather than snippets of early data. The archived data will consist of the calibrated and de-trended flux times series (light curves) for all target stars observed. GOs, however, should receive their data much sooner. We expect to reprocess the data several times based on a better understanding of the nature of the photometry as the mission progresses, including a final reprocessing of the entire data set following the end of flight operations. The data will be maintained at the MAST for up to ten years after the end of the mission.

8.3. *Participating Scientist Program*

The PSP is separate from the GO and DAP programs in that the PSP is reserved for proposals that complement and enhance the core planet detection scientific program of the *Kepler Mission*. These programs might consist of supportive ground based follow-up observing, additional or improved analysis methods for planet detection, etc.

9. Status and Summary

The *Kepler Mission* is progressing through development toward a launch at the end of 2008. The optics have been delivered and tested. All of the flight CCDs have been in hand for over a year. The photometer assembly, integration and test will be completed within the year and delivered for spacecraft integration. The mission is designed to be capable of detecting hundreds of terrestrial planets, if they are common, or provide a significant null result if they are not. Either result would be profound. In addition, the photometric data base will be unique in precision, completeness, duration and number of objects and serve the community in many areas of astrophysical research for years to come.

Acknowledgements

Funding for this mission is provided by NASA's Discovery Program Office.

References

Abazajian, Kevork *et al.* 2003, *ApJ*, The First Data Release Of The Sloan Digital Sky Survey, 126, 2081–2086

Borucki, W. J. & Summers, A.L. 1984, *Icarus* 58, 121

Borucki, W. J., Koch, D., Basri, G., Brown, T., Caldwell, D., DeVore, E., Dunham, E., Gautier, T., Geary, J., Gilliland, R., Gould, A., Howell, S., Jenkins, J., & Latham, D. *Kepler Mission*: Design, Expected Science Results, Opportunities to Participate, 2005, in *A Decade of Extrasolar Planets around Normal Stars* (ed. M. Livio), Cambridge: Cambridge University Press, in preparation

Brown, Timothy M. and Gilliland, Ronald L. *Asterioseismology*, 1994, ARAA, 32, 37–82

Castelli, F. & Kurucz, R. L. 2003, New grids of ATLAS9 model atmospheres, in *IAU Symposium 210, Modelling of Stellar Atmospheres*, (eds. N.E. Piskunov, W.W. Weiss. and D.F. Gray) 2003.

Guinan, E. F., Engle, S. G., & Devinney, E. 2007, in *Solar and Stellar Physics through Eclipses* (eds. O. Demircan, S. O. Selam and B. Albayrak), ASP Conf. Series, in press

Latham, D. 1992, in *IAU Coll. ASPCS*, McAlister and Hartkopf (eds.), 32:135, 110

Kasting, J.F., Whitmire, D.P., & Reynolds, R.T. 1993, *Icarus* 101, 108

Koch, D. G., Borucki, W., Dunham, E., Jenkins, J., Webster, L., & Witteborn, F. 2000, CCD Photometry Tests for a Mission to Detect Earth-Size Planets in the Extended Solar Neighborhood, in *SPIE Conference 4013, UV, Optical and IR Space Telescopes and Instruments*, Munich, Germany

Koch, D., Borucki, W., Dunham, E., Geary, J., Gilliland, R.,, Jenkins, J., Latham, D., Bachtell, E., Berry, D., Deininger, W., Duren, R., Gautier, T. N., Gillis, L., Mayer, D., Miller, C., Shafer, D., Sobeck, C., Stewart, C., & Weiss, M. 2004, Overview and status of the *Kepler Mission*, in *SPIE Conf 5487, Optical, Infrared, and Millimeter Space Telescopes*, Glasgow, Scotland

Krivova, N. A., Solanki, S. K. & Floyd, L. 2006, *A&A*, 452, 631

Niarchos, P.G., Munari, U., & Zwitter, T. 2007, these proceedings, 244

Sahu, K. C. & Gilliland, R. L. 2003, *ApJ*, 584, 1042

Discussion

RALPH GAUME: Given the reprioritizations going on now at NASA Headquarters and the fact that *Kepler* is getting close to launch, do you feel that *Kepler* is safe from budget reprogramming?

KOCH: I don't know of any program that is "safe". We have had to change our baseline about a dozen times, mostly due to external influences. I think all programs are on thin ice and very often one does not know how thin the ice may be at any one time.

CARLSON CHAMBLISS: Low-metallicity stars are probably less likely than others to have Earth-like planets. Are these going to be systematically removed from the 100,000-star sample?

KOCH: No. The *Kepler Mission* is designed to conduct an unbiased search. The mission has enough capacity to include every star in the field of view for which an Earth-size planet can be detected.

Binary Stars as Critical Tools & Tests
in Contemporary Astrophysics
Proceedings IAU Symposium No. 240, 2006
W.I. Hartkopf, E.F. Guinan & P. Harmanec, eds.

Evaluating the Gaia Contribution to the Field of Eclipsing Binaries with Ground-Based Spectroscopy and Hipparcos Photometry

P. G. Niarchos[1]†, U. Munari[2,3] and T. Zwitter[4]

[1]Department of Astrophysics, Astronomy and Mechanics, National and Kapodistrian University of Athens, Athens, Greece
email: pniarcho@phys.uoa.gr

[2]Osservatorio Astronomico di Padova, Sede di Asiago, 36012 Asiago (VI), Italy

[3]Dipartimento di Astronomia dell'Universita di Padova, Osservatorio Astrofisico, 36012 Asiago (VI), Italy

[4]University of Ljubljana, Department of Physics, Jadranska 19, 1000 Ljubljana, Slovenia

Abstract. During its definition phase, ESA's Cornerstone mission Gaia was designed to perform extremely accurate photometry in 10 medium plus 5 broad bands and to collect about 90 epoch spectra with a resolving power of 11500 and a wavelength range 8480–8740 Å (centered on the CaII triplet in the far red). Combining epoch photometry from the ground and the Hipparcos mission with ground-based spectra strictly simulating Gaia ones, we have investigated in a series of papers the performance expected from Gaia on SB2 EBs. We review here the results we obtained. The Gaia design is now under major revision, and its impact on EBs will be briefly addressed.

Keywords. Surveys – stars: fundamental parameters – binaries: eclipsing – binaries: spectroscopic

---

1. Introduction

Gaia is one of the next cornerstones of ESA's science programme (2000). It will be launched around end-2011 with a Soyuz–Fregat launcher and it will be put in a Lissajous-type orbit around the Sun-Earth Lagrangian point L2. The objectives of the Gaia mission are to build a catalogue of $\sim 10^9$ stars with accurate positions, parallaxes, proper motions, magnitudes and radial velocities. The catalogue will be complete up to $V = 20^{\mathrm{th}}$ mag with no input catalogue and therefore no associated bias.

The primary goal of Gaia is to explore the composition, formation and evolution of the Galaxy by studying the dynamics and intrinsic properties of a wide range of stellar types across the whole Galaxy. The quantities complementary to the kinematics can be derived from the spectral energy distribution of the stars by spectral photometry (dispersed images) and spectroscopy. The Gaia core science case requires measurement of: distance, space velocity, spectral energy distribution, duplicity, atmospheric parameters from which absolute quantities can be derived, like: luminosity, temperature, metallicity, galactic orbit.

The spacecraft and payload configuration was re-optimised by the industrial teams in their Phase B2/C/D proposal in response to the mission requirements document issued

† Present address: Dept. of Astrophysics, Astronomy and Mechanics, University of Athens, Greece.

by the ESA project team in 2005. As a result, from early 2006, the final Gaia payload looks somewhat different from the previous design, although all functionality is preserved.

2. Gaia's contribution to the study of eclipsing binaries

Gaia is designed to obtain for a large sample of stars extremely precise micro-arcsec astrometry, spectral photometry with two photometers (Blue and Red) providing low resolution spectrophotometric measurements for each object over the wavelength ranges 330–660 and 650–1000 nm, respectively, and medium resolution spectroscopy. The Radial-Velocity Spectrograph (RVS) will register spectra of all objects brighter than about 17th mag.

It is expected that about 1×10^6 EBs (with $V \leqslant 16^{th}$ mag) will be discovered and some 10^5 of these will be characterized as double-lined in Gaia spectral observations. Moreover, most of the Gaia binaries will be of spectral type G or K (Zwitter & Henden 2003) for which accurate solutions exist for only a small number of systems. Even if for only 1% of the observed EBs reliable parameters are derived, this will be a giant leap in comparison with what has been obtained so far from ground-based observations. The number of photometric points per star in the five-year mission lifetime is estimated to be ~ 70 to 100 and the number of spectra ~ 40. The observing fashion will be quite similar to Hipparcos operational mode.

The aim of the present investigation is to review the results of the investigations made so far aiming to evaluate the Gaia contribution to the field of Eclipsing Binaries. Such investigations for some detached, semi-detached and contact binaries of the basic spectral types A, F, G, K have been done by Munari *et al.* (2001), Niarchos & Manimanis (2003), Zwitter *et al.* (2003), Marrese *et al.* (2004), Niarchos *et al.* (2005) and Milone *et al.* (2005). These investigations are based on ground-based and Hipparcos (Gaia-like) photometry, and on ground-based (Gaia-like) spectroscopy.

The first step for such an evaluation can be accomplished by studying a small sample of EBs, using ground-based and Hipparcos (Gaia-like) observations.

3. Selection of systems

There are crucial questions regarding the reliability of the derived stellar parameters from Gaia observations. Some of them are:

- How Gaia observations can be compared with the state-of-the-art ground-based observations?
- Can Gaia observations permit us to derive fundamental parameters as well or better than previous work?
- Can it add substantially to the list of very well-determined systems from ground-based data?
- What is the accuracy to which Close Eclipsing Binaries can be investigated using Gaia data alone?

The first step in our study was the selection of proper eclipsing systems to be analysed. The following criteria of selection were adopted:

(*a*) EBs observed by the Hipparcos/Tycho mission (H_P observations)

(*b*) Ground-based photometric observations of high quality for the same systems (V observations)

(*c*) Accurate spectroscopic mass ratios for the same systems (determined from radial-velocity measurements using modern techniques)

Table 1. Basic data of the systems of group 1

System	State	Spectral type	Mass ratio q	Reference
RZ Dra	semi-detached	A5 + K2	0.444	1
V1010 Oph	semi-detached	A5 + F6	0.448	1
AB And	contact	G5	0.560	1
ϵ CrA	contact	F2	0.112	1
YY CrB	contact	F8	0.243	2
XY Leo	contact	K0	0.500	2
XZ Leo	contact	A5	0.348	2
V566 Oph	contact	F4 + F8	0.237	2
V839 Oph	contact	G0	0.305	2
AH Vir	contact	K0	0.303	2

1: Niarchos & Manimanis 2003 (and references therein); 2: Niarchos *et al.* 2005 (and references therein).

Table 2. Basic data of the systems of group 2

System	State	Spectral type	Mass ratio q	Reference
V432 Aur	detached	G0	0.560	3
SV Cam	semi-detached	G5	0.303	4
BS Dra	detached	F5	0.500	4
GK Dra	detached	G0	0.112	2
HP Dra	detached	G5	0.500	4
UV Leo	detached	G0	0.237	2
UW LMi	detached	G0	0.305	3
CN Lyn	detached	F5	0.500	3
OO Peg	detached	A2	0.348	1
V505 Per	detached	F5	0.448	1
V570 Per	detached	F5	0.444	1
V781 Tau	contact	G0	0.243	2

1: Munari *et al.* 2001 (and references therein); 2: Zwitter *et al.* 2003 (and references therein); 3: Marrese *et al.* 2004 (and references therein); 4: Milone *et al.* 2005 (and references therein).

(*d*) Main types of EBs (Detached, Semi-detached, Contact)

The systems selected were divided into two groups according to the observations used for their analysis: For the systems of group 1: H_P photometry, ground-based V photometry and accurate mass ratios determined fron ground-based spectroscopy were used; For the systems of group 2: H_P photometry and accurate mass-ratios determined from ground-based Gaia-like spectroscopy (Asiago spectroscopy) were used. For one system of group 2 (V 505 Per) ground-based photometry was also used.

A higher resolving power R $= \lambda/\Delta\lambda = 17000$ for ground-based spectroscopy over the Gaia wavelength interval (8480–8740 Å) has been adopted to compensate for the lower number of spectra that we have secured compared to the twice larger that Gaia will collect at a resolving power 11500. The spectroscopic observations have been obtained with the Echelle+CCD spectrograph on the 1.82-m telescope operated by Osservatorio Astronomico di Padova atop Mt. Ekar (Asiago). A 2″.2 slit width was adopted to match the R $= \lambda/\Delta\lambda = 17000$ requirement. The detector was a UV-coated Thompson CCD 1024×1024 pixel, 19 micron square size. The Gaia spectral range is covered without gaps in a single order by the Asiago Echelle spectrograph. The actual observations however extended over a much larger wavelength interval (4500–9000 Å). Here we will limit the analysis to the Gaia spectral interval.

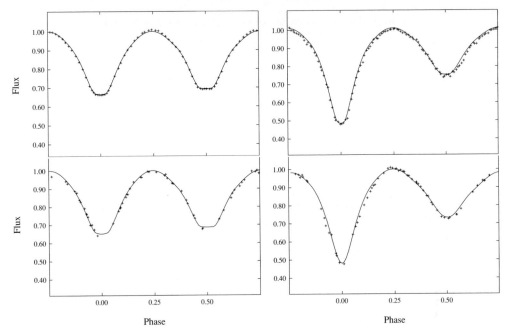

Figure 1. Light curves of the contact system V566 Oph (left) and the semi-detached system V1010 Oph (right). The upper curves are from ground-based V observations and the lower ones from Hipparcos H_P data. The solid lines represent our light-curve solutions.

4. Method of analysis and comparison of absolute elements

The Wilson-Devinney DC program was used for the analysis in the appropriate mode. The free parameters were: $\phi_\circ, i, T_2, \Omega$ (according to the mode used) and L_1, while the fixed ones were: T_1 (from spectral type); theoretical values for g, A and x were assumed based on spectral type. The spectroscopic mass ratio q was used as fixed parameter for systems of group 1, while simultaneous radial velocity and light curve analysis was performed for systems of group 2. The third light was assumed zero in all cases.

The elements derived from the light-curve analyses were combined with the available spectroscopic elements in order to compute the absolute elements (physical parameters) of the systems studied. These parameters are listed in Tables 3 and 4 together with their percentage difference. Four examples of light curves and their solutions are given in Figures 1 and 2.

5. Summary and Conclusions

The agreement between the fundamental stellar parameters, derived from ground-based and Hipparcos (Gaia-like) observations is more than satisfactory. Their difference is within the limits of the combined errors. The stellar parameters could be determined (in many cases) at about 2% accuracy level. Of course there are exceptions for active systems and for those with poor photometric phase coverage.

The strength of the Gaia mission is in the numbers. Gaia will observe $\sim 4 \times 10^5$ eclipsing binaries brighter than $V \leqslant 15$ and $\sim 10^5$ of these will be double-lined (SB2) systems. For $V \leqslant 13$ the number of SB2 will be about 16 000 for which Gaia should provide orbital solutions formally accurate to $\sim 2\%$ (Munari *et al.* 2004). This is a fantastic number compared to $\leqslant 100$ systems studied at similar accuracy by ground-based observations so far. Moreover most of the Gaia binaries will be of G-K spectral type (Zwitter & Henden

P. G. Niarchos *et al.*

Table 3. Absolute elements of group 2 systems in solar units

System	M_1	M_2	R_1	R_2	L_1	L_2
V432 Aur(2)	0.98(2)	1.06(2)	1.39(8)	2.13(14)	2.36(20)	4.86(35)
SV Cam(2)	0.86(10)	0.65(8)	0.98(10)	1.18(12)	0.99(14)	0.33(10)
BS Dra(2)	1.29(8)	1.28(9)	1.46(2)	1.40(7)	3.61(4)	3.34(18)
GK Dra(2)	1.46(7)	1.81(7)	2.43(4)	2.83(5)	13.26(15)	15.84(18)
HP Dra(2)	1.10(2)	1.10(3)	1.17(3)	0.89(9)	1.57(5)	1.16(16)
UV Leo(2)	1.21(9)	1.11(10)	0.97(2)	1.22(4)	1.17(4)	1.43(7)
UW LMi(2)	1.06(2)	1.04(2)	1.23(5)	1.21(6)	2.39(16)	2.31(7)
CN Lyn(2)	1.04(2)	1.04(2)	1.80(21)	1.84(24)	5.11(53)	5.19(60)
OO Peg(2)	1.72(3)	1.69(3)	2.19(8)	1.37(5)	25.07(44)	9.43(24)
V505 Per(1)	1.26(1)	1.25(1)	1.29(3)	1.27(7)	2.60(8)	2.36(15)
V505 (2)	1.30(2)	1.28(2)	1.40(2)	1.14(3)	2.96(5)	1.94(6)
% difference	**3.1**	**2.3**	**7.8**	**11.3**	**12.1**	**21.6**
V570 Per	1.28(3)	1.22(3)	1.64(16)	1.01(25)	4.14(37)	1.33(41)
V781 Tau(2)	0.51(1)	1.15(3)	0.76(1)	1.11(1)	0.85(2)	1.57(2)

(1): ground-based observations (Marschall *et al.* 1997), (2): Gaia (expected) obs.
(Munari *et al.* 2001, Zwitter *et al.* 2003, Marrese *et al.* 2004, Milone *et al.* 2005)

Table 4. Absolute elements of group 1 systems in solar units

System	M_1	M_2	R_1	R_2	L_1	L_2
RZ Dra (1)	1.40(4)	0.62(3)	1.62(1)	1.12(1)	10.1(4)	1.01(6)
RZ Dra (2)	1.42(5)	0.59(4)	1.55(1)	1.11(1)	9.60(47)	0.86(5)
% difference	**1.4**	**4.8**	**4.3**	**0.9**	**5**	**15**
V1010 Oph (1)	1.87(12)	0.90(5)	2.08(6)	1.48(8)	10.7(7)	3.47(35)
V1010 Oph (2)	1.88(14)	0.89(7)	2.08(1)	1.46(1)	11.7(8)	3.07(8)
% difference	**0.5**	**1.1**	**0**	**1.4**	**9.3**	**12**
AB And (1)	1.01(2)	0.56(1)	1.04(4)	0.80(5)	0.87(7)	0.46(5)
AB And (2)	1.00(9)	0.55(5)	1.06(4)	0.83(5)	0.91(7)	0.45(5)
% difference	**1.0**	**1.8**	**1.9**	**3.7**	**4.6**	**2.2**
ϵ CrA (1)	1.75(4)	0.21(2)	2.20(3)	0.80(1)	11.1(1)	1.08(1)
ϵ CrA (2)	1.69(7)	0.22(5)	2.12(12)	0.80(4)	10.3(20)	1.07(14)
% difference	**3.4**	**4.8**	**3.6**	**1.2**	**7.2**	**0.9**
YY CrB (1)	1.41(9)	0.34(2)	1.40(7)	0.77(10)	2.46(26)	0.81(20)
YY CrB (2)	1.37(16)	0.33(4)	1.36(7)	0.74(10)	2.33(25)	0.74(20)
% difference	**2.8**	**2.9**	**2.9**	**3.9**	**5.3**	**8.6**
XY Leo (1)	0.82(4)	0.41(2)	0.87(1)	0.64(2)	0.38(2)	0.14(1)
XY Leo (2)	0.83(21)	0.41(10)	0.86(1)	0.62(2)	0.37(2)	0.17(1)
% difference	**1.2**	**0.7**	**1.1**	**3.1**	**2.6**	**21.4**
XZ Leo (1)	1.83(5)	0.64(2)	1.71(2)	1.07(3)	7.29(29)	2.56(16)
XZ Leo (2)	1.82(13)	0.63(6)	1.71(2)	1.07(3)	7.26(29)	2.69(19)
% difference	**0.5**	**1.6**	**0.2**	**0.2**	**0.4**	**5.1**
V566 Oph (1)	1.40(3)	0.33(1)	1.47(1)	0.79(1)	4.57(4)	1.26(1)
V566 Oph (2)	1.54(11)	0.36(4)	1.518(3)	0.819(4)	4.99(24)	1.26(2)
% difference	**10**	**9.1**	**3.5**	**3.7**	**9.2**	**0.8**
V839 Oph (1)	1.62(4)	0.49(1)	1.53(6)	0.93(8)	3.12(26)	1.32(23)
V839 Oph (2)	1.61(17)	0.49(5)	1.52(6)	0.92(8)	3.10(26)	1.22(21)
% difference	**0.6**	**0.8**	**0.7**	**1.1**	**0.6**	**7.6**
AH Vir (1)	1.45(13)	0.44(4)	1.36(1)	0.77(1)	2.45(8)	0.71(2)
AH Vir (2)	1.33(20)	0.40(6)	1.44(1)	0.88(1)	2.76(9)	0.91(3)
% difference	**8.3**	**9.1**	**5.9**	**14.3**	**12.7**	**28.2**

(1): ground-based observations, (2): Gaia (expected) observations
(Niarchos & Manimanis 2003, Niarchos *et al.* 2005)

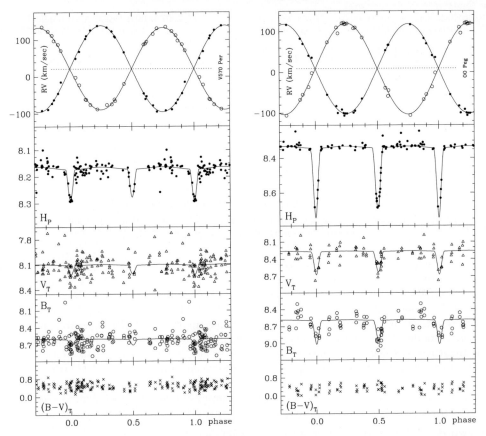

Figure 2. Radial velocity curves (from Asiago spectroscopy) and Hipparcos H_P and Tycho V_T, B_T, $(B - V)_T$ light curves of the detached systems V505 Per (left) and OO Peg (right). The solid lines represent our light curve solutions. (From Munari *et al.* 2001)

2003) for which accurate solutions exist only for a small number of systems. Although the expected accuracy will be moderate, the large amount of data will allow us to look for large deviations from the "normal" mass-radius-luminosity relations. There is no doubt that Gaia observations of EBs will have an immense impact on theories of stellar structure and evolution.

References

Marrese, P. M., Munari, U., Siviero, A., *et al.* 2004, *A&A* 413, 635 (Paper III)

Marschall, L. A., Stefanik, R. P., Lacy, C. H., *et al.* 1997, *AJ* 114, 793

Milone, E. F., Munari, U., Marrese, P. M., *et al.* 2005, *A&A* 441, 605 (Paper IV)

Munari, U., Tomov, T., Zwitter, T. *et al.* 2001, *A&A* 378, 477 (Paper I)

Munari, U., Zwitter, T., & Milone, E. F. 2004, in: Hilditch, R., Hensberge, H. & Pavlovski, K. (ed.) *Spectroscopically and Spatially Resolving the Components of the Close Binary Stars* (ASP Conf. Ser. 318), p. 422

Niarchos, P. & Manimanis, V. 2003, *A&A* 405, 263

Niarchos, P., Manimanis, V., & Gazeas, K. 2005, in: Turon, C., O'Flaherty, K.S., & Perryman, M.A.C. (ed.), *The Three-Dimensional Universe with GAIA* (ESA SP-576), p. 607

Zwitter, T. & Henden, A.A. 2003, in: Munari, U. (ed.) *Gaia Spectroscopy, Science and Technology* (ASP Conf. Ser. 298), p. 489

Zwitter, T., Munari, U., Marrese, P. M., *et al.* 2003, *A&A* 404, 333 (Paper II)

Binary Stars as Critical Tools & Tests
in Contemporary Astrophysics
Proceedings IAU Symposium No. 240, 2006
W.I. Hartkopf, E.F. Guinan & P. Harmanec, eds.

© 2007 International Astronomical Union
doi:10.1017/S1743921307004103

Multiplicity of Herbig Ae/Be Stars

Sandrine J. Thomas[1,2], Nicole S. van der Bliek[2], Bernadette Rodgers[3], Greg Doppmann[3] and Jérôme Bouvier[4]

[1]LAO, UCO/Lick Observatory, 1156 High Street, Santa Cruz, CA 95064, USA.
email: sthomas@ucolick.org

[2]NOAO/CTIO, Casilla 603, La Serena, Chile
email: nvdbliek@ctio.noao.edu

[3]Gemini Observatory, Casilla 603, La Serena, Chile
email: brodgers@gemini.edu, gdoppmann@gemini.edu

[4]University of Grenoble, BP53, F 38041, Grenoble, Cedex 9, France
email: Jerome.Bouvier@obs.ujf-grenoble.fr

Abstract. One of the most interesting constraints on star formation models comes from the study of multiplicity of young stars as a function of mass. While multiplicity studies of low-mass T Tauri stars have been quite exhaustive, an unbiased and systematic investigation of multiplicity among intermediate-mass Herbig Ae/Be (HAEBE) stars is still lacking. We are therefore conducting a photometric and spectroscopic survey of HAEBE stars to detect companions, establish their physical association with the primary and investigate their properties. The frequency and degree of multiplicity of HAEBE systems will provide new constraints on their formation mechanisms. In this paper we present preliminary results of the high resolution imaging part of the survey, carried out with the adaptive optics system Altair-NIRI on Gemini North. Of 72 stars observed, we find 44 possible binaries or multiples, including at least 25 not previously known.

Keywords. Herbig Ae/Be stars, binaries, pre-main sequence

1. Introduction

Herbig Ae/Be (HAEBE) stars are pre main-sequence stars of intermediate mass (2 to 9 solar masses). With masses between those of low-mass T Tauri stars and high mass young stars, HAEBE stars fill an important parameter space in addressing the question of star formation as a function of mass. At the same time, it is very likely that stars of different mass actually form and evolve together, as many stars form in clusters. A promising approach to constrain star formation models is thus to investigate the frequency of multiple pre-main sequence stars and study their properties as a function of mass of the primary stars. Like their low mass counterparts, the T Tauri stars, HAEBE stars are often found in groups: e.g., Bouvier & Corporon (2001) find that HAEBE stars have a binary fraction significantly higher than both field G dwarfs and low mass young stars (both of order 60%), while Testi *et al.* (1997) find a correlation between stellar density and spectral type. However, there are few studies of multiplicity of HAEBE stars and usually based on relatively small samples. A comprehensive, unbiased survey of the immediate environments of HAEBE stars is still lacking. We are therefore conducting a photometric and spectroscopic survey of HAEBE stars to detect multiple systems and to investigate the nature of the companions. The sample consists of all sources listed in Tables 1-3 of Thé *et al.* (1994) supplemented with HAEBE stars from the literature. This survey will provide us with a statistically meaningful data set on multiplicity of HAEBE stars.

250

Here we present the first results of high resolution imaging observations, using Altair–NIRI at Gemini North (program GN-2005B-Q-81), to search for companions to HAEBE stars.

2. High resolution near infrared imaging

Our total sample consists of more than 300 HAEBE stars with spectral types ranging from B to F selected from Tables 1-3 of Thé *et al.* (1994), supplemented with HAEBE stars listed in the literature. Previous studies found that visual companions can be 3 to 5 magnitudes fainter than the primaries [Leinert *et al.* (1997), Testi *et al.* (1997)]. To detect such faint companions at separations of less than $1''$, we obtained high resolution near infrared (NIR) K-band images, using adaptive optics systems on large telescopes: Altair–NIRI at Gemini North and NACO at the VLT. Here we present the results of the observations taken with Altair–NIRI at Gemini North, during the period September 2005 until March 2006. The stars in our sample have K magnitudes ranging from $K = 4$ to $K = 12$, and they are bright enough to be used as guide stars themselves. Depending on the brightness of the source and on the observing conditions, the stars were observed either with the broad-band K-short filter or with a narrow-band K filter centered at Br γ. The field of view is $11''$ and the pixel size is 21.77 mas. Finally, we used the targets that did not have a (resolved) companion as reference point spread functions (PSF).

The data were reduced in a standard way (dark subtraction, flatfielding, sky substraction) with the IRAF Gemini package *niri*. In the case of relatively wide companions the astrometry and relative photometry were obtained by PSF fitting of the reference star, using an IDL version of the DAOPHOT PSF fitting procedure. For closer companions a PSF reconstruction method following Tokovinin *et al.* (2006) was used. Fig. 1 shows an example: HD 37411, where the system is actually a triple star or more.

Figure 1. Example of a discovery, HD 37411.

We determined the detection limit of the system to check the reliability of our discoveries. This limit is defined as the minimum difference of magnitude ΔK observable at a given distance from the primary. First, the intensity radial variance σ_{var} of the image centered on the primary is calculated. Then, the detection limit is set such that the minimum intensity increase due to the presence of a companion is greater than $5\sigma_{rad}$.

Due to seeing variations, this detection limit has been calculated for each night of observation. However, the conditions of observation were chosen to be very similar for each night and therefore we could calculate a fairly accurate model shown on Fig. 2.

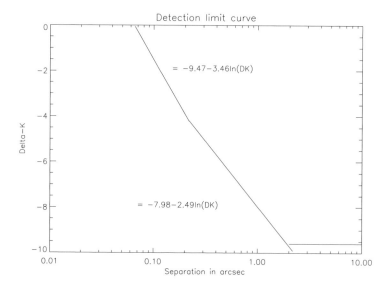

Figure 2. Detection limit for Altair.

3. Preliminary results & future work

Of the 72 sources observed with Altair–NIRI during the period from September 2005 to March 2006, 44 potential binaries were found, of which 31 are not previously listed in the literature and 3 are ambiguous (may be artifacts). Not including these 3, this brings the total number of HAEBE binary candidates to 66, nearly doubling the previously known sample. Table 1 lists the data of the binary candidates observed with Altair–NIRI, regardless of their separation.

Of the 44 HAEBE binaries observed, 23, or 50%, are potentially multiples rather than binaries, suggesting a preliminary binary fraction of HAEBE stars greater than 1. In Table 1 sources with more than one possible companion are marked with an asterisk.

Table 1. Multiplicity. The targets with * have more than one potential companion detected; with [1] are the new discoveries; with [2] are stars included in other studies but not identified as binaries. The 3 targets with (?) are ambiguous ones.

Target	ρ	θ	Δm	Target	ρ	θ	Δm	Target	ρ	θ	Δm
PS Cep[*1]	6.66	310.1	7.9	V1493 Cyg[*1]	0.58	294.8	3.2	VX Cas[*2]	5.34	165.3	4.8
V1429 Aql[*1]	1.19	319.3	7.4	V1982 Cyg[*1]	2.01	302.1	7.7	V628 Cas[*]	0.76	269.6	2.8
LkHA 168[*1]	0.31	208.9	2.6	HD 37258[1]	0.99	30.3	2.8	V633 Cas	5.78	4.4	5.5
V751 Cyg[*1]	0.67	295.0	2.7	BD-06 1259[1]	2.04	46.7	4.0	V1578 Cyg[*]	1.31	309.6	4.6
HD 235495[*1]	2.08	297.0	4.6	V1977 Cyg[2]	4.75	54.5	6.6	IL Cep[*]	7.80	329.2	0.1
HBC 531[*1]	6.61	163.0	5.7	WW Vul[2]	7.29	239.7	6.5	V700 Mon[*]	0.15	55.6	3.8
HBC 324[*1]	8.01	354.0	3.9	V374 Cep[2]	6.17	29.2	8.3	XY Per	1.33	256.7	0.1
HD 37411[*1]	0.46	357.5	3.3	VV Ser[2]	7.47	249.6	7.1	V892 Tau	3.91	23.0	7.6
V350 Ori[1]	0.29	206.8	3.2	HD 179218[*2]	2.54	140.5	6.6	HK Ori[*]	0.35	40.8	1.8
V385 Cep[*1]	2.84	229.9	7.9	V699 Mon[2]	8.50	226.6	4.8	R Mon	0.71	293.6	4.9
HD 245906[1]	0.13	77.1	1.5	V1271 Ori[2]	8.38	294.7	6.7	BD-06 1253	0.12	232.8	1.4
LkHA 201[*1]	2.34	325.8	6.2	CQ Tau[2]	2.09	55.5	8.5	HBC 334[*](?)	0.44	347	4
HBC 535[*1]	3.67	221.3	7.1	V590 Mon[2]	5.07	97.1	6.6	V390 Cep(?)	0.30	248.8	5.2
V1685 Cyg[1]	0.72	175.1	5.4	HD 36112[2]	2.28	311.3	8.3	VY Mon(?)	2.05	240.7	8.1
LkHA 147[*1]	1.63	340.7	3.0	HD 37357[2]	0.14	226.3	1.7				

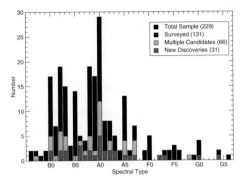

Figure 3. Multiplicity as a function of spectral type. Only stars with well-determined spectral types are included.

Fig. 3 shows multiplicity as a function of spectral type for our sample so far. The detection of possible companions around 50% of the stars surveyed does not appear to be a strong function of spectral type. However, more data are needed to draw significant conclusions.

The next step will be to establish physical association of the companions for which we will use NIR photometry and spectroscopy. For true companions we will determine spectral types and look for evidence of pre-main sequence status, either emission lines or near-infrared excess indicating the presence of circumstellar disks. We also plan to determine rotational and radial velocities. The final goals are to understand the nature of the companion stars and the effect of the nearby primary on their formation, and to study the multiplicity of HAEBE stars as function of mass.

Acknowledgements

We are thankful to Maria-José Cordero, an undergraduate student from the program PIA (Prácticas de Investigación en Astronomía) in Chile, who worked on the first results of the data set. Based on observations obtained at the Gemini Observatory, which is operated by the Association of Universities for Research in Astronomy, Inc., under a cooperative agreement with the NSF on behalf of the Gemini partnership: the National Science Foundation (United States), the Particle Physics and Astronomy Research Council (United Kingdom), the National Research Council (Canada), CONICYT (Chile), the Australian Research Council (Australia), CNPq (Brazil) and CONICET (Argentina).

References

Bouvier J. & Corporon P. 2001, *IAUS* 200, 155
Leinert C., Richichi A., & Haas M. 1997, *A&A* 318, 472
Testi L., Palla F., Prusti T., Natta A., & Maltagliati S. 1997, *A&A* 320, 159
Tokovinin A., Thomas S., Sterzik M., & Udry S. 2006, *A&A*, 450, p681
Thé, P. S., de Winter D., & Perez M. R.1994 *A&AS* 104, 315

Binary Stars as Critical Tools & Tests
in Contemporary Astrophysics
Proceedings IAU Symposium No. 240, 2006
W.I. Hartkopf, E.F. Guinan & P. Harmanec, eds.

A Survey of Stellar Families: Multiplicity Among Solar-type Stars

Deepak Raghavan[1]†, H. A. McAlister[1], T. J. Henry[1], and B. D. Mason[2]

[1]Center for High Angular Resolution Astronomy, Georgia State University, Atlanta GA, USA

[2]U.S. Naval Observatory, 3450 Massachusetts Avenue NW, Washington DC, USA

Abstract. Stellar multiplicity is a fundamental astrophysical property. In addition to being the only physical basis for accurate mass determination, this parameter is believed to influence important aspects such as planet formation and stability. Contrary to earlier expectations, recent studies have shown that even against selection biases, as many as 23% of the planetary systems reside in multiple star systems (Raghavan *et al.* 2006). Leveraging recent efforts in identifying stellar and substellar companions to solar-type stars, and augmenting them with targeted observations, we are conducting a comprehensive survey, aimed at providing a modern update to the seminal work of Duquennoy & Mayor (1991). The details of our sample, survey methods, and some preliminary results are presented here.

Keywords. binaries (including multiples), planetary systems, surveys

1. Introduction

The primary motivation of this effort is to better understand the variety of environments inhabited and fostered by solar-type stars in our Galaxy. We hope to accomplish this via a comprehensive multiplicity survey of solar-type stars in the solar neighborhood. Since the seminal work of Duquennoy & Mayor (1991, hereafter DM), our understanding of solar-type stars has grown substantially. The DM survey predated the *Hipparcos Catalog* (ESA 1997) and hence could not leverage its accurate parallaxes in defining the volume-limited sample. Since DM, several high-precision radial velocity surveys (e.g., Nidever *et al.* 2002, Mayor *et al.* 2004, Marcy *et al.* 2005) have identified companions from stars down to planets. Astrometric efforts such as speckle interferometry (Mason *et al.* 2004), adaptive optics (Luhman & Jayawardhana 2002), and long-baseline interferometry (Bagnuolo *et al.* 2006) have been very useful in identifying and characterizing orbits of binary stars. Multi-epoch archival images from the *Digitized Sky Survey*† (DSS) and the *SuperCOSMOS Sky Survey* (SSS; Hambly *et al.* 2001) allow us to unearth wide Common Proper Motion (CPM) pairs. This work leverages these prior efforts and augments them with new observations with speckle and long-baseline interferometry.

2. The Sample of Solar-Type Stars

We have extracted an unbiased volume-limited sample of 455 primary stars (including our Sun) as representatives of solar-type stars in the Galaxy. Our sample includes stars with a Hipparcos parallax of 40 mas or larger, with an error less than 5%. We further restrict our targets to a proximity band of 2.0 magnitudes above or 1.5 magnitudes below an iterative best-fit main sequence, resulting in the inclusion of luminosity classes IV, V,

---

† email: raghavan@chara.gsu.edu

† See http://stdatu.stsci.edu/cgi-bin/dss_form.

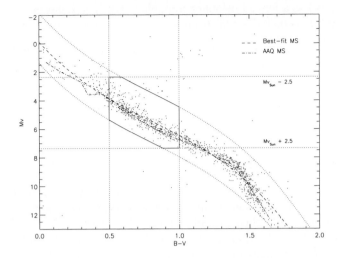

Figure 1. Target list of 455 solar-type stars within 25 pc from Hipparcos. AAQ refers to Allen's *Astrophysical Quantities* (2000).

and VI. The larger band above the main sequence allows for the inclusion of multiple star systems. Finally, we limit targets to $0.5 \leqslant B - V \leqslant 1.0$ in order to select solar-type stars. The above criteria result in the selection of all stars with V–band flux in the range of 0.1–10.0 times the Solar value (see Figure 1), giving us a physical basis for our definition of "solar-type".

2.1. *Comparison with Duquennoy & Mayor (1991)*

This effort is an update to the DM multiplicity survey of solar-type stars. DM used a volume-limited sample of 164 primary stars, selected from Gliese (1969). However, as pointed out in Halbwachs *et al.* (2003), there are substantial differences between a sample selected from Gliese (1969) and the more accurate *Hipparcos Catalog*. We applied the DM selection criteria of spectral types F7 to G9, luminosity classes IV, V, and VI, declination above $-15°$, and parallax greater than 45 mas to the *Hipparcos Catalog*, resulting in the selection of 148 primary stars. Moreover, only 92 of these stars overlap with the DM sample. This implies that 44% of DM's targets are now known to fall outside their parameter space, and their study excluded 38% of the targets now recognized to match their selection criteria. Our sample of 455 primary stars includes 106 of the DM stars. With current information and a larger sample, we hope to present updated multiplicity statistics of better accuracy and precision.

3. Sources Leveraged in the Survey

Every star in our target list is checked against known stellar multiplicity studies. First, we extract information on each target from available catalogs. The *Washington Double Star Catalog* (WDS) is an excellent resource for astrometric pairs. However, it is a catalog of doubles, and hence contains many listings of optical pairs, so we investigate each entry to verify its validity as a gravitationally bound companion. We also polled the *Sixth Catalog of Orbits of Visual Binary Stars* (VB6), the *9th Catalog of Spectroscopic Binary Orbits* (SB9), and the *Hipparcos Catalog*. The *Fourth Catalog of Interferometric Measurements of Binary Stars* (4IF) contains high resolution measures from speckle, adaptive optics (AO) and long-baseline interferometry, and while it substantially overlaps with the WDS, it has a few additional companions (such as lunar occultation measures), and importantly, also includes null results. Finally, we search individual publications to ensure completeness and to identify null results that are not included in the catalogs. The lack of published null results is one frustrating aspect of this search, but we are actively pursuing them, both in publications (e.g., Nidever *et al.* 2002, Luhman & Jayawardhana 2002), as well as through collaborations.

4. Observations

Even with the wealth of available observations, significant gaps exist. Some of these are due to the newness of advanced techniques such as adaptive optics and long-baseline interferometry, while others are simply due to incomplete coverage of prior efforts for our targets. Our observations center around three primary areas. First, we use multi-epoch archival images from the DSS and SSS to blink an area of the sky around each primary to identify wide CPM companions. These are then confirmed or refuted by comparing a photometric distance estimate of the companion to the Hipparcos distance of the primary. Second, we are collecting new speckle observations to help confirm candidates, or identify new ones for unobserved systems. Finally, we plan targeted observations using the CHARA Array's long-baseline interferometric capabilities to discover new companions and more fully characterize known spectroscopic orbits.

5. Preliminary Results & Future Work

We have completed a first pass in assimilating companion information from the WDS, SB9, VB6, 4IF, DM, and Hipparcos catalogs into a central database. We have also blinked the DSS and SSS images to identify CPM companions and to verify the physicality of WDS entries. The WDS lists 459 pairs for 205 of our target stars. We have confirmed 131 (29%) of these to be gravitationally bound companions based on the availability of visual or spectroscopic orbits, or because of matching parallax and proper motions. An additional 52 (11%) remain candidate companions, while 276 (60%) have been confirmed to be optical pairs, based on relative motion between the two stars. Upon blinking archival images, 368 of our 455 targets exhibited detectable proper motions, allowing us to search for CPM candidates, and an additional 43 exhibited a marginal proper motion. We confirmed 52 systems with known CPM companions, two of which are triples. Including candidates, there are 66 systems with CPM candidate companions, 5 of which are triples. Six of the CPM companions detected are potentially new discoveries. Overall, our current percentage of single:double:triple:quadruple is 69:26:4:1. If all of our candidates were to be confirmed, the percentages would be 56:33:9:2, including one possibly sextuple system. In comparison, the DM results were 57:38:4:1 for multiples with orbits, and 51:40:7:2 including candidates. While the larger fraction of singles found in our sample is consistent with the studies of M dwarfs (Henry & McCarthy 1990, Fischer & Marcy 1992), it is too early to jump to that conclusion, because the multiplicity search for our targets is as of yet incomplete, and hence our percentage of singles is an upper limit. Finally, of the 162 planetary systems discovered as of July 2006, 32 (20%) are in our target list. The ratios for this subsample are 78:22:0:0 for confirmed companions, and 69:28:0:3 including candidates. The larger fraction of singles among exoplanet systems is not surprising because planet search programs do not target known binaries. Nascent efforts targeting binaries for planet search (e.g., Konacki 2005) might reveal the answer to this question. Stay tuned!

References

Allen, C.W. 2000, *Astrophysical Quantities*, Fourth Edition, Cox, A. N., Editor, Springer-Verlag, New York

Bagnuolo, W.G., Jr., *et al.* 2006, *AJ*, 131, 2695

Duquennoy, A., & Mayor, M. 1991, *A&A*, 248, 485

ESA 1997, *The Hipparcos and Tycho Catalogues*, ESA SP-1200 (Noordwijk: ESA)

Fischer, D.A. & Marcy, G.W. 1992, *ApJ*, 396, 178

Gliese, W. 1969, *Veroeffentlichungen des Astronomischen Rechen-Instituts Heidelberg*, 22, 1

Halbwachs, J.L., Mayor, M., Udry, S., & Arenou, F. 2003, *A&A*, 397, 159

Hambly, N.C., Davenhall, A.C., Irwin, M.J., & MacGillivray, H.T. 2001, *MNRAS*, 326, 1279

Henry, T.J. & McCarthy, D.W., Jr. 1990, *ApJ*, 350, 334

Konacki, M. 2005, *ApJ*, 626, 431

Luhman, K.L. & Jayawardhana, R. 2002, *ApJ*, 566, 1132

Marcy, G., Butler, R.P., Fischer, D., Vogt, S., Wright, J.T., Tinney, C.G., & Jones, H.R.A. 2005, *Progress of Theoretical Physics Supplement*, 158, 24

Mason, B.D., Hartkopf, W.I., Wycoff, G.L., Rafferty, T.J., Urban, S.E., & Flagg, L. 2004, *AJ*, 128, 3012

Mayor, M., Udry, S., Naef, D., Pepe, F., Queloz, D., Santos, N.C., & Burnet, M. 2004, *A&A*, 415, 391

Nidever, D.L., Marcy, G.W., Butler, R.P., Fischer, D.A., & Vogt, S.S. 2002, *ApJS*, 141, 503

Raghavan, D., Henry, T.J., Mason, B.D., Subasavage, J.P., Jao, W.C., Beaulieu, T.D., & Hambly, N.C. 2006, *ApJ*, 646, 523

Binary Stars as Critical Tools & Tests
in Contemporary Astrophysics
Proceedings IAU Symposium No. 240, 2006
W.I. Hartkopf, E.F. Guinan & P. Harmanec, eds.

Close Binaries in the CoRoT Space Experiment

Carla Maceroni[1] and Ignasi Ribas[2]

[1]INAF - Osservatorio Astronomico di Roma, via Frascati 33, I-00040 Monteporzio (RM) Italy
email: maceroni@oa-roma.inaf.it

[2]CSIC - Institut d'Estudis Espacials de Catalunya, Barcelona, Spain
email: iribas@ieec.uab.es

Abstract. We discuss the impact that the space experiment CoRoT, whose launch is scheduled for late 2006, will have on the field of close binary.

Keywords. binaries:close, binaries:eclipsing, stars:late-type, stars:activity, stars: oscillation, stars: planetary systems

1. Introduction

CoRoT (COnvection, ROtation and planetary Transits) is a french-led international "small" space mission whose launch is scheduled for November 2006. The mission is devoted to the achievement of two parallel "core programs", astroseismology and extra-solar planet search, wich require the same type of observations, i.e. high accuracy photometry and long continuous monitoring of targets †.

CoRoT will provide high accuracy (10^{-3} – 10^{-4}) lightcurves of more than hundred thousand stars. The observations will be performed in two modes, both characterized by continuous monitoring of a preselected field: Long and Short Runs lasting, respectively, 150^d and 30^d. The expected duty cycle is of ~94%.

Many selected binary systems will be observed in the Additional Programme (AP), i.e., research programs outside the core programs. Moreover, about a thousand new binaries are expected as by-product of the exoplanet search and hundreds of them will have extremely accurate light curves.

The exoplanet targets have apparent magnitude in the interval $11.5 > V > 16.5$, and a typical sampling of 8 min. Besides, targets brighter than $V = 15$ will have as well some color information (thanks to a prism that will separate the blue and red part of the spectra with a resolution ~ 4). These will be as well the characteristics of most CoRoT binaries.

Table 1 and Figure 1 (see Maceroni & Ribas 2006) show an estimate of the expected number of binaries per CoRoT Exo-field, by means of the Besançon Galaxy model (Robin *et al.* 2003). The constraints are: spectral type G-M and luminosity class V-IV (as the Exoplanet search will privilege late-type dwarfs), within the magnitude range V=10.5–15.5. Eclipsing binaries are assumed to be 0.5–1% of all monitored stars (as suggested by surveys as Vulcan, STARE, OGLE).The results in Table 1 shall be multiplied by a minimum number of five Long Runs and ten Short Runs.

---

† complete information on the mission is available at `http://corot.oamp.fr`

Table 1. Expected binaries per 3.4 square degree exoplanet field in the Summer (centered at $\alpha = 6^{\mathrm{h}}50^{\mathrm{m}}; \delta = 0°$) and Winter ($\alpha = 18^{\mathrm{h}}50^{\mathrm{m}}; \delta = 0°$) pointing directions

Spectral type	expected EB number
G0–G4	7.2 – 31.1
G5–G9	3.2 – 10.5
K0–K4	2.9 – 11.9
K5–K9	0.35 – 0.8
M0–M4	0.09 – 0.17

2. Close binaries related scientific programs

The community involved in CoRoT and AP on binaries has assumed the (informal) organization of a 'Binary Thematic Team' (BTT) ‡, coordinated by the authors. The BTT has been granted several scientific programs (two applications for data of known binaries and archival data) for the first CoRoT runs (2006/2007) which are aimed to (Maceroni & Ribas 2006 for details):

• derive basic stellar parameters (masses, radii) for *well-behaved* systems, i.e., preferentially detached binaries. Exquisite, better than 1% accuracy is expected,

• study second-order effects in the light curves (limb and gravity darkening),

• study the manifestations of stellar activity in late-type components by eclipse tomography, and derive information on rotational period and differential rotation (from spot migration), and

• perform astroseismology in suitable eclipsing binaries.

In the latter case while the search for (non-radial) oscillation in binary system components is more challenging than in single stars, the advantage is that the masses, radii, inclination, and therefore the rotational velocity can be known. Also, the mutual eclipses of the components could help to identify and assign the different modes. In close binaries tidal frequencies could be excited.

A possible target for a pointed Short Run in the Seismology field is the binary IM Mon, the brightest binary in the CoRoT field of view. As the photometry of seismology target will be at a few ppm level, IM Mon could become the eclipsing system with the best ever measured light curve.

3. Binaries as a concern

The most serious problem for planetary transit detection in CoRoT will arise from contamination by background eclipsing binaries (BEBs).

The CoRoT exoplanet program essentially performs aperture photometry on a crowded stellar field. In spite of the definition of specific masks, conceived to limit contamination from nearby stars, a contribution of faint background objects to the light curve of the target is unavoidable. The high frequency of false alarms from BEBs was already identified from the experience of OGLE planetary transit search: only five true planets out of 177 alarms were confirmed by follow-up observations (Udalski *et al.* 2004), a result in agreement with the theoretical estimation of Brown (2003).

Blind tests have been done by Moutou, Pont, Barge *et al.* (2005) to evaluate the ability of detrending the lightcurves from instrumental and environmental biases in the specific context of CoRoT, and of eventually discriminating real planets from false alarms. A complex light curve simulator, has been realized (Auvergne, Boisnard & Buey 2003),

‡ The BTT webpage can be found at `http://thor.ieec.uab.es/binteam/`

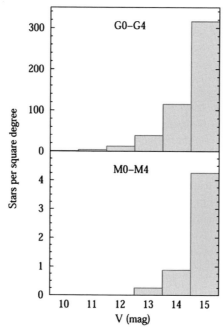

Figure 1. The distribution in apparent magnitude in the winter field of binaries with late-type components (and luminosity class IV-V) as obtained from the Besançon galaxy model and the hypothesis that eclipsing binaries are $\sim 0.5 - 1.0$ % of all monitored similar stars.

which includes most sources of noise and environmental perturbations. The blind test consisted in analyzing, without prior knowledge, a set of simulated CoRoT lightcurves containing signal from planets and, as well, from other stellar sources of noise, such as stellar activity and background binaries.

While BEBs were confirmed to be the major source of false alarms, many cases could be discriminated just by detailed analysis of the folded lightcurve, on the basis of color behavior of the transit and by detection of a secondary minimum. Only $\sim$10% of alarms will need more time and resources such as follow-up spectroscopy†.

Acknowledgements

CM acknowledges funding of this research project by MIUR/Cofin and F-INAF programs and a grant from the IAU General Secretary.

References

Auvergne, M., Boisnard, L., & Buey, J-T. 2003, *SPIE* 4853, 170

Brown, T. M. 2003, *ApJ* 593, L125

Maceroni, C. & Ribas, I. 2006, *astro-ph/0511171*

Moutou, C., *et al.* 2005, *A&A* 437, 355

Robin, A. C., Reylé, C., Derrière, S., & S. Picaud 2003, *A&A* 409, 523

Udalski, A., Szymanski, M. K., Kubiak, M.,Pietrzynski, G., Soszynski, I., Zebrun, K., Szewczyk, O. & Wyrzykowski, L. 2004, *AcA* 54, 313

† see the presentation of F. Pont (2006) available from the website of the 10th CoRoT Week: www.obs-nice.fr/cassiopee/COROTWeek10/presentations.html

Binary Stars as Critical Tools & Tests
in Contemporary Astrophysics
Proceedings IAU Symposium No. 240, 2006
W.I. Hartkopf, E.F. Guinan & P. Harmanec, eds.

© 2007 International Astronomical Union
doi:10.1017/S1743921307004139

Detectability of Planets in Wide Binaries by Ground-Based Relative Astrometry with AO

R. Neuhäuser[1] †, A. Seifahrt[1,2], T. Röll[1], A. Bedalov[1], and M. Mugrauer[1]

[1]Astrophysikalisches Institut, Universität Jena, Schillergäßchen 2-3, 07745 Jena, Germany
[2]European Southern Observatory, Karl-Schwarzschild-Str. 2, 85748 Garching, Germany

Abstract. Many planet candidates have been detected by radial-velocity variations of the primary star; they are planet *candidates*, because of the unknown orbit inclination. Detection of the wobble in the two other dimensions, to be measured by astrometry, would yield the inclination and, hence, true mass of the companions. We aim to show that planets can be confirmed or discovered in a close visual stellar binary system by measuring the astrometric wobble of the exoplanet host star as a periodic variation of the separation, even from the ground. We test the feasibility with HD 19994, a visual binary with one radial velocity planet candidate. We use the adaptive optics camera NACO at the VLT with its smallest pixel scale ($\sim$ 13 mas) for high-precision astrometric measurements. The separations measured in 120 single images taken within one night are shown to follow white noise, so that the standard deviation can be divided by the square root of the number of images to obtain the precision. In this paper we present the first results and investigate the achievable precision in relative astrometry with adaptive optics. With careful data reduction it is possible to achieve a relative astrometric precision as low as 50 μas for a 0$''$.6 binary with VLT/NACO observations in one hour, the best relative astrometric precision ever achieved with a single telescope from the ground. The relative astrometric precision demonstrated here with AO at an 8-m mirror is sufficient to detect the astrometric signal of the planet HD 19994 Ab as periodic variation of the separation between HD 19994 A and B.

1. Introduction

Since the radial velocity technique can yield only lower mass limits, all planets (or planet candidates) found by this technique have to be confirmed by other methods. Of $\sim$ 200 radial velocity planet candidates found so far, 14 have been confirmed by transit and two by astrometry (see, e.g., exoplanet.eu.) The two planets GJ 876b and 55 Cancri d have been confirmed by astrometry using the Hubble Space Telescope Fine Guidance Sensor (Benedict *et al.* 2002, McArthur *et al.* 2004), with a precision down to 0.04 mas (milliarcsec), sufficient to detect the wobble of the host star in the plane of the sky.

The astrometric displacement is given by

$$\theta \, [\mathrm{mas}] = 0.960 \times \frac{a}{[5\,\mathrm{AU}]} \times \frac{[10\,\mathrm{pc}]}{d} \times \frac{M_{\mathrm{pl}}}{[M_{\mathrm{jup}}]} \times \frac{[M_\odot]}{M_\star}$$

with planet mass M_{pl} in Jupiter masses at a semi-major axis a (in units of 5 AU) in a circular orbit around a host star with mass $M_\star$ in solar masses at a distance d (in units of 10 pc).

Here, we present both a feasibility study to determine the astrometric precision of the Adaptive Optics (AO) camera NAos-COnica (NACO) of the ESO Very Large Telescope

† rne@astro.uni-jena.de

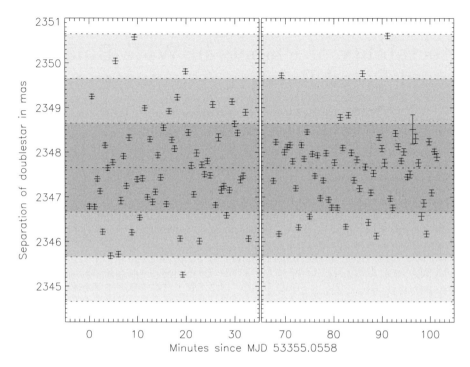

Figure 1. Measured separation for HD 19994 (similar for HD 19063). Error bars denote 1σ error for individual measurements. The standard deviation of the complete dataset is shown as dashed lines and grey-shaded areas for 1, 2, and 3σ, respectively. The standard deviation of the mean, thus the precision of the measured binary separation, is $0.998/\sqrt{120}\ mas\ =\ 91.2\ \mu as$.

(VLT) as well as first measurements in order to determine the mass of HD 19994 Ab, a RV planet candidate in orbit around HD 19994: The RV data suggest a lower mass limit of the planet candidate of $m \cdot \sin i = 1.68\ M_{\rm Jup}$ in an orbit with semi-major axis $a = 1.42$ AU with an eccentricity of $e = 0.30 \pm 0.04$, i.e., a 535.7 ± 3.1 day orbital period (Mayor *et al.* 2004). The full (peak-to-peak) astrometric wobble, from Kepler's $3^{\rm rd}$ Law, is at least 0.155 mas (for a circular orbit) and 0.131 mas by taking the observed eccentricity into account – this is assuming the minimum mass $m \cdot \sin i$ as the true mass.

2. Observations and data reduction

We observed HD 19994 and a calibration binary (HD 19063) with the S13 camera (13.26 mas/pixel pixel scale, $14'' \times 14''$ FOV) in a NB_2.17 narrow-band filter. Short exposures of both target binaries were taken for one hour each, equally split up into two time slots before and after meridian passage. We obtained a total of 120 frames for HD 19994 and 117 frames for HD 19063. Data reduction followed the standard technique of dark subtraction, division by a flat field and the application of a bad pixel mask with the *eclipse* software. In each single exposure (0.347 sec), we measured the separation between A and B of both binaries. We have then made sure that the separations follow white noise by a K-S test (see a future paper by Neuhäuser *et al.* for details). Hence, we can divide the standard deviation of the mean by the square root of the number of measurements to obtain the precision of the separation measurement, the *precision*, not the *accuracy*. For the latter, we would need to know the true pixel scale. See Figure 1 and Table 1.

Table 1. Astrometric results from the measurements of the binaries HD 19994 and HD 19063. K-S test values higher than 0.6 are interpreted as indicating white noise.

Target	Mean separation (mas)	Standard deviation	N	K-S test	Precision achieved
HD 19994	2347.652	0.998 mas	120	0.69	91.2 μas
	2347.632	1.110 mas	60	0.85	143.3 μas
	2347.673	0.882 mas	60	0.64	113.9 μas
HD 19063					
	649.628	0.377 mas	57	0.74	49.9 μas
	649.596	0.391 mas	60	0.48	(failed)

Differential chromatic refraction (DRC) or Allan noise (see Pravdo & Shaklan 1996) do not matter here, because we observe in the near IR with a narrow-band filter (hence no DCR) and with AO (hence no Allan noise).

3. Results and conclusions

We have observed two bright visual binaries (0.6 to 2.4 arc sec separation) with VLT/NACOs smallest pixel scale (13.26 mas/pixel) by taking $\sim$ 120 short (0.347 s) exposures per binary within a few hours, separated in four bins of $\sim$ 60 exposures each (15 December 2004, ESO program 075.C-0288.A). We could confirm that the separations measured follow white noise (in three out of four bins), so that we are allowed to divide the mean of the separations by the square root of the number of measurements to obtain the precision of the separation measurement. By doing so, we measured the separation between HD 19994 A and B to be 2347.652 mas (milliarcseconds) with the relative precision of 91.2μas (microarcseconds). For HD 19063, we obtained 649.628 mas with a precision being as low as 49.9μas. These are the most precise measurements in relative astrometry obtained with single–aperture telescopes from the ground, sufficient to detect the astrometric wobble due to a planet, to be seen as periodic change in the binary separation.

Such a high precision as shown here can be applied not only to measurements of masses of previously-detected RV planets (or candidates), but also to search for new planets in wide binaries, or even single stars with a bright (back- or foreground) star within the isoplanatic angle. Other possible applications are orbit determinations in stellar multiples, ground-based parallax and proper motion measurements, observations of expanding, contracting, or rotating star clusters, etc.

The main limitation of this technique is the need for a bright star near the target, namely within the isoplanatic angle. And, as long as no pixel scale calibrator is available with a precision down to 1/100000, the measurement can achieve this high precision only for relative astrometry, but not in absolute terms.

References

Benedict, G.F., McArthur, B.E., Forveille, T., *et al.*, 2002, ApJ 581, L115

Mayor, M., Udry, S., Naef, D., Pepe, F., Queloz, D., Santos, N.C., & Burnet, M. 2004, A&A 415, 391

McArthur B.E., Endl M., Cochran W.D., *et al.*, 2004, ApJ 614, L81

Pravdo, S.H., & Shaklan, S.B. 1996, ApJ 465, 264

Binary Stars as Critical Tools & Tests
in Contemporary Astrophysics
Proceedings IAU Symposium No. 240, 2006
W.I. Hartkopf, E.F. Guinan & P. Harmanec, eds.

Poster Abstracts (Session 4)

(Full Posters are available at http://www.journals.cambridge.org/jid_IAU)

CCD Astrometry and Photometry of Visual Double Stars VI. Northern Hipparcos Wide Pairs Measured in the Years 2003–2005

D. Sinachopoulos[1], P. Gavras[2], Th. Medupe[3], Ch. Ducourant[4], and O. Dionatos[2]

[1] *National Observatory of Athens, Athens, Greece,*
[2] *National & Kapodestrian University of Athens, Athens, Greece,*
[3] *South African Astronomical Observatory, Cape town, South Africa,*
[4] *Observatory of the University of Bordeaux, Bordeaux, France*

The relative positions of the Hipparcos visual double star components are currently known with a precision around fifty mas. Modern CCD astrometric observations of these objects achieve an accuracy of their angular separation between 10 and 20 mas per observation. New CCD measurements have been obtained at Kryonerion Observatory in the north hemisphere. They provide current relative positions of visual double stars, which are at least twice as accurate as the ones provided by Hipparcos. The new measurements will permit us to extract the physical pairs from the sample, and the double stars, which have components of common origin. Final statistics of these systems will improve our understanding of stellar formation and evolution rates, of wide binaries in the solar neighborhood.

Eclipsing Binaries In Open Clusters

J. Southworth

University of Warwick, Coventry, United Kingdom

Detached eclipsing binary stars are of fundamental importance in providing constraints on theoretical stellar models and the physics they contain. They have been used to calibrate theoretical models and to constrain the amount of convective core overshooting and mixing length used by the models. However, a major limitation in this work is that the age and chemical composition are not in general known for each eclipsing binary, allowing these quantities to be freely adjusted when comparing to theoretical models. I describe an observational program to study eclipsing binaries in a sample open clusters with a range of ages and chemical compositions. In this situation, theoretical predictions are required to simultaneously match the masses, radii and luminosities of the two eclipsing stars and the radiative properties of all the other stars in the cluster. These more detailed constraints mean that subtle physical phenomena, such as convective overshooting and mixing length, can be disentangled from the effects of age and metallicity. I present newly acquired light curves for several eclipsing binaries in the young open cluster NGC 7128. This will uniquely allow mass-radius and mass-luminosity plots to be constructed which contain many stars with the same age and metallicity but a wide range of masses. In addition, the age, chemical composition and distance of NGC 7128 can be obtained without the complications inherent in the comparison of theoretical isochrones to the cluster main sequence in colour-magnitude diagrams.

A Hundred New Preliminary Orbits and Masses from Hipparcos, Ground-based Astrometry and Radial Velocities

G.A. Gontcharov

Pulkovo Observatory, Saint Petersburg, Russia

Stellar pairs with large magnitude difference and period of several decades are hard to investigate by spectroscopic, interferometric and visual methods. A combination of Hipparcos and ground-based astrometry (see Gontcharov & Kiyaeva, 2002 A&A, 391, 647) can help to derive orbits and masses for such astrometric binaries. A direct combination of the Hipparcos/Tycho data with 400 astrometric ground-based catalogues of the 20th century (including all ground-based catalogues used for Tycho-2) reduced to the ICRF/Hipparcos system is used to detect the variations of transversal velocity with median precision of 1 mas/year, which corresponds to 0.5 km/s at 100 pc. The radial velocity variations are detected with a median precision of 0.7 km/s as a by-product of a new compilation of 500 sources of precise radial velocities (including WEB, Barbier-Brossat and Figon, Geneva-Copenhagen survey (2004 A&A, 418, 989) and CORAVEL K-M giants (2005 A&A, 430, 165)) reduced to a common new standard of 837 stars, the Pulkovo RAdial VELOcities catalogue of 35000 stars (PRAVELO, www.geocities.com/orionspiral). Similar precision allows us to use transversal and radial velocities together. However, the data are still rather heterogeneous and have small S/N ratio mainly because the accuracy of observations improved with time. Therefore, least square solution uncertainties show several minima. A new approach is developed to solve it. As a first result, preliminary orbits and component masses are calculated for 100 astrometric binaries with no previous orbit calculation and 20 binaries with known visual or spectroscopic orbit. Many new orbital pairs appear to be A, F, G stars with white dwarfs. The duplicity/multiplicity of many famous stars is discussed in detail, such as Polaris, Mizar and Arcturus.

This research is a part of Orion Spiral Arm CAtalogue project (OSACA) supported by Russian Foundation for basic research grant 05-02-17047.

Procedure for the Classification of Eclipsing Binaries

E. Oblak[1], O.Yu. Malkov[2], E.A. Avvakumova[3], and J. Torra[4]

[1] *Observatoire de Besancon, Besancon, France,*
[2] *Institute of Astronomy, Moscow, Russia,*
[3] *Ural State University, Ekaterinbourg, Russia,*
[4] *Universitat de Barcelona, Barcelona, Spain*

In this work we present a procedure for the automatic classification of eclipsing binaries. The procedure is based on the data from a new catalogue of 6330 eclipsing variable stars, compiled by the authors and representing the largest list of eclipsing binaries classified from observations. The procedure allows the classification of a given system basing on a set of observational parameters even if the set is incomplete. Results of an application of the procedure to a test sample of 1029 classified systems and to a sample of 5301 unclassified systems are discussed. The classified eclipsing binaries will be used for determination of astrophysical parameters of their components.

Frontiers of Transient Phenomena in X-ray Binaries and Cataclysmic Variables Investigated by a High-speed CCD Camera and an Automated Monitor Telescope

D. Nogami[1] and S. Mineshige[2]

[1] *Hida Observatory, Kyoto University, Kamitakara, Gifu, Japan,*
[2] *Yukawa Institute, Kyoto University, Kyoto, Japan*

We are developing two new systems of a high-speed CCD camera, and an automated monitor telescope. This camera using a frame transfer-type CCD enables us to take images each 27.3 msec at the highest speed. We try to investigate accretion and eruption phenomena around compact stars by optical light. The automated monitor system of X-ray binaries and cataclysmic variables we are developing is a small system of a 30-cm reflector, a CCD camera, and a computer. It is a cheap system, but can monitor more than 150 systems each night. It will reveal long-term light curve of programmed stars of those transient systems, and catch sudden outbursts/decays. We will be able to start follow-up observations best to clarify the mechanism of these activities, as early as possible.

We here report the current status of these projects, the target physics, and the future development.

Search for Binary Systems from the SDSS

K.-W. Lee, B.-C. Lee, and M.-G. Park

Kyungpook National University, Daegu, Korea, South

Some cool stars exhibiting a strong Hα emission line in their spectra are actually binary systems with an undiscovered component such as RS CVn types, AM Hers, symbiotic systems and so forth. We therefore have searched candidates of binary systems from the *Sloan Digital Sky Survey Fourth Data Release* (SDSS DR4) for the spectra showing strong Balmer emission lines in late-type stars cooler than K spectral type. These categoric stars include such as M, L, and cool C-types having a low effective temperature ($<3,500$ K), hence, in general, exhibiting absorption features of Balmer series. Of 23,936 sample classified as late-type in the SDSS DR4, we selected 312 stars satisfying our constraints on the flux of Hα emission line in the continuum removed spectra, greater than 20 times level compared with continuum removed one. By hand inspection of those spectra, we find four candidates of binary systems, presumably all cataclysmic variables: SDSS J012018.49−102536.1, J133336.22+491157.7, J152831.00+380305.8, and J215101.33+123739.9 including two previously identified ones. In order to confirm these objects as binary systems and to determine their variable types, we are planning to perform follow-up observations.

Gaia Treatment of Astrometric Binaries with a Variable Component: VIM, VIMA, VIMO

J.-L. Halbwachs[1], D. Pourbaix[2], M. Mayor[3], and S. Udry[3]

[1] *Observatoire Astronomique de Strasbourg, Strasbourg, France,*
[2] *Institut d'Astronomie et d'Astrophysique, Bruxelles, Belgium,*
[3] *Observatoire Astronomique de l'Université de Genève, Sauverny, Switzerland*

Gaia is an ESA project that will conduct a census of one thousand million stars in our Galaxy. Each target star will receive about 100 1-D high-precision astrometric measurements over a five-year period. The magnitude of the stars will be accurately estimated in the same time.

Among the huge quantity of stars observed by Gaia, some will be unresolved binaries hosting a photometric variable; since the astrometric abscissa and the total luminosity of these stars will be recorded simultaneously, three distinct models will be considered to describe these observations:

(a) the VIM model (VIM=Variability Induced Movers), is dedicated to binaries with components having fixed relative positions;

(b) the VIMA model introduces acceleration terms in order to describe a short part of an orbit;

(c) the VIMO model is used to derive the orbital parameters of binaries with orbital period up to about 10 years.

The VIM model was already used in the Hipparcos data reduction. Each astrometric measurement is fitted with a linear function of the parameters of the stars, and their derivation is therefore straightforward.

The VIMA model is more complicated. It includes a parameter $h = ((1+q)/q)10^{-0.4 \text{mC}}$, which is a function of the mass ratio of the system and of the magnitude of the non-variable component. When "h" is fixed, the model remains linear.

The VIMO model is still more complicated, since, additionally to h, it is necessary to fix P, e, T_0 and ω to obtain a linear system.

A method of computation was tried with simulated data, and its results look fairly good.

In conclusion, unresolved binaries with a variable component will receive accurate astrometric parameters in the final Gaia catalogue, and other properties of these systems will still be derived with the software that we are preparing.

Using MECI to Mine Eclipsing Binaries from Photometric Exoplanet Surveys

J. Devor and D. Charbonneau

Harvard–Smithsonian Center for Astrophysics, Cambridge, MA, United States

We describe the Method for Eclipsing Component Identification (MECI), which is an automated method for assigning the most likely absolute physical parameters to the components of an eclipsing binary. MECI is unique in that it requires only the photometric light curve and combined color of the eclipsing binaries. We have implemented this method using published theoretical isochrones and limb-darkening coefficients, and publicly released its source code. MECI lends itself to creating large catalogues through the systematic analyses of datasets consisting of photometric time series, such as those produced by OGLE, MACHO, HAT, and many others surveys. We will be presenting results of data mining the Trans-Atlantic Exoplanet Survey (TrES). This sort of mining technique may be used for both characterizing stellar populations and for discovering rare and interesting binary systems. Of particular interest are the lower main-sequence stars, for which models underestimate their sizes by as much as 20%. Progress in this area has been hampered by the small number of suitable M-dwarf binary systems with accurately determined stellar properties. Finding additional systems by mining Exoplanet Surveys may provide significant benefits for our understanding of such low-mass stars.

WIRE Satellite Light Curves of Bright Eclipsing Binary Stars

H. Bruntt[1] and J. Southworth[2]

[1]*School of Physics, Department of Physics, Sydney, Australia,*
[2]*University of Warwick, Coventry, United Kingdom*

We are undertaking a project to obtain extensive high-precision photometry of bright eclipsing binary stars using the star tracker aboard the WIRE satellite. Our immediate aim is to measure the masses and radii of our target systems to the highest possible accuracy.

The resulting physical properties will be excellent tests and calibrators of theoretical models, in particular because several of our targets have very low mass ratios. These data will also allow us to investigate the reliability of eclipsing binary light curve models and limb darkening coefficients, to see just how far such analyses can be pushed. We present a 29-day light curve (sampling rate 15 seconds, with coverage of 30% of each 90-minute Earth orbit of the satellite) of the totally-eclipsing binary ψ Centauri with point-to-point scatter of only 2 mmag.

We measure the relative radii of the stars to a precision of 0.2% (random error) from these data, and also find clear evidence for g-mode pulsations in the primary star with amplitudes of only 0.2 mmag. We have also obtained a 25-day light curve of AR Cassiopeiae which will allow us to accurately measure its apsidal motion and the relative radii of the components. In addition to six modulation at the rotational period of the primary star, and several oscillation modes with periods in the range 1–2 days.

Session 5: Binary Stars as Critical Tools:

The need to improve basic calibration;
Increasing possibilities of classical methods

Figure 1. (top left to bottom right) Speakers Jason Aufdenberg, David Valls-Gabaud, Alvaro Giménez, Todd Henry, Ben Lane, and Terry Oswalt.

Figure 2. (top left to bottom right) Speakers Andrei Tokovinin, Juan Echevarría, M. Virginia McSwain, and Julio Chanamé.

Figure 3. Evening on the Vltava.

Binary Stars as Critical Tools & Tests
in Contemporary Astrophysics
Proceedings IAU Symposium No. 240, 2006
W.I. Hartkopf, E.F. Guinan & P. Harmanec, eds.

Interferometric Constraints on Gravity Darkening with Application to the Modeling of Spica A & B

J.P. Aufdenberg[1]†, M. J. Ireland[2], A. Mérand[3] V. Coudé du Foresto[4],
O. Absil[5], E. Di Folco[6], P. Kervella[4], W. G. Bagnuolo[7], D. R. Gies[7],
S. T. Ridgway[1], D. H. Berger[8], T. A. ten Brummelaar[3],
H. A. McAlister[7], J. Sturmann[3], L. Sturmann[3], N. H. Turner[3]
and A. P. Jacob[9]

[1]National Optical Astronomy Observatory, 950 N. Cherry Ave, Tucson, AZ 85719, USA
email: aufded93@erau.edu
[2]Planetary Science, California Institude of Technology, 1200 E. California Blvd, Mail Code
150-21, Pasadena CA 91125, USA
[3]The CHARA Array, Mount Wilson Observatory, Mount Wilson, CA 91023, USA
[4]LESIA, UMR 8109, Observatoire de Paris, 5 place J. Janssen, 92195 Meudon, France
[5]Insitut d'Astrophysique et de Géophysique, University of Liège, 17 Allée du Six Août,
B-40000 Liège, Belgium
[6]Observatoire de Genève, Switzerland
[7]Center for High Angular Resolution Astronomy, Department of Physics and Astronomy,
Georgia State University, P.O. Box 3969, Atlanta, GA 30302, USA
[8]University of Michigan, Department of Astronomy, 500 Church St, 917 Dennison Bldg., Ann
Arbor, MI 48109, USA
[9]School of Physics, University of Sydney, NSW 2006, Austrailia

Abstract. In 2005 we obtained very precise interferometric measurements of the pole-on rapid rotator Vega (A0 V) with the longest baselines of the Center for High Angular Angular Resolution (CHARA) Array and the Fiber Linked Unit for Optical Recombination (FLUOR). For the analysis of these data, we developed a code for mapping sophisticated PHOENIX model atmospheres on to the surface of rotationally distorted stars described by a Roche-von Zeipel formalism. Given a set of input parameters for a star or binary pair, this code predicts the interferometric visibility, spectral energy distribution and high-resolution line spectrum expected for the system. For the gravity-darkened Vega, our model provides a very good match to the K-band interferometric data, a good match to the spectral energy distribution – except below 160 nm – and a rather poor match to weak lines in the high dispersion spectrum where the model appears overly gravity darkened. In 2006, we used the CHARA Array and FLUOR to obtain high precision measurements of the massive, non-eclipsing, double-line spectroscopic binary Spica, a 4-day period system where both components are gravity darkened rapid rotators. These data supplement recent data obtained with the Sydney University Stellar Interferometer (SUSI). Our study follows the classic 1971 study by Herbison-Evans *et al.* who resolved Spica as a binary with the Narrabri Stellar Intensity Interferometer (NSII). We will report on our progress modeling the new interferometric and archival spectroscopic data, with the goal towards better constraining the apsidal constant.

Keywords. binaries: close, binaries: spectroscopic, methods: numerical, stars: atmospheres, stars: fundamental parameters, stars: individual (Spica, Vega), stars: interiors, stars: oscillations, stars: rotation, techniques: interferometric

---

† Present address: Physical Sciences Department, Embry-Riddle Aeronautical University, 600 S. Clyde Morris Blvd, Daytona Beach, FL 32114, USA

1. Introduction

The double-line spectroscopic binary Spica (α Vir = HR 5056 = HD 116658) was the second star, after Capella (see Anderson 1920 and Merrill 1922), to have its "visual" orbital elements measured interferometrically by Herbison-Evans et al. (1971). Spica is important astrophysically because it is a close ($P = 4.01$ d), non-eclipsing massive binary ($\sim 11\ M_{\odot} + \sim 7 M_{\odot}$) in an eccentric orbit ($e \simeq 0.1$) whose components' tidal and rotational distortions lead to the advance of its longitude of periastron with time – in other words Spica exhibits apsidal motion. The advance of periastron depends on the internal structure of the stars and therefore observationally constraining Spica's apsidal motion provides a measurement of mass distribution inside its components. The primary, Spica A, is a β Cephei non-radial pulsator (with a prominent 4-hr period) providing the potential to compare its interior structure as revealed by both apsidal motion and asteroseismology.

The interferometric orbit for Spica from the Narrabri Stellar Intensity Interferometer (NSII) in 1971 provided an apsidal constant for this non-eclipsing system and immediately sparked theoretical investigations of Spica A's interior by Mathis & Odell (1973) and Odell (1974). These studies found that the mass distribution of Spica A from stellar interior models was less centrally concentrated than indicated by the apsidal motion constant. At the time of these theoretical studies this result held true not only for Spica, but also for many eclipsing eccentric double-line systems with more accurate radii. Over the next two decades, improvements in stellar interior models (better opacities, core overshooting, rotation effects) significantly reduced the discrepancies between theory and observation for most systems, but not for Spica (see Claret & Gimenez 1993). More recent work from Claret & Willems (2002) and Claret (2003) indicates that Spica remains one of the massive binaries for which there is still disagreement with theory.

The refinement of Spica's interferometric orbit, together with tighter constraints on the angular diameters of its components, promise to more tightly constrain our knowledge of the system's apsidal constant and provide an important check on theoretical expectations for stars in this mass range just off the main sequence. The apsidal constant is a function of the ratio of the primary radius to the semi-major axis of the orbit to the fifth power,

$$k_{2,\mathrm{obs}} \propto \left(\frac{\theta_A}{\theta_{SMA}} \right)^5, \qquad (1.1)$$

thus sensitive to the angular diameter of the primary θ_A and the angular measure of the semi-major axis, parameters probed by an interferometer. Advances in visibility calibration for long-baseline interferometry make possible visibility measurements up to five times more precise than used to establish the original Spica interferometric orbit. Here we present a report of our first steps toward better constraining Spica's orbit and fundamental parameters using new interferometric data from the Sydney University Stellar Interferometer (SUSI) and from the Center for High Angular Resolution (CHARA) Array (see ten Brummelaar et al. 2005) using the Fiber-Link Unit for Optical Recombination (FLUOR, see Coudé du Foresto et al. 2003). The analysis of these data is tackled with state-of-the-art model stellar atmospheres for the components including rotational distortion and gravity darkening. Our model aims to predict self-consistently the interferometry, high-resolution spectroscopy and mean spectrophotometric properties of the system as an aid in constraining the system's fundamental parameters.

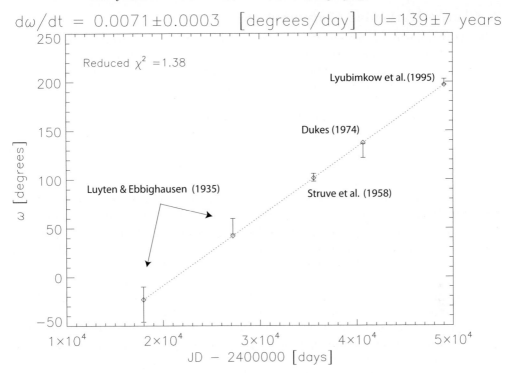

$d\omega/dt = 0.0071 \pm 0.0003$ [degrees/day] $U = 139 \pm 7$ years

Figure 1. A linear fit to evolution of Spica's longitude of periastron with time. ω values from Luyten & Ebbighausen (1935), Struve et al. (1958), Dukes (1974) and Lyubimkov et al. (1995). The corresponding apsidal period, U, is 139 ± 7 years.

2. Constraining Spica's Apsidal-Motion Constant

Our computation of Spica's mean observed apsidal motion is based on parameters from the literature and the formulae of Claret & Willems (2002). Our fit to the advance of periastron, which corresponds to $\dot{\omega} = 0.0071 \pm 0.0003$ degrees/day†, is shown in Figure 1. For computing $\log k_{2,\mathrm{obs}}$, we adopted values and uncertainties for the component masses, the orbital inclination, the component angular diameters, and the angular size of the semi-major axis from Herbison-Evans et al. (1971). The semi-amplitudes $K_{1,2}$ of the components and the eccentricity were taken from spectroscopic orbit of Shobbrook et al. (1972). The component $v \sin i$ values are from Lyubimkov et al. (1995). We find $\log k_{2,\mathrm{obs}} = -2.66 \pm 0.19$. Our value is consistent with that from Claret & Gimenez (1993). Our error bar together with Claret and Gimenez's Figure 9 shows that Spica's $\log k_{2,\mathrm{obs}}$ value is marginally consistent with theory, $\log k_{2,\mathrm{theo}} \simeq -2.4$, just outside 1 sigma. Our analysis of the error budget indicates that the uncertainty in $\log k_{2\ \mathrm{obs}}$ can be reduced from 0.19 to 0.11 if the uncertainties in the primary's (equatorial) radius, the semi-major axis, the inclination and the eccentricity can all be reduced to 1%. An additional reward from an improved orbital solution will be a precise distance estimate (via orbital parallax) independent from *HIPPARCOS*. Lastly, more tightly constraining Spica's fundamental parameters will help to better identify the primary's β Cephei pulsational modes as discussed by Smith (1985a) and Smith (1985b). Spica may prove a valuable target for on-going and upcoming space-based asteroseismology missions.

† This corresponds to $\dot{\omega} = 0.0284 \pm 0.0014$ degrees/cycle (periastron to periaston).

3. Gravity Darkening Considerations

Accurate angular diameters for Spica A & B will be model dependent because even with the CHARA Array's 313-m E1–W1 baseline the component stars themselves are not fully resolved. The components are expected to be gravity darkened, particularly Spica A with a $v \sin i$ of ~ 150 km s^{-1}, 50% of the angular break-up speed for an inclination of 63.7°. Therefore the darkening is expected to deviate from standard limb darkening models and will be dependent on the orientation of the orbit and the rotation axes of the stars relative to the orbit. We assume here that the angular momentum vectors of the stars and the orbit are aligned.

Perhaps the best interferometric gravity darkening measurements on a star thus far are those obtained by CHARA/FLUOR on Vega (see Aufdenberg et al. 2006). To analyze these data we developed a code to produce synthetic visibilities for rapidly rotating stars assuming purely radiative Von Zeipel darkening (where the β exponent in the Von Zeipel relation $T_{\mathrm{eff}}/T_{\mathrm{eff}}^{\mathrm{pole}} = (g/g_{\mathrm{pole}})^{\beta}$ is 0.25). Such a model provides a very good fit to the CHARA/FLUOR visibility data for Vega, a model independently confirmed by observations from Navy Prototype Optical Interferometer at optical wavelengths (see Peterson et al. 2006). Based only on these independent interferometric data, our gravity darkening model for purely radiative stars would seem to be sufficient to apply to the Spica problem. However a closer look reveals that this model fails to reproduce aspects of the high quality spectrophotometric and high resolution spectroscopic data available on Vega. As pointed out in Aufdenberg et al. (2006), the model is too bright in the far ultraviolet, below 160 nm, suggesting a deviation from Von Zeipel.

Far more puzzling is the failure of this interferometric Vega model to reproduce Vega's high dispersion spectrum. Work to fit Vega's high-dispersion spectrum by Hill et al. (2004) finds a much more slowly rotating Vega: $V_{\mathrm{eq}} = 115$ km $^{-1}$ (spectroscopic) versus $V_{\mathrm{eq}} = 270$ km s^{-1} (interferometric). Since Aufdenberg et al. (2006), we have upgraded our code to generate synthetic high dispersion spectra for rotating stars. Our own independent high-dispersion synthesis for Vega (see Figure 2) confirms the much lower equatorial speed found by Hill et al. (2004) assuming standard Von Zeipel darkening. The higher of the two equatorial speeds is favored to solve Vega's mass-luminosity problem (a bright pole explains the high apparent absolute magnitude), yet such a model implies a pole-to-equator effective temperature difference of $\sim$2000 K. Such a temperature difference is enough to expect that Vega's atmosphere near the equator may be convectively unstable, further complicating the darkening law. Furthermore, the effects of differential rotation and meridional circulation may come into play.

For the hotter early B-type atmospheres in the Spica system we assume a purely radiative von Zeipel darkening holds, but we recognize this may not be good assumption in the presence of mutual illumination and tidal distortion. We hope that including rotational distortion and gravity darkening is better than assuming spherical stars with regular limb darkening.

4. SUSI and CHARA/FLUOR Observations of Spica

The first interferometric measurements of Spica obtained with the Narrabri Stellar Intensity Interferometer (NSII) have not yet been followed up with modern long-baseline interferometry, as least in the published literature. As noted above, the uncertainty in the apsidal constant, $k_{2,\mathrm{obs}}$, is very sensitive to uncertainties in both the equatorial angular diameter of the primary star θ_{A} and the angular diameter of the semi-major axis θ_{SMA}. Estimates for these parameters will be tied to the precision of the visibility measurements

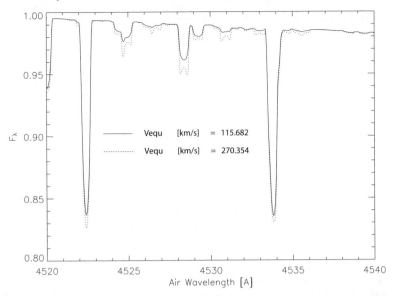

Figure 2. Two synthetic spectra for Vega with different equatorial velocities. The higher velocity model shows deeper line cores for the stronger lines and distorted, inverted line cores in the weaker lines. This distortion is due to strong darkening near the limb (Vega's equator in our pole-on view) where the local effective temperature enhances formation of these lines and the modest projected rotation velocity ($v \sin i \simeq 21$ km s^{-1}) shifts the line redward and blueward on opposite regions of the limb causing the inverted line core. The 270 km s^{-1} equatorial velocity is consistent with interferometric data (see Aufdenberg et al. 2006 and Peterson et al. 2006), while the 115 km s^{-1} best matches Vega's observed high-resolution spectrum (see Hill et al. 2004).

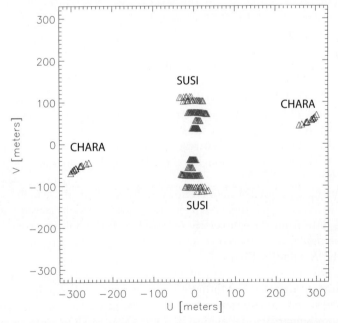

Figure 3. Baseline and position angle coverage of Spica from the CHARA/FLUOR and SUSI in-terferometers mapped on the U-V plane. CHARA's long east-west projected baselines at K-band have equivalent resolving power to SUSI's longest baselines at 700 nm.

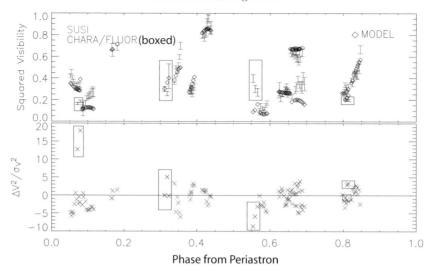

Figure 4. A preliminary comparison of model visibilities (diamonds) to the SUSI data (error bars) and CHARA/FLUOR data (error bars in boxes) on Spica. The residuals are shown in the lower panel. Our preliminary model matches these data best near phases 0.3 and 0.8 (relative to periastron), while the largest deviations occur near phases 0.1 and 0.6. At these latter pair of phases the two stars are near their minimum angular separation on the sky according the model.

and the model chosen to interpret them. While the Herbison-Evans et al. (1971) visibilities have uncertainties of $\simeq \pm 10\%$, the best data from our first season (2006) of CHARA/FLUOR observations are at the level of 5%. Visibilities at the level of 1-2% have been demonstrated with FLUOR on Vega by Aufdenberg et al. (2006), and we hope to obtain data of this precision on Spica with CHARA/FLUOR in the coming season. For our initial analysis we are combining our CHARA/FLUOR data with 112 squared visibility points obtained by Ireland (2005) at SUSI. These data, on a north-south baseline, nicely complement the CHARA data obtained on its longest (313 meter) east-west baseline. While the CHARA baseline is at least three times longer than the SUSI baselines, the SUSI data were obtained at a wavelength of 700 nm, roughly three times shorter than the FLUOR K-band. Figure 3 shows the baseline coverage in the U-V plane for CHARA and SUSI data sets. The corresponding squared visibility data are shown in Figure 4, along with preliminary model points. The model points are based on synthetic images of the Spica system which are Fourier transformed to yield synthetic visiblities. One synthetic image and Fourier map is shown in Figure 5 at phase 0.30 from periastron. For this same model, two high-dispersion spectra have been synthesized and compared to archival echelle data from the University of Toledo's Ritter Observatory (see Figure 6). Lastly, the same model is compared to archival spectrophotometry of Spica from the far ultraviolet to the K-band in Figure 7.

At this time we can say that present constraints on Spica's orbit and components are sufficient to yield reasonable synthetic matches to high dispersion spectroscopy, the spectrophotometry, and the interferometry at several phases at two wavelengths. A comparison of our preliminary model against the CHARA/FLUOR data at phases 0.1 and 0.6 (see Figure 4) indicates some important deficiency or deficiencies. Whether the resolution to this problem lies in further iteration of the system's orbital parameters or more realistic models for the system's components remains to be seen. High precision interferometry appears to be key to revising our model for the system.

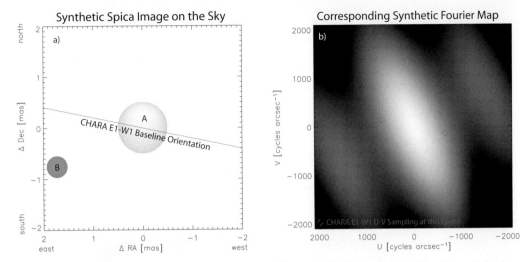

Figure 5. Panel a) shows a synthetic image of Spica A & B at phase 0.30, along with the orientation of the CHARA baseline. Both stars are limb and gravity darkened, however the dynamic range in the image makes Spica B appear as a uniform disk. Panel b) shows the Fourier map of the image; the brightness corresponds to the value of the squared visibility. The system is most resolved in a direction along a line connecting the two stars, and least resolved perpendicular to this line. As a result the CHARA/FLUOR squared visibility is ~ 0.4 as seen in Figure 4 at phase 0.30. The diamonds show the snapshot U-V sampling of the two telescope E1-W1 baseline at this single epoch.

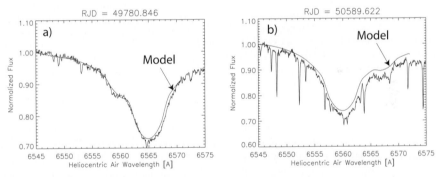

Figure 6. Two archival spectra of Spica from Ritter Observatory at phases a) 0.415 and b) 0.873 compared with the same model used to generate the synthetic squared visibilities shown in Figure 4. Strong telluric lines affect the spectrum in panel b). The Hα components of both the primary and the secondary can been discerned on opposite sides in the two panels. The model comparison isn't too bad considering the model was not fit to these data (apart from the flux normalization), only to the interferometry. Simultaneous fitting to both the spectroscopy and interferometry will hopefully lead to a more consistent model for Spica.

5. Future Work

In addition to pursuing the collection of more precise interferometric measurements on Spica with CHARA/FLUOR, we will be comparing our code to the large database of echelle spectra from Georgia State University first analyzed by Riddle et al. (2001). Improvements to our model for Spica are now underway. Our code to generate synthetic interferometry, synthetic spectrophotometry, and synthetic high-resolution line profiles for Spica does not yet include the effects of tidal distortions or the effects of mutual illumination and heating. Tidal distortion is evident in ellipsoidal light curve variations

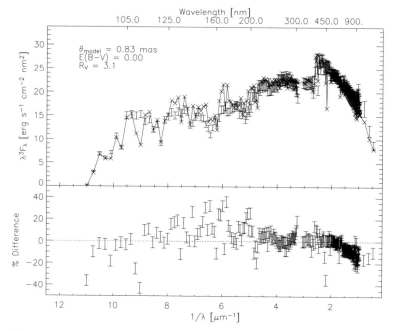

Figure 7. Top panel shows non-contemporaneous absolute spectrophotometry of Spica (error bars) compared with the model used to generate the synthetic squared visibilities shown in Figure 4. θ_{model} refers to the model angular diameter of Spica A. Data in the far ultraviolet from Morales et al. (2001), in the ultraviolet from the *International Ultraviolet Explorer* (rebinned high dispersion spectra SWP33082RL and LWP12841RL), optical data from Glushneva et al. (1992) and Glushneva et al. (1998), and J, H, K fluxes from Ducati (2002). The lower panel shows the percentage difference between the model and data. The comparison in the near-IR suggests the temperature and/or radius of the secondary may need to be adjusted. We seek to simultaneously fit these spectrophotometric data along with the interferometric and high-dispersion spectroscopic data to arrive at a best model for Spica components.

($\Delta V \simeq 0.03$, see Dukes 1974) due to the changing projection of the tri-axial primary on the sky throughout the orbit. Our plan is to take the tidally distorted Roche surfaces from D.H. Bradstreet's Binary Maker 3.0 software and map our PHOENIX model radiation field grids onto this geometry. The reflection effect will change the brightness ratio of the components as a function of phase an thus effect the squared visibility. To treat this effect in the code we plan to use the approach of Cranmer (1993). While the observations and modeling of Spica which lie ahead are quite challenging, we hope to better constrain the system's fundamental parameters, thus further constraining Spica's apsidal-motion constant, thus providing a probe of the primary's interior which can be independently probed in turn via asteroseismology.

Acknowledgements

JPA thanks the Michelson Fellowship Program (under contract with the Jet Propulsion Laboratory, which is funded by NASA, and managed by the California Institute of Technology) for three years of generous support. JPA also thanks NOAO (operated by AURA, Inc, under cooperative agreement with the National Science Foundation), for hosting my Michelson Postdoctoral Fellowship in Tucson. The CHARA Array has been supported by National Science Foundation, Georgia State University, and W. M. Keck Foundation.

References

Anderson, J.A. 1920, *ApJ*, 51, 263

Aufdenberg, J.P., Mérand, A., Foresto, V.C.d., Absil, O., Di Folco, E., Kervella, P., Ridgway, S.T., Berger, D.H., ten Brummelaar, T.A., McAlister, H.A., Sturmann, J., Sturmann, L., & Turner, N.H. 2006, *ApJ*, 645, 664

ten Brummelaar, T.A., McAlister, H.A., Ridgway, S.T., Bagnuolo, W.G., Turner, N.H., Sturmann, L., Sturmann, J., Berger, D.H., Ogden, C.E., Cadman, R., Hartkopf, W.I., Hopper, C.H., & Shure, M.A. 2005, *ApJ*, 628, 453

Claret, A. 2003, *A&A*, 399, 1115

Claret, A. & Gimenez, A. 1993, *A&A*, 277, 487

Claret, A. & Willems, B. 2002, *A&A*, 388, 518

Coudé du Foresto, V., Borde, P.J., Merand, A., Baudouin, C., Remond, A., Perrin, G.S., Ridgway, S.T., ten Brummelaar, T.A., & McAlister, H.A. 2003, in Interferometry for Optical Astronomy II. Edited by Wesley A. Traub. Proceedings of the SPIE, Volume 4838, 280–285

Cranmer, S.R. 1993, *MNRAS*, 263, 989

Ducati, J.R. 2002, VizieR Online Data Catalog, 2237, 0

Dukes, R.J. 1974, *ApJ*, 192, 81

Glushneva, I.N., Doroshenko, V.T., Fetisova, T.S., Khruzina, T.S., Kolotilov, E.A., Mossakovskaya, L.V., Ovchinnikov, S.L., & Voloshina, I.B. 1998, VizieR Online Data Catalog, 3208, 0

Glushneva, I.N., Kharitonov, A.V., Knyazeva, L.N., & Shenavrin, V.I. 1992, *A&AS*, 92, 1

Herbison-Evans, D., Hanbury Brown, R., Davis, J., & Allen, L.R. 1971, *MNRAS*, 151, 161

Hill, G., Gulliver, A.F., & Adelman, S.J. 2004, in: J. Zverko *et al.* (eds.), *The A-Star Puzzle.*, Proc. IAU Symposium No. 224 (San Francisco: ASP), p.35

Ireland, M. 2005, PhD thesis, University of Sydney

Luyten, W.J. & Ebbighausen, E. 1935, *ApJ*, 81, 305

Lyubimkov, L.S., Rachkovskaya, T.M., Rostopchin, S.I., & Tarasov, A.E. 1995, *Astron. Rep.*, 39, 186

Mathis, J.S. & Odell, A.P. 1973, *ApJ*, 180, 517

Merrill, P.W. 1922, *ApJ*, 56, 40

Morales, C., Orozco, V., Gómez, J.F., Trapero, J., Talavera, A., Bowyer, S., Edelstein, J., Korpela, E., Lampton, M., & Drake, J.J. 2001, *ApJ*, 552, 278

Odell, A.P. 1974, *ApJ*, 192, 417

Peterson, D.M., Hummel, C.A., Pauls, T.A., Armstrong, J.T., Benson, J.A., Gilbreath, G.C., Hindsley, R.B., Hutter, D.J., Johnston, K.J., Mozurkewich, D., & Schmitt, H.R. 2006, *Nature*, 440, 896

Riddle, R.L., Bagnuolo, W.G., & Gies, D.R. 2001, American Astronomical Society Meeting, 199

Shobbrook, R.R., Lomb, N.R., & Herbison Evans, D. 1972, *MNRAS*, 156, 165

Smith, M.A. 1985a, *ApJ*, 297, 206

—. 1985b, *ApJ*, 297, 224

Struve, O., Sahade, J., Huang, S.-S., & Zebergs, V. 1958, *ApJ*, 128, 310

Discussion

IZOLD PUSTYLNIK: It is known from classical papers that for rapid rotators the meridional circulation can be important. Do your models take this effect into account, and do you see from CHARA data any evidence on that?

AUFDENBERG: Our models do not take meridional circulation into account. From the CHARA data alone, a standard (purely radiative) von Zeipel gravity darkening law fits the data very well in the case of Vega. This also appears to be the case with the optical interferometric data from the Navy Prototype Optical Interferometer, the work of Peterson *et al.*. What we don't understand is why the model fits the interferometric data well, yet poorly reproduces aspects of the spectrophotometric data and very high resolution spectral line profile data. Meridional circulation may well play a role in the resolution of the problem.

Juan Manuel Echevarria: Can you further comment on the rotational velocity difference from spectroscopy and interferometry in Vega?

Aufdenberg: The work of Hill *et al.* shows that particular weak metal lines can be fit with a von Zeipel darkened model not exceeding an equatorial velocity of 115 km/s. We have independently confirmed this with our model. Above this velocity, in particular a velocity of 270 km/s which best matches our interferometric data, these lines show a reversed core (as in Figure 2) indicating the darkening is too strong near the equator. A more sophisticated model is needed to find a consistent set of parameters to match both the spectroscopic and the interferometric data. Such a model might include differential rotation such as in the recent work by Jackson, MacGregor, and Skumanich on rapidly rotating main-sequence stars.

Binary Stars as Critical Tools & Tests
in Contemporary Astrophysics
Proceedings IAU Symposium No. 240, 2006
W.I. Hartkopf, E.F. Guinan & P. Harmanec, eds.

The Distance to the Pleiades Revisited

David Valls-Gabaud

GEPI, CNRS UMR 8111, Observatoire de Paris, France
email: david.valls-gabaud@obspm.fr

Abstract. The parallax of the Pleiades has been mired in controversy ever since the very first astrometric measures in the late 1880s. Over a century later, the measures from the HIPPARCOS catalogue gave results which were inconsistent with the distance inferred from the fitting of the colour-magnitude diagram. We briefly review here the debate and focus on the various attempts made at solving the problem, and especially those using binary stars. The only double-lined eclipsing binary found so far in the Pleiades, HD 23642, provides not only the final answer to the problem but also, through detailed state-of-the-art analyses, the fundamental calibration for binaries in more distant clusters and hence in the Local Group. We discuss some of the various sources of systematic uncertainties that limit, so far, the accuracy of the measured stellar parameters to about 1%, and the progress that is required to break this barrier.

Keywords. Stars: fundamental parameters, Stars: binaries: eclipsing, Stars: binaries: spectroscopic, Galaxy: open clusters, Methods: statistical

1. Introduction

> [...] *the project of determining the parallax*
> *of the Pleiades is not altogether hopeless.*
>
> Agnes M. Clerke (1893)

The distance to the Pleiades has a long, distinguished and painful history. Even though the first attempts at measuring the parallax of its members with photographic plates resulted in widely different results, the optimism of Clerke (1893) seemed reasonable at the time. However, over a century later, and in spite of the tremendous progress made, some pessimism seems to remain:

> *I have been in astronomy so long that I don't*
> *really believe astronomical distances are much good.*
>
> Donald Lynden-Bell (1998)

In the context of the cosmological distance scale, pinning down the distances of nearby clusters is essential to calibrate the primary, and then the secondary, distance indicators. The largest contribution to the *systematic* uncertainty in the measurement of the Hubble parameter H_0 comes from the uncertainties in the distance to the Large Magellanic Cloud, which in turn reflect, in part, the systematics in the primary calibrations. Measuring locally H_0 to within a systematic uncertainty below 3% would allow us to constrain, for instance, the equation of state of the dark energy field that seems to accelerate the expansion of the universe.

Over time, and quite remarkably, the trigonometric parallaxes of the Pleiades have systematically decreased, yielding historically increasing distances from 27 pc (Schouten 1919), 83 pc (Alden 1923), 101 pc (Binnendijk 1946) and 130 pc (van Leeuwen 1983), a distance which appeared to agree with the photometric parallaxes inferred from its colour-magnitude diagram.

2. HIPPARCOS *vs* Stellar Physics ?

The parallax of the Hyades inferred from HIPPARCOS (Perryman *et al.* 1998) showed that the half-mass radius was a significant fraction (~12%) of the distance, and hence depth effects are important when interpreting the colour-magnitude diagram, and especially when testing stellar evolution models. More distant similar clusters should provide much better tests, and it came as a surprise that the HIPPARCOS parallax was significantly larger than the one inferred from photometric parallaxes (van Leeuwen & Hansen-Ruiz 1998, Mermilliod *et al.* 1997, Robichon *et al.* 1999). The difference is very significant, and amounts to some 13 pc, that is, around 1 milli-arcsecond or -0.3 magnitudes (Pinsonneault *et al.* 1998). The error in the HIPPARCOS zero point amounting to less than 0.1 mas (Arenou *et al.* 1995, Lindegren 1995), the controversy started in earnest. Either stellar evolution had to be modified to account for this new distance, or there are unaccounted problems in the HIPPARCOS catalogue, or both.

Changes suggested for the stellar tracks included an increase in the helium abundance, up to $Y = 0.34$ (Belikov *et al.* 1998) and a decrease in metallicity, down to $Z = 0.012$ (Castellani *et al.* (2002). Although a photometric abundance of [Fe/H]$= -0.1$ was claimed (Grenon 2002), precise spectroscopic abundances range from 0.026 (Boesgaard & Friel 1990) to 0.13 (Cayrel *et al.* 1988) through 0.04 (King *et al.* 2000). It would therefore be highly surprising that the helium abundance would be as large as 0.34, given that the Hyades, at [Fe/H]$=0.14$, has an inferred $Y = 0.26 \pm 0.02$. On the other hand, there are hardly any spectroscopic *measures* of the helium and iron abundance in clusters, and hence the dispersion may be very large. As van Leeuwen (1999) observed, in clusters with well-measured HIPPARCOS parallaxes, there seems to be a trend in age in their colour-magnitude diagram that appears to be unexplained by current stellar evolutionary tracks. The most recent, and accurate, colour-magnitude diagram (CMD) for the Pleiades has been made by An *et al.* (2006). In the $B - V$ diagram, the lower main-sequence stars appear far too blue in comparison with the predicted isochrone. This may perhaps be explained by their chromospheric activity (even though in the $V - I_c$ and $V - K_s$ diagrams they appear normal). In contrast, it is in the $V - I_c$ diagram of M67, where stars appear to be bluer than predicted. To make different CMDs consistent with the same distance, an *empirical* correction has to be applied to each CMD and to each cluster. It is therefore only the upper, brighter half of the CMD that seems safe from strong systematic trends, a conclusion also reached by Percival *et al.* (2005), who showed that the optical CMDs required an implausible low [Fe/H]$=-0.4$, while the IR CMDs are consistent above $M_V = 6$ with a distance of 134(± 3) pc. Changes in the shapes of the main sequence may also arise from variations in the mixing length parameter, which usually is assumed to be constant at all masses, ages and metallicities, while there may be tentative evidence of the contrary (Lastennet *et al.* 2003).

There are also more indirect ways of constraining the distance. For instance, Fox Machado *et al.* (2006) use measured seismological modes along with assumed ages and metallicities to produce consistent values for the distance. Unless better constraints for each star are used, this purely astrophysical method does not give yet strong constraints to be compared to the photometric parallax.

Ever since the very first HR diagram, by Hans Rosenberg (1911)† the Pleiades have been used as a calibrator and many of the distances inferred have found the value of the calibration. There is therefore much space for a revised distance, and the HIPPARCOS parallax has no reason to be, *a priori*, rejected. This is especially true given that the other method to measure distances, using the convergent point, is not ideal in the Pleiades,

---

† and *not* by Hertzsprung or Russell.

since the radial velocity is small and hence the expansion-related proper motion at large distances from the centre is small and there are few stars to measure. At any rate, the effect is roughly similar to the intrinsic velocity dispersion and makes any distance estimate highly uncertain.

Ground-based trigonometric parallaxes, on the other hand, yield values which appear systematically smaller than the ones from HIPPARCOS, even though the differences are within 1σ (Gatewood *et al.* 2000).

Could the HIPPARCOS parallaxes be in error, as claimed in various papers (e.g., Makarov 2002)? Part of the problem resides in the fact that the HIPPARCOS-inferred distance does not come from the straight average of the distances of stars which are members of the clusters, but rather comes as a *global* solution to parallaxes and proper motions based on the individual abscissa residuals. Correlations between abscissae measured within each data set, reduced independently by the two consortia (NDAC and FAST), are now taken into account (van Leeuwen 2004) but the inferred proper motions are inconsistent with accurate ground-based measures. This indicates that even the revised parallax from Hipparcos, at about 8.0 mas, is subject to caution. It may well be that the global solution for the cluster is actually not properly attached to the overall solution, and that there may be *local* zero-point offsets. The stars of the Pleiades being so bright, the error in the abscissa residuals come from the attitude of the satellite, rather than photon noise (which dominates at the fainter magnitudes). A full, new reduction taking into account all these effects is currently been completed (van Leeuwen 2004).

Could a different trigonometric-based parallax be made? For instance, using the Fine Guidance Sensor onboard HST, Soderblom *et al.* (2005) found that the 3 stars they observed had parallaxes inconsistent (within 1.5σ) with HIPPARCOS. Note however that the HST-based parallaxes are differential, that is, measured with respect to background stars which have their own parallaxes. One has to assume their distances based on spectral types and absolute magnitudes, and iteratively find a consistent solution for both background and foreground stars. In other words, one relies on calibrations *using* stellar physics, and so the argument is somewhat circular, as it is not very surprising to find that the 3 Pleiades stars measured by Soderblom *et al.* (2005) do agree with the distance inferred from the fitting of the CMD with stellar isochrones.

3. Binaries to the Rescue

Remarkably, binaries *can* solve the controversy, as they may bring a fully stellar physics independent way of measuring distances. The first of class of binaries which comes to mind that can yield distance measurements is the interferometric one. Atlas, one of the "seven" (or is it six?) bright stars that make the naked-eye Pleiades, was known to be an interferometric binary and observations carried out at the Mark III interferometer in 1989–1992 and then at the PTI (1996–1999) by Pan *et al.* (2004) gave a distance between 133 and 137 pc. However, they had to assume a mass–luminosity relation to get these values, and we are again in a case of circular reasoning: astrophysical inputs are likely to yield the photometric parallax... Using 12 further astrometric measures at the Mark III and NPOI, and, crucially, spectra, Zwahlen *et al.* (2004) succeeded in getting a purely geometrical distance to Atlas of 132±4 pc. A note of caution comes from the fact that the spectra had to be disentangled in order to produce a radial velocity curve for both components, and the low amplitude of the curve (37 and 26 km/s) means that there is room for improvement. Nevertheless, this was a superb achievement and further measures on other such binaries would be welcome, especially to assess the depth of the cluster. The individual HIPPARCOS parallax for Atlas is not that discrepant with this value.

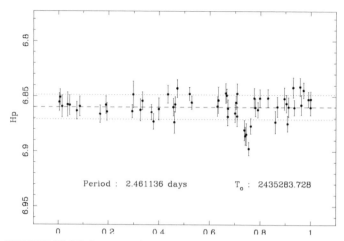

Figure 1. The HIPPARCOS light curve, folded with the period found by Griffin (1995), showing the secondary eclipse. HIPPARCOS managed to miss the primary eclipse completely, while the secondary eclipse is sampled with 5 points.

The second class of binaries that are important in this context is the one formed by double-lined (SB2), eclipsing binaries. In this case, not only the full orbital solution can be computed, but also the physical one, yielding the distance with no astrophysical calibrations whatsoever. Peering through the catalogues of SB2 binaries, there are only four of them in the Pleiades. Using the periods measured from the radial velocity curves, we folded, back in November 1997, the HIPPARCOS light curves and found, much to our delight, that HD 23642 had eclipses, as predicted by Griffin (1995). Unfortunately the sampling of HIPPARCOS was less than ideal in that it missed completely the primary eclipse (Figure 1) ! This was independently found by Torres (2003) as well.

Even though the eclipse is very shallow, denoting a grazing configuration, the star is very bright and serious amateurs could produce detailed light curves. An international observing campaign was launched, resulting in the detection of the primary eclipse with a depth of 0.08 mag, and the confirmation of the 0.05 mag-deep secondary eclipse (Miles 1999) both with CCD and photoelectric detectors. The star is so bright ($H_p = 6.84$) that professional instruments are unable to monitor it, and the role of amateurs has been essential in securing fully-sampled multi-band light curves.

Incidentally, the first "professional" observations of HD 23642 were made by Galileo (1610) in January–February 1610 (Figure 2). Given the importance of this binary for the distance scale determination, we have been securing high-precision light curves and spectra to make this binary the primary calibrator for eclipsing SB2s.

4. HD 23642 and the distance of the Pleiades

HD 23642 was already singled out by Giannuzzi (1985) who attempted to infer its distance through astrophysical calibrations, assuming an age and composition. This approach has a number of problems (e.g., Lastennet & Valls-Gabaud 2002). Munari *et al.* (2005) took five spectra and combined these new radial velocities with the old values measured in photographic plates, along with some 500 and 430 photometric measurements in the B and V bands respectively. They measured the effective temperature of the primary to be 9671±46 K, and the secondary at 8023±544 K, which yielded the

PLEIADUM CONSTELLATIO

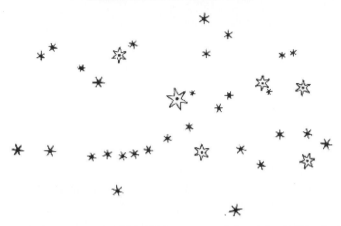

Figure 2. The first observations of HD 23642 were carried out by Galileo in 1610 as part of his quick survey of the winter sky.

absolute magnitudes and hence a distance of 132±2 pc. This determination, with a precision of 1.5%, depends critically on the adopted reddening of E(B−V)=0.012. This is an implausibly small value, even knowing the strongly differential reddening of the cluster. Southworth *et al.* (2005) analysed the same data set as Munari *et al.* (2005), and, even adopting the same reddening, found more realistic values for the effective temperatures (9750±250 K and 7600±400 K) yielding a larger distance of 139±3 pc and widening the difference with respect to the HIPPARCOS distance of 111±12 pc for this binary. A crucial aspect of their work was to realise that systematics may dominate the error budget and they explored this problem using extensive Monte Carlo simulations of synthetic light curves at their best solution. Unfortunately they kept fixed the mass ratio to the value found spectroscopically, so the cross-talk between the radial velocity curves and the light curves was not explored. In addition, the peculiar nature of the secundary, an Am star, makes both the interpretation of the light curves and effective temperature more uncertain (Burkhart & Coupry 2000, Hui-Bon-Hoa 2000, Böhm-Vitense 2006).

Our approach has been to keep under control as much as feasible the systematic errors. Following the well-established HIPPARCOS tradition, we will not give preliminary results, but will describe part of the work that we have been doing over the past eight years on this unique binary. First the photometry. Systematics being different at different telescopes, we secured light curves in the same photometric filters at different locations and telescopes. Next, we covered the optical light curves observing in the *UBVRI* filters, with a fourfold redundancy in *V* and twice in *I*. In total, we made 12382 photometric observations (Figure 3). This dataset allows us to constrain both model atmospheres and limb darkening (e.g., Claret & Hauschildt 2003).

Second, we also took spectra at different telescopes/instruments/resolutions and made a careful analysis of the zero point offsets, along with the influence of weigths given to each observation. The resulting radial velocity curve has over 80 new spectra with well understood systematics (Figure 4).

Third, we secured very high resolution and signal-to-noise spectra at both the CFHT (using Gecko and Espadons) and VLT (UVES) so that a detailed abundance analysis of each component can be carried out (Figure 5). Similarly, the spectra taken with different telescopes allow us to understand the observed variations in the line profiles through Fourier analysis. The realisation that Vega is a rapid, pole-on rotator (Peterson *et al.*

HD 23642   UBVRI light curves

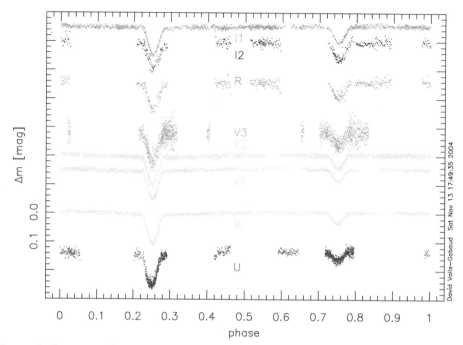

Figure 3. Summary of our 12382 photometric observations over the $UBVRI$ filters carried out over the past 7 years at different locations/telescopes to assess systematics properly.

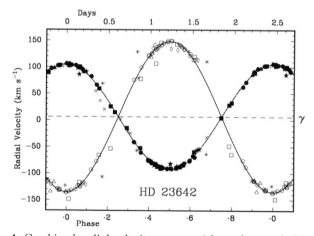

Figure 4. Combined radial velocity curves with modern and old spectra.

2006) provides yet another cautionary tale. The interstellar absorption lines over this wide wavelength range, combined with spectra of other stars in the Pleiades, allow us to map the gas that may be tracing the dust content along the line of sight (e.g., White *et al.* 2001).

Finally, we combined the optical spectra with 2MASS IR photometry and IUE SWP and LWP spectra to set constraints on the full spectral energy distribution spanning the 1000 Å– 2.2 μm range (Figure 6).

This unique dataset allows us to address a number of problems in quantifying the systematic errors: (1) zero points in radial velocities, (2) differential reddening due to the

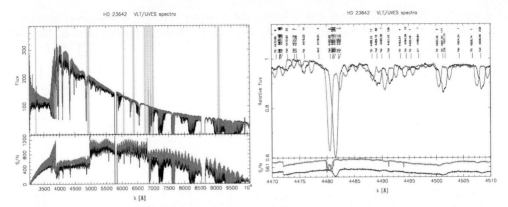

Figure 5. (*Left*) UVES spectra taken at the VLT spanning the range 3400 – 9,500 Å with an overall signal-to-noise ratio larger than 400. The shaded regions mark areas with problems in the detector. Note also the gaps at 5800 and 8600 Å. (*Right*) Zoom in the 4490 Å with an average S/N=560. The identifications and EWs in the rest frame are marked.

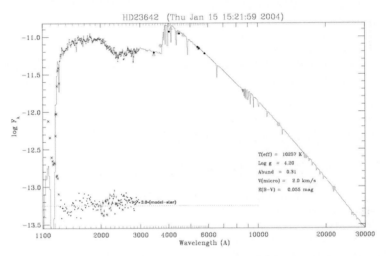

Figure 6. Overall spectral energy distribution, using IUE, optical and 2MASS data.

patchy extinction, (3) the effect of diffuse light due to the reflection nebula, and possible third light contribution, (4) the effects of the assumed limb darkening, and (5) the effects of assumptions regarding model atmospheres.

5. Summary

Although the distance to HD 23642 is not quite settled yet, we are on the right path to quantify properly both the statistical and systematic errors with unprecedented detail. This binary will prove to be the fundamental calibration for eclipsing, double-lined spectroscopic binaries, and will be essential for studies of these binaries in the Local Group. The parallax of the Pleiades is not an hopeless project, as Agnes Clerke correctly pointed out, but truly breaking the 1% barrier in precision and accuracy does require more hard work than anticipated.

Acknowledgements

I would like to thank my collaborators in this 8 year-long project: I. Ribas, G. Torres, R.F. Griffin, E. Fitzpatrick, F. Royer, J.M. Gómez Forrellad, E. García Melendo, C.H.

Lacy and D. Terrell. Thanks are also due to the members of the Variable Star Section of the British Astronomical Association, and especially R. Miles, J. Saxton, and K. Holland.

References

Alden, H.L. 1923, *AJ* 35, 61
An, D., *et al.* 2006, *ApJ* in press
Arenou, F., *et al.* 1995, *A&A* 304, 52
Belikov, A.N. *et al.* 1998, *A&A* 332, 575
Binnendijk, L. 1946, *Ann. Sterrew. Leiden* 19, 2
Boesgaard, A.M., Friel, E.D. 1990, *ApJ* 351, 467
Böhm-Vitense, E. 2006, *PASP* 118, 419
Burkhart, C., Coupry, M.F. 2000, *A&A* 354, 216
Castellani, V. *et al.* 2002, *MNRAS* 334, 193
Cayrel, R. *et al.* 1988, in *IAU Symp* 132, 449
Claret, A., Hauschildt, P.H. 2003, *A&A* 412, 241
Clerke, A.M. 1893, *Observatory* 16, 198
Fox Machado, L. *et al.* 2006, *A&A* 446, 611
Galileo, G. 1610, *Sidereus Nuncius* (Florence)
Gatewood, G. *et al.* 2000, *ApJ* 533, 938
Giannuzzi, M.A. 1995, *A&A* 293, 360
Grenon, M. 2002, *Highlights in Astronomy* 12, 680
Griffin, R.F. 1995, *JRASC* 89, 53
Hui-Bon-Hoa, A. 2000, *A&AS* 144, 203
King, J.R. *et al.* 2000, *ApJ* 533, 944
Lastennet, E. *et al.* 2003, *A&A* 409, 611
Lastennet, E. & Valls-Gabaud 2002, *A&A* 396, 551
Lindegren, L. 1995, *A&A* 304, 61
Lynden-Bell, D. 1998, in *IAU Symp.* 192, 15
Makarov, V.V. 2002, *AJ* 124, 3299
Mermilliod, J.C. *et al.* 1997, in *Hipparcos Venice '97*, ESA SP-402, 643
Miles, R. 1999, *J. Brit. Astron. Assoc.* 109, 106
Munari, U., *et al.* 2004, *A&A* 418, L31
Pan, X., Shao, M., Kulkarni, S.R. 2004, *Nature* 427, 326
Percival, S.M. *et al.* 2005, *A&A* 429, 887
Perryman, M.A.C. *et al.* 1998, *A&A* 331, 81
Peterson, D.M. *et al.* 2006, *Nature* 440, 896
Pinsonneault, M.H., *et al.* 1998, *ApJ* 504, 170
Robichon, N., *et al.* 1999, *A&A* 345, 471
Rosenberg, H. 1911, *Ast. Nach.* 186, 71
Schouten, W.J.A. 1919, *Observatory* 42, 240
Soderblom, D.R. *et al.* 2005, *AJ* 129, 1616
Southworth, J. *et al.* 2005, *A&A* 429, 645
Tomkin, J. 2005, in *ASP Conf. Series*, 336, 199
Torres, G. 2003, *IBVS* 5402, 1
Van Leeuwen, F. 1983, *Ph.D. Thesis*, Leiden Univ.
Van Leeuwen, F. 1999, *A&A* 341, L71
Van Leeuwen, F. 2004, in *Transits of Venus*, IAU Coll. 196, 347
Van Leeuwen, F., Hansen-Ruiz, C.S. 1997, in *Hipparcos Venice '97*, ESA SP-402, 689
White, R.E. *et al.* 2001, *ApJ* 132, 253
Zwahlen, N. *et al.* 2004, *A&A* 425, L45

Discussion

VIRGINIA TRIMBLE: So I guess the question is: What went wrong with the Hipparcos parallaxes ?

VALLS–GABAUD: There is nothing fundamentally wrong with the parallaxes provided by HIPPARCOS! As a matter of fact, the comparison of accurate, parallax- and stellar physics- independent distances and the Hipparcos-inferred ones yield a perfect correlation within 50 pc (Tomkin 2005) and at larger distances the dispersion increases, exactly as one expects. The trouble may arise when computing average distances of extended clusters, from a handful of bright stars (Van Leeuwen 2004).

Binary Stars as Critical Tools & Tests
in Contemporary Astrophysics
Proceedings IAU Symposium No. 240, 2006
W.I. Hartkopf, E.F. Guinan & P. Harmanec, eds.

The Apsidal Motion Test in Eclipsing Binaries

Alvaro Giménez

Research and Scientific Support Department, ESA, ESTEC, Noordwijk, the Netherlands

Abstract. The use of eccentric eclipsing binaries to test stellar internal structure models, as well as the equations of motion provided by General Relativity, is reviewed. Close to 80 years have elapsed since the first ideas were produced in this field and many results have been obtained since then. It appears that, in general, a good understanding of stellar structure within the main sequence is available while the same level of knowledge can not be claimed beyond the termination age. The equations of general relativity could not be disproved with observational data though some systems cannot still be fully explained. In the near future, the analysis of evolved systems, very low mass stars, the effects of tidal resonances and the presence of third bodies has to be further explored. In addition, the analysis of large data bases obtained by means of extensive photometric surveys will certainly change the picture from an observational point of view.

Keywords. Eclipsing binaries, Eccentric Binaries, Apsidal Motion, Internal Structure, General Relativity

1. Introduction

The purpose of this review is to present briefly the history of the apsidal motion test in the field of close binaries as well as the current status of the subject, latest results and prospects for the future. The initial idea to measure the motion of the periastron of binary stars in order to have some insight into their internal structure was given by Russell (1928). His problem was the accumulation of observational data on eclipsing binaries with spectroscopic orbits leading to values of the average density of the stars well below that needed to achieve configurations in hydrostatic equilibrium. It was clear that the central density concentration had to be much higher than the average value. In eccentric binaries, the behavior of the orbit is ruled by the distortion of the components and this is a function of the internal density concentration as well as the mass ratio and the separation of the components. Stellar distortion, or deviation from the point-like behavior, is responsible for the secular movement of the periastron and its observational measurement should obviously lead to an empirical measurement of the level of the density concentration.

The idea was right but the proposed equations for the apsidal motion were not fully developed. A debate was initiated about the proper treatment of the problem and Kopal (1936) proposed to measure the stellar distortion directly by means of the photometric variations outside of eclipse, but the existence of additional effects, like reflection, precluded to reach reliable conclusions. On the other hand, stellar models were not available to predict the rate of motion of the periastron though Chandrasekhar (1933) contributed to solve this by computing the internal density concentration, and corresponding distortions, for polytropic models. The equations of motion were revised by Cowling (1938) who identified the internal structure parameter k_2 to be related to the apsidal motion rate through the fifth power of the relative radii of the stars. Accurate absolute dimensions were therefore necessary for any reliable interpretation of the observational data. Kopal (1938) contributed to the development of the equations by introducing the effects of libration and Sterne (1939a) extended the work by Cowling (1938) using terms up to r^9 while producing the necessary tools to analyze the observations of eclipse timings under the assumption that the line of sight remains within the orbital plane.

Sterne (1939b) could thus deliver the first comparison of polytropic models with observations, collected for 5 binary systems. The best agreement was found for a polytropic index $n = 3$, also known as the standard model. Sterne was nevertheless well aware that the comparison of non-accurate data with unrealistic models could not lead to reliable conclusions and the method did not advance further. In parallel, Levi-Civita (1937) had found the equations for the relativistic motion of eccentric binary stars to be equivalent to those corresponding to the perihelion motion of the planet Mercury. The results were probably academic at the time but proved to be applicable years later as described in Section 3.

Schwarzschild (1958) revitalized the apsidal motion test by making a comparison using reliable observational data for 8 systems and new internal structure models based on realistic physical assumptions. The result then was that the observations were indicative of much more centrally condensed stars than actually predicted. Schwarzschild already noticed that the situation improved, but was far from solved, taking into account stellar evolution. New detailed equations for the apsidal motion were given by Kopal (1959) including their modification when non-synchronized rotation, inclined axes, or nodal variations are taken into account. Kopal (1965) also revised the situation of the comparison between models and observations increasing the number of binary systems to 14 but being only able confirm the disagreement, even considering cases just within the main sequence.

In the following decade, the apsidal motion test was revisited by several authors. It is interesting to note the critical review by Batten (1973) and the evolutionary considerations by Petty (1973). Stothers (1974) introduced in the discussion the effects of the adopted opacities and stellar rotation, from the point of view of the models computations, both in fact leading to more centrally concentrated stellar mass distribution, or lower values of the internal structure constants k_2, as actually observed. On the other hand, O'Dell (1974) questioned the predicted size of the cores of massive stars. Finally, Kopal (1978) reviewed the dynamical behavior of close binaries with no definitive solution to the disagreement between models and observations.

The interest for further analysis of eccentric eclipsing binaries showing apsidal motion increased considerably as a consequence of all these studies. Giménez (1981a) tried to put together all the accumulated data and made a critical discussion of the comparison with the best available theoretical models. The new approach included a revision of the selection of the data sample. Only well-detached systems with well-determined eccentricities and clean light curves were used. From an observational point of view, the determination of apsidal motion rates was purely based on the measurement of the changing relative position of deep eclipses. Further, in order to make the comparison with the relevant models, accurate absolute dimensions for the component stars were used. Even with these severe constraints, the number of systems used in the comparison could be increased to 20 as discussed in detail by Giménez (1981b).

The results, also shown by Giménez & Garcia-Pelayo (1982), clearly indicated the strong dependence of the variation of k_2 with age and, to a lower degree, with the adopted opacities. It was then clear that new modern evolutionary models were needed. For the first time, the observational data were precise enough to identify the source of the problem in the apsidal motion test.

2. The quadrupole term

During the eighties the effort was concentrated in the computation of new models. The old opacities by Cox & Stewart (1970), adopted by Hejlesen (1980), were very useful for the comparison with absolute dimensions obtained from well-detached eclipsing binaries (see Andersen 1991), though they presented a problem with the fit of the effective temperatures for low mass stars and the internal structure constants were only published years later (Hejlesen 1987). The solution to this problem came with the use of better opacity tables. Jeffery (1984) computed models using those by Carson (1976) that predicted more centrally concentrated stars, but proved later to be based on unrealistic assumptions. More detailed computations of the internal structure constants were published by Claret & Giménez (1989) but the real improvement was made when the new Los Alamos opacities by Rogers & Iglesias (1992) were available and the corresponding computations were made by Claret & Giménez (1992), taking also into account the effect of a moderate amount of convective core overshooting.

In parallel, the equations for the determination of apsidal motion rates by means of the analysis of the position of the eclipses were revised by Giménez & Garcia-Pelayo (1983) and new cases of double-lined eclipsing binaries with eccentric orbit could be added to the data sample available for the comparison with the models. During the same period of time, several studies were carried out to assess tidal effects in the rotation of the component stars. This parameter was known to play an important role in the determination of the internal structure constants as well as in their modelling.

Using 24 binary systems and the latest models, a detailed comparison of the k_2 internal structure constants was published by Claret & Giménez (1993). The adopted procedure first verified that the observed effective temperatures of the component stars were in good agreement with those predicted by the models and these, in addition, provided the same evolutionary age for the primary and the secondary component of each system. Finally, the observed rotational velocities were compared with the orbital period. As a result the quality of the data, and the goodness of the models to reproduce them, could be assessed.

The comparison of the internal structure constants showed a much better agreement than in previous studies, certainly as a result of the consideration of stellar evolution. Nevertheless, a systematic difference could still be seen, again in the sense of stars being more centrally concentrated than predicted. The difference was found also to be a function of evolution but with a much smaller amplitude than that indicated by Giménez & Garcia-Pelayo (1982). In any case, the disagreement was reduced to just three cases with the lowest data quality (which were also the more evolved ones as indicated by Giménez 1984) and the main part of the data sample was very well reproduced, with no systematic effect when the models were corrected for the effect of rotation. It has already been mentioned that rotation has to be taken into account since models, generally computed for non-rotating configurations, are compared to real, rotating, stars in the apsidal motion test. Claret (1999) has analyzed the issue in detail, computing rotating models, and provided a good confirmation of the correction term given by Stothers (1974). Moreover, the situation with the more evolved systems could be improved by assuming some degree of convective core overshooting, fully consistent with independent tests of the width of the main sequence by Andersen, Nordström & Clausen (1990).

Further studies were obviously required and were developed in three different areas. New lists of candidates to be monitored superseded those by Kopal (1978) and Giménez & Delgado (1980), and were given by Hegedus (1988), Giménez (1994), Petrova & Orlov (1999) and Hegedus, Giménez & Claret (2005). They have allowed the systematic monitoring of more than 100 eclipsing binaries in order to determine accurate apsidal motion rates and absolute dimensions. The second area of development was an improvement in the methods used to analyze the data. Lacy (1992) proposed an exact iterative method while Giménez & Bastero (1995) extended the earlier equations by Giménez & Garcia-Pelayo (1983), and van Hamme & Wilson (1998) added new ideas using the simultaneous solution of all the observational evidence rather than just the position of the eclipses. As a third area of research, new detailed theoretical models were computed in the 1995–2005 decade for a wide range of masses and chemical compositions by Claret (see e.g. Claret 1995), allowing the comparison with observational data for different degrees of metallicity, rotation, and convective overshooting.

The latest general comparison of models with observational data has been published by Claret & Willems (2003). Allowing the fit of models for the best chemical composition combination led to an excellent comparison of the predicted and observed effective temperatures as well as the evolutionary ages of the two components of each system. The comparison of the internal structure constants showed no significant difference at all when a moderate degree of convective core overshooting was assumed and thus the long fight to reconcile models and observations could be closed. A reasonable agreement was definitely found provided that a) the initial chemical composition is taken into account, b) the empirical values of the internal structure are corrected for rotational effects, and c) some amount of convective core overshooting is considered (see also Giménez *et al.* 1999). The determination of the initial chemical composition for a sample of binary stars allowed in addition the estimation of the fundamental chemical enrichment law in our Galaxy. This had already been shown by Ribas *et al.* (2000a) for a wider sample of binary systems and showed an excellent agreement with results obtained through other approaches. A similar use of the test for the required degree of convective core overshooting allowed Ribas, Jordi & Giménez (2000) to show a first attempt to look into its variation with mass.

Two possible complications were nevertheless identified that could bias the apsidal motion test assumptions and give unexpected results in some individual cases. The presence of an unseen third body in a given system could affect the observational data and the assumption of equilibrium tides made by Cowling and Sterne could not be valid. In fact, the importance of dynamical and resonance tides was already pointed out by Papaloizou & Pringle (1980) and discussed by Quataert, Kumar & Ao (1996), Smeyers, Willems & van Hoolst (1998), and Smeyers & Willems (2001). Nevertheless, when the modified equations are applied to eccentric binaries, like those currently used in the apsidal motion test, Claret & Willems (2003) showed that no significant effect should be expected and the classical approach can be safely used within the level of the observational uncertainties. On the other hand, the discovery of third bodies in the adopted sample of eccentric binaries, mostly young systems, increased considerably with the precision of the observational data. Light-time effect was mainly used together with the need for third-light in the light curve, to detect the presence of these companions. The equations by Martynov (1948) could be applied to correct the observed apsidal motion rates. Nevertheless, the existence of unseen bodies remains a source of uncertainty in the test. A classical example is AS Cam, an anomalous case that could be explained with a non-coplanar third body, as discussed by Khodykin & Vedenesev (1994) and Kozyreva *et al.* (1999).

3. The relativistic term

Studies looking for the internal density concentration of the stars corrected the observed apsidal motion rates for the contribution due to the relativistic equations of motion, as provided in early times by Levi-Civita (1937). But the test of general relativity is an interesting subject on its own and apsidal motion measurements can also be used for this purpose. Rudkjobing (1959) already suggested that the massive eccentric binary DI Her, with an orbital period of around 10 days, could be an excellent observational target. Unfortunately, despite many attempts, the detection of the relativistic term in the apsidal motion of DI Her failed to give the expected results. In 1982, Giménez & Scaltriti (1982) could measure reliably the apsidal motion of the system V889 Aql to be in good agreement with the equations of General Relativity. However, the system was less massive than the original target suggested by Rudkjobing (1959) or those mentioned by Koch (1973). Koch indeed proposed 5 candidates, 2 of them with good absolute dimensions, but none with a clear determination of the apsidal motion rate. Giménez (1985) subsequently showed that the key parameter is not the total mass of the system but the relative radii of the component stars since the newtonian apsidal motion rate decreases much faster than the relativistic term with this parameter. Following this argument, a new list of 23 candidate systems for the relativistic test was put together by Giménez (1985) leading necessarily to very slow rotation periods and thus difficulties in their measurement, as discussed by Claret (1997). By 1985, already 5 systems had been studied with a relativistic term dominating the total apsidal motion rate: 4 in good agreement with General Relativity (V889 Aql, EK Cep, V1143 Cyg, and VV Pyx) and one in clear disagreement (DI Her), as found by Guinan & Maloney (1985). A new approach to formalize gravitation by Moffat (1989) was then used to explain the problem though, when applied to all systems with apsidal motion, the use of the same parametrization failed to fit the otherwise good agreement in the predicted and observed internal structure constants as shown by Claret (1997).

Dynamical tides are less important in the well-detached systems selected for the relativistic test than in closer binaries so that they could not be used to explain the disagreement of DI Her. Non-synchronization of the rotational axis with that perpendicular to the orbit was proposed by Shakura (1985) but only the presence of a third body in an almost orthogonal orbit with respect to the orbital plane of the eclipsing system could lead to the observed apsidal motion rate, as indicated by Khodykin & Vedenesev (1994) in the case of AS Cam, and discussed by Hsuan & Mardling (2006) for DI Her. The effect of a third body is an important issue in the study of apsidal motions and several systems in the list of Giménez (1985) have been later found to have companions. The expected effect is nevertheless to increase the apsidal motion rate rather than to slow it, as indicated by the equations of Martynov (1948) in the case of coplanar orbits. A simple guess indicates that at least 1/3 of all binaries in the referred list of candidates may have a third body, and possibly 1/2.

A revision of the situation has been carried out with inputs from the literature collected in the last two decades. The list of candidate systems now include 36 binaries, instead of 23, with 25 of them having good absolute dimensions, instead of 9. Realistic apsidal motion determinations are available for 16 binaries in the sample, compared to the previous 5 and a new comparison with theoretical predictions could be thus attempted. Preliminary results show that a good agreement is found for 12 out of the mentioned 16 cases and they are shown in Figure 1.

Giménez (1985) had found that 4 systems were in agreement with the theory and 1 in disagreement. The situation now is that 12 systems are in agreement and 4 failed the comparison (DI Her, BW Aqr, SW CMa and ES Lac). Although the statistics of failures remains more or less the same, the interesting point is that all except DI Her present a faster than expected apsidal motion; i.e., they can be easily explained with a, not yet seen, coplanar third body. The case of DI Her remains the only odd case requiring a third

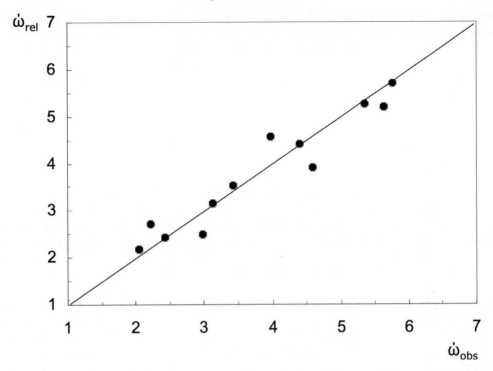

Figure 1. Comparison of the predicted and observed relativistic term in the apsidal motion rate of selected eclipsing binaries.

body with an orbit perpendicular to that of the eclipsing system. One case is of course not enough to disprove a solid theory and, in addition, other cases are known to have non coplanar third bodies, like that of SS Lac that stopped showing eclipses due to the induced change in the inclination of the orbit (Torres 2001) or the previously mentioned AS Cam.

4. The future

The field of apsidal motion studies in eclipsing binaries is not over despite the advance of the last years. The first point requiring further analysis is linked to the fact that most of the systems studied up to now cover medium mass stars within the main sequence and close to solar chemical composition. An extension of the test has therefore to be carried out in other domains of the HR diagram. The study of very low mass stars, like CM Dra, OGLE-TR-122 (Pont *et al.* 2005) or the recently discovered brown-dwarf binary (Stassun, Mathieu & Valenti 2006), will certainly help understanding the structure of purely, or almost so, convective configurations. On the other side, the analysis of evolved massive systems, like V380 Cyg or V453 Cyg, recently studied by Southworth, Maxted & Smalley (2004), will provide a better knowledge of convective core overshooting, including its variation with mass and age. Finally, binaries formed in a different environment, or metallicity, like HV 2274 in the LMC studied by Ribas *et al.* (2000b), will provide an insight into the internal structure of extragalactic stars.

In these studies it is clear that increased efforts have to be devoted to the determination of accurate absolute dimensions for all systems showing apsidal motion and, if possible, for all eccentric eclipsing binaries, as potential candidates for the test of the models.

Of course, new minima should be obtained for the most interesting cases, in order to determine accurate apsidal motion periods. One of the most important aspects in the coming years for the comparison between observations and stellar structure models will be to assess the effect of third bodies in the systems studied. To look carefully for any indication of a third body, either directly, through their effect in the light curve or in the variation of the eclipse timings, will be critical.

The observational constraints imposed by Giménez (1981a) for the apsidal motion test, can now be partly lifted thanks to the much better observational data available. An example is the clean light curve of the eccentric system RZ Eri shown by Burki, Kviz & North (1992) with heavily spotted components. But the future of the test will be mainly driven by the large increase in the discovery of new eclipsing binaries. The first way to go is certainly the systematic and in depth exploitation of existing data bases of light curves of eclipsing binaries. The use of the Hipparcos data has shown, despite its low photometric precision, how the determination of additional eclipse timings is critical in the determination of accurate apsidal motion rates (Otero 2003). Large surveys in crowded fields, looking for microlensing events, like OGLE, MACHO and EROS, have discovered thousands of new eclipsing binaries, many of them showing eccentric orbits and thus providing many more candidate systems than those in classical catalogues. Identifying those that best suit the kind of fundamental tests mentioned above will be the main task in the coming years. A good example of the determination of apsidal motion periods from binaries discovered in a survey of the LMC, using automated algorithms, has been shown by Michalska & Pigulski (2006). The newly discovered eccentric eclipsing binaries in the Magellanic Clouds and beyond, like the system in M33 studied by Bonanos *et al.* (2006), will open further possibilities for testing stellar structure outside our own Galaxy. But another area of interest that I would like to emphasize is the study of dynamical tides and resonant oscillations in eccentric systems known to have pulsating components. The precision and amount of data accumulated for this type of eclipsing systems, neglected for some time, has been increasing recently and the apsidal motion test can provide very important new pieces of information.

References

Andersen, J. 1991, *Astron. & Astrophys. Reviews*, 3, 91
Andersen, J., Nordström, B., & Clausen, J.V. 1990, *ApJ*, 363, L33
Batten, A.H. 1973, *Binary and Multiple Systems of Stars*, Pergamon Press, New York
Bonanos, A.Z. *et al.* 2006, , preprint in astro-ph/0606279
Burki, G., Kviz, Z., & North, P. 1992 *A&A*, 256, 463
Carson, T.R. 1976, *Ann. Rev. Astron. Astrophys.*, 14, 95
Chandrasekhar, S. 1933, *MNRAS*, 93, 449
Claret, A. 1995, *A&A Suppl.*, 109, 441
Claret, A. 1997, *A&A*, 327, 11
Claret, A. 1999, *A&A*, 350, 56
Claret, A. & Giménez, A. 1989, *A&A*, 81, 1
Claret, A. & Giménez, A. 1992, *A&A Suppl.*, 96, 255
Claret, A. & Giménez, A. 1993, *A&A*, 277, 487
Claret, A. & Willems, B. 2003, *A&A*, 410, 289
Cowling, T.G. 1938 *MNRAS*, 98, 734
Cox, A.N. & Stewart, J.N 1970, *ApJ Suppl.*, 19, 243
Giménez, A. 1981a, *in Photometric and Spectroscopic Binary Systems*, NATO Advanced Institute Series, Eds. E.B. Carling & Z. Kopal, Kluwer, Dordrecht, 511
Giménez, A. 1981b, *PhD Thesis, University of Granada*

Giménez, A. 1984, *Observational Tests of the Stellar Evolution Theory*, IAU Symp. 105, Eds. A, Maeder & A. Renzini, Reidel, Dordrecht, 419

Giménez, A. 1985, *ApJ*, 297, 405

Giménez, A. 1994, *Exp. Astron.*, 5, 91

Giménez, A. & Bastero, M. 1995, *Astrophys. Space Sci.*, 226, 99

Giménez, A. & Delgado, A. 1980, *I.B.V.S., IAU Comm. 27*, No. 1815

Giménez, A. & Garcia-Pelayo, J.M. 1982, *Binary and Multiple Stars as Tracers of Stellar Evolution*, Eds. Z. Kopal & J. Rahe, Reidel, Dordrecht, 37

Giménez, A. & Garcia-Pelayo, J.M. 1983, *Astrophys. Space Science*, 92, 203

Giménez, A. & Scaltriti, F. 1982, *A&Ap*, 115, 321

Giménez, A., Claret, A., Ribas, I., & Jordi, C. 1999, *Theory and Tests of Convection in Stellar Structure*, Eds. A. Giménez, E.F. Guinan & B. Montesinos. ASP Conf. Series, 173, 41

Guinan, E.F. & Maloney, F.P. 1985, *AJ*, 90, 1519

van Hamme, W.V. & Wilson, R.E. 1998, *BAAS*, 30, 1402

Hegedus, T. 1988, *Bull. Inf. CDS*, 35, 15

Hegedus, T., Giménez, A., & Claret, A. 1985, *Tidal Evolution and Oscillations in Binary Stars*, Eds. A. Claret, A. Giménez & J.P. Zahn, ASP Conf. Series, Vol. 333, 88

Hejlesen, H.E. 1980, *A&A Suppl.*, 39, 347

Hejlesen, H.E. 1987, *A&A Suppl.*, 69, 251

Hsuan, K. & Mardling, R.A. 2006, *Astrophys. Space Sci.*, in press

Jeffery, S.F. 1984 *MNRAS*, 207, 323

Khodykin, S.A. & Vedenesev, V.G. 1994, *ApJ*, 475, 798

Koch, R.H. 1973, *ApJ*, 183, 275

Kopal, Z. 1936, *MNRAS*, 96, 854

Kopal, Z. 1938, *MNRAS*, 98, 448

Kopal, Z. 1959, *Close Binary Systems*, Chapman & Hall, London

Kopal, Z. 1965, *Advances in Astron. & Astrophys.*, 3, 89

Kopal, Z. 1978, *Dynamics of Close Binary Systems*, Reidel Publ., Dordrecht

Kozyreva, V.S., Zakharov, A.I., & Khaliullin, Kh.F. 1999, *IBVS*, No. 4690

Lacy, C. 1992, *AJ*, 104, 2213

Levi-Civita, T. 1937, *Amer. J. Mathem.*, 59, 225

Martynov, D. Ya. 1948, *Izv. Engelhardt Obs. Kazan*, No. 25

Michalska, G. & Pigulski, A. 2006, *A&A*, preprint in astro-ph/0501380

Moffat 1989, *Phys. Rev. D*, 39, 474

O'Dell, A.P. 1974, *ApJ*, 192, 417

Otero 2003, *IBVS*, 5482

Papaloizou, J. & Pringle, J.E. 1980, *MNRAS*, 193, 603

Petrova, A.V. & Orlov, V.V. 1999, *AJ*, 117, 587

Petty, A.F. 1973, *Astrophys. Space Sci.*, 21, 189

Pont, F., Melo, C.H.F., Bouchy, F., Udry, S., Queloz, D., Mayor, M., & Santos, N.C. 2005, *A&A*, 433, L21

Quataert, E.J., Kumar, P., & Ao, C.O. 1993, *ApJ*, 463, 284

Ribas, I., Jordi, C., & Giménez, A. 2000, *MNRAS*, 318, L55

Ribas, I., Jordi, C., Torra, J., & Giménez, A. 2000a, *MNRAS*, 313, 99

Ribas, I., Guinan, E.F., Fitzpatrick, E.L., de Warf, L.E., Maloney, F.P., Maurone, P.A., Bradstreet, E.H., Giménez, A., & Pritchard, J.D. 2000b, *ApJ*, 528, 692

Rogers, F.J. & Iglesias, C.A. 1992, *ApJ Suppl.*, 79, 507

Rudkjobing, M. 1959, *Ann. d'Astroph.*, 22, 111

Russell, H.N. 1928, *MNRAS*, 88, 642

Schwarzschild, M. 1958, *Structure and Evolution of the Stars*, Princeton University Press

Shakura, N.I. 1985 *Sov. Astr. Lett.*, 11, 225

Smeyers, P., Willems, B., & van Hoolst, T. 1998, *A&A*, 335, 622

Smeyers, P. & Willems, B. 2001, *A&A*, 373, 173

Southworth, J., Maxted, P.F.L., & Smalley, B. 2004, *MNRAS*, 351, 1277

Stassun, K.G., Mathieu, R.D., & Valenti, J.A. 2006, *Nature*, 440, 311

Sterne, T.E. 1939a, *MNRAS*, 99, 451
Sterne, T.E. 1939b, *MNRAS*, 99, 662
Stothers, R. 1974, *ApJ*, 194, 651
Torres, G. 2001, *AJ*, 121, 2227

Discussion

GUILLEM ANGLADA-ESCUDÉ: What is the timing accuracy required for your studies? Does it really depend very strongly on color information?

GIMÉNEZ: It depends, of course, on the period of each particular system. The photometry and the deepness of the light curve and the eclipses ultimately determine your timing precision. Multicolor minima are used to accurately calibrate minima's asymmetries that can potentially perturb the instant of minima estimation.

NANCY EVANS: Just to let you know about a little more information we will provide to let you interpret your very interesting results. We (with Guinan, Ribas, Fitzpatrick, etc.) have got the FUSE spectra of some of the hottest systems to study the temperatures.

GIMÉNEZ: This is excellent news.

ROBERT WILSON: Could differential rotation be a significant factor? Tidal locking makes the outer envelope rotate at the orbital rate, but stay at another rate in the somewhat deeper interior.

GIMÉNEZ: Indeed some observational results of the apsidal motion test can be interpreted as an effect of convective core overshooting but could also be understood by any physical process that produces more massive cores. Differential rotation, if it is faster inside the star, may provide one of these processes.

Binary Stars as Critical Tools & Tests
in Contemporary Astrophysics
Proceedings IAU Symposium No. 240, 2006
W.I. Hartkopf, E.F. Guinan & P. Harmanec, eds.

© 2007 International Astronomical Union
doi:10.1017/S174392130700419X

The Sun's Smaller Cousins are Running the Universe — The Masses of Red and Brown Dwarfs

Todd J. Henry

Department of Physics and Astronomy, Georgia State University, Atlanta GA, USA

Abstract. Although not one red dwarf can be seen with the naked eye, they dominate the solar neighborhood, accounting for no less than 70% of all stars. The large numbers of these wee denizens of the night actually translate into a surprisingly large amount of mass – in fact, more mass is found in M dwarfs than in any other stellar spectral type.

To determine just how important red dwarfs are to the nature of the Universe, an accurate mass-luminosity relation (MLR) must be determined so that a relatively easily determined characteristic, luminosity, can be converted into the critical parameter, mass. Results from a decade-long observational effort to calibrate the MLR for red dwarfs using the Fine Guidance Sensors (FGSs) on the Hubble Space Telescope (HST) will be highlighted. For many of the binary systems targeted, the interferometric data from HST are combined with radial velocity data to further improve the mass measurements, which often have errors less than 5%. Related results from a large southern sky parallax program to determine accurate distances to red dwarfs, and a search for companions orbiting them, will also be discussed.

The state of mass determinations for M dwarfs' smaller cousins, the L and T dwarfs, will also be reviewed. Although we do not yet know the size of the true population of these (primarily) substellar objects, only by mapping out the interplay of their masses and luminosities (which change drastically with time) can we understand their contribution to the mass budget of the Universe.

Discussion

JUAN ECHEVARRIA: With so many new M dwarf masses derived from observations, what improvements have been made on the M–R relation?

HENRY: We are using the CHARA Array to measure directly the radii of M dwarfs. We find that between 0.5 and 0.7 $R_{\odot}$, the models do not fit the data well. Between 0.2 and 0.5 $R_{\odot}$, all seems to be well. There is clearly more observational work to be done, but our first-order improvements in the M–R relation indicate that there is quite a bit of understanding yet to be had.

HANS ZINNECKER: How many of your M stars in the 'Hood actually live in binary systems (i.e., what is the binary fraction)?

HENRY: Of M dwarf systems, ∼30% are multiples.

Binary Stars as Critical Tools & Tests
in Contemporary Astrophysics
Proceedings IAU Symposium No. 240, 2006
W.I. Hartkopf, E.F. Guinan & P. Harmanec, eds.

© 2007 International Astronomical Union
doi:10.1017/S1743921307004206

Fragile Binaries: Observational Leverage on Difficult Astrophysical Problems

T.D. Oswalt[1], K.B. Johnston[1], M. Rudkin[1], T. Vaccaro[1], and D. Valls-Gabaud[2]

[1]Dept. of Physics and Astronomy, Florida Institute of Technology
150 West University Blvd., Melbourne, FL 32901
email: toswalt@fit.edu, kyjohnst@fit.edu, mrudkin@fit.edu, tvaccaro@fit.edu

[2]Laboratoire d'Astrophysique UMR CNRS 5572, Observatoire Midi-Pyrénées
14 Avenue Edouard Belin, F-31400 Toulouse Cdx, France.
email: David.Valls-Gabaud@obspm.fr

Abstract. Loosely bound, *fragile* binary stars, whose separations may reach ~ 0.1 pc, are like open clusters with two coeval components. They provide a largely overlooked avenue for the investigation of many astrophysical questions. For example, the orbital distribution of fragile binaries with two long-lived main-sequence components provides a sensitive test of the cumulative effects of the Galactic environment. In pairs where one component is evolved, the orbits have been amplified by post-main-sequence mass loss, potentially providing useful constraints on the initial-to-final mass relation for white dwarfs. The nearly featureless spectra of cool white dwarfs usually provide little information about intrinsic radial velocity, full space motion, population membership, metallicity, etc. However, distant main sequence companions provide benchmarks against which those properties can be determined. In addition, the cooling ages of white dwarf components provide useful limits on the ages of their main sequence companions, independent of other stellar age determination methods. This paper summarizes some of the ways fragile binaries provide useful leverage on these and other problems of interest.

Keywords. binaries: common proper motion - stars: white dwarfs

1. Fragile Binaries: A Definition

During the first large-scale proper motion surveys of the mid-twentieth century, Luyten (1969 et seq) and Giclas *et al.* (1971) identified over 6000 very wide common-proper-motion binaries ($CPMBs$) based upon angular proximity in the sky ($< 300''$) and similar proper motion. Most are likely to be low-mass main-sequence pairs (MS+MS). About 10% of these CPMBs were identified as potentially containing at least one white dwarf (WD) component based upon their location in reduced proper motion diagrams constructed from a color index, apparent magnitudes and proper motions. Luyten chose the symbol H ($= m + log\mu + 5$) for reduced proper motion as a tribute to Enjar Hertzsprung, who he credited with the first use of the technique.

The separations of CPMBs range from 10 to 10^4 AU, averaging about 10^3 AU (Oswalt *et al.* 1998). CPMBs may not represent the large separation tail of a continuous distribution of binary separations that includes the closer classical visual and spectroscopic binary classes. The latter are believed to form via fragmentation or fission processes; the former may arise from multi-body capture or cluster disintegration processes (see Zinnecker 1984 for an early review). In any case, CPMBs are almost certainly the most common type of binary system (see Luyten 1969). Oswalt & Smith (1995) and Smith & Oswalt (1995) have shown that even the current large sample of known CPMBs within 100 pc of the Sun is at most 20% complete and that many new CPMBs remain to be found among the new large-scale surveys such as the *Sloan Digital Sky Survey* (see Smith *et al.* 2005) and by combining existing surveys (see Chanamé & Gould 2004, Lépine & Shara 2005). We suggest here that the term *fragile binary* (FB) is more physically descriptive of such loosely bound pairs than the traditional term CPMB, because they are exquisitely sensitive to the influence of the Galactic environment (see below).

2. Importance of Fragile Binaries with Evolved Components

FBs provide unique observational leverage on important astrophysical problems that are difficult to address using samples of single stars or clusters. The importance of wide unevolved (MS+MS) pairs is outlined elsewhere in this volume and we defer that discussion to the papers by Allen *et al.*, Chanamé, Sinachopolous *et al.*, Kiyaeva, and Poveda *et al.*. For the remainder of this paper we will examine only those areas of opportunity afforded by FBs containing at least one evolved component, i.e., WD+MS or WD+WD pairs. FBs with more highly evolved companions (neutron stars or black holes) experienced a supernova explosion that almost certainly disrupted them.

As the end stage of over 90% of all stars the Galaxy has ever spawned, the observed physical properties of WDs provide important boundary conditions on evolutionary models for stars less massive than $\sim 8\,M_\odot$. For example, they constrain the mass-radius relation for degenerate matter and provided one of the first tests of relativity theory, via gravitational redshift measurements. They comprise an accumulating history of the Galaxy containing clues to the initial mass function (IMF), star formation rate (SFR) and initial-to-final mass relation (IFMR) for degenerate stars. Moreover, they provide information about the various components of the Galaxy, e.g., the thick and thin disk, halo and dark matter.

WDs have a very narrow mass distribution with a mean of $\sim 0.6 \pm 0.1\,M_\odot$ (Silvestri *et al.* 2001). Lacking significant energy sources, the evolution of WDs is a simple cooling process in which there is an inverse relation between the age and luminosity. The *cooling ages* of the oldest WDs played a key role in resolving the discrepancy between the age of the Universe derived from the Hubble constant and globular clusters (Liebert *et al.* 1988, Oswalt *et al.* 1996, Bergeron *et al.* 1997; see also the review paper by Lineweaver 1998).

At present about 500 FBs consisting of WD+MS components have been spectroscopically identified. Most are among the Luyten and Giclas proper motion surveys. Such wide noninteracting pairs are of special interest because the WD component gives information about its MS companion that would otherwise be unobtainable if it were a single star and vice versa. Because a faint star like a WD is many times more likely to be found when it is near a brighter MS star of high proper motion, FBs are likely to harbor a more complete sample of WDs, all other things being equal (search volume, magnitude limit, proper motion limit, etc.). Moreover, in an H-R diagram, the natural dispersion about the WD cooling track is about half the scatter of the MS. Thus, for pairs without trigonometric parallax determinations, a WD component provides a more precise photometric parallax for its MS companion than one based on fits to the MS. Even if trigonometric parallaxes are available, a pair provides two independent distance and luminosity estimates that will be consistent if they are bound. Oswalt *et al.* (1996) used this leverage to probe a deeper sample of WDs to obtain a new WD luminosity function that set a firm lower limit of $\sim 10\,$Gyr to the age of the Galaxy. It also indicated that WDs contribute only a few percent of the dark matter content in the Galaxy.

Among WD+MS pairs, the MS components provide a benchmark for the intrinsic radial velocity and, when trigonometric or photometric parallaxes are available, the full space motion of each pair can be computed. With such large orbits, the orbital velocity of both components is $<1\,\mathrm{km\,s^{-1}}$, usually below the precision of measurement in faint stars. Thus, the observed difference in radial velocity between a WD and MS component is essentially the gravitational redshift of the WD. FBs have provided several hundred independent WD mass determinations via gravitational redshift measurements (see Silvestri *et al.* 2001 and references therein).

Age is one of the most difficult to determine physical properties of single stars. In the years since Skumanich (1972) first introduced it, Barnes (2001), Lachaume *et al.* (1999), Soderblom *et al.* (1991) and others have shown that Ca II H&K emission, a proxy for chromospheric activity, provides a reliable age estimate for F, G and K MS stars. For WD components that are warm enough to exhibit Balmer lines, it is relatively easy to estimate cooling ages, temperatures, gravities and *final* masses from line profile fits, as in Kawka & Vennes (2006). In a wide non-interacting binary, the cooling age of a white dwarf (WD) component provides a firm lower limit to the age of any distant main sequence (MS) companion. The difference between the apparent ages of the WD+MS components is the time the WD originally spent as a MS star. Using the canonical mass vs. lifetime relation for MS stars (e.g., Cox 2000), one can then estimate the *initial mass* of the WD, i.e., its mass before post-MS evolution mass loss occurred. The relation between this and a WD's current (i.e., *final*) mass, the so-called *initial-to-final mass relation* (IFMR) is one of the weakest links in all of stellar evolution (Jeffries 1997, Weidemann 2000, Catalán 2007). FBs offer a unique opportunity to improve this situation that we are currently exploiting.

Silvestri *et al.* (2005) attempted to extend the chromospheric activity vs. age relation to M stars using the ages derived from distant WD companions in 116 FBs. In M stars, Hα emission serves as a better proxy for chromospheric activity than Ca II H&K. The nearby Ca H$_2$ and Ti O$_5$ bands are useful indicators of effective temperature and metallicity (Reid *et al.* 1995). Silvestri *et al.* (2005) found that in general a higher fraction of early M stars are active (i.e., have strong Hα emission) in accord with studies of M dwarfs whose ages were derived from cluster membership (Reid *et al.* 1995). Clearly a different excitation mechanism applies to M stars than the canonical self-sustaining dynamo in F, G and K stars. A much larger sample of WD+dM stars will be needed to fully explore where the transition to a proposed 'turbulent dynamo' process occurs, and to identify any additional variables, such as metallicity, that influence whether a star maintains long-lived chromospheric activity or not.

All but a handful of the currently known FBs are members of the Galactic disk (Silvestri *et al.* 2005, Montiero *et al.* 2006). Chanamé (2007) is conducting a search of the *Sloan Digital Sky Survey* to find FBs in the halo. Most will be long-lived low-mass MS+MS pairs whose orbits can set constraints on the dark matter content in the halo (see Chanamé & Gould 2004). The number of WD+MS pairs found to have large space motions is likely to be sufficient to provide the first robust determination of the WD luminosity function and age determination of the halo. As outlined above, the full space motions of these pairs will be provided by their MS companions regardless of whether their old cool WD companions exhibit measurable lines. Moreover, the metallicities of the WD progenitors is easily measured via their MS companions' spectra and this provides an independent indicator of population membership that single WDs cannot provide.

During post-MS evolution much mass is lost and the orbit of a FB expands. Greenstein (1986) and Oswalt & Sion (1988) were among the first to present observational evidence for such expansion. Oswalt & Strunk (1994) and Sterzik & Durisen (2004) noticed that FBs with MS components of early spectral type tend to have much wider mean separations than those of later spectral class. Valls-Gabaud (1988) and Wood &

Oswalt (1998) showed that expansion of up to an order of magnitude can occur with the right combination of initial primary mass and binary mass ratio. Fahiri (2006) showed observational evidence that a gap in the orbital separation distribution may occur near 5 AU. Closer pairs experienced a common envelope phase and their orbits contracted, while wider pairs' orbits expanded as the primary became a WD.

Recently, Johnston *et al.* (2007) have taken up the challenge of modeling the orbital evolution of FBs in greater detail, including not only the effects of post-MS mass loss, but eventually including the perturbations caused by the Galactic tidal potential, giant molecular cloud encounters, and stellar interactions. Using large samples of FBs drawn from the SDSS, model separation distributions will be constructed for observed samples of evolved (WD+MS, WD+WD) pairs and unevolved (MS+MS) pairs. By matching observed and computed angular orbit separation distributions for Galactic disk FBs, Johnston *et al.* are attempting to derive average mass loss as a function of average initial mass, thereby achieving a new constraint on the WD $IFMR$ that is independent of other techniques. As the known sample of FBs in the halo becomes large enough, an attempt will be made to model orbit amplification in these ancient pairs as well.

Among the rarest of all FBs are those pairs consisting of two WD components. Luyten (1969) identified about two dozen such pairs in his original proper motion survey. Even today, only about 50 WD+WD pairs are known. Probably this is because both components tend to be intrinsically faint, but also because the Galaxy is not old enough for many of them to have formed. Sion *et al.* (1991) showed that WD+WD pairs tend to have slightly smaller average physical separations than WD+MS pairs. This could plausibly have arisen from initial mass ratios that were closer to unity in the progenitors of WD+WD pairs. Because they contain some of the oldest known WDs, such pairs have also played an important role in determining the age of the Galaxy (Oswalt *et al.* 1996, Bergeron *et al.* 1997). Holley–Bockelmann examined the possibility that the closest WD+WD pairs will pose a significant source of foreground noise for future gravitational wave detectors.

It came as a surprise to the WD community when Saumon & Jacobsen (1999) and Hanson (1999) independently predicted that the coolest WDs would not be red in color, but blue, due to the onset of collisionally induced absorption by H_2 molecules. It was not long after that the first cool blue degenerate star was actually discovered (Hodgkin *et al.* 2000). Oppenheimer *et al.* (2001) stirred up a controversy by finding several dozen such objects in a deep proper motion survey and suggesting that most or all of the dark matter content in the Galactic halo might be cool blue degenerate stars. FBs played an important role in rejecting this conclusion. Silvestri *et al.* (2002) showed that among $\sim$100 FBs consisting of WD+MS stars, all but one pair were high velocity members of the thick disk, not the halo. This was a firm conclusion because the MS companions provided radial velocities, gravitational redshift masses, full UVW space motions and metallicities for their WD companions (see Silvestri *et al.* 2005 for details). Silvestri *et al.* (2002) showed that the velocity histograms for the WD+MS sample, look like those of Oppenheimer *et al.* (2001), when degraded by the *zero radial velocity* assumption necessary for single WDs lacking radial velocity measurements. Recently, Monteiro *et al.* (2006) and Chanamé (2007) have begun looking for FBs of high space velocity. Such pairs will be very important probes of the age and dark matter content in the halo.

3. Conclusions

Much science remains to be gleaned from the many thousands of FBs awaiting study in the large new imaging and proper motion surveys now underway. It is not much of a stretch to assert FBs are comparable in importance to open clusters in terms of the leverage they potentially provide on astrophysical problems. Although each has only 2-3 coeval components, FBs span nearly a continuous range of ages over the entire history of the Galaxy, (~10 Gyr). By comparison only a half dozen or so age milestones are provided by the nearest clusters that contain comparably bright stars. Finally, the shear number of FBs (at least 12,000 WD+MS pairs in the SDSS alone, according to Smith *et al.* 2005) provide a golden opportunity to attack various problems using subsets that isolate such variables as mass, spectral type, metallicity, age, etc.

Luyten (1969) was right in declaring that we have been *ignoring the common man in space* by focusing our attention on the more flashy close interacting binaries, OB associations and other rare members of the stellar zoo. He would be pleased to know that several groups are now beginning to use these pairs in ways he could not foresee 40 years ago.

Acknowledgements

We would like to thank the Scientific Organizing Committee of IAU Symposium 240 for the invitation to present this review and we are grateful to our hosts in Prague for a warm welcome and a productive meeting. One of us (TDO) acknowledges support provided by the U.S. National Science Foundation via grant AST-0206115.

References

Allen, C., Poveda, A., & Hemández. Alcántara, A. 2007, these proceedings, 405.

Barnes, S. 2001, "An Assessment of the Rotation Rates of the Host Stars of Extrasolar Planets", *ApJ* 561, 1095

Bergeron, P., Ruiz, M-T., & Leggett, S.K. 1997, "The Chemical Evolution of Cool White Dwarfs and the Age of the Local Galactic Disk", *ApJS* 108, 339

Catalán, S., Ribas, I., Isern, J., García-Berro, E., & Allende, Prieto, C. 2007, these proceedings, 380.

Chanamé, J. 2007, these proceedings, 316.

Chanamé, J. & Gould, A. 2004, "Disk and Halo Wide Binaries from the Revised Luyten Catalog: Probes of Star Formation and MACHO Dark Matter", *ApJ* 601, 289

Cox, A. 2000 (ed.), *Allen's Astrophysical Quantities, 4th ed.* (New York: AIP Press), Springer

Fahiri, J. 2006, "White Dwarf-Red Dwarf Systems Resolved with the Hubble Space Telescope. I. First Results", *ApJ* 646, 480

Giclas, H.L., Burnham, R., & Thomas, N.G. 1971, *Lowell Proper Motion Survey: Northern Hemisphere: The G-Numbered Stars.* (Flagstaff: Lowell Observatory)

Greenstein, J.L., 1986, "White Dwarfs in Wide Binaries I. Physical Properties", *AJ*, 92, 859

Hanson, B.M.S. 1999, "Cooling Models for Old White Dwarfs", *ApJ* 520, 680

Hodgkin, S., Oppenheimer, B., Hambly, N. Jameson, R. Smartt, S., & Steele, I. 2000, "Infrared Spectrum of an Extremely Cool White Dwarf Star", *Nature* 403, 57

Jeffries, R.D. 1997, "On the Initial-Final Mass Relation and the Maximum Mass of White Dwarf Progenitors", *MNRAS* 288, 585

Johnston, K., Oswalt, T., & Valls-Gabaud, D. 2007 (this volume)

Kawka, A. & Vennes, S. 2006, "Spectroscopic Identification of Cool White Dwarfs in the Solar Neighborhood", *ApJ*, 643, 402

Kiyaeva, O. 2007, these proceedings, 131.

Lachaume, R. *et al.*, 1999, "Age Determinations of Main-Sequence Stars: Combining Different Methods", *A&A* 348, 897

Lépine & Shara, M. 2005, "A Catalog of Northern Stars with Annual Proper Motions Larger than 0''.15 (LSPM-NORTH Catalog)", *AJ* 129, 1483-1522

Liebert, J., Dahn, C., & Monet, D. 1988, "Luminosity Function of White Dwarfs", *ApJ* 332, 891

Lineweaver, C.H. 1999, "A Younger Age for the Universe", *Science* 284, 1503

Luyten, W.J., 1969, *Proper Motion Survey with the Forty-Eight Inch Schmidt Telescope XXI: Double Stars with Common Proper Motion*, et seq. (Univ. Minn. Press: Minneapolis)

Monteiro, H., Jao, W-C, Henry, T., Subasavage, J., & Beaulieu, T. 2006, "Ages of White Dwarf – Red Subdwarf Systems", *ApJ* 638, 446

Oppenheimer, B. Saumon, D., Hodgkin, S.T., Jameson, R.F., Hambly, N.C., Chabrier, G., Filippenko, A.V., Coil, A.L., & Brown, M.E. 2001, "Observations of Ultra-Cool White Dwarfs", *ApJ* 550, 448

Oswalt, T.D., Sion, E.M. 1989, "On the Physical Separations of Wide White Dwarf Binaries", in IAU Coll. 114, *White Dwarfs*, Springer-Verlag: Berlin, p. 454

Oswalt, T.D. & Smith, J.A. 1995, "On the Luminosity Function of White Dwarfs in Wide Binaries", in *White Dwarfs*, eds. D. Koester & K. Werner, (Springer-Verlag: Heidelberg), p. 113

Oswalt, T., Smith, J., & Wood, M. 1998, "Wide Binaries: Probes of the Galaxy's Dark Matter Content", Transactions 23rd IAU G/A, Joint Disc. 10, *Low Luminosity Stars*, (Kluwer: Dordrecht)

Oswalt, T., Smith, J., Wood, M., & Hintzen, P. 1996, "New Limits to the Galactic Disk Age from the Luminosity Function of White Dwarfs in Wide Binaries", *Nature (Letters)*, 382, 692

Oswalt, T. & Strunk, D. 1994, "A Catalog of White Dwarfs in Wide Binaries", *BAAS* 26, 901

Poveda, A., Allen, C., & Hernández-Aleántara, A., 2007, these proceedings, 417

Reid, I.N., Hawley, S.L., & Mateo, M. 1995, "Chromospheric and Coronal Activity in Low-Mass Hyades Dwarfs", *MNRAS* 272, 828

Saumon, D. & Jacobson, S.B. 1999, "Pure Hydrogen Model Atmospheres for Very Cool White Dwarfs", *ApJ* 511, L107

Silvestri, N., Oswalt, T., Smith, J.A., Wood, M., Reid, N., & Sion, E., 2001, "White Dwarfs in Common Proper Motion Binary System: Mass Distribution & Kinematics", *AJ* 121, 503

Silvestri, N., Oswalt, T., & Hawley, S. 2002, "Wide Binary Systems and the Nature of High-Velocity White Dwarfs", *AJ* 124, 1118

Silvestri, N., Hawley, S., & Oswalt, T. 2005, "The Chromospheric Activity and Ages of M Dwarf Stars in Wide Binary Systems", *ApJ* 129, 2428

Sinachopoulos, D., Gauras, P., Medupe, Th., Ducourant, Ch., & Dionatos, O. 2007, these proceedings, 264

Sion, E. *et al.* 1991, "The Physical Properties of Double Degenerate Wide Common Proper Motion Systems", *AJ* 101, 1476

Skumanich, A. 1972, "Time Scales for CA II Emission Decay, Rotational Braking, and Lithium Depletion", *ApJ* 171, 565

Smith, J.A. & Oswalt, T. 1995, "Exploration of the Lower Main Sequence Among Wide Binaries", in *The Bottom of the Main Sequence and Beyond*, ed. C.G. Tinney, (Springer–Verlag: Heidelberg), p. 24

Smith, J.A., Silvestri, N.M., Oswalt, T.D., Harris, H.C., Kleinman, S.J., Munn, J.A., Nitta, A., & Rudkin, M.A. 2005, "Sloan Digital Sky Survey: Proper Motion Systems Containing White Dwarfs", *ASP Conf.* 334, 127

Soderblom, D.R., Duncan, D.K., & Johnson, D.R.H. 1991, "The Chromospheric Emission-Age Relation for Stars of the Lower Main Sequence and Its Implications for the Star Formation Rate", *ApJ* 375, 722

Sterzik M. & Durisen, R. 2004, "Are Binary Separations Related to Their System Mass?", in *The Environment & Evolution of Double & Multiple Stars,* Proc. IAU Coll. 191, eds. C. Allen & C. Scarfe. *Rev. Mex. de Astron. y Astrofisica (Conf.)* 21, 58

Valls-Gabaud, D. 1988, "Evidence for Mass Loss in Visual Binary Stars", *Ap&SS* 142, 289

Weidemann, V. 2000, "Revision of the Initial-to-Final Mass Relation", *A&A* 363, 647

Wood, M. & Oswalt, T. 1998, "White Dwarf Cosmochronology I. Monte Carlo Simulations of Proper Motion and Magnitude Limited Samples Using Schmidt's $\frac{1}{V_{max}}$ Estimator", *ApJ* 497, 870

Zinnecker, H. 1984, *Ap&SS* 99, 41

Binary Stars as Critical Tools & Tests
in Contemporary Astrophysics
Proceedings IAU Symposium No. 240, 2006
W.I. Hartkopf, E.F. Guinan & P. Harmanec, eds.

© 2007 International Astronomical Union
doi:10.1017/S1743921307004218

Measurement of the Tidal Dissipation in Multiple Stars

Andrei Tokovinin

Cerro Tololo Inter-American Observatory, Casilla 603 La Serena, Chile
e-mail: atokovinin@ctio.noao.edu

Abstract. Considerable effort has been spent to determine the period of tidal circularization in close binaries as a function of age, in order to constrain the tidal dissipation theory. A new, direct method of measuring the tidal dissipation by precise timings of periastron passages in eccentric binaries undergoing circularization is proposed. Such binaries with components just leaving the Main Sequence can be found as inner systems in multiple stars. Three examples are given.

1. Introduction

Tidal interaction in close binary stars is important in many respects. It provides a dissipative mechanism of orbit circularization, converting eccentric binaries with long periods into short-period systems with circular orbits. This orbital *circularization* is accompanied by the *synchronization* of component's axial rotation with the orbit. Both processes can act when stars are just forming, at the pre-Main Sequence (PMS) stage, on the Main Sequence (MS), or in systems with evolved components. In all cases the knowledge of the tidal forces is needed to understand the physics and evolution of binary systems, including some special important cases such as X-ray binaries or pulsars. Here I propose a new method that can potentially lead to a direct measurement of the dissipation.

Traditionally, the tidal dissipation in Main Sequence (MS) stars is constrained from the circularization periods $P_{\rm circ}$ in open clusters. Binaries with $P < P_{\rm circ} \approx 10^{\rm d}$ have circular orbits while orbits with longer periods are eccentric. The $P_{\rm circ}$ is only an indirect measure of tidal circularization, being dependent on unknown initial conditions and evolutionary history. Its determination is affected by the small number of short-period binaries in each cluster. Meibom & Mathieu (2005) reach a conclusion that current theories of tidal interaction explain circularization and synchronization qualitatively, but not quantitatively.

The circularization with constant angular momentum follows the track $P(1 - e^2)^{3/2} =$ const, i.e. is accompanied by the shortening of the orbital period. The period can be determined with a much better precision than eccentricity. A detection of period changes will lead to the measurement of the tidal dissipation rate. In interpreting these variations, we will have to account for the period changes caused by other effects such as by the exchange of the angular momentum with components, dynamical tides, etc.

A measurable period change caused by tidal dissipation happens in binaries with a short circularization time-scale. Evidently, such binaries are rare. They can be found among short-period PMS binaries with eccentric (still circularizing) orbits. Another option is to study inner sub-systems in multiple stars with sub-giant components, as explained below.

2. Origin and destiny of very eccentric binaries

The origin of very high eccentricities is naturally explained by Kozai cycles – strong eccentricity modulation in triple systems with high mutual orbit inclinations (Eggleton 2006). These cycles are softened by the relativistic apsidal motion (AM): a short-period sub-system cannot acquire very large eccentricities by the Kozai mechanism if its tertiary companion is too distant. The Kozai cycles are also perturbed by the tidal friction, so that eventually (i) the inner binary becomes locked into the high-e state of the cycle, and then (ii) the eccentricity decreases on the frictional timescale, with angular momentum being roughly constant, till circularity is reached. The orbit evolves on a path like the dash-dotted line of Fig. 1. The product of such evolution will be a spectroscopic binary with giant components and near-circular orbit. There are many such systems among known multiple stars (Tokovinin 1997). The existence of highly eccentric sub-systems in multiples is not a rare event but rather a natural outcome of the dynamical evolution.

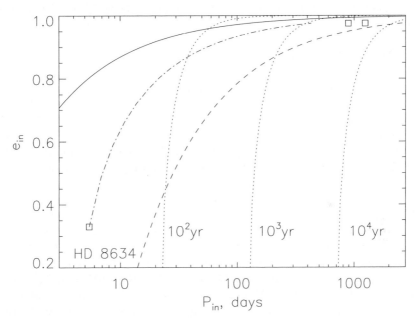

Figure 1. Eccentricity of inner sub-systems in triple stars vs. their period. The full line is the P, e relation when the two components just touch each other at periastron for a sum of radii equal 3.6 $R_\odot$ (like in 41 Dra). The dashed line is the P, e locus for orbits which have the same periastron distance as a 10-day circular orbit, $P(1 - e)^{3/2} = 10^{\rm d}$. The dotted lines are the maximum eccentricities of Kozai cycles softened by the relativistic AM, for 3 representative tertiary periods $P_{\rm out}^*$. The squares denote three systems discussed in the text. The dash-dot line is the line along which HD 8634 must have evolved at constant angular momentum but decreasing e to reach its present position.

The squares in Fig. 1 are the real systems illustrating this scenario. **Gliese 586A** (Duquennoy et al. 1992) held for a while the world record of eccentricity, $e = 0.9752$. It is a pair of unevolved low-mass dwarfs accompanied by a physical component B and an even more distant common proper motion companion C at 20.2′ (24 000 AU). The circularization of this system was studied by Goldman & Mazeh (1994). Rapid circularization is not expected in this system. **41 Dra** (Tokovinin et al. 2003) has an even larger eccentricity $e = 0.9754$ and forms a physical quadruple with a 10.5$^{\rm d}$ spectroscopic binary

40 Dra. These F-type stars are moderately evolved, the age is estimated as 2.5 Gyr by isochrone fitting. As the component's radius increases, so does the tidal dissipation. Thus 41 Dra may have already started its phase of rapid circularization.

HD 8634 = HR 407 is a binary with $P = 5.4^d < P_{\text{circ}}$ and $e = 0.33$. The orbit was computed by Wright & Pugh (1954) and later re-determined by Mayor & Mazeh (1987) in search of precession caused by a tertiary companion. No precession was found, yet a faint tertiary at $1.5''$ has been discovered with adaptive optics by Tokovinin *et al.* (2006). The primary F5III component is slightly evolved. It was a late-A type star on the MS. It is likely that HD 8634 is now undergoing rapid circularization, evolving from a high-eccentricity state.

3. Can we observe the tidal circularization?

Uncertainties in the tidal theory do not permit reliable estimation of the circularization time scale for very eccentric binaries (Goldman & Mazeh 1994), hence a direct measurement of the orbital evolution in a system like 41 Dra will be of great value. The radial velocity at periastron changes by 3 km s^{-1} per hour. Presently we measure the periastron time (and period) to 0.1 h with radial velocities accurate to 0.3 km s^{-1}. With a conservative 10-fold increase in precision we hope to bring the timing error down to 0.01 h. This gives a measurable effect $dP/P = 3.34 \times 10^{-7}$ (time scale of 3 Myr). The next periastron passage will occur on April 1, 2008.

The case of HD 8634 is tantalizing. Application of Eq. 2 of Verbunt & Phinney (1995) to a similar system with a convective primary gives the circularization time of ~ 4 Myr. If a new, precise periastron timing is made now, the period variation on a time scale of 10 Myr or shorter is detectable. Complications such as motion around the tertiary and spin-orbit interaction must be addressed when interpreting future timings.

4. Conclusions

It is shown that binary stars undergoing presently rapid tidal orbital evolution are formed naturally within multiple systems. Period changes in such systems can be detected and will provide a direct measurement of the tidal dissipation. Monitoring a sample of (otherwise uninteresting) spectroscopic binaries with precise RV techniques over a long period of time will eventually lead to the successful detection of orbital tidal evolution.

References

Duquenoy, A., Mayor, M., Andersen, J. *et al.* 1992, *A&A* 254, L13
Eggleton P. 2006, *Evolutionary Processes in Binary and Multiple Stars.* (Cambridge, UK: Cambridge Univ. Press)
Goldman, I. & Mazeh, T. 1994, *ApJ* 429, 362
Mayor, M. & Mazeh, T. 1987, *A&A* 171, 157
Meibom, S. & Mathieu, M. 2005, *AJ* 620, 970
Tokovinin, A. 1997, *A&AS* 124, 75
Tokovinin, A., Balega, Y.Y., Pluzhnik, E.A. *et al.* 2003, *A&A* 409, 245
Tokovinin, A., Thomas, S., Sterzik, M., & Udry, S. 2006, *A&A* 450, 681
Verbunt, F. & Phinney, E.S. 1995, *A&A* 296, 709
Wright, K.O. & Pugh, R.E. 1954, *Publ. DAO* 9, 407

Binary Stars as Critical Tools & Tests
in Contemporary Astrophysics
Proceedings IAU Symposium No. 240, 2006
W.I. Hartkopf, E.F. Guinan & P. Harmanec, eds.

U Geminorum: a Test Case for Orbital Parameters Determination

Juan Echevarría, Eduardo de la Fuente, and Rafael Costero

Instituto de Astronomía, *Universidad Nacional Autónoma de México,*
Apartado Postal 70-264, México, D.F., México

Abstract. High-resolution spectroscopy of U Gem was obtained during quiescence. We did not find a hot spot or gas stream around the outer boundaries of the accretion disk. Instead, we detected a strong narrow emission near the location of the secondary star. We measured the radial velocity curve from the wings of the double-peaked Hα emission line, and obtained a semi-amplitude value in excellent agreement with the ultraviolet results by Long & Gilliland (1999). We present also a new method to obtain K_2, which enhances the detection of absorption or emission features arising in the late-type companion. Our results are compared with published values derived from the near-infrared NaI line doublet. From a comparison of the TiO band with those of late type M stars, we find that a best fit is obtained for a M6 V star, contributing 5% of the total light at that spectral region. Assuming that the radial velocity semi-amplitudes reflect accurately the motion of the binary components, then from our results: $K_{em} = 108 \pm 2$ km s^{-1}; $K_{abs} = 310 \pm 5$ km s^{-1}, and using the inclination angle by Zhang & Robinson (1987); $i = 69°.7 \pm 0.7$, the system parameters become: $M_{WD} = 1.20 \pm 0.05\,M_\odot$; $M_{RD} = 0.42 \pm 0.04\,M_\odot$; and $a = 1.55 \pm 0.02\,R_\odot$. Based on the separation of the double emission peaks, we calculate an outer disk radius of $R_{out}/a \sim 0.63$, close to the distance of the inner Lagrangian point $L_1/a \sim 0.63$. Therefore we suggest that, at the time of observations, the accretion disk was filling the Roche-Lobe of the primary, and that the matter leaving the L_1 point was colliding with the disc directly, producing the hot spot at this location. Specific details not included in the printed version can be found in the Electronic Poster (EP).

Keywords. binaries: close – stars: nova, cataclysmic variable – stars: individual (U Geminorum)

1. Observations

U Geminorum was observed in 1999 January 15 with the Echelle spectrograph at the f/7.5 Cassegrain focus of the 2.1-m telescope of the Observatorio Astrónomico Nacional at San Pedro Mártir, B.C., México. The Thompson 1024×1024 CCD was used to cover a spectral range from λ5200 to λ9100 Å with a spectral resolution of R=18,000. An echellette grating of 150 l/mm, with Blaze around 7000 Å was used. The spectra shows a strong Hα emission line. No absorption features were detected from the secondary star. A complete orbital cycle was obtained with twenty-one spectra with an exposure time of 600s each. Thirteen further spectra were then acquired with an exposure of 300s each. The flux standard HD 17520 and the late spectral M star HR 3950 were observed on the same night.

2. Radial Velocities

We derive radial velocities of the primary star from the prominent Hα emission line by using a method based on a cross-correlating technique and also using the standard method of measuring the wings of the line. In the case of the secondary star, we were unable to detect any single absorption line in the individual spectra, and therefore it was not possible to use any standard method. However, we propose and use a new method

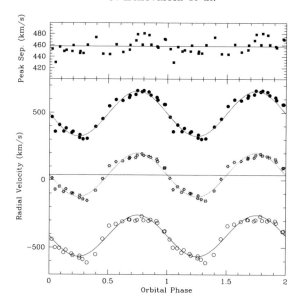

Figure 1. Radial velocity curve of the double peaks. The separation of the peaks has a mean value of about 460 km s^{-1}.

to derive the semi-amplitude of the companion star, based on a co-adding technique (see EP for full details).

2.1. *The Primary Star*

In this section we compare three methods for determining the radial velocity of the primary star, based on measurements of the Hα emission line. We adopt the ephemeris: HJD $= 2,437,638.82566(4) + 0.1769061911(28)$ E, for the inferior conjunction of the secondary star. These ephemeris are used throughout this paper for all our phase folded diagrams and Doppler Tomography. Full details can be found in Echevarría, de la Fuente & Costero (2007). To match the signal-to-noise ratio of the first twenty-one spectra, we have co-added (in pairs) the last thirteen exposures. The last three spectra were added to form two different spectra. This is to avoid losing the last single spectrum. This adds to a total of twenty-eight 600 s spectra. A handicap to this approach is that due to the large read-out time of the Thompson CCD we are effectively smearing the phase coverage of these last co-added spectra to nearly 900 s. However, the mean heliocentric time was corrected for each sum.

We have measured the position of the peaks using a double–Gaussian fit. The results are shown in Figure 1. The mean semi-amplitude and peak separation are $K_{\rm em} = 168 \pm 8$ km s^{-1} and 460 km s^{-1} respectively.

We have cross-correlated the Hα spectra with a specially constructed template. We selected a spectrum close to orbital phase 0.75, a phase at which we should expect a minimum distortion in the peaks due to asymmetric components. The selected spectrum was highly smoothed to minimize high-frequency correlations. This template is shown in the EP. The semi-amplitude of the radial velocity curve is $K_{\rm em} = 129 \pm 6$ km s^{-1}. The radial velocity curve is shown in the EP.

The Hα emission lines were measured using the standard double–Gaussian technique and its diagnostic diagrams as described in Shafter, Szkody & Thorstensen (1986). For full details see Echevarría, de la Fuente & Costero (2007) and the EP. The semi-amplitude of the radial velocity curve is $K_{\rm em} = 108 \pm 2$ km s^{-1}. The radial velocity curve is shown in the EP.

2.2. *The Secondary Star: a new method to determine K_2*

We were unable to detect any absorption features from the secondary star in any single spectra either visually or using the standard cross-correlation technique. We have been able, however, to detect the NaI $\lambda\lambda$ 8183.3, 8194.8 Å doublet and the TiO Head band around λ 7050 Å with a new technique, as well as $H\alpha$ emission as well, which enables us to derive the semi-amplitude K_{abs} of the secondary. The method is fully explained in Echevarría, de la Fuente & Costero (2007) and the EP.

We have applied our criteria to U Gem. The time of the inferior conjunction of the secondary and the orbital period were taken from section 2.1. The results are shown in the EP Figures.

For the NaI doublet $\lambda\lambda$ 8183, 8195 Å, the spectra were *co-phased*, varying K_{pr} between 250 and 450 km s^{-1}. The line depth of the blue and red components of the doublet (stars and open circles respectively), as well as the mean value of both lines (dots) are shown in the diagram. In both lines we find a best solution for $K_2 = 310$ km s^{-1}. As it approaches its maximum value, the line depth oscillates slightly but in the same way for both lines. We found a similar behavior on the artificial spectra process described above for low signal-to-noise features. The figure (see EP) shows the *co-phased* spectrum of the NaI doublet for our best solution of K_2. Similar results were found for the TiO band and the $H\alpha$ emission with the best solution for $K_2 = 310$ km s^{-1}.

3. Doppler Tomography

Doppler Tomography is a useful and powerful tool to study the material orbiting the white dwarf, including the gas stream coming from the secondary star as well as emission regions arising from the companion itself. A detailed formulation of this technique can be found in Marsh & Horne (1988). The Doppler Tomography results derived here from the $H\alpha$ emission line were constructed using the code developed by Spruit (1998).

Since our observations of the object cover 1.5 orbital cycles, and with the intention to avoid disparities on the intensity of the trailed and reconstructed spectra and on the tomographic map, we have carefully selected spectra covering a full cycle only. In addition to this, we excluded from the calculations of the Tomography map, the spectra taken during the partial eclipse of the accretion disc (phases between 0.95 and 0.05). The results are shown in the EP figure. Here, we present a blow-up of the $H\alpha$ emission near the secondary.

4. Basic system parameters

Assuming that the radial velocity semi-amplitudes reflect accurately the motion of the binary components, then from our results: $K_{em} = 108 \pm 2$ km s^{-1}; $K_{abs} = 310 \pm 5$ km s^{-1}, and adopting $P = 0.1769061911$ and using the inclination angle by Zhang & Robinson (1987); $i = 69°.7 \pm 0.7$, the system parameters become: $M_{WD} = 1.20 \pm 0.05\,M_\odot$; $M_{RD} = 0.42 \pm 0.04\,M_\odot$; and $a = 1.55 \pm 0.02\,R_\odot$.

4.1. *The inner and outer size of the disc*

The dimensions of the disc –the inner and outer radius– can be derived from the observed Balmer emission line. Its peak-to-peak velocity separation is related to its outer radius, while the wings of the line, coming from the high velocity regions of the disc, can give an estimate of the inner radius (e.g., Smak 2001). The peak-to-peak velocity separation, as well as the velocity of the blue and red wings of $H\alpha$ (10% above the continuum), of the 28 individual spectra were measured. From these measurements we derive a mean value of $V_{peak} = 460$ km s^{-1} and $V_{wings} = 1200$ km s^{-1}.

These velocities are related to the disc radii, which can be obtained from numerical disc simulations, tidal limitations and analytical approximations (see Warner 1995 and

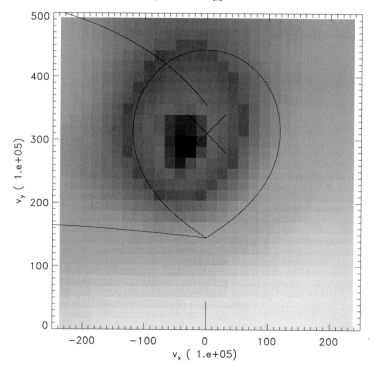

references therein). Here we assume that the material in the disc at radius r is moving with Keplerian rotational velocity V. Then the radius, in units of the binary separation is given by: $r/a = (K_{em} + K_{abs})K_{abs}/V^2$, (e.g., Horne, Wade, & Szkody 1986).

Smak (1981) has shown that the observed maximum intensity of the double-peak emission in Keplerian discs occurs close to the velocity of its outer radius. From the observed value we obtain an outer radius of $R_{out}/a = 0.61$. An inner radius $R_{in}/a = 0.09$ is derived from the wing measurements and the assumption of the Keplerian approximation. These outer radii can be compared with the inner Lagrangian point and the mean R of the primary. The inner radius can be compared with the radius of a white dwarf and with the expected boundary layer radius. The distance R_{L_1}/a from the center of the primary to the inner Lagrangian point is: $R_{L_1}/a = 1 - w + 1/3w^2 + 1/9W^3$, where $w^3 = q/(3(1+q))$ (see Kopal 1959). Using $q = 0.35$ we obtain $R_{L_1}/a = 0.63$. The disc, therefore, appears to be large, almost filling the Roche lobe of the primary, with the matter leaving the L_1 point colliding with the disc directly, producing the hot spot at this location. Full details can be found in the Electronic Poster (EP) and in Echevarría, de la Fuente & Costero (2007).

References

Echevarría, J., de la Fuente, E., & Costero, R. 2007, *AJ*, in preparation
Horne, K., Wade, R.A., & Szkody, P. 1986, *MNRAS*, 219, 791
Kopal, Z., 1959, *Close Binary Systems*, Champan & Hall, London.
Long, K.S. & Gilliland, R.L. 1999, *ApJ*, 511, 916
Marsh, T.R. & Horne, K. 1988, *MNRAS*, 235, 269
Shafter, A.W., Szkody, P., & Thorstensen, J.R. 1986, *ApJ*, 308, 765
Smak, J. 2001, *Acta Astr.*, 51, 279
Spruit, H.C. 1998, preprint, astro-ph/9806141
Zhang, E.H. & Robinson, E.L., 1987 *ApJ*, 321, 813

Binary Stars as Critical Tools & Tests
in Contemporary Astrophysics
Proceedings IAU Symposium No. 240, 2006
W.I. Hartkopf, E.F. Guinan & P. Harmanec, eds.

© 2007 International Astronomical Union
doi:10.1017/S1743921307004231

The Ejection of Runaway Massive Binaries

M. Virginia McSwain[1], Scott M. Ransom[2], Tabetha S. Boyajian[3], Erika D. Grundstrom[3], and Mallory S.E. Roberts[4]

[1]NSF Astronomy and Astrophysics Postdoctoral Fellow;
Department of Astronomy, Yale University, P.O. Box 208101,
New Haven, CT 06520-8101, USA, email: mcswain@astro.yale.edu
[2]National Radio Astronomy Observatory, 520 Edgemont Road,
Charlottesville, VA 22903, USA, email: sransom@nrao.edu
[3]Department of Physics and Astronomy, Georgia State University,
P.O. Box 4106, Atlanta, GA 30302-4106, USA,
email: tabetha@chara.gsu.edu, erika@chara.gsu.edu, gies@chara.gsu.edu
[4]Eureka Scientific, Inc., 2452 Delmer Street Suite 100,
Oakland, CA 94602-3017, USA, email: malloryr@gmail.com

Abstract. The runaway O-type stars HD 14633 and HD 15137 are both SB1 systems that were probably ejected from the open cluster NGC 654. Were these stars dynamically ejected by close gravitational encounters in the dense cluster, or did the binaries each receive a kick from a supernova in one member? We present new results from our investigation of the optical, X-ray, and radio properties of these binary systems to discuss the probable ejection scenarios. We argue that these binaries may have been ejected via dynamical interactions in the dense cluster environment.

Keywords. binaries: spectroscopic, stars: early-type, stars: kinematics, stars: winds, stars: individual (HD 14633, HD 15137), pulsars: general, X-rays: binaries

1. Introduction

Most O- and B-type stars form in open clusters and stellar associations, but a small fraction are observed at high galactic latitudes and with large peculiar space velocities. These runaway stars were likely ejected from the clusters of their birth, either by close multi-body interactions in the dense cluster environment or by supernovae explosions in close binaries. Identifying the dominant scenario producing runaway stars can offer important clues to the evolution of close binary stars and open clusters.

Distinguishing between the dynamical or supernova ejection mechanisms among runaway spectroscopic binaries can be a difficult task since both are expected to produce eccentric, relatively short period binaries (Leonard & Duncan 1990; Portegies Zwart 2000). Only dynamical interactions are expected to produce double-lined spectroscopic binaries (SB2s), while a single-lined system (SB1) may be formed either way. It can be nearly impossible to detect a cool, low mass, optical companion with an O- or B-type primary, while a neutron star companion might be identified either as a pulsar or by X-ray emission produced during mass accretion.

Spectroscopic investigations of HD 14633 and HD 15137 have found short orbital periods and low mass companions (Boyajian *et al.* 2005; McSwain *et al.* 2006). HD 14633 was ejected from the open cluster NGC 654 about 14 Myr ago, and HD 15137 was ejected from the same cluster about 10 Myr ago (Boyajian *et al.* 2005). Presumably, both systems obtained their high runaway velocities during a supernova explosion in a close binary, and the O stars remain bound to the stellar remnant. In this work, we

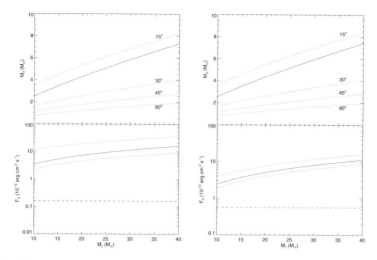

Figure 1. (top) Mass diagram for a range of inclination angles (*dotted lines*) and the most probable companion mass (*solid line*), based on the statistical method of Mazeh & Goldberg (1992). (bottom) The predicted time-averaged X-ray flux (*solid line*) compared to the observed upper limit of F_X from *ROSAT*/PSPC observations (*dashed line*). This prediction assumes the minimum value of M_2, and a higher mass companion will produce an even greater F_X. The left and right plots show the relationships for HD 14633 and HD 15137, respectively.

test this hypothesis using a collection of new observations and archival data across the electromagnetic spectrum.

2. Pulsar Search Results

Although previous searches have failed to detect radio pulsars in runaway OB stars (Philp *et al.* 1996; Sayer *et al.* 1996), we performed a new, more sensitive search to investigate the companions of both HD 14633 and HD 15137. Our search offers several advantages over these earlier searches: our new orbital ephemerides (McSwain *et al.* 2006) allowed us to schedule observations near the time of apastron, we used higher frequencies that are more likely to detect radio pulses dispersed by the stellar winds, and we obtained a better flux density sensitivity than previous searches.

We observed both targets using the National Radio Astronomy Observatory's 100-m Green Bank Telescope and the Pulsar Spigot back-end, and the data were reduced using the PRESTO software package (Ransom 2001). No pulsars were detected in either HD 14633 or HD 15137.

3. Predicted X-ray Luminosity from a Wind Accretion Model

If HD 14633 and HD 15137 do contain compact companions, should these stars have detectable X-ray emission? We predict the X-ray flux of each quiet MXRB candidate using the wind accretion model of Lamers *et al.* (1976), modified for eccentric orbits. In Figure 1, the predicted fluxes are compared to the estimated upper limits for their unabsorbed X-ray fluxes using other detected sources from the *ROSAT* All-Sky Survey and pointed *ROSAT* observations within a 30' radius of each star (White *et al.* 2000; Voges *et al.* 2000). We predict X-ray fluxes of HD 14633 and HD 15137 that are an order of magnitude greater than the observed upper limits. However, the lack of X-ray detections does not conclusively rule out neutron stars.

The conditions for wind accretion onto a neutron star depend strongly upon the spin rate and the magnetic field (Lipunov 1992). A young neutron star in the ejector or propellor regimes will not accrete significant amounts of material because its fast rotation and/or large magnetic field sweep material out of the system at a distance larger than the accretion capture radius, R_a. As the neutron star spins down over time, the corotation radius, R_c, becomes larger than both the magnetospheric radius, R_{mag}, and R_a. We estimate these values for the critical distances in our systems: $R_a \sim 10^{10}$ cm, $R_c \sim 10^8 - 10^9$ cm, and $R_{mag} \sim 10^9$ cm. Unrestrained accretion requires $R_{mag} < R_a$ and $R_{mag} < R_c$, which may be possible in these systems. However, even a non-accreting neutron star could be detected by its thermal or non-thermal X-ray spectrum (Lipunov 1992). Further X-ray observations are required to confirm or refute the presence of neutron stars in HD 14633 and HD 15137.

4. Conclusions

If HD 14633 and HD 15137 were ejected by supernovae in these close binary systems, they should contain neutron stars detectable as either radio pulsars and/or X-ray sources. Our search for pulsars with the Green Bank Telescope revealed no detections, and the predicted X-ray emission from wind accretion is much larger than the observed limits. Furthermore, neither system exhibits an infrared excess from an accretion disk around a compact companion (McSwain *et al.* 2006). While a neutron star cannot be ruled out, the probable companion masses (Figure 1) support higher mass objects. Therefore we argue that these binaries do not contain neutron stars, and we suggest that they may have been ejected from NGC 654 by dynamical interactions in the dense cluster environment.

Acknowledgements

M. V. M. gratefully acknowledges travel support from NRAO, and she is supported by an NSF Astronomy and Astrophysics Postdoctoral Fellowship under award AST-0401460. This material is based on work supported by the National Science Foundation under Grants No. AST-0205297 and AST-0506573 (D. R. G.). Institutional support has been provided from the GSU College of Arts and Sciences and from the Research Program Enhancement fund of the Board of Regents of the University System of Georgia, administered through the GSU Office of the Vice President for Research.

References

Boyajian, T.S., Beaulieu, T.D., Gies, D.R., Huang, W., McSwain, M.V., Riddle, R.L., Wingert, D.W., & De Becker, M. 2005, *ApJ*, 621, 978

Lamers, H.J.G.L.M., van den Heuvel, E.P.J., & Petterson, J.A. 1976, *A&A*, 49, 327

Leonard, P.J.T. & Duncan, M.J. 1990, *AJ*, 99, 608

Lipunov, V.M. 1992, *Astrophysics of Neutron Stars* (Berlin: Springer-Verlag)

Mazeh, T. & Goldberg, D. 1992, *ApJ*, 394, 592

McSwain, M.V., Boyajian, T.S., Grundstrom, E.D., & Gies, D.R. 2006, *ApJ*, in press (astro-ph/0608270)

Philp, C.J., Evans, C.R., Leonard, P.J.T., & Frail, D.A. 1996, *AJ*, 111, 1220

Portegies Zwart, S.F. 2000, *ApJ*, 544, 437

Ransom, S.M. 2001, Ph.D. thesis, Harvard University

Sayer, R.W., Nice, D.J., & Kaspi, V.M. 1996, *ApJ*, 461, 357

Voges, W. *et al.* 2000, *IAU Circ.*, 7432, 1

White N.E., Giommi P., & Angelini L. 2000, *The WGACAT version of the ROSAT PSPC Catalogue*, Rev. 1, (VizieR No. IX/31)

Binary Stars as Critical Tools & Tests
in Contemporary Astrophysics
Proceedings IAU Symposium No. 240, 2006
W.I. Hartkopf, E.F. Guinan & P. Harmanec, eds.

Catalogs of Wide Binaries:
Impact on Galactic Astronomy

Julio Chanamé

Space Telescope Science Institute, Baltimore, MD, USA
email: jchaname@stsci.edu

Abstract. Wide binaries, particularly in large numbers and as free from selection biases as possible, constitute a largely overlooked tool for studying the Galaxy. The goal of this review is to highlight the potential inherent to large samples of field wide binaries for research on problems as varied as star formation in the early Galaxy, the nature of halo dark matter, the evolution of the stellar halo, new geometric distances, metallicities, masses, and ages of field stars and white dwarfs, and much more. Using the Revised NLTT as an illustrative example, I review the main steps in the assembly of a large catalog of wide binaries useful for multiple applications. The capability of cleanly separating between the Galactic disk and halo populations using good colors and proper motions is emphasized. The critical role of large surveys for research on wide binaries as well as for the better understanding of the Galaxy in general is stressed throughout. Finally, I point out the potential for assembling new samples of wide binaries from available proper-motion surveys, and report on current efforts of using the SDSS towards this goal.

Keywords. surveys, catalogs, (stars:) binaries: general, stars: fundamental parameters, stars: kinematics, Galaxy: general, (cosmology:) dark matter

1. Introduction

Knowledge obtained from the study of stars in binary systems has been fundamental for astronomy, and influences almost all of its branches. Their impact is easily illustrated with just a few examples: the masses of individual stars (the single most important parameter in stellar evolution) are measured with confidence only for stars in binary systems†; eclipsing binaries provide a primary method for measuring distances (Paczyński 1997) and are playing a major role in securing the base of the cosmological distance ladder (Ribas *et al.* 2005; Bonanos *et al.* 2006); close binary systems harboring white dwarfs, as likely progenitors of type-Ia supernovae, are thought to be main players driving the chemical evolution of galaxies and the intergalactic medium (Pagel 1997). And the list can go on and on.

Most of that knowledge, however, has come from the study of binary systems in close orbits, while the population of binaries at the wide end of the distribution of orbital separations (semimajor axes $a \gtrsim 100 \, \text{AU}$) remains poorly explored. Clearly, the main reason behind this historical bias is the much longer orbital timescale inherent to very wide binary systems (of the order of hundreds of years and longer), making their unambiguous identification a task that also requires a relatively long timescale (though, fortunately, not as long as their complete periods!). In contrast to the case of close binaries, where some kind of variability is typically the property that helps uncover them, a reasonably large number of very wide binaries can not be identified via the typical photometric or

---

† Stellar mass measurements for single stars have been achieved thanks to gravitational microlensing as well (Drake *et al.* 2004; Jiang *et al.* 2004). However, the errors in this technique are not yet competitive with those for stars in binary systems.

spectroscopic campaign lasting from several months to a few years. Instead, wide binaries require accurate astrometry over timescales of the order of decades.

Nevertheless, wide binaries hold great potential for a large variety of studies in Galactic astronomy. A summary list of today's applications of wide binaries can include, in no particular order:

• probing for the processes and conditions of star formation as a function of age and metallicity in star-forming regions (White & Ghez 2001), and as a function of environment (disk vs. halo) during the assembly of the Galaxy (Chanamé & Gould 2004);

• to obtain new and accurate distances to faint low-mass stars in the field (Gould & Chanamé 2004; Lépine & Bongiorno 2006), allowing high-precision studies of stellar properties near the bottom of the main sequence (Patten *et al.* 2006);

• determination of the metallicities of field M-dwarfs (Bonfils *et al.* 2005) and the study of the age-metallicity relation in the Galactic disk (Monteiro *et al.* 2006);

• tests of not well constrained stellar evolutionary processes such as internal mixing and diffusion via the study of the surface abundances of the members of wide binaries with twin components (Martín *et al.* 2002);

• exploration of the age and chemical evolution of the stellar halo and its underlying substructure via the study of halo wide binaries with evolved components (Ivans, Chanamé, & Gould, in preparation);

• measurement of the masses of white dwarfs and constraints on the (dark) mass density of stellar remnants in the local Galactic disk (Silvestri *et al.* 2001);

• to place severe constraints on the nature of halo dark matter that complement the decades-long microlensing campaigns (Yoo, Chanamé, & Gould 2004), and to explore the dynamical and merger history of the Galaxy (Allen & Poveda 2007);

• studies of the initial-to-final mass relation of white dwarfs via the determination of the mass of their main-sequence progenitors (Catalán *et al.* 2007; Silvestri *et al.* 2005);

• constraining mass loss during post main sequence stages, white dwarf progenitor masses, and the density of matter returned to the interstellar medium via the investigation of the evolution of orbital separations of wide binaries harboring evolved components (Johnston, Oswalt, & Valls-Gabaud 2007);

• studying the impact of planetary systems on the evolution of their host stars via the detailed surface abundances of stars in wide binaries (Desidera *et al.* 2004).

The main purpose of this review is therefore to highlight the large potential inherent to *large samples* of wide binaries† for the study of the Galaxy. The emphasis on the words *large samples* is not accidental: various of the applications mentioned above are not necessarily the product of studies of wide binaries on a system by system basis but, rather, statistically as a population. They have only been possible thanks to the relatively large numbers of *genuine* binaries that became available when astronomers changed their strategies and started to look for them in a systematic way, making big efforts to avoid possible selection biases. This change of approach, in turn, only became possible with the construction and exploitation of large photometric and astrometric databases such as the NLTT, USNO, *Hipparcos*, 2MASS, etc. Therefore, the critical role of *large surveys* in this revolution, not only in the field of wide binaries but for Galactic astronomy in general, cannot be emphasized enough, and this crucial point is stressed again in § 2 and throughout the present contribution.

In § 3 a brief report is presented of the status of the currently available catalogs of wide binaries, using the case of the Revised NLTT (rNLTT; Chanamé & Gould 2004) catalog

---

† The discussion in the present contribution is restricted to wide binaries in the field, i.e., does not include systems in clusters and/or star forming-regions.

to illustrate the steps involved in their construction. A description of the ongoing efforts and prospects of using the Sloan Digital Sky Survey (SDSS) to build an even larger catalog of wide binaries is presented in § 3.2. Finally, a few of the various applications of catalogs of wide binaries to problems of Galactic astronomy are briefly outlined in § 4.

2. Surveys: the crucial starting point

Progress in astronomy over the last 25 or so years has been deeply influenced by the steady increase in importance of the role of large and systematic surveys of all types. Several of these even prompted the deployment of instrumentation beyond Earth's atmosphere. The impact of these strategies is now seen across areas that range from the cosmological (HST, 2dF, WMAP, ...) to that of extrasolar planets (SuperWasp, Kepler). In the Galactic context, the extraordinary explosion of new science and discoveries that the SDSS (a survey mainly designed to address cosmological questions) is allowing at this very moment has its most immediate predecessors, both scientific as well as technological, in surveys such as *Hipparcos* and 2MASS. Today, the case for staying on the same track is evidenced by the efforts (and funding!) put in upcoming surveys such as GAIA.

Research on wide binaries is no exception to this trend and progress in this area is also intimately tied to the development and use of large databases. Commonly referred to as common proper-motion systems, a large number of wide binaries is a natural and relatively straightforward product of astrometric campaigns and proper-motion surveys.

Moreover, the combination of good photometry with good astrometry, either coming from a single survey or from the combination of separate ones, has allowed in the last few years the clean separation between large numbers of stars unambiguously belonging to the disk and halo of the Galaxy, thus opening a new avenue for studying Galactic structure and evolution with high statistical significance. On the subject of our interest here, the rNLTT catalog of disk and halo wide binaries, assembled from the combination of surveys such as the NLTT, USNO-A, *Hipparcos*, and 2MASS, constitutes an excellent example of this synergy, and its capacity of telling whether a single star belongs to the disk or the halo of the Galaxy constitutes a powerful tool for the unambiguous identification of pairs of stars that, even though having proper motions sharing the same magnitude and direction on the sky, are not really bound to each other, and would contaminate a sample of wide binaries selected only on common proper-motion criteria (see § 3.1).

The big advantage of the strategy of surveys lies not only in the greatly improved statistics but, most importantly, in the degree of completeness of the resulting databases, which, with well understood selection effects, if any, permits a relatively unbiased view of whatever the particular survey aims at. It is this crucial characteristic that, in the case of wide binaries, makes possible two of their most interesting applications to astronomy, as are their use as probes of the nature of dark matter and of the processes and conditions of star formation (see § 4).

The approach of large surveys has proven to be of great power and is likely to stay as one of the favored ways to advance in astrophysics. People interested in wide binaries and their applications must therefore consider carefully what can be done with the already existing databases as well as prepare in advance for an efficient use and exploitation of the surveys to come. Joint Discussion 13 on this very same IAU General Assembly, entitled *Exploiting Large Surveys for Galactic Astronomy*, dealt with this subject in detail, and readers are encouraged to consult those proceedings (see also their website at http://clavius.as.arizona.edu/vo/jd13/).

3. Catalogs of field wide binaries

As noted before, being the characteristic fingerprint of wide binaries the almost identical proper-motion vectors of their two stars (up to differences due to measurement uncertainty and relative orbital motion only), they are intimately associated to proper-motion surveys. The rNLTT wide binaries, extracted from the Luyten survey of high proper-motion stars (Luyten 1979, 1980), constitute likely the most homogeneous and complete catalog of such objects available today, although not the only one.

Before the rNLTT, the most successful efforts to assemble large numbers of wide binaries were those of Poveda *et al.* (1994) and Allen *et al.* (2000), who based their searches on databases such as Gliese's Catalog of Nearby Stars and the NLTT itself, though not exploiting the latter in its full capacity. See the contributions by A. Poveda and C. Allen on these proceedings for details on their samples and some of their uses in astronomy. The efforts by Allen *et al.* (2000) and Zapatero Osorio & Martín (2004) concentrated on the search for wide companions to metal-poor stars selected from databases such as the Carney-Latham surveys. A few earlier attempts to identify smaller samples of field wide binaries are reported in the introductory section of Chanamé & Gould (2004).

Other proper-motion catalogs available today that hold potential for a search of wide binaries include UCAC (Rafferty *et al.* 2001), SuperCOSMOS (Hambly *et al.* 2001), LSPM (Lépine & Shara 2005), and SDSS ∩ USNO-B (Gould & Kollmeier 2004; Munn *et al.* 2004). Although a few new common proper-motion pairs of intrinsic interest have been identified from these databases (Scholz *et al.* 2002, 2005; Seifahrt *et al.* 2005; Monteiro *et al.* 2006), only one report of a systematic search for wide binaries using these catalogs exists, which appeared the very same day this contribution was to be submitted. In a manner analogous to the search of Gould & Chanamé (2004) in the rNLTT, Lépine & Bongiorno (2006) searched in the LSPM catalog for faint common proper-motion companions of *Hipparcos* stars, uncovering 521 systems, of which 130 are new (see § 4).

The optimal construction of a catalog of field wide binaries is outlined in § 3.1, using the case of the rNLTT as an illustrative example. It will be seen how, with appropriate photometric and astrometric data, it is possible to not only identify genuine field wide binaries, but also to determine with high confidence, on a pair by pair basis, their membership in either the disk or the halo populations of the Galaxy. Disentangling between pairs belonging to the Galactic thin and thick disks, however, would require more information, such as the full three-dimensional velocities and metallicities of the stars. Finally, in § 3.2, I describe an ongoing search for wide binaries in SDSS ∩ USNO-B.

3.1. *Building a catalog: the rNLTT wide binaries*

The NLTT proper-motion survey was available long before the Chanamé & Gould (2004) work, and Luyten himself identified and recorded a substantial fraction of the NLTT wide binaries in his Luyten Double-Star Catalog (LDS, Luyten 1940-87). However, two factors worked against the construction of a *complete* sample of wide binaries from the original NLTT. First, the crude photographic colors did not allow the construction of a reliable reduced proper motion (RPM) diagram, thus preventing the identification of chance alignments (i.e., pairs of unrelated stars whose proper-motion vectors are aligned on the sky just by chance) via the non-consistent colors of the stars in such pairs (more below). Second, the large proper-motion errors in the original NLTT did not allow Luyten to identify a fair number of genuine binaries at large angular separations (the *non-Luyten* binaries of Chanamé & Gould 2004) because it was impossible to separate them from the numerous unrelated optical pairs at those separations.

Only with the construction of the Revised NLTT (Gould & Salim 2003; Salim & Gould 2003) could these problems be overcome. An upgraded version of the original NLTT,

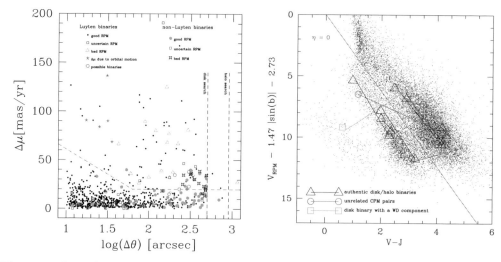

Figure 1. (a; left) Initial selection of candidate binaries in rNLTT: vector proper motion difference vs. angular separation for systems with $\Delta\theta > 10''$. Pairs below the dashed line were automatically accepted as candidate binaries for their subsequent classification. The dashed vertical lines indicate the upper limits of the homogeneous search for disk and halo binaries. (b; right) Reduced proper motion diagram of rNLTT stars, clearly separating the disk, halo, and white dwarf tracks. A few examples of how this diagram is used to test and classify binaries are shown (see text).

the rNLTT catalog provides optical and (CCD) infrared photometry for the majority of the NLTT stars by matching them to the USNO-A and 2MASS surveys. Thanks to the better colors and large color baseline, this proved sufficient to permit a clean separation between disk and halo stars using a RPM diagram (Salim & Gould 2002). Furthermore, the better temporal baseline of the rNLTT relative to the NLTT translated into a substantial improvement in the accuracy of the proper-motion measurements, which allowed Chanamé & Gould (2004) to perform a uniform and complete search of wide common proper-motion pairs up to angular separations much larger than what Luyten was able to do. It is worth noticing, here again, the fundamental importance of large surveys and the potential inherent in their combination and adequate exploitation.

Still, with the main shortcomings of the original NLTT overcome with the publication of the rNLTT, it was by no means straightforward to dig out a *uniformly selected* catalog of wide binaries from it. In order to produce a sample of wide binaries as free from selection effects as possible, a number of complications had to be dealt with, such as blending, saturation, missing independent proper-motion measurements, and others. See Chanamé & Gould (2004) for a detailed account of the solutions to these problems. Here only a brief outline of the selection and classification of the rNLTT wide binaries is given.

The first step is to select all pairs of stars that seem to be moving together on the sky, i.e., all pairs with similar proper-motion vectors. This was done by plotting the magnitude of the vector proper motion difference (denoted by $\Delta\mu = |\Delta\boldsymbol{\mu}|$) of the components of these pairs against their angular separation ($\Delta\theta$), as shown in Figure 1a. Pairs with separations smaller than $\Delta\theta = 10''$ as measured by Luyten were not subjected to this selection because of the small chance of two unassociated stars lying so close in both position and velocity space. Thus the plot in Figure 1a begins at $\Delta\theta = 10''$ and for pairs with smaller separations all efforts were focused on their classification (which, based on the consistency of their colors, constitutes another test for their reality).

For the rest of common proper-motion systems wider than $10''$, good candidates were considered to be those falling below the dashed line in Figure 1a, set at wide separations at $\Delta\mu = 20\,\mathrm{mas\,yr}^{-1}$. Note that all the NLTT stars are moving faster than $180\,\mathrm{mas\,yr}^{-1}$, so the above maximum-$\Delta\mu$ requirement is already a strong test, and pairs that satisfy it have already a high likelihood of being real binaries. The $20\,\mathrm{mas\,yr}^{-1}$ cutoff was somewhat relaxed at closer separations because the chance of contamination by unassociated pairs is lower and also because some real pairs at close separations can have significant orbital motions (pairs indicated by red asterisks in Figure 1a).

This search for rNLTT wide binaries was systematically done up to separations of $\Delta\theta = 500''$ for disk binaries and $\Delta\theta = 900''$ for halo binaries, thus avoiding any possible selection effects as a function of angular separation. In the same spirit, the exclusion from the analysis of pairs outside the allowed region of the $\Delta\theta - \Delta\mu$ plane does not mean that all these systems should be regarded as false binaries. Rather, it simply reflects the extreme care put in being as complete and free from selection effects as possible, but at the same time rigorous in not being contaminated by unrelated optical pairs. It is only thanks to this characteristic that the rNLTT wide binaries (or any other catalog similarly constructed) can be used as probes of star formation as well as of the nature of dark matter in the Galaxy.

The second test for the reality of the (up to this point) candidate binaries is related with their classification as either disk or halo pairs. This is done via examination of each pair in a RPM diagram, and all the possibilities are illustrated in Figure 1b, where the clear separation between the tracks of disk and halo stars (at both sides of the line labeled $\eta = 0$) is evident. When classifying the binaries, one expects not only that both components belong to the same population, but also that their positions on the RPM diagram are consistent with both stars having the same age and metallicity. Therefore, for real binaries (blue triangles in Fig. 1b) one expects that the line connecting their positions on the RPM diagram should be approximately parallel, within measurement errors, to the corresponding disk or halo track. Candidate pairs composed of one disk and one halo member are rejected as chance alignments, and so are pairs composed of two disk or two halo stars if the line connecting them is inconsistent with being parallel to the respective sequence (red circles in Figure 1b). The only cases permitted not to follow this parallel rule are those involving a white dwarf companion (as the green pair of squares in Figure 1b).

The above two-step procedure yielded 999 genuine wide binaries: 883 pairs belonging to the Galactic disk and 116 belonging to the Galactic halo. Among the disk binaries, 82 pairs contained a white dwarf, while no halo stars with clear white dwarf companions were found.

3.2. *Wide binaries from the SDSS∩ USNO-B*

The Sloan Digital Sky Survey, providing superb optical CCD photometry for almost a quarter of the sky, is already revolutionizing our understanding of the Galaxy on several fronts. Although a single-epoch survey, it can be combined with older photometric and/or proper-motion surveys in order to produce the two epochs necessary to compute new proper motions for its stars. Gould & Kollmeier (2004) and Munn *et al.* (2004) already accomplished this for the SDSS First Data Release by cross-correlating it with the USNO-B catalog, producing nearly 400,000 stars down to $r' < 20$, moving faster than $20\,\mathrm{mas\,yr}^{-1}$, and with proper-motion errors of about $4\,\mathrm{mas\,yr}^{-1}$.

Given the depth and sky coverage of SDSS, there is little doubt that it will provide a significantly larger number of binaries than the rNLTT, and efforts toward this goal are already underway (see Smith *et al.* 2005 and Greaves 2005 for recent searches in

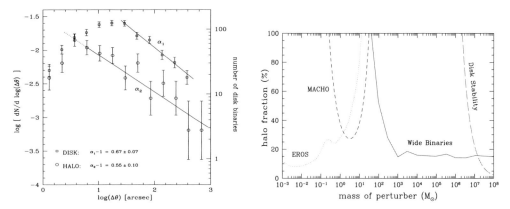

Figure 2. (a; left) Distribution of angular separations for the rNLTT disk and halo wide binaries (Chanamé & Gould 2004). Note that disk and halo binaries show basically the same distributions, and that the data for both populations of binaries are consistent with a single power-law out to the widest separations, with no hint of a break or cutoff. (b; right) Exclusion contours for MACHO dark matter at the 2σ level of the microlensing experiments and the rNLTT halo wide binaries (Yoo, Chanamé, & Gould 2004).

SDSS of pairs with white dwarf components). The challenge, however, lies in being able to distinguish the truly bound systems from among the large number of spurious pairs inevitably present in such a large collection of stars with very similar proper motions. For the nearby Luyten stars, with proper motions $|\boldsymbol{\mu}| > 180\,\mathrm{mas\,yr}^{-1}$, just the combination of photometry and astrometry was enough to do the selection. But with the more distant SDSS stars, moving at $20 - 40\,\mathrm{mas\,yr}^{-1}$, the measured $\Delta\vec{\mu}$ alone does not reject all the apparent binaries, even though the photometry is excellent. Therefore, it is necessary to obtain radial velocities of the components of the common proper-motion candidates that pass the first tests (consistent photometry and proper motions) to clean the initial sample and select the real binaries. Binary candidates have been selected and the radial velocities of the brightest pairs have already been measured using the SMARTS 1.5-m telescope at CTIO (Chanamé, van der Marel, van de Voort, & Gould, in preparation). The large majority of targets, however, with magnitudes in the range $16 < r < 20$, require larger apertures. Our plans to complete these observations using telescopes at the MDM and Kitt Peak Observatories are already underway.

4. Impact of wide binary catalogs on Galactic astronomy

Here, I only point out a few of the various applications of wide binaries to Galactic astronomy that are outlined in § 1. See T. Oswalt's contribution for the several uses of wide binaries involving white dwarfs, and those of C. Allen and A. Poveda for ways of exploiting their potential as probes of dynamical processes in the Galactic disk.

One of the most useful characteristics of wide binaries is that both components share properties such as age, metallicity, distance, etc., and, therefore, when any of these is measured for one of them, typically the primary, it automatically can be assigned to the fainter secondary. This has been exploited to obtain parallaxes to faint low-mass stars in the field by Gould & Chanamé (2004) and Lépine & Bongiorno (2006), who searched for wide companions of *Hipparcos* stars in the rNLTT and LSPM catalogs, respectively. Other applications listed in § 1 apply this same principle to get ages, metallicities, etc.

Nevertheless, two of the most exciting applications of wide binaries are the result of statistical studies for which not only the large size but, perhaps most importantly, the completeness of any sample of these objects is critical. In particular, their distribution of (physical or angular) separations can be used to gain insights into the processes of star formation, and also to place strong limits to the mass and density of compact dark matter (MACHOs) in the Galactic halo.

Using the rNLTT, Chanamé & Gould (2004) were able to perform the first comparison of the distributions of separations of large and clean samples of disk and halo wide binaries, as shown in Figure 2a. Somewhat surprisingly, they found both distributions to be the same within the uncertainties, despite the radically different Galactic environments to which they belong today. This provides a unique insight on the environmental conditions in which star formation proceeded in the early Galaxy, and any successful model of the formation of the Galaxy should be able to reproduce this finding.

Finally, the very wide binaries ($a \gtrsim 0.1$ pc) are so weakly bound that they can be significantly disturbed, even disrupted, by gravitational interactions as they orbit the Galaxy, including those with giant molecular clouds, large-scale tides, passing stars, as well as massive dark objects (MACHOs). If such encounters were frequent during the lifetime of any given binary, then one would expect the widest systems to be relatively depleted. Early attempts to learn about disk dark matter in the solar neighborhood via wide binaries (Bahcall, Hut & Tremaine 1985) did not achieve useful conclusions due to the small number of wide binaries available then, as well as the intrinsic complexity of the disk potential (Weinberg, Shapiro & Wasserman 1987).

The situation in the Galactic halo was easier to model, but an adequate sample of halo wide binaries did not exist until the rNLTT binaries were published. Only then, using detailed Monte Carlo simulations of the evolution of a population of wide binaries immersed in a halo filled with compact objects with various combinations of mass and density, Yoo, Chanamé, & Gould (2004) were able to exclude a Milky Way halo mostly composed of MACHOs with masses $M > 43\,\mathrm{M_\odot}$ at the standard local halo density and at 95% confidence level, as shown in Figure 2b. In this way, halo wide binaries were shown to complement the results of the microlensing campaigns, mostly sensitive to lower masses, and provided us with an alternative experiment for addressing the same key question.

Acknowledgements

I wish to thank Bill Hartkopf for inviting me to this Symposium, and Roeland van der Marel for his interest and support of my work on wide binaries. Thanks as well to Andy Gould and Inese Ivans for a critical reading of this manuscript, and to Freeke van de Voort for her help and enthusiasm in digging into noisy spectra in the search for new binaries in SDSS. Finally, I acknowledge the support of both the American Astronomical Society and the International Astronomical Union through grants to travel and attend the IAU General Assembly.

References

Allen, C., Poveda, A. & Hernández-Alcántara, A. 2007, IAU Symposium 240, these proceedings, 405
Allen, C., Poveda, A., & Herrera, M. 2000, *A&A*, 356, 529
Bahcall, J. N., Hut, P., & Tremaine, S. 1985, *ApJ*, 290, 15
Bonanos, A. Z., *et al.* 2006, *Ap&SS*, 79
Bonfils, X., *et al.* 2005, *A&A*, 442, 635
Catalán, S., *et al.* 2007, these proceedings, 380
Chanamé, J. & Gould, A. 2004, *ApJ*, 601, 289
Desidera, S., *et al.* 2004, *A&A*, 420, 683
Drake, A.J., Cook, K.H., & Keller, S.C. 2004, *ApJ*, 607, L29

Gould, A., & Chanamé, J. 2004, *ApJS*, 150, 455

Gould, A., & Kollmeier, J. 2004, *ApJS*, 152, 103

Gould, A., & Salim, S. 2003, *ApJ*, 582, 1001

Greaves, J. 2005, *Journal of Double Star Observations*, 1, 41

Hambly, N. C., *et al.* 2001, *MNRAS*, 326, 1315

Jiang, G., *et al.* 2004, *ApJ*, 617, 1307

Johnston, K., Oswalt, T., & Valls-Gabaud, D. 2007, these proceedings (Paster), 429

Lépine, S., & Shara, M. M. 2005, *AJ*, 129, 1483

Lépine, S., & Bongiorno, B. 2005, AJ 133, 889

Luyten, W. J. 1979, 1980, *New Luyten Catalog of Stars with Proper Motions Larger than Two Tenths of an Arcsecond* (Minneapolis: University of Minnesota Press)

Luyten, W.J. 1940-87, *The LDS Catalogue: Double Stars with Common Proper Motion* (Minneapolis: University of Minnesota Press)

Martín, E.L., 2002, *ApJ*, 579, 437

Monteiro, H., Jao, W.-C., Henry, T., Subasavage, J., & Beaulieu, T. 2006, *ApJ*, 638, 446

Munn, J. A., *et al.* 2004, *AJ*, 127, 3034

Paczyński, B. 1997, *The Extragalactic Distance Scale*, 273 (astro-ph/9608094)

Pagel, B. 1997, *Nucleosynthesis and Chemical Evolution of Galaxies* (Cambridge Univ Press)

Patten, B.M., *et al.* 2006, ApJ 651, 502

Poveda, A., *et al.* 1994, *Revista Mexicana de Astronomía y Astrofísica*, 28, 43

Rafferty, T. J., *et al.* 2001, ASP Conf. 232: *The New Era of Wide Field Astronomy*, 232, 308

Ribas, I., *et al.* 2005, *ApJ*, 635, L37

Salim, S., & Gould, A. 2002, *ApJ*, 575, L83

Salim, S. & Gould, A. 2003, *ApJ*, 582, 1011

Scholz, R.-D., *et al.* 2002, *ApJ*, 565, 539

Scholz, R.-D., *et al.* 2005, *A&A*, 430, L49

Seifahrt, A., *et al.* 2005, *A&A*, 440, 967

Silvestri, N., *et al.* 2001, *AJ*, 121, 503

Silvestri, N., Hawley, S. L., & Oswalt, T. D. 2005, *AJ*, 129, 2428

Smith, J.A., Silvestri, N.M., Oswalt, T.D., *et al.* 2005, ASP Conf. Ser. 334: *14th European Workshop on White Dwarfs*, 334, 127

Weinberg, M.D., Shapiro, S.L., & Wasserman, I. 1987, *ApJ*, 312, 367

White, R. J., & Ghez, A. M. 2001, *ApJ*, 556, 265

Yoo, J., Chanamé, J., & Gould, A. 2004, *ApJ*, 601, 311

Zapatero Osorio, M. R., & Martín, E. L. 2004, *A&A*, 419, 167

Discussion

PETR HARMANEC: Is there some explanation why metallicity depends on eccentricity?

CHANAME: The metallicity of a star does not depend on the eccentricity of its orbit, nor viceversa. It is true, however, that the oldest and most metal-poor stars in the Galaxy are halo stars, and these in general have large eccentricities in comparison with the more nearly circular orbits exhibited by disk stars, which, in turn, have higher metal abundances than halo stars. These are statistical facts applicable to these two distinct stellar populations in general, and should not be used in a star-by-star basis. The plot that showed a correlation between metallicity (as indicated by the UV excess) and the eccentricity of the orbits of the stars around the Galaxy is from the 1962 paper by Eggen, Lynden-Bell, & Sandage (their Figure 4). What I intended to convey by showing this plot is simply how the interplay between the kinematics and abundances of large numbers of stars can be used to infer things about the history of the Galaxy at large.

NANCY EVANS: You should be able to get a distribution of mass ratios. How does this work out?

CHANAME: We did compute these distributions, and they are available in Chaname & Gould (2004). Although this requires the adoption of distances and then color-magnitude relations (therefore introducing model-dependent assumptions and uncertainties), my recollection is that disk and halo wide binaries are also similar in their distributions of mass-ratio. But please take a look at the paper itself for a thorough discussion on this point.

VALERI MAKAROV: Do you see the bifurcation of halo stars and disc stars on the color-magnitude diagram (M_V vs. $(V - J)$) if you use trigomometric parallaxes instead of reduced proper motions?

CHANAME: No, the difference is not as clear as with the reduced proper motions, as shown in the quoted paper. It is a useful property of the reduced proper motion method that it reflects the different kinematics of the halo and the disc.

Binary Stars as Critical Tools & Tests
in Contemporary Astrophysics
Proceedings IAU Symposium No. 240, 2006
W.I. Hartkopf, E.F. Guinan & P. Harmanec, eds.

Poster Abstracts (Session 5)

(Full Posters are available at http://www.journals.cambridge.org/jid_IAU)

Absolute Parameters of the O-type Eclipsing Binary V1007 Sco

S. Nesslinger[1], H. Drechsel[1], R. Lorenz[1], P. Harmanec[3], P. Mayer[2], and M. Wolf[2]

[1] *Dr. Remeis Observatory, University of Erlangen-Nuremberg, Bamberg, Germany,*
[2] *Astronomical Institute of the Charles University, Prague, Czech Republic,*
[3] *Astronomical Institute, Academy of Sciences, Ondrejov, Czech Republic*

A definitive set of orbital and physical parameters of the binary system V1007 Sco is presented. Both stars are classified as O7 giants. Light and RV curves were simultaneously analyzed with PHOEBE, a modern WD-based solution code. A statistical approach yielded realistic error estimates. Orbital eccentricity is 0.12 with an apsidal motion period of 112 years.

CM Draconis: Masses and Radii of Very Low Mass Stars

J.C. Morales[1,3], I. Ribas[1,2], C. Jordi[1,3], G. Torres[4], E.F. Guinan[5], and D. Charbonneau[4]

[1] *Institut d'Estudis Espacials de Catalunya, Barcelona, Spain*
[2] *Institut de Ciències de l'Espai – CSIC, Bellaterra, Spain,*
[3] *Dept. d'Astronomia i Meteorologia, Universitat de Barcelona, Barcelona, Spain,*
[4] *Harvard-Smithsonian Center for Astrophysics, Cambridge, MA, United States,*
[5] *Dept. of Astronomy, Villanova University, Villanova, PA, United States*

Eclipsing binaries provide the only high-accuracy ($\sim$1-2%) measures of stellar masses and radii, making them suitable for relevant tests of the models of structure and evolution of stars. At the low-mass end of the main sequence, current results have shown that models predict radii smaller by $\sim$10% and effective temperatures larger by $\sim$5% than observed. Although no definitive explanation of such differences has been yet reached, some evidence suggests that the disagreement could be related to the strong activity and intense magnetic fields in these stars. In this work we study CM Draconis, one of the least massive eclipsing binaries known. Their components are very similar, both with dM4 spectral type, and masses and radii of about $0.23\,M_\odot$ and $0.25\,R_\odot$, respectively. We have analyzed light curves in the R and I bands taken between 1975 and 2004 with a total of 25051 measurements. To calculate the fundamental properties of this system with accuracies better than 1% we have considered spots on the surface of the stars and modeled the light curve variations accordingly. From the resulting accurate masses and radii we plan to carry out a thorough test of the models for these fully convective stars. This will be especially interesting since the mechanism driving magnetic activity is thought to be different from that of more massive stars. In addition, the extended time-span of the observations has led to the detection of apsidal motion in this system, which has a slightly eccentric orbit. This provides a further check on models through the determination of the internal structure of the stars

A Mass Estimate in a Sample of Double Stars

S. Ninkovic and Z. Cvetkovic

Astronomical Observatory, Belgrade, Yugoslavia (Serbia and Montenegro)

The total masses of binaries are calculated on the basis of their orbital elements from the *Sixth Catalog of Orbits of Visual Binary Stars*. They are then compared with the values resulted from the mass-luminosity relation for the Main Sequence where as the input data are used: trigonometric parallax, total apparent magnitude of the pair (source *Hipparcos Catalog*) and magnitude difference (sources *Hipparcos Catalog* and *Photometric Magnitude Difference Catalog*). It seems that for the pairs indicated as having qualitative orbital elements the agreement between the total-mass values obtained in these two ways is satisfactory.

GQ Lup, 2M1207, and AB Pic: Planet Companion Candidates Imaged Directly and their Relevance in Orbital Dynamics and Mass Estimation via Theoretical Models

R. Neuhaeuser[1], M. Mugrauer[1], and A. Seifahrt[2]

[1] *AIU University Jena, Jena, Germany,*
[2] *ESO Garching, Garching, Germany*

In 2005, evidence was presented for three exo-planets imaged directly: GQ Lup, 2M 1207, and AB Pic. In all three cases, a faint red object is co-moving with a young nearby star. The masses of these companions are determined through theoretical models, which are under dispute and have not yet been tested successfully in the relevant parameter range of young ages and low masses. We show that being co-moving with another star and having a late spectral type is necessary, but not sufficient for being gravitationally bound. We discuss the relevance of these three wide visual binary systems for orbital dynamics and testing theoretical models. We will also present new images from 2005 and 2006 to investigate orbital motion of GQ Lup b around GQ Lup A.

Low-mass Eclipsing Binaries to Refine Barnes-Evans-like Relations

P.J. Amado

Univ. Granada-IAA(CSIC), Granada, Spain

The relation of the surface brightness, a parameter related to the apparent magnitude and the angular diameter of a star, with a colour index was first calibrated by Wesselink (1969) and later refined by Barnes & Evans (1976). This calibration has a very wide range of applicability amongst which is the evaluation of various stellar parameters. In this work, we use a number of low-mass eclipsing binaries whose parameters have been accurately determined to refine the Barnes-Evans-like relation $F_V - I_c - K)$ introduced by Amado *et al.* (1999), a relation more suitable for the low temperature of cool stars.

Eclipsing Binaries as a Test for Synthetic Photometry

U. Heiter[1], B. Smalley[2], Ch. Stütz[3], F. Kupka[4], and O. Kochukhov[1]

[1] *Department of Astronomy and Space Physics, Uppsala University, Uppsala, Sweden,*
[2] *Astrophysics Group, School of Chemistry and Physics, Keele University, Keele, United Kingdom,*
[3] *Institute for Astronomy, University of Vienna, Vienna, Austria,*
[4] *Max-Planck-Institute for Astrophysics, Garching, Germany*

Narrow-band photometry is a viable tool to characterize large numbers of stars. The connection between observed colors and astrophysical parameters has to rely on synthetic photometry calculated from stellar atmosphere models. Here, we present synthetic $H\beta$ indices calculated from 1D model atmospheres, which implement various treatments of convection. The calculated indices are transformed to the standard system using observed medium-resolution spectra from recently published stellar libraries. We test how well the synthetic photometry reproduces observed indices by using a number of eclipsing binary systems. For these stars, atmospheric parameters can be determined independently from the models with highest possible accuracy. As a preliminary conclusion, the computed indices deviate from the observed ones by an amount expected from the observational errors and the accuracy of the atmospheric parameters.

Note: The full text of this paper will be published in the proceedings of IAU Symposium 239, *Convection in Astrophysics*.

Looking for Exoplanets in Bright Stars with Small Field-of-View Detectors

Eder Martioli and Francisco Jablonski

Instituo Nacional de Pesquisas Espaciais – Divisâo de Astrofisica, Sâ José dos Campos, SP, Brasil

Differential photometry is a robust technique for ground-based observations of transits since it sorts out slow variations of sky transparency as well as other first order effects that are common to all stars in the field-of-view (FOV) of the imaging detector. To work properly, differential photometry has to obey a few requirements like similar brightness of the target and reference stars, similar colors and a relative proximity in the plane of the sky to avoid sensitivity variations like those caused by vignetting. It happens that for bright stars these conditions are hardly met. Typical CCDs in a $\sim$60-cm class telescope give a FOV of $\sim$10 arcmin and this is not enough to have suitable reference stars in the same image frame. Also, bright ($V < 7$) stars tend to saturate the detector for the shortest practical integration times. To minimize these problems, we tested an instrumental setup in which half of the detector is covered with a neutral density ($D = 2.3$) filter. We report CCD observations on which we achieved mmag precision for bright systems that are not known to show transits, like τ Boo, 55 Cnc and HD 162020, as well as the known case of HD 209458.

Multiplicity Study of Exoplanet Host Stars: The HD 3651 AB System

M. Mugrauer[1], A. Seifahrt[2], R. Neuhaeuser[1], T. Mazeh[3], and T. Schmidt[1]

[1] Astrophysikalisches Institut, Universität Jena, Jena, Germany,
[3] European Southern Observatory, Garching, Germany,
[3] Tel Aviv University, Tel Aviv, Israel

We present new results from our ongoing multiplicity study of exoplanet host stars. We found new stellar companions of the exoplanet host stars GJ 3021 and HD 40979 and present our imaging and spectroscopic data of the wide companion of the exoplanet host star HD 27442. GJ 3021 is a new close planet hosting binary system with a M3–M5 stellar companion ($\sim$0.125 $M_\odot$) which is separated from its primary by only 70AU. In contrast, HD 40979 is one of the widest planet hosting stellar systems known today with a projected separation of $\sim$6400 AU. We present our observations of the wide companion HD 40979B, which turned out to be a stellar pair composed of a $\sim$0.8 $M_\odot$ and a $\sim$0.4 $M_\odot$ dwarf with a projected separation of $\sim$130 AU. Hence, HD 40979 is a new member of the small group of planet hosting triple star systems known today. Finally, we present our observations of the planet hosting subgiant HD 27442, which has a co-moving companion with a projected separation of $\sim$240 AU. The V- and H-band magnitudes of this faint companion are fully consistent with a relatively young, hot white dwarf, with an effective temperature of $\sim$14400K, and cooling age of $\sim$220 Myr. With follow-up spectroscopy which shows Hydrogen absorption features in its optical and infrared spectra, we confirm the white dwarf nature of this companion. With the subgiant exoplanet host star and its white-dwarf companion, HD 27442AB is the most evolved planet hosting stellar system presently known.

Preliminary Orbit and Masses of the Nearby Binary L Dwarf GJ 1001BC

D.A. Golimowski[1], D. Minniti[2], T.J. Henry[3], and H.C. Ford[1]

[1] Johns Hopkins University, Baltimore, MD, United States,
[2] Pontificia Universidad Catolica de Chile, Santiago, Chile,
[3] Georgia State University, Atlanta, GA, United States

We present preliminary results of a continuing VLT program to map the orbit of the nearby binary L4.5 dwarf GJ 1001BC (LHS 102BC). Since discovering its duplicity in 2002 and 2003 using HST's NICMOS and ACS, we have obtained high-resolution images of GJ 1001BC at three epochs between October 2004 and November 2005, using the NAOS/CONICA system at VLT-UT4 (Yepun). Our HST and VLT images cover $\sim$75% of GJ 1001BC's 4-year orbit. A least-squares fit of a Keplerian orbit yields a combined binary mass of 0.100 $\pm$ 0.026 $M_\odot$ for a tentative parallactic distance measurement of 13.0 $\pm$ 0.7 pc to the M dwarf GJ 1001A. Hypothetically assuming a 3:2 mass ratio for the nearly equal-luminosity L dwarfs, we estimate masses of 0.060 $\pm$ 0.016 $M_\odot$ and 0.040 $\pm$ 0.010 $M_\odot$ for GJ 1001B and C, respectively. If these preliminary values are sustained by our continuing orbit and parallax measurements, then GJ 1001C will be the least massive L dwarf for which a dynamical mass has been measured.

Near-Infrared Light Curves of a Young, Eclipsing Binary of Brown Dwarfs

Y. Gómez Maqueo Chew[1], K.G. Stassun[1], R. Mathieu[2], and L.P. Vaz[3],

[1] *Vanderbilt University, Nashville, TN, United States,*
[2] *University of Wisconsin, Madison, WI, United States,*
[3] *Universidade Federal de Minas Gerais, Belo Horizonte, Brazil*

We present the near-infrared light curves for an eclipsing binary system, 2MASS J05352184−0546085, in which both components are brown dwarfs (Stassun *et al.* 2006). The system is a member of the Orion Nebula Cluster, and therefore has a likely age of only a few million years.

We model light curves using a Wilson-Devinney based code to derive fundamental system properties, including effective temperatures and radii. Our analysis includes *JHK* light curves obtained with the 1.3-m SMARTS telescope at CTIO in Chile.

A thorough spectroscopic and photometric analysis of eclipsing binary systems yields highly accurate properties of the system and of its components, in such a manner that they are determined independently of their distance and other assumptions. The masses of both components, 0.054 and 0.034 $M_\odot$, have been measured with accuracies better than 10%.

This system is of particular interest because it provides with the measurements of the two least massive of pre-main-sequence objects known to date, and consequently provides useful data for testing the predictions of current early evolution and star formation models.

RZ Cassiopeia: an Eclipsing Binary with a Pulsating Component

Alex Golovin[1] and Elena Pavlenko[2]

[1] *Kiyv National Taras Shevchenko University; visiting astronomer of the Crimean Astrophysical Observatory, Kyiv, Ukraine,*
[2] *Crimean Astrophysical Observatory, Nauchny, Ukraine*

We report time-resolved *VR*-band CCD photometry of the eclipsing binary RZ Cas obtained with a 38-cm Cassegrain telescope at the Crimean Astrophysical Observatory during July 2004 – October 2005.

Obtained lightcurves clearly demonstrates rapid pulsations with a period about 22 minutes. Periodogram analysis of such oscillations also is reported. On the 12, January, 2005 we observed rapid variability with higher amplitude ($\sim 0^m.1$) that, perhaps, may be interpreted as high-mass-transfer-rate event and inhomogeneity of accretion stream.

Follow-up observations (both, photometric and spectroscopic) of RZ Cas are strictly desirable for more detailed study of such event.

Orbital Period Changes of OB-type Contact Binaries and Their Implications for the Triplicity, Formation and Evolution of This Type of Binary Star

S.-B. Qian[1], J.M. Kreiner[2], L. Liu[1], J.-J. He[1], L.-Y. Zhu[1], J.-Z. Yuan[1], and Z.-B. Dai[1]

[1] *Yunnan Observatory, Kunming, China,*
[2] *Cracow Pedagogical University, Cracow, Poland*

Orbital period variations of 9 well-observed OB-type contact binary stars, LY Aur, BH Cen, V382 Cyg, V729 Cyg, AW Lac, TU Mus, RZ Pyx, V701 Sco and CT Tau, are investigated in detail. Of the nine systems, V701 Sco and CT Tau are two contact binaries containing twin components with a mass ratio of unity, LY Aur and V729 Cyg have the longest period among contact binary stars (P=4.0 and 6.6 days, respectively), and BH Cen and V701 Sco are members of the two extremely young galactic clusters IC 2994 and NGC 6383. It was discovered that, apart from the two systems with twin components (V701 Sco and CT Tau), the orbital periods of the remaining 7 binary stars show a long-term increase. This is different from the situations of the late-type (W UMa-type) contact binaries where both secular period increase and decrease are usually encountered, indicating that magnetic fields may play an important role in causing the long-term period decrease of W UMa-type contact binary stars. The fact that no long-term continuous period variations were found for V701 Sco and CT Tau may suggest that contact binaries with twin components can be in an equilibrium. Based on the rates of period changes (dP/dt) of the 7 sample binary stars, statistical relations between dP/dt and orbital period (P) and the mean density of the secondary component were found. Our results suggest that the period increases of the short-period systems (P<2 days) may be mainly caused by a mass transfer from the less massive component to the more massive one, while for the long-period ones (P>2 days), LY Aur and V729 Cyg, their period increases may be resulted from a combination of stellar wind and mass transfer from the secondary to the primary.

Meanwhile, cyclic period changes are found for all of the nine binary systems. Those periodic variations can be plausibly explained as the results of light-travel time effects suggesting that they are triple systems. The astrophysical parameters of the tertiary components in the nine systems have been determined. The tertiary components in the seven binaries, BH Cen, V382 Cyg, AW Lac, TU Mus, RZ Pyx, V701 Sco and CT Tau, may be invisible, while those in LY Aur and V729 Cyg may be the fainter visual companions in the two systems. It is possible that the tertiary components in those binaries played an important role for the formations and evolutions of the contact configurations by bringing angular momentum out from the central systems. Thus they have short initial periods and can evolve into contact configurations in a short timescale.

Quasi-Molecular K-H$_2$ Absorption as an Alternative to the Resurgence of CaH Bands in the Spectra of T-Type Dwarfs: is the Cloud-Clearing Scheme at Stake?

F. Allard[1], N.F. Allard[2], C.M.S. Johnas[3], P.H. Hauschildt[3], D. Homeier[4], J.K. Kielkopf[5], and F. Spiegelman[6]

[1] *Centre de Recherche Astronomique de Lyon, Lyon, France,*
[2] *Institut d'Astrophysique de Paris, Paris, France,*
[3] *Hamburger Strenwarte, Hamburg, Germany,*
[4] *Institut für Astrophysik Götingen, Götingen, Germany,*
[5] *University of Louisville, Louisville, KY, United States,*
[6] *Universite Paul Sabatier, Toulouse, France*

As brown dwarfs cool off with time, their atmospheres become denser and more transparent, allowing the emitted thermal flux to escape from deeper atmospheric layers. Burgasser *et al.* (2002) have investigated and classified the red spectra of T dwarfs in a spectral sequence where a resurgence of the hydride bands, after disappearing in the M to L spectral transition, occur between the late L to T before disappearing again in the late T dwarfs. CaH for example is identified in mid-T dwarfs at around 0.7μm (Burgassser 2003). The authors explain this resurgence by a cloud-clearing scheme where holes would allow to see the CaH from deeper enriched layers, while it is settled out from the uppermost atmospheric layers seen on the rest of the brown dwarf surface.

We present the first synthetic spectra of T dwarfs including a semi-classical modelling of the pressure broadening of alkalis lines (Na I D, Li I, K I, Rb I, and Cs I fundamental resonance doublets) by molecular hydrogen and helium, the most important species in these atmospheres. We compare the models to the T dwarfs red optical spectra of Burgasser *et al.* (2003) and we find that the 0.7μm feature has been wrongly identified to CaH. In particular, the very strong KI resonance transition doublet at 0.77μm explains by itself this absorption feature by producing a quasi-molecular satellite absorption feature at this wavelength. The strength of this satellite is very sensitive to the density of perturbers in the lower photosphere and to the background opacity provided by the Na I D red wing, which explains naturally both its apparition in late L dwarfs and its vanishing in late T dwarfs.

We find in conclusion that no cloud-clearing scheme or non-equilibrium processes is necessary to explain this absorption feature, and the evolution of the red optical spectrum of T dwarfs. And this should teach us caution about these atmospheres often too enthusiastly considered planetary. MHR 3D convection models are nevertheless underway to estimate the likelihood of cloud-clearings in late L and T dwarfs.

Search for Exoplanets and Variables in the Open Cluster NGC 381

J.-H. Hu and W.-H. Ip

Institute of Astronomy, National Central University, Jhongli City, Taiwan

We present the result of a search for exoplanets and variable stars in the open cluster star fields, NGC 381, with the Lulin One-meter Telescope (LOT), Taiwan. The main scientific goal is to use time series CCD photometry measurements to detect exoplanets via transit effects. Observations of open clusters would give us important information on the formation of planets in different stellar environments. The secondary scientific goal is to discover and study variable stars with the similar data analysis process. Several program stellar clusters have been observed. Four variable candidates were discovered in the NGC 381 star field. One of them was identified as one of the suspected cluster member, CT 123, by Crinklaw & Talbert, (1988).

The Binary Star Gamma Persei — Bright, but Ill-Understood

R.E.M. Griffin

HIA/DAO, Victoria, Canada

Gamma Per looks like an unexceptional composite-spectrum binary, most often classified as G8 III + A3 V. Nevertheless, early studies indicated that the masses of the component stars were on the high side, until Popper showed that the component stars were not over-massive so much as over-luminous. Pourbaix then went further, constraining the masses on the basis of evolutionary theory and largely removing any discrepancies in mass. However, recent CCD spectroscopy, from which the secondary spectrum has been fully isolated by a spectrum subtraction technique, finds that the masses of the individual components ($3.9\ M_\odot$ for the giant and $2.5\ M_\odot$ for the dwarf) are indeed greater than would be expected from their spectral types, thus re-opening the question of the origin and evolution of this somewhat unusual system.

Analyses of more than 25 composite-spectrum systems by spectrum subtraction have now produced enough results that a picture is beginning to emerge — and it is an unexpected one: like γ Per, many other systems also have secondary stars that have started evolving away from the main-sequence, thus challenging us to explain how the two stars can have attained that state of rapid evolution and yet be of the same age. The importance of the new analysis of gamma Per is the high precision with which the observed stellar parameters of one of these systems can be determined.

The Population of Close Binaries Dynamically Formed in Hierarchical Triple Systems, with Application to Extrasolar Planets

Daniel C. Fabrycky

Princeton University, Princeton, NJ, United States

Because of the large radii of pre-main-sequence stars, the current separation of close binaries was not likely established at their formation. The secular perturbation of a third star orbiting a binary at high inclination can cause its eccentricity to grow close to unity as angular momentum is gained by the third star. Close pericenter passages at high eccentricity cause dissipative tides, and the loss of orbital energy can result in a close binary. A population of triple stars is considered, of which some will undergo this evolution. The distribution of binary period and of relative inclination of the third star's orbit to the binary plane is derived using a Monte-Carlo method. The averaged (secular) equations of motion are integrated, including a model for tidal dissipation and extra precession due to the tidal and spin distortion of the stars and general relativity. The predicted period distribution matches reasonably to a set of spectroscopic binaries with known companions: there is an excess of such binaries with periods below about 5 days relative to the population of true binary systems. Reasonable statistics for the relative inclinations of such triple stars will be built up by optical interferometers in the near future, and these observations can check the predicted distribution.

This mechanism is also applicable to extrasolar planets hosted by binary stars, causing them to migrate to short periods. For planets with periods less than 20 days, there is a paucity of massive (minimum mass > 2 Jupiter mass) planets orbiting single stars, yet there are three such planets (Tau Boo b, HD 195019b, Gl 86b) in binaries. Therefore the single-star migration mechanism is apparently sensitive to planetary mass, whereas the binary-star migration mechanism is not; the latter matches the present study. Furthermore, HD 195019b and Gl 86b have circularized orbits, whereas single-star planets with comparable periods show a range of eccentricities; planetary radius inflation resulting from the epoch of high eccentricity may have led to this efficient circularization.

The UV Spectrum of the Binary Star 88 Her: Activity Cycles in the Circumstellar Envelope

A. Granada[1], L. Cidale[1,2], and C. Quiroga[1,2]

[1] *Falcultad de Ciencias Astronómicas y Geofísicas, UNLP, Argentina,*
[2] *Instituto de Astrofísica La Plata, CONICET*

Since its discovery as a variable star, 88 Her has undergone three long-term photometric variation cycles with transitions between Be-shell and normal B phases. From the spectroscopic study of fifteen high resolution spectra obtained by the IUE satellite between 1981 and 1992 we were able to set parameters such as optical depths and location of line forming regions. We also found that the periodic radial velocity variations of UV Fe II lines agree with the binary orbital period of 86.7 days (Harmanec *et al.*, 1974) and that the line absorption depth variations have a cycle of about 1560 days. Our aims are to relate the properties of the circumstellar envelope of 88 Her to the spectroscopic variability observed in Fe II and Mg II UV lines, and to understand the mechanisms which cause them, as well as the influence that binarity has on them.

Session 6: Binary Stars as Critical Tools:
Evolutionary models for binary and multiple stars

Figure 1. (top left to bottom right) Speakers Cathie Clarke, Peter Eggleton, Dmitry Bisikalo, Hans Zinnecker, Aliz Derekas, and Silvia Catalán.

Figure 2. (top) Panos Niarchos talking with Patricia Lampens. (lower left) Detail of the Habsburg monument to Franciscus II. (lower right) Týnský church with the tomb of Tycho Brahe (1546-1601) at Staroměstské Square.

Binary Stars as Critical Tools & Tests
in Contemporary Astrophysics
Proceedings IAU Symposium No. 240, 2006
W.I. Hartkopf, E.F. Guinan & P. Harmanec, eds.

© 2007 International Astronomical Union
doi:10.1017/S1743921307004279

The Formation of Binary Stars

C.J. Clarke

Institute of Astronomy, Madingley Road, Cambridge, CB3 OHA, U.K.
email: cclarke@ast.cam.ac.uk

Abstract. I argue that binary star statistics offer the best observational constraints on current hydrodynamical simulations of star forming clusters. In these simulations, clusters form hierarchically from the bottom up, and dynamical interactions, mediated by the presence of circumstellar material, play a vital role at the lowest (few body) level of the hierarchy. Such a scenario produces a rich array of complex multiple systems whose properties are in many respects consistent with observations. I however highlight two areas of current disagreement: the simulations over-produce low mass single stars and under-produce binaries with low mass ratios. It is currently unclear to what extent these shortcomings reflect numerical issues and to what extent the omission of relevant physical processes. I conclude with a theorist's wish list for observational diagnostics that would most meaningfully constrain future modeling efforts.

1. Introduction

A decade ago, any review of binary formation mechanisms would have started by setting out a list of possible options (fission, prompt initial fragmentation, disc fragmentation, capture) and would have assessed the feasibility of each, with some speculative attempt to map each of the scenarios onto a set of observational characteristics (see, e.g., Clarke 1995)., We can now say with some confidence that fission is unlikely to be viable (see Tohline & Durisen 2001 for the most authoritative recent investigation of this issue). Binary formation through star-disc capture almost certainly occurs in dense young clusters like the Orion Nebula Cluster, but since such encounters are suffered by less than ~10% of all stars, even in the very densest regions of the cluster core (see Scally & Clarke 2001, Olczak *et al.* 2006), this cannot be a major route for binary production. We are therefore left with fragmentation at various stages of the collapse process, plus star-disc (or dynamical) capture in the very early (deeply embedded) stage of star cluster formation.

The reason why it is not fruitful to try and isolate the observational properties of each of these production routes is that the field has moved on from the situation where simulations started with idealised initial conditions, set up in order to create a binary through one of the aforementioned methods. Instead, we are in era where it is possible to simulate the formation of *entire clusters*. In this case, the creation route of any particular binary is not guided by the prejudices of the simulator, but emerges from the complicated non-linear development of the Jeans instability in a turbulent star forming cloud. Examination, *a posteriori* of how a particular binary came into being shows that a combination of mechanisms is usually involved in the creation of a single system. Thus we are fast abandoning the search for *the* binary formation mechanism and are instead interrogating the simulations, in order to see whether they can reproduce a range of observational statistics on binary stars.

It is not difficult to understand why the subject has advanced in this way. In addition to the obvious advances in the speed of computational hardware, there have been some algorithmic advances which are of particular importance to the binary formation

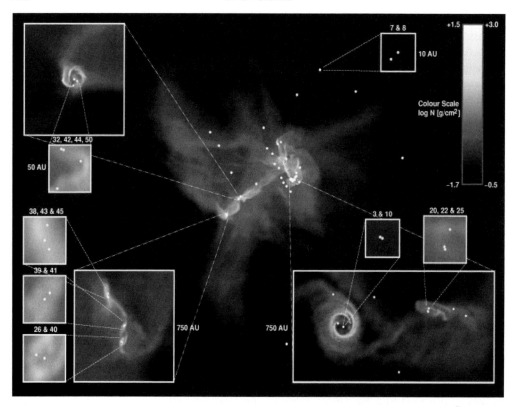

Figure 1. The locations and states of seven close binary systems created after 4×10^5 years in the simulations of Bate *et al.* (2002). The insets (labeled with the dimensions of the binaries in AU) show the variety of multiple systems and distributions of circumstellar material produced.

problem. Chief amongst these is the development of 'sink particles', these being point masses that interact gravitationally and through appropriate hydrodynamical boundary conditions with the rest of the flow but are themselves excised from the domain of detailed computation. This is an essential measure in such simulations, which produce a large number of collapsed regions ('stars') whose internal structure is not itself of interest but which, if not treated in this way, would impose a prohibitive computational load. The implementation of such particles is relatively straightforward in the case of a Lagrangian scheme such as SPH (Bate *et al.* 1995) but has only recently been successfully deployed in Eulerian (grid based) codes (Krumholz *et al.* 2005).

Figure 1 is a snapshot from the simulations of Bate *et al.* (2002) of the fragmentation of a turbulent molecular cloud which singles out some of the binaries and multiple systems created after $\sim 10^5$ years. It is notable that the basic star formation mode is one of small N clusters, so there is plenty of scope for the kind of small N dynamical interactions (exchange, binary hardening, ejection of singles) first explored in the N body context by van Albada (1968). It is, however, also obvious that gas dynamics plays an important role; we see tidal structures in the inset images that evidence disc-disc interactions, as well as massive circumbinary discs which are themselves subject to further fragmentation. The reader is directed to Bate *et al.* (2002, 2003) for a detailed analysis of the creation routes of the multiple systems formed.

Although such simulations generate eye-catching and broadly observationally credible results, it is obviously desirable to check their fidelity through as many routes as possible.

Potential sources of error include the use of incorrect initial conditions or, more likely, the omission of important physical effects such as magnetic fields and a realistic treatment of radiative transfer (see, however, Whitehouse *et al.* 2005, Dale *et al.* 2005, Price & Monaghan 2005 for recent progress in incorporating such effects in SPH codes). Numerical errors are an ever present possibility, so it is encouraging that there is the prospect that Adaptive Mesh Refinement (AMR) codes may develop to the point where they can provide a credible cross-check on results obtained with existing SPH simulations.

The most obvious zeroth order observational check is whether the simulations can reproduce the observed initial mass function (IMF), but this they do easily. In fact the rather featureless form of the IMF makes it too easy to replicate (as evidenced by the plethora of 'successful' IMF theories over the years). Binary statistics are much harder to replicate by any theory, however, on account of the richness of the diagnostic information they contain. It would appear unlikely that any model could accidentally replicate such a sweep of observational diagnostics as the binary and multiplicity fraction, the mass ratio distribution, the eccentricity and period distribution, all as a function of primary mass; indeed, we should be reassured by the fact that current models are conspicuously lacking in several of these areas. *Thus all those engaged in the painstaking task of accumulating binary star statistics should be reassured that, ultimately, their work will provide the most stringent test of star formation theories.*

In this contribution, I will focus on the two properties of binary stars which are set during the main accretion phase of star formation i.e., over the period (< 1 Myr) covered by the simulations: the multiplicity (as a function of primary mass) and distribution of binary mass ratios. Further important information is of course also preserved in the period and eccentricity distributions of binaries. However, these latter quantities can be substantially modified during later pre-main sequence stages. i.e., long after the masses of the constituent bodies have been set. Even at this late stage, it is well known (e.g., Artymowicz *et al.* 1991) that a small quantity of gas in circumbinary orbit can extract significant angular momentum from the binary and hence modify the orbital elements significantly.

2. Multiplicity as a function of stellar mass

Figure 2 provides a schematic illustration of the sorts of 'system architectures' commonly produced by turbulent fragmentation calculations at the end of the accretion phase. These derive from the study of Delgado *et al.* (2004), which simulated multiple realisations of turbulent $5M_\odot$ cores. This focus on small scale cores implies that of course one misses the interactions on larger scales that are captured by simulations such as Bate *et al.* (2002), which model a $50M_\odot$ cloud (see Figure 1). Nevertheless, the computational savings afforded by such small scale simulations means that it is possible, at the same numerical resolution as used by Bate *et al.*, to obtain a statistically meaningful set of multiple star systems. In contrast to the Bate *et al.* calculations, these multiples can then be integrated to the point where a large fraction of the gas has accreted on to the stars and are thereafter integrated to the point of dynamical stability as a pure N-body system.

From Figure 2, it is immediately obvious that the initial star formation process involves the creation of high order quasi-hierarchical multiples. The 13 multiples generated in the simulations have membership numbers roughly uniformly distributed in the range $N = 2 - 7$, where the membership number excludes 'outliers' (see below). A remarkable feature of the multiples shown in Figure 2 is the tendency for each constituent subsystem to apportion its mass roughly equally between components (for example, the

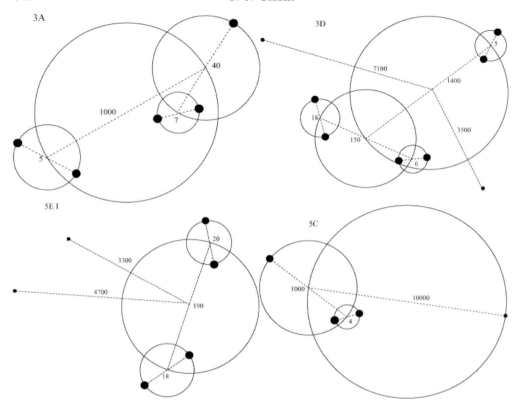

Figure 2. Schematic diagrams (not to scale) of some common system architectures produced in the simulations of Delgado *et al.* (2004) after ~ 1 Myr. The numbers refer to separations in AU.

system shown in the upper left consists of a nearly equal mass 7 AU binary with total mass nearly equal to its triple companion at 40 AU, whilst the total mass of the triple is roughly the same as the total mass of the (roughly equal mass ratio) 5 AU binary at 1000 AU). Evidently, some process (see Section 3) is ensuring a roughly equitable mass distribution at every level of the hierarchy.

The exception to this rule lies in the most distant members of the hierarchy, i.e., the so-called 'outliers' which are single stars located at $\sim 3000 - 10000$ AU from the centre of mass of the multiple. There are typically $2 - 3$ such outliers per multiple at the end of the accretion stage, and these outliers are always of very low mass (usually in the brown dwarf regime). Their creation is easily understood in terms of the dynamics of small N multiple systems: low mass objects acquire high kinetic energy at the expense of high mass stars in bound subsystems and are often ejected from their natal cores (Reipurth & Clarke 2001). Sometimes such interactions however impart insufficient energy to unbind the low mass object completely and they thus end up in weakly bound orbits with apocentres at many times the typical size of the parent multiple.

The above description refers to the end of the hydrodynamical evolution stage (i.e., after about a Myr) and is thus applicable to the predicted multiplicity statistics of the youngest pre-main sequence stars. After a further ~ 10 Myr of purely N-body integration, the situation looks very different: although there are still some high order multiples (up to $N = 6$) which are apparently dynamically stable hierarchies (according to the criteria of Eggleton & Kiseleva 1995), the majority of systems have been broken up into their constituent binary stars. It is also notable that only 10% of the 'outliers' present at ~ 1 Myr remain bound at 10 Myr: as expected, such low mass and weakly bound objects are liable to escape during the orbital reconfiguration of the parent multiples.

In terms of commonly used descriptors of multiple star properties, Delgado *et al.* (2004) found that the multiplicity fraction (defined as the number of multiples divided by the number of multiples plus singles) remains roughly constant (at ~ 0.2) during the phase of dynamical break up (i.e., between $\sim 1 - 10$ Myr). The reason that this is the case — despite the drastic reconfiguration of the multiples over that period — is that the breaking down of multiples into (mainly) singles and binaries increases the number of multiples and singles by about the same factor. The companion fraction, on the other hand, (i.e., the mean number of companions per system) changes strongly (from about 1 to about 0.3) through the same reconfiguration. This illustrates the care that is needed when interpreting these quantities: although the companion fraction better captures the fact that high order multiples are being broken up, the values are misleading unless one thinks carefully about what is involved (i.e., a companion frequency of 1, which might suggest that most systems contain two stars, in this case represents the situation where the majority of stars are either in high order multiples or are singles).

The general outcome of these models, therefore, is that higher order multiple systems are abundant among young stars, but that these tend to break up as a result of dynamical interactions on timescales of ~ 10 Myr. This is in broad qualitative agreement with the recent multiplicity studies of young stars by Correia *et al.* (2006), which find some evidence that higher order systems are favoured in pre-main sequence, compared with main sequence, binaries. Correia *et al.* however point out that their observed abundances of higher order systems (compared with binaries) are somewhat lower than predicted by the simulations.

An obvious discrepancy, as pointed out by Goodwin & Kroupa (2005) is that the multiplicity fraction produced by the simulations is much lower than that observed, i.e., about 0.2 compared with values of $0.5 - 0.6$ in the observational literature (see Duquennoy & Mayor 1991, Tokovinin & Smekhov 2002). This relatively low multiplicity fraction is a consequence of the large numbers of stars produced per core (typically > 3). [Note, however, that early estimates (McDonald & Clarke 1993) relating number of stars per core to multiplicity fraction were overly pessimistic, since they assumed, as is the case for purely N-body interactions, that only one binary is formed per cluster. In the hydrodynamic simulations, we have seen that the creation of high order multiples, which then decay dynamically, actually produces several binaries per star forming core].

If we scrutinise the simulations more closely (Figure 3) we see that the discrepancy is entirely at the low mass end (indeed, the simulations somewhat over-produce binaries at the high mass end (around $1 M_\odot$), although this would be remedied by modeling larger cores so that higher mass stars would be able to disrupt some of the solar mass binaries). However, the model results that best fit the binary fraction in the stellar domain (those where 60% of the core mass is accreted onto the stars before the multiples are evolved as N-body systems), produce *no* binaries in the brown dwarf regime. The observational value is however in the range $10 - 20\%$ (Martin *et al.* 2003). Since stars and brown dwarfs are formed in roughly equal numbers, one can understand why the simulations can match the stellar binary data and yet produce a global multiplicity fraction which is so low.

The reason for the over-production of single brown dwarfs in the simulations is probably due to the fact that the discs in the simulations can fragment too easily. Despite some unpublished claims that such fragmentation was a result of under-resolution in the SPH calculations, further convergence tests have not supported this assertion (Bonnell & Bate,

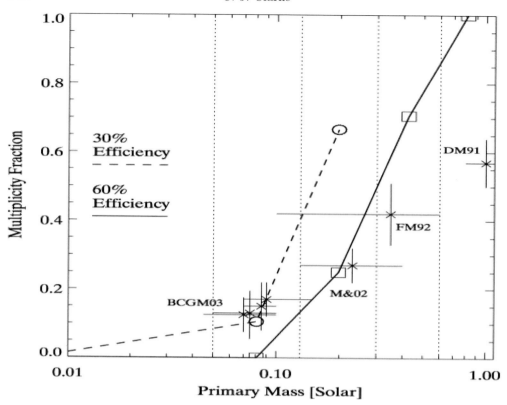

Figure 3. Multiplicity fraction as a function of primary mass produced by the simulations of Delgado *et al.* (2004). Observational data is labeled according to its source, i.e., DM91 = Duquennoy & Mayor (1991), FM92 = Fischer & Marcy (1992), M&02 = Marchal *et al.* (2003), BCGM03 = Bouy *et al.* (2003), Close *et al.* (2003), Gizis *et al.* (2003), Martin *et al.* (2003).

private communication). Instead, it is more likely that excessive disc fragmentation is a result of the unrealistic treatment of the thermal physics. Following Gammie (2001) is it well known that discs should only fragment if the local cooling timescale is less than around three times the local dynamical timescale. Given the cooling rates to be expected in circumstellar discs when radiative cooling with a realistic opacity law is assumed, this criterion is only expected to be met at large radii in the disc (i.e., at > 100 AU; Rafikov 2002) †. The binary formation simulations discussed here (i.e., those of Bate *et al.* 2002 and Delgado *et al.* 2004; see also Wadsley *et al.* 2006) instead employ an isothermal equation of state, until the gas has collapsed to very high densities. An isothermal equation of state however implies an effectively *zero* cooling timescale, since *pdV* work can be radiated away instantly; thus it is unsurprising that the discs in these simulations are able to fragment at essentially all radii. In future, it will be desirable to explore binary formation in simulations which, instead of using an isothermal equation of state, treat cooling of the disc gas via radiative diffusion (see Whitehouse *et al.* 2005).

---

† Indeed, according to the recent simulations of Lodato *et al.* (2006), discs with longer cooling timescales do not fragment even when subject to large amplitude impulsive interactions with passing stars.

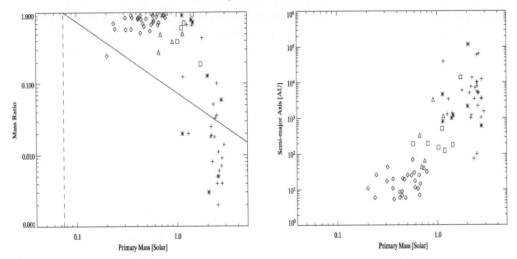

Figure 4. Mass ratios and separations of multiples as a function of 'primary' mass, from the simulations of Delgado *et al.* (2004). Here mass ratio q is the ratio of the mass of the outermost object of the subsystem under consideration to the total mass of all the objects interior to it (i.e., for a triple [(1-2)-3], q=m3/(m1 +m2)). Primary mass refers to the total mass of all the objects interior to that for which the mass ratio is being calculated. Diamonds correspond to binaries (either independent or bound to larger structures), triangles denote triples, squares quadruples, asterisks quintuples and crosses higher-order multiples.

3. The binary mass ratio distribution.

Figure 4 illustrates an obvious deficiency of all hydrodynamical models to date - the tendency not to produce binaries with low mass ratios. This tendency is shown at every level of the clustering hierarchy, as was noted in Section 2 above. Figure 4 shows that binaries are almost never produced in the simulations with $q = M_2/M_1 < 0.5$, with the exception of the very low mass 'outliers' which remain in wide orbits (~ 5000 AU) for about 10 Myr or so, before being dynamically ejected from the system. The under-production of low q pairs appears to be a generic property of turbulent fragmentation calculations (see also Goodwin *et al.* 2004) and appears to be a consequence of accretion of gas upon the proto-binary. For any plausible history of binary assembly, the material falling in upon a newly formed binary is generally of high specific angular momentum compared with the binary itself, and in these circumstances it is widely claimed (Artymowicz 1983, Bate & Bonnell 1997) that such gas will preferentially accrete upon the secondary. In this case, binaries evolve rapidly towards unit mass ratio, regardless of their initial mass ratio (Bate 2000). In the SPH simulations, this certainly appears to be the case and implies that any solution to the problem should relate to this accretion phase rather than to the creation of the binary.

Before examining possible solutions, it is important to be clear that there is indeed an observational problem. It is often cited that among the closer binaries (i.e., those with a period of less than 2000 days), the q distribution rises gently towards unit mass ratio (Mazeh *et al.* 1992) and is therefore not grossly inconsistent with the output of the simulations. However, these systems represent a small fraction of all binaries (recall that the median binary period is ~ 200 years) and *for the binary population as a whole* it is clear that the mass ratio distribution rises quite steeply towards low q (Duquennoy & Mayor 1991, Evans 1995, Woitas *et al.* 2001, Prato *et al.* 2002, Halbwachs *et al.* 2003). Given that in the case of wider pairs, this dataset consists of binaries culled from the literature, and which can therefore be expected to be selectively incomplete at *low q*, it is hard to escape the conclusion that low q pairs are abundant in nature.

It has recently been suggested, however, that the lack of low q binaries in the simulations might be an artifact of the SPH calculations. Ochi *et al.* (2005) used a grid-based code to re-examine the (SPH) calculations of Bate & Bonnell (1997), which modeled gaseous accretion onto a protobinary. Both codes agree that in the case of accretion of gas whose specific angular momentum is larger than that of the protobinary, the accreted gas falls preferentially into the Roche lobe of the *secondary*, which is to be expected, given the larger displacement of the secondary from the system centre of mass. In the SPH calculations, this gas is retained in the secondary's Roche lobe and then accretes onto the secondary star. In the grid-based calculation, however, gas that enters the secondary's Roche lobe near the L2 point then circulates half an orbit around the secondary and then passes through the L1 point into the primary's Roche lobe, where it is then accreted. Thus in these latter calculations, the mass ratio of the binary can decline even when the inflowing material has high specific angular momentum.

It is not easy to assess which of these simulations is correct. Proponents of the grid-based code can legitimately question whether it is not the large viscosity in the SPH calculations which artificially shrinks the orbits of gas in the secondary's Roche lobe and thus prevents the transfer of material into the primary's Roche lobe. On the other hand, it can also be reasonably pointed out that the grid-based code models rather warm gas which can be deflected too readily by pressure gradients in the low gravity environs of the L1 point. Detailed follow up of these issues is under way, but is not currently pointing to an obvious resolution. If the lack of low q systems turns out *not* to be a numerical artifact, it will be necessary to take a harder look at the assumptions (initial conditions, physics modeled) behind the simulations.

4. Summary

Binary star statistics offer the best opportunities to check the veracity of star formation simulations. Current simulations, in which star formation proceeds in a highly interactive small N cluster mode, provide the right ingredients for generating a wealth of binaries and higher order multiple systems, in broad agreement with observations. This allows one to focus on areas of disagreement, and here we have highlighted the over-production of low mass single stars and the under-production of binaries with low mass ratios.

There are also a number of areas where current observational datasets are too incomplete to test the simulations. A theorist's wish list in this regard would run as follows.

i) The assembly of the sort of binary statistics that are currently available for G dwarfs to a range of lower mass bins,

ii) A better collation of the high order multiplicity statistics and of how mass is distributed at each level of the hierarchy, and

iii) An understanding of how the latter properties vary as a function of age. According to the models, the pre-main sequence stage should involve significant reconfiguration (breaking up) of high order multiple systems. In particular, it is predicted that rich young multiple systems should be accompanied by low mass outliers in weakly bound orbits. Evidently, deep imaging of the environs of pre-main sequence multiple systems is required to check out this prediction.

References

Artymowicz, P., 1983, *Acta Astron.* 33, 223

Artymowicz, P. Clarke, C., Lubow, S., & Pringle, J., 1991, *ApJL* 370, 35

Bate, M.R., 2000, *MNRAS* 314, 33.

Bate, M. & Bonnell, I., 1997, *MNRAS* 285, 33

Bate, M., Bonnell, I., & Price, N., 1995, *MNRAS* 277, 362

Bate, M., Bonnell, I., & Bromm, V., 2002, *MNRAS* 336, 705.

Bate, M., Bonnell, I., & Bromm, V., 2003, *MNRAS* 339, 577.

Bouy H., Brandner W., Martn E. L., Delfosse X., Allard F., & Basri G., 2003, *AJ*, 126, 1526

Clarke, C., 1995, *Ap&SS* 223, 73

Close, L. M., Siegler N., Freed M., & Biller B., 2003, *ApJ*, 587, 407

Correia, S., Zinnecker, H., Ratzka, T., & Sterzik, M., 2006. *A&A* 459, 909.

Dale, J., Bonnell, I., Clarke, C., & Bate, M., 2005, *MNRAS* 358, 291

Delgado, E., Clarke, C., Bate, M., & Hodgkin, S., 2004, *MNRAS* 351, 671

Duquennoy, A. & Mayor, M., 1991, *A&A* 248, 485

Eggleton, P. & Kiseleva, L., 1995, *ApJ* 455, 640

Evans, N., 1995, *ApJ* 445, 393

Gammie, C., 2001, *ApJ* 553, 174

Gizis J.E., Reid I. N., Knapp G.R., Liebert J., Kirkpatrick J.D., Koerner D.W., & Burgasser A.J., 2003, *AJ*, 125, 3302

Goodwin, S. & Kroupa, P., 2005, *A&A* 439, 565

Goodwin, S., Whitworth, A., & Ward-Thompson, D., 2004, *A&A* 423, 169

Halbwachs, J., Mayor, M., Udry, S., & Arenau, F., 2003, *A&A* 397, 159

Krumholz, M., McKee, C., & Klein, R., 2004, *ApJ* 611, 399

Lodato, G., Meru, F., Clarke, C., & Rice, W.K., 2006, *MNRAS* in press

Martin. E., Barrado y Navescues, D., Baraffe, I., Bouy, H., & Dahm, S., 2003, *ApJ* 594, 525

Mazeh, T., Goldberg, D., Duquennoy, A., & Mayor, M., 1992, *ApJ* 401, 365

McDonald, J. & Clarke, C., 1993, *MNRAS* 262, 800

Ochi, Y., Sugimoto, K., & Hanawa, T., 2005, *ApJ* 623, 922

Olczak, C., Pfalzner, S., & Spurzem, R., 2005, *ApJ* 642, 1140

Prato, L., Simon, M., Mazeh, T., McLean, I., Norman, D., & Zucker, S., 2002, *ApJ* 569, 863

Price, D. & Monaghan, J., 2005, *MNRAS* 364, 384

Rafikov, R., 2005, *ApJ* 621, L69

Reipurth, B. & Clarke, C., 2001, *AJ* 122, 432

Scally, A. & Clarke, C., 2001, *MNRAS* 325, 449

Tohline, J. & Durisen, R., 2001, *The Formation of Binary Stars*, Proceedings of IAU Symp. 200, held 10-15 April 2000, in Potsdam, Germany, Edited by Hans Zinnecker and Robert D. Mathieu, 2001, p. 40.

Tokovinin, A. & Smekhov, M., 2002, *A&A* 382, 118

van Albada, T., 1968, *Bull. Astron. Inst. Neth.* 19, 479

Shen, S. & Wadsley, J., 2006, *ApJ* 651, L145.

Whitehouse, S., Bate, M., & Monaghan, J., 2005, *MNRAS* 364, 1367

Woitas, J., Leinert, C., & Kohler, R., 2001, *A&A* 376, 982

Discussion

ANDREI TOKOVININ: What is the influence of high angular momentum (high beta) on the stability of resulting multiples and ejections? Did you model accretion of matter with different $\vec{J}$ during formation of a multiple system?

CLARKE: The simulations do not impose any net rotation in the star forming core, but the initial gas has non-zero vorticity and thus has some degree of rotational support. From the size scales of the resulting multiples (Figure 1) one can judge that, since the initial core size is $\sim 10^4$ AU, $\beta \sim 0.1$. We have not attempted to *impose* a range of β values on our initial conditions, so cannot comment on any systematic dependence of the multiple star properties on β. Likewise, the accretion of material with different $\vec{J}$ is a natural outcome of the simulations. As we discuss, there is a tendency for late arriving material to have high $\vec{J}$ and, in these SPH simulations, for the mass ratios to increase at every level of the hierarchy as a result.

Binary Stars as Critical Tools & Tests
in Contemporary Astrophysics
Proceedings IAU Symposium No. 240, 2006
W.I. Hartkopf, E.F. Guinan & P. Harmanec, eds.

The Incidence of Multiplicity Among Bright Stellar Systems

Peter P. Eggleton[1], Ludmila Kisseleva-Eggleton[2] and Xander Dearborn[3]

[1]Institute of Geophysics and Planetary Physics,
Lawrence Livermore National Laboratory, 7000 East Ave, Livermore, CA94551, USA
email: ppe@igpp.ucllnl.org

[2]U. C. Berkeley Extension, 1995 University Ave, Berkeley, CA 94720-7002
email: ludmilake@yahoo.com

[3]1770 Walnut Ave, Livermore, CA 94551, USA
email: powerhawk56@hotmail.com

Abstract. We consider the multiplicity of stellar systems with (combined) magnitude brighter than 6.00 in Hipparcos magnitudes. We identify 4555 such bright systems, and the frequencies of multiplicities 1, 2, ..., 7 are found to be 2722, 1412, 299, 86, 22, 12 and 2. We also consider the distributions of periods of orbits and sub-orbits. For the even more restricted set of 474 systems with $V_H \leqslant 4.00$ the proportions of higher multiples up to sextuple are progressively larger, suggesting incompleteness in even the relatively well-studied larger sample.

We construct a Monte-Carlo algorithm that will generate systems with roughly the observed multiplicities and orbital parameters, taking account of selection effects.

Keywords. stars: binaries, including multiples

1. Introduction

The *Bright Star Catalogue* (BSC: Hoffleit & Jaschek 1983) is a fundamental resource when considering the stellar population of the Galaxy, or at any rate the nearer parts of the Galaxy. It lists multiplicities, but these are often visual multiplicities that may include line-of-sight, or 'optical', coincidences. Although roughly limited to magnitude 6.5, it is not entirely complete to this magnitude, and also includes several fainter stars. The *Multiple Star Catalog* (MSC: Tokovinin 1997) carefully identifies many multiple systems, but restricts itself to multiplicity $\geqslant 3$. The *Hipparcos Catalog* (HIP: Perryman *et al.* 1997) has useful data such as parallaxes and proper motions that can help to distinguish optical from physical systems.

The great majority of multiple systems are 'hierarchical', with, for example, a wide 'binary' containing two closer 'binaries'. However non-hierarchical systems exist, in small numbers. We identify 17 non-hierarchical systems in our sample.

For 'brightness', we choose the Hipparcos magnitude system, and we include in the brightness *all* physical components even if their separation is quite wide. We make the cutoff at $V_H = 6.00$. The Hipparcos data are particularly useful because variable stars are averaged in a systematic way. A handful of stars which fail to be listed in the BSC, even down to magnitude 6.5, are Miras whose Hipparcos averaged magnitudes are brighter than 6.0.

Note that we use the word 'system', and even the words 'multiple system', to include the possibility of multiplicity one, i.e., single stars; and we avoid the term 'star' as ambiguous. In this paper we count systems, and within systems we count components, so far as we are able.

2. Multiplicity

In addition to the BSC, MSC and HIP we also used used the *Eighth Spectroscopic Binary Catalog* (BFM; Batten, Fletcher & McCarthy 1989), the *Ninth...* (SB9; Pourbaix *et al.*2004), the *Sixth USNO Catalog of Visual Binary Orbits* (Hartkopf et al 2000), the *CCDM* catalog (Dommanget & Nys 2002) of close multiples (many of which however are optical rather than physical multiples), the GCVS (Samus *et al.* 2004) on eclipsers and ellipsoidal variables, and the CHARA survey of speckle observations (McAlister *et al.* 1989).

Several more catalogs have been scanned, either electronically or by eye. We do not list them here, for brevity, but they will be listed later in a more detailed publication. However, we mention one, Makarov & Kaplan (2005), who considered systems whose Tycho and Hipparcos parallaxes and proper motions showed nonlinear behavior with time, i.e., acceleration, suggesting astrometric binaries. There are 349 such 'astrometric accelerators' in our catalog, denoted by 'A' in the reference column of Table 1. Occasionally one, or even more, is in a system already known to be a visual or spectroscopic binary, and we estimated whether the astrometric acceleration might or might not be due to the known companion.

In addition, several thousand papers were read, and are referenced directly if the data was different from (and, as we judged, better than) data from the principal catalogs mentioned above. Our overarching criterion was 'on the balance of probabilities', rather than 'beyond reasonable doubt'.

Following McClure (1983) and Boffin & Jorissen (1988), we assume that *all* Ba stars are binaries, with a white-dwarf component. Many Ba stars have indeed been determined to have spectroscopic orbits, and in a small number a white dwarf has actually been detected in the UV. However the case for binarity is not just that some are confirmed binaries, but more strongly that a physical mechanism exists to explain the Ba anomaly in terms of binarity, specifically with a white-dwarf companion, and that the anomaly is very hard to explain otherwise. An example where we assume binarity, despite the absence of an orbit or a white-dwarf spectrum, is given in Table 1 (HR 459).

Table 1 lists a sample of our results. The main body of the Table is being prepared for electronic publication, and in the printed version here we include only a few examples, with a range of multiplicities. Many systems consist of two or more HR entries; we list them under the largest relevant HR number (Column 1). For 10 systems at the end which have no HR number, a 'pseudo-HR' number ($\geqslant 9201$) is given, prefixed by 'P'. Column 2 lists what we consider to be the most reasonable multiplicity. Column 3 contains some reference letters (see below). Column 4 contains the parallax from Hipparcos. Column 5 contains our description of the configuration of the system using nested parentheses in roughly the format suggested by Evans (1977). For each individual component we give a magnitude and a spectral type, where we can find them, and for each pair of components we give either a period or a separation in arcseconds, where we can find them. Our reason for preferring this notation here is that, from experience, it can summarise a system sufficiently clearly that one can readily see where each component is in the hierarchy, and yet it only takes one line per system. To convey more information, both about the sub-components and the sub-orbits, the elegant notation of Tokovinin (1997) in the MSC is excellent, but this either takes $n-1$ lines for an n-tuple system, or else would be rather confusing to follow.

Reference letters in upper-case refer to particular catalogs, as listed in an annex to the on-line version. The *absence* of letters also indicates particular catalog sources: periods in days are from SB9, periods in years from USNO, spectral types from BSC, magnitudes (if

Table 1. Sample Configurations of Bright Multiple Systems

HR	n	ref.	plx	configuration
4:	1		0.009	5.71G5III
91:	3		0.002	5.55(5.95[B5IV + ?; 27.80d, e=.20) + 6.84; 152.7y, e=.10)
120:	2?	A	0.022	5.75(F2V + ?; ?)
136:	6		0.021	3.42(3.68(4.33(B9V + 13.5; 2.4″) + 4.55(A2V + A7V; 44.66y, e=.74); 27.060″) + 5.09(A0V + A0V; .1″); 540″)
152:	2	R	0.005	5.26(K5III + ?; 576.2d, e=.30)
165:	3	R A	0.032	3.43((K3III + ?; 20158d, e=.34) + 13.0M2; 28.7″bin)
233:	3	G M	0.004	5.47(G8IIa + (B9V + ?; 226.d, e=.14); 2091d, e=.53)
439:	2	C	0.002	5.82(K0Ib + B9V; .11″)
459:	2?		0.006	5.58(wd + K2IIIaBa; ?)
553:	2	s	0.055	2.60(A5V + G0:; 107.0d, e=.88)
629:	3?	H M	0.002	5.67(6.07B9V + ?; ?) + 6.95A1Vn; 16.690″)
958:	2	R G s	0.004	5.68(K0II + A7III; 115.0d)
1556:	2	s	0.006	4.74(WDA3 + S3.5/1-; ?)
1564:	4	R R	0.031	5.28(5.67F0IV + ?; ?) + 6.59(F4V + ?; ?); 12.500″)
1788:	6?	m E E	0.004	3.29(3.58((B1V + B2e; 7.990d, e=.01) + (B: + ?; 0.864d); 9.50y, e=.2) + 4.89B2V; 1.695″) + 9.4; 115″)
2788:	3?	M A E	0.023	5.79((F1V + G8IV-V; 1.136d, SD) + ?; ?)
4621:	5?	M A A	0.008	2.32(2.52(B2IVne + ?; ?) + 4.40(B6IIIe + ?; ?) @ 325°, 267″ + 6.5B9 @ 227°, 220″)
4908:	2?	L	0.002	5.37(O9Ib + 11.8K0III; 29.1″)
5340:	1	s	0.089	0.11K1.5IIIFe-0.5
6046:	2	R A s	0.005	5.77(M3II + G7III; 2201d, e=.69)
7776:	6	P	0.009	3.14(3.21((G8II + 7.2(B8V + ?; 8.68d, e=.36); 137d, e=.42) + 8.3; .8″) + 6.09(6.16A0III + 9.14A1; .68″); 205.3″)
8387:	3	m	0.276	4.83(K4.5V + (T1 + T6; .73″); 402.3″bin)
9072:	1?	s	0.031	4.12F4IV
P9203:	3?	H M	0.004	5.79(6.44(B5/6V + ?; ?) + 6.67B8/9V; 129.490″)
P9207:	1		0.007	5.99M8IIIvar

Note. —

1. The full Table will be placed on-line in due course, with a cross-reference Table for ID and a file of references.
2. For systems containing more than one HR component, we use the largest HR no. If there is no genuine HR no. in the system, we give a 'pseudo-HR' no., $\geq$ 9201, prefixed by P. The corresponding HIP and/or HD numbers can then be found in the cross-reference Table. One example shown here, P9203, is HIP 32256 and 32269. We identify only 9 pseudo-HR systems that qualify for our sample.
3. In col. 3, the letters refer to various sources, as mentioned in the text. The absence of a reference letter also implies particular catalog sources: BSC, HIP. MSC. SB9. USNO.

quoted to 2 d.p.), separations (if quoted to 3 d.p.) and parallaxes from HIP. Lower-case reference letters s or m point to specific papers listed in the on-line version.

Table 2 gives the distribution over periods and spectral types. It is noticeable that the period distribution is severely bimodal at early types, becoming roughly unimodal by type F. The statistics on multiplicity are given later, Table 4, where they are compared with a theoretical model.

Table 2. Period Distribution in Systems and Subsystems

$\log P(\mathrm{yr})$	-3.0	-2.0	-1.0	0.0	1.0	2.0	3.0	4.0	5.0	6.0	7.0	8.0	total
sp.													
O	0	5	11	4	0	5	3	6	12	4	1	1	0	52
eB	0	24	40	19	14	20	28	26	32	23	6	2	1	235
lB	0	24	52	22	21	45	50	56	46	18	6	0	0	340
A	0	27	61	24	45	78	78	69	43	26	4	1	0	456
F	0	14	30	24	38	49	68	38	24	14	2	1	0	302
G	1	6	10	22	41	39	54	42	25	20	4	1	1	266
K	0	4	4	12	57	35	40	43	35	19	5	0	1	255
M	1	0	0	3	7	6	5	8	9	3	3	0	0	45
sum	2	104	208	130	223	277	326	288	226	127	31	6	3	1951

Notes – The first column gives the spectral type of the dominant body in the system: eB means early B, i.e., B0 – B3.5, and lB means later B. Wolf-Rayet stars (2) are included under O; S and C stars (4) are included under M. The first column of integers gives the number of systems and sub-systems with $\log P(\mathrm{yr}) \leqslant -3.0$; the second for $-3.0 < \log P \leqslant -2.0$, etc. The two shortest periods are a contact binary subsystem of the G dwarf 44 Boo (HR 5618), and a cataclysmic binary subsystem of the M giant CQ Dra (HR 4765). Long periods are estimates from the angular separation, distance, and Kepler's law.

3. A Monte Carlo Model

We attempt to model the data with a Monte Carlo procedure, based on that of Eggleton *et al.* (1989), selecting masses, mass ratios and periods from distributions with as few parameters as possible. We firstly draw a total system mass from a roughly Salpeter-like distribution, and then we divide it and sub-divide it according to a distribution of mass ratios, allowing a finite probability for the possibility that the mass ratio is zero, i.e., the system or subsystem does not in fact divide further. We allow a maximum of three subdivisions, so that the highest multiplicity we can obtain is 8.

By postulating successive subdivisions, we do not intend to imply that actual multiple systems are produced by a physical mechanism of successive fragmentation. 'Successive bifurcation' is simply a convenient description of what hierarchical binaries actually are, regardless of how they are formed.

Let X be a random number drawn from a uniform distribution on the interval [0,1]. If we take the system mass to be $M = f(X)$, we are in effect postulating a distribution such that $X(M) \equiv f^{-1}(M)$ is the normalised cumulative distribution. The distribution given by

$$M = M_0 \frac{X^{\alpha'}}{(1-X)^{\alpha}} , \qquad (1)$$

has median $M_0 2^{(\alpha-\alpha')}$, and is roughly Salpeter-like at large M ($X \sim 1$) if $\alpha \sim 0.75$.

Our attempts to model the distribution of bright systems will clearly not cast any light on the distribution of K/M dwarfs, since hardly any appear in Table 1 except as secondaries or tertiaries. Consequently we adopt the simplification that $\alpha' = \alpha$, i.e., that the distribution is symmetrical (in $\log M$) about the median, which we expect to be at $M_0 \sim 0.3 - 0.5$, appropriate to early M dwarfs. We adopt provisionally $M_0 = 0.30\,M_\odot$, $\alpha = 0.83$.

The bifurcation probability is determined by another X, a new selection from the random-number generator. We choose a function $X_{bf}(l, M)$ defined on the interval $[0,1]$, and require that the system bifurcate at hierarchical level l if $X \leqslant X_{bf}$, and not bifurcate if $X > X_{bf}$. We supply X_{bf} in tabular form, using as independent variable

$$m = 5 + 2\log\left(\frac{1 + 100M}{100 + M}\right) \tag{2}$$

Of course, the bright-system sample tests the algorithm seriously only for $M \gtrsim 1\,M_\odot$ ($m \gtrsim 5$), i.e., the 15 values at the right.

Table 3. Bifurcation Probability X_{bf} as a Function of Mass and Hierarchical Level

m=	1	2	3	4	5	6	7	8	9
M=	0	.01	.09	.32	1	3.2	11	32	∞
l=0	0.50	0.50	0.50	0.50	0.50	0.65	0.80	0.87	0.92
1	0.10	0.15	0.15	0.15	0.15	0.15	0.20	0.45	0.65
2	0.00	0.00	0.00	0.00	0.00	0.20	0.50	0.70	0.85

The periods are chosen from further distributions, which at the highest hierarchical level ($l = 0$) we take to be somewhat similar to (1):

$$P = P_0\,\frac{X^{\beta'}}{(1 - X)^\beta} \; . \tag{3}$$

We adopt $P_0 \sim 270\,\mathrm{yrs}$, $\beta = 2.5$, $\beta' = 2.0$.

For subsystems or subsubsystems, if any, we choose a period ratio, relative to the next level up, of

$$\frac{P'}{P} = A \cdot 10^{-\gamma X} \; . \tag{4}$$

We provisionally adopt $A = 0.2$, $\gamma = 5$.

The mass ratio q at each bifurcation, if it occurs, is taken from another distribution:

$$q = \max(0.01, X^\delta) \; , \tag{5}$$

where the exponent δ might in principle be a function of both M and P'. We provisionally adopt $\delta = 0.8$. We disallow $q < 0.01$.

For the present we assume that all orbits are circular. It would not be difficult to generate a distribution of eccentricities, but we feel that there are too many wide orbits, of unknown eccentricity, to make this refinement necessary at present.

The age is chosen from a very simple distribution:

$$t = t_0\,X \; , \tag{6}$$

where t_0 is an estimate of the age of the Galaxy, $t_0 \sim 10^{10}\,\mathrm{yr}$. This assumes a uniform production-rate of systems. It is unlikely to be a good assumption over the whole range of possible ages, but most of the systems in Table 1 are massive enough that they must have been formed in the last 10% of the Galaxy's lifetime, and so the assumption of constancy may not be very critical.

The evolution is treated so far in a rather cursory manner, which we hope to improve on in a future paper. In particular, we ignore such binary-specific evolutionary processes as Roche-lobe overflow (RLOF), although we note systems in which we can expect it to have happened. We also ignore the fact that neutron stars are typically blasted out of their parent system by a slightly anisotropic supernova explosion; thus many multiples that we create contain one or more neutron stars. But for the present we evolve each component independently according to its mass and age, following roughly the simple procedure of Eggleton et al.(1989).

Finally, we need to generate a distribution over distance d. Given a set of component masses, as above, we can evolve them to the selected age and thus determine (a) the total luminosity, and (b) the distance d_{max} out to which the system will be visible above the set limit ($V_H = 6$, say). Suppose that we generate a total of $N = 10^6$ systems. To a first approximation, they might populate uniformly a sphere of radius $D \sim 100$ pc. We can think of the given system as being 'cloned' $n \equiv (d_{max}/D)^3$ times, with the rth clone placed randomly within a spherical shell of inner and outer radii $(r-1)^{1/3}D$ and $r^{1/3}D$, i.e., it would be at distance

$$d_r = (r - 1 + X_r)^{1/3}D \ , \ r = 1, 2, \ldots, n \ , \tag{7}$$

where the X_r are a sequence of independent random numbers in [0,1]. Of course, if $d_{max} < D$ the system will only be seen at most once, with probability $(d_{max}/D)^3$; but that is in fact what Equation 7 implies with $n = 1$. Since we have defined D for 10^6 systems, we can write $D = 10^2 d_{nbr}$ where d_{nbr} is the mean distance between nearest neighbors. We adopt $d_{nbr} = 1.39$ pc.

To model a more realistic density distribution (that does not experience Olbers' paradox), suppose that the distance to the rth clone is

$$d_r = x^{1/3} D \sqrt{1 + x^{1/3}D/h} \ , \ x = r - 1 + X_r \ . \tag{8}$$

This approximation is roughly what we expect for a uniform distribution within a disk of thickness h. We make the vertical scale-height h age-dependent, $h = 200 t^{0.3}$, with h in pc and t in units of 10^{10} yr.

A further small refinement that we put in is interstellar absorption, so that the apparent luminosity l of a system at distance d whose intrinsic luminosity is L will be given as

$$l = \frac{L}{d^2} e^{-kd} \ , \tag{10}$$

with $k \sim 0.001$ (d in parsecs), corresponding to roughly 1 magnitude per kiloparsec. This means that the total number of observable clones will in practice be less than the crude number n estimated above.

We put in one further procedure which is convenient numerically though it has no astrophysical significance. A given component is at its brightest, as a giant, for only a short period of time compared with the Galactic age. Given a mass and age selected at random as above (and considering, for the purpose of exposition, only single stars) the chance is rather small, even among 10^6 samples, that a given star will be in this rather bright stage; but on the other hand it will require to be cloned many times since it will be visible to a large distance. This introduces considerable scatter as between one simulation and another that is identical except insofar as the random numbers differ. We can mitigate this by using a non-uniform distribution of X, Y over the unit square, and using the Jacobian of the transformation to modulate the density. Thus the sample of 10^6 systems, which might produce only say 10 giants cloned 100 times each, can produce instead say 500 giants cloned 2 times each.

Once we have a Monte Carlo selection of multiples, we can then examine them in a 'theoretical observatory', to determine how many of the components and sub-binaries will be actually recognisable. We look for spectroscopic binaries, assuming that the velocity amplitude has to be greater than some threshold which we take to depend mainly on spectral type: fairly low for G/K components ($\sim 2\,\mathrm{km/s}$), increasing to $\sim 30\,\mathrm{km/s}$ for O components, which tend to have both few and broad lines, and also to be intrinsically variable in radial velocity at this level. We further required the period to be less than 30 years, to simulate the patience of observers.

For a visual binary to be theoretically observable, we require that the angular separation be greater than some threshold that depends on the difference in magnitude of the (sub)components:

$$\rho > 0.4\,|\Delta V_{\mathrm{H}}| + 0.1 \;, \tag{11}$$

where ρ is the angular separation in arcsecs. We further require that the combined magnitude of a detectable component or subsytem be $\leqslant 14$. Although we can expect that there are many companions that are fainter, Table 1 has only a handful fainter than this limit.

Table 4. Multiplicities in Theoretical and actual Samples

	total	1	2	3	4	5	6	7	8	av.
$V_{\mathrm{H}} \leqslant 6$										
raw	4649	1596	2041	531	285	118	43	31	4	2.05
apparent	4649	2918	1323	279	93	24	9	3	0	1.50
actual	4555	2722	1412	299	86	22	12	2	0	1.53
$V_{\mathrm{H}} \leqslant 4$										
raw	566	178	256	60	43	16	9	3	1	2.13
apparent	566	313	189	43	13	4	2	2	0	1.62
actual	474	213	176	52	18	9	6	0	0	1.84

4. Discussion

The result of the above selection effects is to transform the raw multiplicities into apparent multiplicities. Table 4 is an example, for both the larger sample of $V_{\mathrm{H}} \leqslant 6$ and the smaller sample $V_{\mathrm{H}} \leqslant 4$: the top row gives the raw multiplicities, the second row the apparent multiplicities after theoretical observation, and the third row the observed multiplicities. It can be seen that in the larger sample the frequencies of multiplicities 1 to 7 are reasonably matched, with the main discrepancy being a $\sim 6\%$ shortage of binaries in the 'apparent' sample. The overall average multiplicity is reasonably matched within the limits of small-number statistics. For the smaller sample there is less good agreement in the average, although it has moved in the right direction. This discrepancy, although fairly modest, confirms our view that the smaller sample has simply been more thoroughly studied than the larger. Our model, however, is based on trying to fit the larger sample (including the data of Table 2), because of its greater statistical weight.

The bimodality of the period distribution at early types seen in Table 2, shading to unimodality at later types, is in fact quite well-modeled by our 'apparent' data. This supports, though it does not prove, the view that it is largely a selection effect.

Data on multiplicity is important as a constraint on (a) the star-formation problem, (b) the problem of the evolution of the Galactic stellar population, (c) N-body simulations of dynamics and evolution of clusters and (d) the interaction of dynamics and evolution through the effect of Kozai cycles. We discuss one sub-topic each of (b), (d) briefly.

A likely effect of triples under (b) is the production of 'anomalous binaries'. In short-period binaries we can expect that a merger of the two components is a fairly common event. Case A systems can evolve conservatively only if the initial mass ratio is fairly mild ($q \gtrsim 0.6$; Nelson & Eggleton 2001), and if q is not this mild then a merger seems quite a likely event. It would be hard to determine that a particular currently-single star is a merged remnant of a former binary. But within a primordial triple system, it is possible that such a merged remnant would be identifiable, because the wide binary that remains after the merger of the close sub-binary would be expected, in at least some circumstances, to show an anomaly where the two components appear to be of different ages. R.E.M. Griffin (to be published) has found a number of such apparently anomalous systems, of which γ Per (HR 915) is an example. Although the mass ratio, obtained from careful deconvolution of the two spectra (G8III + A2IV; 5330 d, $e = 0.79$), is 1.53 (M_G/M_A), the A component seems surprisingly large and luminous compared with what it should be on the ZAMS; and it ought to be very close to the ZAMS if it is coeval with the G component. A possible explanation is that the giant is the merged product of a former sub-binary with a mass-ratio of ~ 0.5, since this could allow the more massive two of the original three components to evolve at roughly the same rate (Eggleton & Kiseleva 1996, Eggleton 2006). We hope to test shortly the possibility that the appropriate primordial triple parameters, from our Monte Carlo model, will give an acceptable number of potential progenitors. Alcock *et al.* (1999) and Evans *et al.* (2005) have noted that a similar process might lead to Cepheid binaries of an anomalous character, such as may be required to reconcile observed Cepheids with the theoretical models of the Cepheid pulsation phenomenon.

Under (d), triple stars in which the two orbits are misaligned can be subject to the dynamical effect of Kozai cycles, and these in turn can allow tidal friction to become important in the course of $10^6 - 10^9$ yrs and cause the inner orbit to become smaller. Tokovinin & Smekhov (2002) have noted that periods above ~ 10 days are rarer (relative to shorter periods) among the inner binaries of triples than in the population as a whole, and possibly this is accounted for by the combination of Kozai cycles and tidal friction (KCTF). Pribulla & Rucinski (2006) have noted that as many as 42% of contact binaries appear to be in triples (and arguably 59% in a more thoroughly examined subsample), and it could be that KCTF has contributed to this; although we probably need the additional influence of magnetic braking to drive fairly close low-mass binaries generated by KCTF to contact on a timscale of $\lesssim 10^9$.

Acknowledgements

This study has been carried out partly under the auspices of the U.S. Department of Energy, National Nuclear Security Administration, by the University of California, Lawrence Livermore National Laboratory, under contract No. W-7405-Eng-48.

References

Alcock, C. *et al.* (1999) *AJ*, 117, 920
Batten, A.H., Fletcher, J.M., & McCarthy, D.G. (1989; BFM) *PDAO*, 17, 1
Boffin, H.M.J. & Jorissen, A. (1988) *A&A*, 205, 155
Dommanget, J. & Nys O. (2002; CCDM) *Observations et Travaux* 54, 5
Eggleton, P.P. (2006) in *A Life with Stars,* eds. Kaper, L., van der Klis, M., & Wijers, R.A.M.J., *New Astron. Rev.*, in press
Eggleton, P.P., Fitchett, M.J., & Tout, C.A. (1989) *ApJ*, 347, 998
Eggleton, P.P. & Kiseleva, L.G. (1996) in '*Evolutionary Processes in Binary Stars*' ed. Wijers, R. A. M. J., Davies, M. B., & Tout, C.A., NATO ASI Series C, 477, p345
Evans, D.S. (1977) *Rev. Mex A&A*, 3, 13
Evans, N.R., Carpenter, K.G., Robinson, R., Kienzle, F., & Dekas, A.E. (2005), *AJ*, 130, 789
Hartkopf, W.I., Mason, B.D., McAlister, H.A., *et al.* (2000; USNO) *AJ*, 119, 3084

Hoffleit, D. & Jaschek, C. (1983; BSC) *The Bright Star Catalogue*, 4th ed. New Haven: Yale University Observatory

McAlister, H.A., Hartkopf, W.I., Sowell, J.R., *et al.* (1989; CHARA) *AJ*, 97, 510

McClure, R.D. (1983) *ApJ*, 268, 264

Makarov, V.V. & Kaplan, G.H. (2005) *AJ*, 129, 2424

Nelson, C.A. & Eggleton, P.P. (2001) *ApJ*, 552, 664

Perryman, M.A.C., Lindegren, L., Kovalevsky, J., *et al.* (1997; HIP) *A&A*, 323, 49

Pourbaix, D., Tokovinin, A.A., Batten, A.H., *et al.* (2004; SB9) *A&A*, 424, 727

Pribulla, T. & Rucinski, S.M. (2006) *AJ*, 131, 2986

Samus, N.N., Durlevich, O.V. *et al.*, (2004;GCVS) Institute of Astronomy of Russian Academy of Sciences and Sternberg State Astronomical Institute of the Moscow State University

Tokovinin, A.A. (1997; MSC) *A&AS*, 124, 75

Tokovinin, A.A. & Smekhov, M.G. (2002) *A&A*, 382, 118

Discussion

JUAN MANUEL ECHEVARRIA: From your sample, which is close to the number of stars seen by the naked eye (at least of an 18 year old person), what is the percentage of multiple stars in the 'classical' sense (i.e., 2 or more)?

EGGLETON: The proportion of 2 or more is about 40% for the 6th mag. sample, and about 55% for the 4th mag. subsample. In our theoretical 6th mag sample, before theoretically observing them, the proportion is about 65%.

THEO TEN BRUMMELAAR: You seem to have a large number of free parameters to fit what seems to be a rather small sample. Could you please comment on this?

EGGLETON: There are about 20 parameters that matter. We consider this to be quite a small number considering the range of multiplicities we attempt to encompass. Supposing that one could divide about 5000 stars into 20 bins, each bin being dominated by one parameter, we might reasonably hope that the parameters could be individually estimated to 10%. Of course in the real world we would find some parameters less tightly constrained and some more tightly.

ROGER GRIFFIN: I think you may have inflated your figures for multiplicity by your 'balance of probability' criterion. In the case of HR 233 you particularly commented on the period of 226 days for the inner orbit. That period does not exist - the orbit concerned was withdrawn by the author and repudiated by the editor of the journal concerned (see the first page of the May 1984 Astronomical Journal).

EGGLETON: I am aware of the inadequacy of that particular claim. However R.E.M. Griffin tells me that nevertheless the hot star appears to have greater r.v. variation, and on a shorter timescale, than is likely for just a binary with your well-established longer period. I think it is therefore still reasonable to conclude on a balance of probabilities that the system is at least triple. I hope you will be pleased to see that Arcturus (HR 5340) is listed by us as probably single, although Hipparcos claimed to resolve it into 2. We noted your analysis (*Obs*, 118, 299, 1998) that showed that the very accurate radial velocities over ~ 50 years do not support such binarity.

Binary Stars as Critical Tools & Tests
in Contemporary Astrophysics
Proceedings IAU Symposium No. 240, 2006
W.I. Hartkopf, E.F. Guinan & P. Harmanec, eds.

Testing and Improving the Dynamical Theory of Mass Exchange

Dmitry Bisikalo[1] and Takuya Matsuda[2]

[1]Institute of Astronomy of the Russian Academy of Sciences, Moscow, Russia
email: bisikalo@inasan.ru

[2]Department of Earth and Planetary Sciences, Kobe University, Kobe, Japan
email: tmatsuda312@yahoo.co.jp

Abstract. The study of the flow structure is of great importance, and the results can be used both for consideration of the evolutionary status of binary stars and for the interpretation of observational data. In this report we present the review of 3D gas dynamic models used for the description of the mass exchange in close binaries.

Main features of the flow structure in steady-state close binaries are summarized. It is shown that in self-consistent considerations the interaction between the stream from the inner Lagrangian point and the forming accretion disk is shock-free, and, hence, a "hot spot" does not form at the outer edge of the disk. To explain the presence of the observed zones of high luminosity in close binaries a self-consistent "hot line" model was proposed according to which the excess energy is released in a shock wave formed due to interaction between the circumdisk halo and the stream. The "hot line" model was confronted with observations and confirmed by virtue of comparison of synthetic and observational light curves for cataclysmic variables and by the analysis of Doppler tomograms.

The special attention is paid to the physics of accretion disks in binary systems and particularly to waves in disks. The possible observational manifestations of the "hot line" wave and two arms of the tidal shocks are discussed. We also suggest that an additional spiral density wave can exist in inner parts of the cold accretion disk. This spiral wave is due to the retrograde precession of flow lines in the binary system. The results of 3D gas dynamic simulation have shown that a considerable increase in the accretion rate (by an order of magnitude) is associated with the formation of the "precessional" spiral wave. Based on this fact we suggest a new mechanism for the superoutbursts and superhumps in close binaries.

Keywords. binaries: close, mass loss, accretion, accretion disks

1. Introduction

The overwhelming majority of stars (some researchers believe up to 80%) are binaries. The presence of a gravity-connected companion of the star can affect the physical processes in the star and appreciably change its evolution. The fact is that belonging to a binary system limits the maximum size of the star. As early as 1848, E.A. Roche studied the motion of probe particles in the vicinity of a binary within a restricted problem of three bodies. He found that a certain space close to every component can be chosen, and within this space the gravity field of this component mainly affects the motion of a probe particle. In the course of the evolution of the star — a component of a binary — the expansion out of the limits of this space which is now referred to as the Roche lobe results in mass loss of the star. A number of observations prove the complex flow structure in close binary stars (CBSs) caused by mass transfer in a system. Starting from the study of Struve (1941) who first conceived the idea of a gas stream appearing between components in β Lyrae to explain the peculiar behaviour of the spectrum at eclipse, the

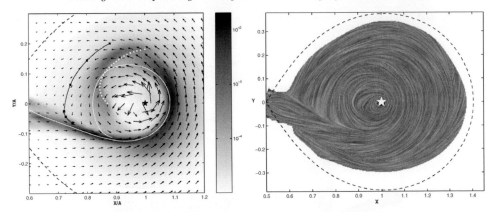

Figure 1. *Left panel*: The distribution of density over the equatorial plane. The X and Y coordinates are expressed in terms of the separation A. Arrows are the velocity vectors. The black asterisk is the accretor. The dashed line bounds the Roche lobe. The dotted line is the tidally induced spiral shock. Gas dynamic trajectory of a particle moving from L_1 to accretor is shown by a white line with circles. Another gas dynamic trajectory is shown by a black line with squares. Symbols of larger size correspond to passing of trajectory through shock wave located on the edge of the stream. *Right panel*: The visualization of the velocity vectors field using line integral convolution method. The X and Y coordinates are expressed in terms of the separation A. The white asterisk is the accretor, the dashed line is the Roche equipotential.

effects of mass transfer resulting in formation of gas flows, streams, disks, circumbinary envelopes and other structures were observed in a number of CBSs.

The main objective of this paper is the study of gas dynamics of mass transfer in CBSs. The study of the flow structure is of great importance, and the results can be used both for consideration of the evolutionary status and for the interpretation of observational data.

2. Numerical modeling of mass transfer in close binaries

To describe gas flows correctly, we should solve a complete system of gas dynamic equations. This system can be solved only within the framework of rather complex mathematical models. In the last fifteen years gas dynamics of mass transfer was numerically studied with the help of more realistic 3D models (see, e.g. pioneering works by Nagasawa *et al.*, 1991; Hirose *et al.*, 1991; Molteni *et al.*, 1991). In this paper we summarize the results of 3D numerical simulation of mass transfer in semidetached binaries that were mainly obtained by Bisikalo *et al.* (1998 - 2006), Molteni *et al.* (2001), Harmanec *et al.* (2002), Boyarchuk *et al.* (2002), Fridman *et al.* (2003), Kaigorodov *et al.* (2006).

Analysis conducted in Bisikalo *et al.* (2003) has shown that for realistic values of parameters ($\dot{M} \simeq 10^{-12} \div 10^{-7} M_\odot \, \mathrm{y}^{-1}$ and $\alpha \simeq 10^{-1} \div 10^{-2}$) the gas temperature in outer parts of the disk is between $\sim 10^4$K and $\sim 2 \cdot 10^5$ K. This implies that both hot and cold accretion disks can form in close binaries.

Let us consider the morphology of steady-state gaseous flows in a system with hot accretion disk. These solutions were obtained for temperatures of the outer parts of the accretion disk of 100 000 $\div$ 200 000 K.

The morphology of gaseous flows in considered binary system can be evaluated from Figure 1 (left panel). In this figure the distribution of density over the equatorial plane and velocity vectors are presented. We also show the gas dynamic trajectory of a particle moving from inner Lagrangian point L_1 to accretor (white line with circles) and the gas

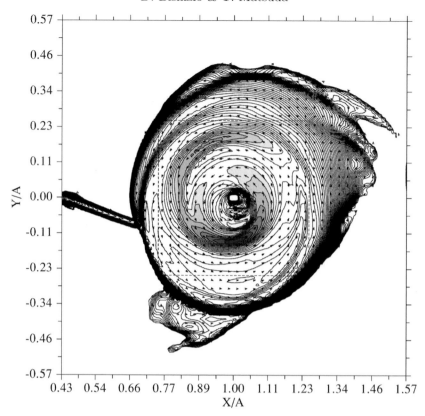

Figure 2. Contours of constant density and velocity vectors in the equatorial plane.

dynamic trajectory passing through the shock wave along the stream edge (black line with squares). The morphology of gaseous flows in semidetached binaries is governed by the stream of matter from L_1, quasi-elliptical accretion disk, circumdisk halo and circumbinary envelope. This classification of the main constituents is based on their physical properties: (i) if the motion of a gas particle is not determined by the gravitational field of accretor then this particle belongs to the circumbinary envelope filling the space between the components of binary; (ii) if a gas particle revolves around the accretor and after that mixes with the matter of the stream then it does not belong to the accretion disk but forms the circumdisk halo; (iii) the accretion disk is formed by the matter of the stream that is gravitationally captured by the accretor and hereinafter does not interact with the stream but moves to the accretor losing the angular momentum.

The analysis of the gas parameters along the flow lines reveals that the flow is smooth for all lines belonging to the disk up to the boundary flow line (white line with circles in Figure 1, left panel). The absence of breaks of the gas parameters suggests a shock-free interaction between the gas flow and the matter of the disk that in turn implies the absence of a "hot spot" on the edge of the disk. The uniform morphology of the flow results in the stream deflecting under the effect of the circumbinary envelope gas, approaches the disk along a tangent line and does not cause any shock perturbation of the disk edge. This fact is also obvious from the so-called texture figure (Figure 1, right panel) where the visualization of the velocity vector field using line integral convolution method is shown. At the same time analysis of the results proves that the interaction of

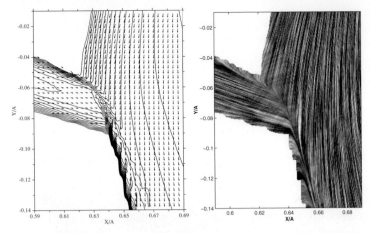

Figure 3. Contours of constant density and velocity vectors (left panel) and visualization of the velocity field (right panel) in the zone of interaction between the stream and disk.

a stream with the circumbinary envelope causes the formation an extended shock wave located along the stream edge. This shock is referred hereafter as "hot line".

From Figure 1 it is also seen that the spiral shock tidally induced by the donor-star appears in the disk. The position of this shock is shown in Figure 1 (left panel) by a white dotted line. Appearance of the tidally induced two-armed spiral shock was discovered by Sawada *et al.* (1986). Our 3D gas dynamic simulations for the "hot" case have shown only one-armed spiral shock while in the place where the second arm should be the flow structure is determined by the stream from L_1. It also should be stressed that in the "hot" case the spiral shock penetrates deeply to the inner part of the disk.

Let us consider the morphology of gaseous flows in a system with cold accretion disk. In the model used the temperature decreases to $\sim 1.4 \cdot 10^4$ K over the entire computation domain due to the radiative cooling. The basic problem here is whether the interaction between the stream and the disk remains shockless, as was shown for relatively hot disks. Figure 2 depicts density distribution and velocity vectors in the equatorial plane of the system (the XY plane). The shocks, which are formed in the disk, are seen as condensed isolines. The latter on the edge of the disk halo correspond to sharp decrease of density up to the background value. We can see the dense circular disk as well as the compact halo. The interaction of gas of the halo with the stream generates the shock outside the disk "hot line". The two-armed spiral shock wave forms in the disk. The both arms are located in the outer part of the disk.

Two panels of Figure 3 show density distribution and velocity vectors (left panel) and visualization of velocity field (right panel) in the zone of stream-disk interaction. Figure 3 shows that in the "cold" case the interaction between the circumdisk halo and the stream displays all features typical of an oblique collision of two streams. We can clearly see two shock waves and a tangential discontinuity between them. The gases forming the halo and stream pass through the shocks corresponding to their flows, mix, and move along the tangential discontinuity between the two shocks. Further, this material forms the disk itself, the halo, and the envelope.

Let us consider the changes occurring during the transition from the hot accretion disk to the cold one. The sketch of main peculiarities of the morphology of gaseous flows in semidetached binaries for "hot" and "cold" cases is given in Figure 4. These schemes are based on the results of 3D gas dynamic simulations. In Figure 4 the fragment of mass-losing star that fills its Roche lobe, the location of the inner Lagrangian point L_1,

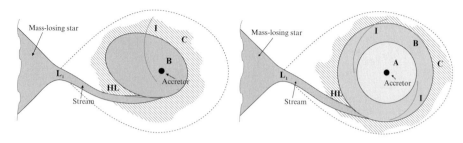

Figure 4. The sketch of main peculiarities of the morphology of gaseous flows in semidetached
binaries for the case of high (left panel) and low (right panel) gas temperature.

the stream of matter from L_1, as well as the location of the accretor are shown. The
dashed line marks the Roche lobe. The accretion disk is shown in Figure 4 as "A" and
"B" zones, the near disk halo is shown as zone "C", the "hot line" is marked by "HL",
and arms of tidal spiral shock are marked by "I". Our 3D gas dynamic simulations have
shown that for the "cold" case when the radiative cooling decreases the gas temperature
to $\sim 1.4 \cdot 10^4$ K the solution has the same qualitative features as that for the "hot"
case, namely: the interaction between the stream and disk is shockless, the region of
enhanced energy release is formed due to the interaction between the circumdisk halo
and the stream and is located beyond the disk. The resulting shock – the "hot line"
("HL" in Figure 4) is fairly extended, that is particularly important for explaining the
observations. However, unlike the solution with a high temperature in the outer regions
of the disk, in the "cold" case, the shape of the zone of shock interaction between the
stream and halo is more complex than a simple line. This is due to the sharp increase
of the halo density as the disk is approached. Those parts of the halo that are far from
the disk have low density, and the shock, due to their interaction with the stream, is
situated along the edge of the stream. As the halo density increases, the shock bends,
and eventually stretches along the edge of the disk. In the "cold" case the accretion disk
(zones "A" and "B" in Figure 4) is significantly more dense as compared to the matter of
the stream, the disk is thinner and has not quasi-elliptical but circular form. The size of
the circumdisk halo is smaller as well. The second arm of the tidal spiral shock is formed,
the both arms do not reach the accretor but are located in the outer part of the disk.

3. The structure of the cold accretion disk

Taking into account that the stream influences the dense inner part of the disk weakly
as well as that all the shocks ("hot line" and two arms of tidal wave) are located in the
outer part of the disk we can introduce a new element of flow structure for the "cold"
case: the inner region of accretion disk (zone "A" in Figure 4) where the influence of gas
dynamic perturbations mentioned above is negligible. Formation of non-perturbed region
in the inner part of the disk allows to consider the latter as a slightly elliptical disk with
the typical size of $\sim 0.2 - 0.3A$ embedded in the gravitational field of binary. It is known
(see, e.g., Warner, 1995), that the influence of companion star results in precession of
orbits of particles rotating around of the binary's component. The precession is retrograde
and its period increases with approaching the accretor.

For the accretion disk the orbits must be replaced by flow lines. Flow lines cannot
intersect and can only touch by being tangent to each other. If the orbits precess such
that the precession of distant flow lines tends to be faster, these distant flow lines will
constantly overtake those with smaller semimajor axes. Since the flow lines in a disk

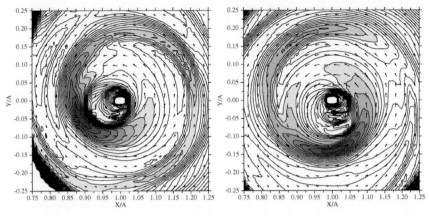

Figure 5. Density contours and velocity vectors in the inner part of the accretion disk for two moments of time. Enlarged image of the region delimited by the dashed curve in Figure 2.

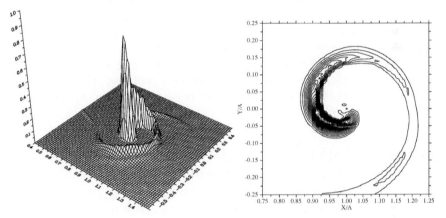

Figure 6. Distribution of the radial flux of matter in the equatorial plane of the disk. The flux is normalized to its maximum value.

cannot intersect, an equilibrium solution is established with time and all the flow lines begin to precess with the same angular velocity, i.e., to display rigid-body rotation. Let us consider a solution with the semimajor axes of the flow lines misaligned with respect to some chosen direction by an angle (turn angle) that is proportional to the semimajor axis of the orbit. It is obvious that such a solution should contain spiral structures. In particular, due to the nonuniformity of the motion along the flow line and the formation of a maximum density at apastron, the curve connecting the apastrons will form a spiral density wave. Indeed we have discovered (Bisikalo *et al.*, 2004a) that the retrograde precession with specified law of precession rate results in formation of the density spiral wave of a new, "precessional" type in the inner part of the disk. This wave is clearly seen in Figure 2 in the region that was supposed to be unaffected by gas-dynamical perturbations. Figure 5 shows an enlarged view of the density distribution and velocity vectors in the inner part of the disk for two moments of time. Figure 5 shows the flows in a coordinate frame that rotates with the orbital period of the binary. The two-armed spiral wave is at rest in this coordinate system, as is quite natural for a tidal wave, while the inner spiral wave moves. Numerical results show that the "precessional" wave moves as a single entity, and its period is about ten times larger than the orbital one. Figure 6 shows the distribution of the radial flux of matter in the equatorial plane of the disk

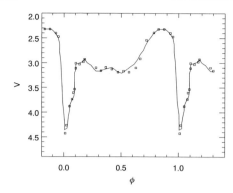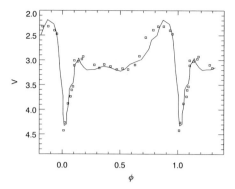

Figure 7. Mean light curve (points) of IP Peg in V filter for quiescent state and synthetic (solid line) light curves obtained in the "hot line" (left panel) and in the "hot spot" (right panel) models.

in two ways; the radial flux is normalized to its maximum value. We can clearly see the spiral shape of the curve passing through the flux peaks. Due to the increase of the radial flux of matter behind the "precessional" density wave, the rate of accretion increases by approximately an order of magnitude compared to the wave-free solution.

Analysis of the results of a 3D gas dynamic simulation fully confirms the possibility of the generation of a spiral wave in the inner parts of a cold disk. The agreement between the qualitative analysis and the computational results makes us confident that the wave has a precessional origin.

4. Astronomical applications of the "hot line" model

4.1. "Hot line" model versus "hot spot" model

One of the best sources of information about the flow structure is the analysis of light curves of cataclysmic variables (CVs). The vast majority of light curves of normal CVs (U Gem type) display a so-called "orbital hump" around orbital phase ~ 0.8. Gorbatskii (1967) and Smak (1970) suggested that this is due to a "hot spot" at the edge of the accretion disk, where the disk collides with the stream from L_1. For many years, the "hot spot" model has been widely used to interpret light curves of cataclysmic binaries.

In terms of gas dynamics one can hardly understand the presence of a shock interaction between the flow and the steady-state accretion disk. Even if at the initial moment the stream from L_1 collides with the earlier–formed disk, in the course of time the morphology of the flow changes so that the stream–disk interaction has become shock-free, since the stream is the only source of matter for the disk. Three-dimensional simulations of gas dynamics of mass transfer prove that in a steady-state self-consistent solution the stream arrives at a tangent line to the disk, and the shock interaction of the disk and the stream (i.e., the standard "hot spot") does not appear. Meantime, the interaction of the circumbinary envelope with the stream was found to form an extended shock wave located along the edge of the stream. In spite of the fact that in this model the region of shock energy release is located outside of the accretion disk, there are good grounds to believe that this region ("hot line") can be considered in observation as an equivalent of a "hot spot" in the disk. In order to make certain that this assumption is valid and, hence, the gas dynamical model without a "hot spot" is adequate, we synthesize the light curves and compare them with observational data.

When applied to the interpretation of optical and infrared light curves of dwarf novae (Khruzina *et al.*, 2001; 2003b), the "hot line" model shows certain advantages over the

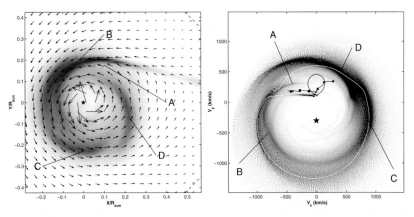

Figure 8. *Left panel*: Intensity map in the equatorial plane of the system in its quiescent state. A, B, C, and D denote zones with the highest intensity of radiation: A - "hot line", B - stream, C -apogee of the quasi-elliptical disk, D - single arm of the tidal shock. *Right panel*: Synthetic Doppler tomogram for the system in its quiescent state. The Roche lobe of the donor star (thick black curve) and the accretor (star) are indicated. The white curve with circles and black curve with squares display gas dynamic trajectories in velocity space (see also Figure 1, left panel). A, B, C, and D denote the same zones of maximum radiation intensity as in the left panel.

"hot spot" model (see Figure 7). Application of the "hot line" model to the light curves of the X-ray nova GU Mus (Khruzina *et al.*, 2003a) demonstrated that this model is able to adequately describe the whole range of features of this X-ray nova's optical and IR orbital variability in quiescence. The "hot spot" model cannot reproduce the unusual shape of the light curve of GU Mus or, in some cases, the quiescent amplitude of the orbital light curves. Comparison between "hot spot" and "hot line" models for interpretation of the light curve of another X-ray nova in quiescence, XTE J1118+480 (Khruzina *et al.*, 2005), has presented conclusive evidence in favor of the latter type of model.

The obtained results show that the "hot line" gas dynamic model can explain rather well the observed light curves of cataclysmic and X-ray binaries. Moreover, the use of the "hot line" model to interpret the light curves in systems where the application of standard models with the "spot" were unsuccessful, also results in rather good agreement of the synthetic and observed light curves.

4.2. *The model of superoutburst in binaries of SU UMa type*

Superhumps are modulations of the light curves of binary systems with periods that differ from the orbital periods by several percent, and are observed mainly during super-outbursts in SU UMa systems. Currently, the most popular model explains these light variations as an effect of the precession of the accretion disk. The presence of the Lindblad $3:1$ resonance in the disk results in an instability that leads to the precession of the outer part of the disk, with a period that is appreciably longer than the orbital period. The beating of the orbital and precessional periods gives rise to the periodic variations that are identified with superhumps. This model has several shortcomings, the most important being that it implies a limit on the maximum component mass ratio. In order for $3:1$ resonance to be located inside the accretion disk, the ratio of the donor and accretor masses q must be lower than ~ 0.33, while there are several observed systems with superhumps where the mass ratio is well above the critical value. The estimates of the mass ratio for TV Col show that q can be as much as $0.6 \div 0.9$ (Retter *et al.*, 2003).

A new mechanism for the formation of superhumps in SU UMa systems was suggested in Bisikalo *et al.* (2004b). The basis of this mechanism is the idea that a precessional-type

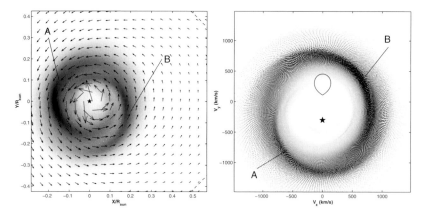

Figure 9. *Left panel*: Intensity map in the equatorial plane of the system in its active state. A and B denote zones with maximum intensity of radiation corresponding to the two arms of the tidal shock. The arrows denote velocity vectors in the laboratory coordinate system. The star marks the accretor. *Right panel*: Synthetic Doppler tomogram for the system in its active state. The Roche lobe of the donor star (thick black curve) and the accretor (star) are shown. A and B denote the same zones with maximum radiation as in the left panel.

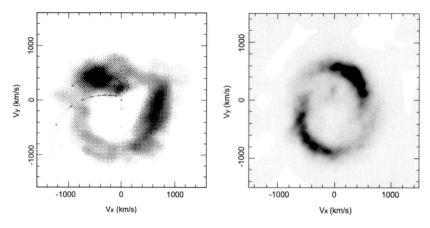

Figure 10. Doppler map of IP Peg for quiescent (left panel) and active (right panel) states of the system (Morales-Rueda *et al.*, 2000; Wolf *et al.*, 1998).

density wave can form in the cold accretion disk (see Section 3). After the formation of the precessional density wave in the disk, the rate of accretion grows sharply (by up to an order of magnitude). Matter approaches the surface of the accretor along the precessional wave, and the region of accretion is localized in azimuth, and, hence, forms a radiating spot at the surface of accretor. The increase in the accretion rate due to the density wave explains both the development of a superoutburst and the amplitude of the superhump (Bisikalo *et al.*, 2004b, 2006; Kaigorodov *et al.*, 2006). The beating of the orbital period and the precessional period of the wave results in the superhump period, which is slightly larger than the orbital period.

If outburst is associated with the formation of a "precessional" wave we will see the typical structure of the cold disk during the outburst (see Figure 4, right panel). On the other hand, being in the frame of this mechanism, during the quiescent state the disk can not have the spiral "precessional" wave. It means that in this state inner parts of the disk

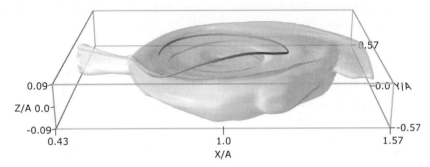

Figure 11. Three-dimensional surface of constant density and fragment of streamline emerging from the neighborhood of L_1.

are gasdynamically disturbed. At the beginning of the quiescent state the gas is heated up to the high temperature, so the solution will be close to one obtained for the hot disk (see Figure 4, left panel). Using 3D gas dynamic simulations alongside with the Doppler tomography technique allows to identify the main features of the flow on the Doppler maps without solution of an ill-posed inverse problem (Kuznetsov *et al.*, 2003). Figures 8 and 9 show intensity maps and synthetic Doppler tomograms for system in quiescent and in outburst state. Comparison of the synthetic tomograms with observations (Figure 10) shows that, in the quiescent state of the system, the most luminous components are (i) the "hot line" (zone A), and (ii) the gas condensation at the apogee of the quasi-elliptical disk (zone C). Both the single arm of spiral shock wave (zone D) and the stream itself (zone B) contribute little to the luminosity. In the active state of the system, when the stream does not play an appreciable role and the disk dominates, both areas of enhanced luminosity in the observational tomograms are associated with the two arms of the spiral shock wave in the disk. Comparison of synthetic Doppler tomograms with observations shows that the considered model does not contradict observations.

This superoutburst model based on the presence of a precessional spiral wave in the accretion disk made it possible for the first time to explain all important observational manifestations of superoutbursts and superhumps in SU UMa systems, including their period, duration, and energy; the anticorrelation of the brightness and color temperature at the maximum of an ordinary superhump; the late-superhump phenomenon, etc. Moreover, this model does not place strict constraints on the component mass ratio, and so can be applied to the superhumps in systems of large mass ratio.

4.3. *On possible nature of pre-eclipse dips in light curves of semidetached systems with steady-state disks*

Observations of low-mass X-ray binaries (LMXBs) have revealed pre-eclipse dips in the X-ray light curves for several systems. Similar light-curve features in various wavelength ranges have also been recorded for a number of cataclysmic binaries in outburst. Further observations showed that light curve dips can also appear when a system is in a stationary state. Studies of the ultraviolet light curves of the eclipsing nova-like cataclysmic binaries UX UMa and RW Tri (Mason *et al.*, 1997) confirmed this result, and suggested that this phenomenon was universal in semi-detached binaries with accretion disks. It is interesting that, in contrast to the cataclysmic systems, systems with stationary disks display pre-eclipse dips at much earlier phases, $\sim 0.6 \div 0.7$ (Mason *et al.*, 1997; Froning *et al.*, 2003).

In order to explain pre-eclipse dips at phases ~ 0.8 in light curves of LMXBs and CVs, the idea that during the outburst the stream collides with the disk and ricochets off the disk's edge is considered as the best way to explain the presence of matter at heights considerably exceeding the disk thickness (see, e.g., Armitage & Livio, 1996). Shock-free nature of the interaction between the stream and the steady-state disk in the stationary solution poses the question on the cause of the of matter presence at significant height above the accretion disk in systems of this type. The results of 3D modeling have shown that in the absence of direct collision between the stream and the disk the formation of thickening of the halo above the disk is also possible (Bisikalo *et al.*, 2005).

In the framework of the "hot line" model the significant part of the matter gets the acceleration in vertical direction. Gas movement in the vertical direction together with its movement along the outer edge of the disk leads to the gradual increase of the near-disk halo width. Maximum of the calculated thickening located above the outer part of the disk corresponds to the ~ 0.7 phase (see Figure 11) that is in agreement with the observed values for systems with steady-state disks. This fact confirms the "hot line" model suggested earlier for description of the flow structure in semi-detached binaries and gives new opportunities for the interpretation of the light curves of such systems.

5. Conclusions

Results of gas dynamic modeling allow us to understand better the structure of gaseous flows in close binaries. For systems with stationary accretion disks the flow structure is described by the "hot line" model that is in a good agreement with observations.

Acknowledgements

The work was supported by the Russian Foundation for Basic Research (projects nos. 05-02-16123, 05-02-17070, 05-02-17874, 06-02-16097), the Scientific Schools Program (grant no. NSh-4820.2006.2), and the basic research programs "Mathematical modeling and intellectual systems" and "Origin and evolution of stars and galaxies" of the RAS.

References

Armitage, P.J. & Livio, M. 1996, *ApJ* 470, 1024

Bisikalo, D.V., Boyarchuk, A.A., Chechetkin, V.M., Kuznetsov, O.A., & Molteni, D. 1998a, *MNRAS*, 300, 39

Bisikalo, D.V., Boyarchuk, A.A., Kuznetsov, O.A., Khruzina, T.S., & Cherepashchuk, A.M. 1998b, *Astron. Rep.* 42, 33

Bisikalo, D.V., Boyarchuk, A.A., Kuznetsov, O.A., & Chechetkin, V.M. 1999a, *Astron. Rep.* 43, 229

Bisikalo, D.V., Boyarchuk, A.A., Chechetkin, V.M., Kuznetsov, O.A., & Molteni, D. 1999b, *Astron. Rep.* 43, 797

Bisikalo, D.V., Harmanec, P., Boyarchuk, A.A., Kuznetsov, O.A., & Hadrava, P. 2000a, *A&A* 353, 1009

Bisikalo, D.V., Boyarchuk, A.A., Kuznetsov, O.A., & Chechetkin, V.M. 2000a, *Astron. Rep.* 44, 26

Bisikalo, D.V., Boyarchuk, A.A., Kilpio, A.A., Kuznetsov, O.A., & Chechetkin, V.M. 2001, *Astron. Rep.* 45, 611

Bisikalo, D.V., Boyarchuk, A.A., Kaigorodov P.V., & Kuznetsov, O.A. 2003, *Astron. Rep.* 47, 809

Bisikalo, D.V., Boyarchuk, A.A., Kaigorodov P.V., Kuznetsov, O.A., & Matsuda T. 2004a, *Astron. Rep.* 48, 449

Bisikalo, D.V., Boyarchuk, A.A., Kaigorodov P.V., Kuznetsov, O.A., & Matsuda T. 2004b, *Astron. Rep.* 48, 588

Bisikalo, D.V. 2005, *Ap&SS* 296, 391

Bisikalo, D.V., Kaigorodov P.V., Boyarchuk, A.A., & Kuznetsov, O.A. 2005, *Astron. Rep.* 49, 701

Bisikalo, D.V., Boyarchuk, A.A., Kaigorodov P.V., Kuznetsov, O.A., & Matsuda T. 2006, *Chin. J. Astron. Astrophys.*, vol. 6, p. 159

Boyarchuk, A.A., Bisikalo, D.V., Kuznetsov, O.A., & Chechetkin, V.M. 2002, *Mass transfer in close binary stars*, Taylor and Francis, London

Fridman, A. M., Boyarchuk, A. A., Bisikalo, D. V., Kuznetsov, O. A., Khoruzhii, O. V., Torgashin, Yu. M., & Kilpio, A. A. 2003, *Phys. Lett.* A 317, 181

Froning, C.S., Long, K.S., & Knigge, C. 2003, *ApJ* 584, 433

Gorbatskii V.G. 1967, *Astrofisica* 3, 245

Hirose, M., Osaki, Y., & Mineshige, S. 1991, *Publ. Astron. Soc. Japan*, vol. 43, p. 809

Harmanec, P., Bisikalo, D. V., Boyarchuk, A. A., & Kuznetsov, O. A. 2002, *A&A*. 396, 937

Kaigorodov P.V., Bisikalo, D.V., Kuznetsov, O.A., & Boyarchuk, A.A. 2006, *Astron. Rep.* 50, 537

Khruzina T.S., Cherepashchuk, A.M., Bisikalo, D.V., Boyarchuk, A.A., & Kuznetsov O.A. 2001, *Astron. Rep.* 45, 538

Khruzina T.S., Cherepashchuk, A.M., Bisikalo, D.V., Boyarchuk, A.A., & Kuznetsov O.A. 2003a, *Astron. Rep.* 47, 621

Khruzina T.S., Cherepashchuk, A.M., Bisikalo, D.V., Boyarchuk, A.A., & Kuznetsov O.A. 2003a, *Astron. Rep.* 47, 848

Khruzina T.S., Cherepashchuk, A.M., Bisikalo, D.V., Boyarchuk, A.A., & Kuznetsov O.A. 2005, *Astron. Rep.* 49, 79

Kuznetsov, O.A., Bisikalo, D.V., Boyarchuk, A.A., Khruzina, T.S., & Cherepashchuk, A.M. 2001, *Astron. Rep.* 45, 872

Mason, K.O., Drew, J.E., & Knigge, C. 1997, *MNRAS* 290, L23

Molteni, D., Belvedere, G., & Lanzafame, G. 1991, *MNRAS*, 249, 748

Molteni, D., Kuznetsov, O. A., Bisikalo, D. V., & Boyarchuk, A. A. 2001, *MNRAS* 327, 1103

Morales-Rueda, L., Marsh, T. R., & Billington, I. 2000, *MNRAS* 313, 454

Nagasawa, M., Matsuda, T., & Kuwahara, K. 1991, *Numer. Astrophys. in Japan*, vol. 2, p. 27

Retter, A., Hellier, C., Augusteijn, T., *et al.* 2003, *MNRAS* 340, 679

Sawada K., Matsuda T., & Hachisu I. 1986, *MNRAS* 219, 75

Smak J. 1970, *AcA* 20, 312

Struve, O. 1941, *ApJ* 93, 104

Warner, B. 1995, *Cataclysmic Variable Stars*, Cambridge University Press, Cambridge

Wolf, S., Barwig, H., Bobinger, A., *et al.* 1998, *A&A* 332, 984

Discussion

JOHN SOUTHWORTH: There are some very good light curves of eclipsing CV where a bright spot model fits well but not perfectly. Would you be able to improve upon this with your model?

BISIKALO: Hopefully, yes.

MERCEDES RICHARDS: What are the inner boundary conditions for your simulations (i.e., do your simulations allow you to follow the flow down to the star)?

BISIKALO: The inner boundary conditions is set close to the surface of the WD. Using the supercomputer allows us to consider the flow structure up to the surface of WD.

SLAVEK RUCINSKI: Can your hydrodynamic model be used for close main-sequence stars when one star fills the Roche lobe and sends the matter straight into the surface of the second star? We see many EB stars where one side is directly heated by the colliding stream (good example: V361 Lyrae).

BISIKALO: Yes, this model is suitable for consideration of the flow structure in such systems. Actually we are now working on simulations of a system where streams collide directly onto the surface of the second star.

MERCEDES RICHARDS: Can you distinguish between the thickened part of the disk and the projection of the gas stream at phase 0.6-0.7 in the synthesized light curve?

BISIKALO: Yes, the synthetic light curve was obtained from the gas dynamic solution, where we can see the real thickenning of the circumdisk halo.

MERCEDES RICHARDS: This is very nice work! Thank you.

Binary Stars as Critical Tools & Tests
in Contemporary Astrophysics
Proceedings IAU Symposium No. 240, 2006
W.I. Hartkopf, E.F. Guinan & P. Harmanec, eds.

© 2007 International Astronomical Union
doi:10.1017/S1743921307004309

Young Binaries as a Test for
Pre-Main Sequence Evolutionary Tracks

Hans Zinnecker

Astrophysikalisches Institut Potsdam, An der Sternwarte 16, D-14482 Potsdam, Germany
email: hzinnecker@aip.de

Abstract. Observations of young low-mass binaries ($t \lesssim 10^7$ yr, $M \lesssim 3\,M_\odot$) can be used to calibrate pre-Main Sequence (pre-MS) evolutionary tracks. Recent high angular resolution HST/FGS, speckle, and long-baseline interferometry have resolved the astrometric orbits of a few SB2 pre-MS binaries and have provided the individual dynamical masses of their components as well as the system orbital parallaxes. Spectroscopic fits and filter photometry have permitted to determine SpT (temperatures) and a good estimate of the absolute magnitude (bolometric luminosity) of the components, which in turn allows one to place the components on a theoretical HR-diagram. In this way, one can check (a) whether the measured dynamical masses agree with the predicted masses on the tracks and (b) whether both components lie on an isochrone, as they should for a coeval physical pair of stars.

With a sufficiently large sample of different masses and ages of resolved SB2 systems, most of the parameter space of pre-MS tracks can be tested, even for very low stellar masses ($M < 0.5\,M_\odot$) and very young ages (< 2 Myr). This is a prerequisite in order to derive the IMF and star formation history in very young clusters and associations.

Keywords. techniques: high angular resolution, techniques: interferometric, techniques: spectroscopic telescopes, astrometry, binaries: spectroscopic, binaries: visual, stars: fundamental parameters, Hertzsprung-Russell diagram, stars: late-type, stars: pre–main-sequence

1. Introduction

I would like to begin my contribution by paying tribute to Czech born Prof. Zdeněk Kopal, one of the fathers of close binary star astrophysics, who died in 1993 and is buried around the corner in the Prague Vyšehrad cemetery. I met him during the total solar eclipse in June 1983 in Indonesia (cf. Kopal 1986) following IAU Coll. No. 80, of which he is the co-editor. It was he (and the late Dr. Rahe from Bamberg) who invited me to present a first review on "binary statistics and star formation" (Zinnecker 1984). This was also my first major paper for an IAU-related event, and I am still proud of it, although to this day hardly anyone took note of it (except T. Oswalt in his talk yesterday). Indeed, this paper now provides an interesting look back in time to what we knew in ~1980 about binary frequency and multiplicity, mass ratio and period distribution. Those were the days when star formation started to take off as a subject, and an early confrontation between observations and theory became possible.

Ten years later, an excellent ARAA review on pre-MS binaries was published (Mathieu 1994). Then, in the year 2000, I hope some of you remember, IAU-Symp. 200 "The Formation of Binary Stars" was held in Potsdam (eds. Zinnecker & Mathieu 2001).

In today's meeting the topic that I want to address is not so much the formation of binary stars but more the use of young pre-MS binaries for checking that we have reliable masses and ages for low-mass pre-MS stars and brown dwarfs. This context already played a role in Mathieu's (1994) review, but in the meantime more and better data are becoming available, mainly through the advent of long-baseline interferometry. A rationale for

interferometry and a summary of results on dynamical mass determinations are given in
Zinnecker & Correia (2004) and Hillenbrand & White (2004), respectively. Here I will try
to explain the logic behind the use of young binaries for testing early stellar evolution. In
addition, the latest results on dynamical mass measurement of young stellar objects will
be presented. A more comprehensive review of this topic is the article by Mathieu *et al.*
(2006) in the proceedings of *Protostars and Planets V* (highly recommended to read).

2. Pre-MS tracks

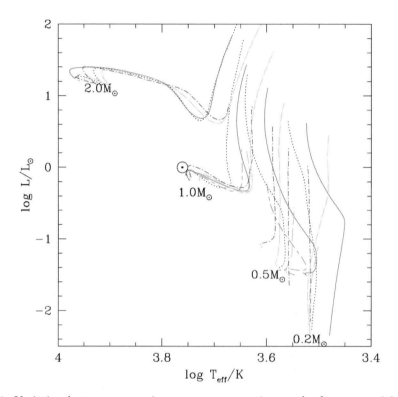

Figure 1. Variation between pre-main-sequence contraction tracks for masses 0.2, 0.5, 1.0,
and 2.0 $M_\odot$; for the following models: S93 – Siess (solid line), DM97 – d'Antona & Mazzitelli,
with 1998 correction (dotted line), B98, $\alpha = 1.9$ – Baraffe *et al.* (long-dashed line), PS99 – Palla
& Stahler (dot-short-dashed line), S00 – Siess (dot-long-dashed line), and Y^2 – Yi/Yale (long–
dash-short-dashed line). Note that the PS99 models, for which no 0.5 $M_\odot$ track is available,
have both the 0.4 and the 0.6 $M_\odot$ tracks plotted instead. Note also that the Y^2 models do not
extend as low as 0.2 $M_\odot$ (figure & caption from Hillenbrand & White 2004).

Figure 1 shows pre-MS tracks for 0.2, 0.5, 1.0, and 2.0 $M_\odot$ as calculated by various
authors (given in the figure caption). The near vertical parts of the curves correspond to
the so-called Hayashi tracks (fully convective quasi-statically contracting stars), while the
more horizontal parts leading on to the Main Sequence correspond to a stellar structure
where the energy transport in the interior is mainly radiative (so-called Henyey tracks).
A good description of pre-MS evolution and a reference set of pre-MS evolutionary tracks
with isomasses and isochrones can be found in the new textbook on 'The Formation of
Stars' by Stahler & Palla (2004; see their Figure 1.18).

One can easily see from Figure 1 that the pre-MS models differ quite substantially for
low masses, particularly below 0.5 $M_\odot$ where molecular opacities start playing a role in

the cool atmospheres. It can happen that someone's $0.2\,M_\odot$ star is another one's $0.5\,M_\odot$ star, for a given point in the bottom right corner of the HR-diagram. Such a big error is clearly unacceptable, and the model input must be fixed or calibrated by appropriate observations.

Table 1. Important young spectroscopic binaries for dynamical mass determinations

References	Objects	Type
Covino *et al.* 2000, *A&A Lett.*	RXJ 0529.4+0041	SB2E
Steffen *et al.* 2001, *AJ*	045251+3016	SB2A
Boden *et al.* 2005, *ApJ*	HD 98800B	SB2A
Schaefer, Simon, Prato 2006, *ESO conf.*	Haro 1-14c	SB2A
Stassun, Mathieu, Valenti 2006, *Nature*	2MASS 0535−05	SB2E

3. Logical steps for testing pre-MS tracks

The first step is, of course, to calculate a set of pre-MS tracks on the computer, using certain input physics (e.g., convection described by a mixing-length parameter or otherwise, opacity sources, metallicity, or even magnetic star spots). Different assumptions on the main input physics lead to substantially different evolutionary tracks.

The next step then is to select those tracks that best describe the early evolution of real stars rather than model stars. How to do this? The trick is to use young low-mass spectroscopic binaries in a variety of star-forming regions (young clusters and associations) with measured orbital periods (and inferred component separations) such that they can be spatially resolved and their orbits followed by high-angular resolution interferometric observations. The relevant periods are around 1 yr, give or take a factor of $3-5$ (i.e., semi-major axes of the order of 1 AU within a factor of $2-3$), depending on the actual distance ($50\,\mathrm{pc}-150\,\mathrm{pc}$). Young eclipsing binaries would be even better but they are very rare and only very few have been found (see the results section). Note that a semi-major axis of 1 AU at 50 pc (e.g., TW Hydrae association) or 150 pc (e.g., the Sco Cen association) correspond to angular separations of 20 mas and 7 mas, respectively, fairly easy to do with Keck or VLTI measurements in the near-IR JHK bands. Because long-baseline interferometers such as Keck and VLTI have only recently begun to operate, it is no wonder that only very few examples have been observed.

In more detail, the useful young binary objects have to meet two more requirements:

a) they had better be weak-line or naked T Tauri stars rather than classical T-Tauri stars. The former have the advantage of possessing no circumstellar or circumbinary accretion disks, hence their bolometric luminosity need not be corrected for accretion luminosity (which in turn requires observing their spectral energy distribution and an appropriate subtraction). The only correction required is for foreground extinction and reddening.

b) they had better be SB2 systems rather than SB1s. Only the combination of SB2 radial velocity curves and a resolved astrometric orbit yields the orbital inclination ($\sin i$), thus giving the individual masses of the components and, importantly too, the so-called orbital parallax. In case we have an SB1, all is not lost, as there is every chance now to transform an SB1 into an SB2 using near-IR rather than optical spectroscopy (Mazeh *et al.* 2003, Prato *et al.* 2003). The reason is that the faint secondary component tends to be cooler and hence relatively brighter w.r.t. the primary in the near-IR, hence the brightness is more nearly equal to unity in the near-IR than in the optical.

The third step is to place the resolved binary components into a theoretical pre-MS HR-diagram or color-magnitude diagram. This is not easy, as we need to get separate SpT and stellar luminosities for the components (being careful to avoid hidden triple systems). How to achieve this? Stellar luminosities, particularly for the fainter secondary, can be estimated from the total luminosity of the system and the brightness ratio in at least one filter; the more filters, the better (note here that the VLTI/AMBER instrument provides measurements in all the JHK filters simultaneously). With more than one filter, we also obtain useful component color information. The decomposition of the spectral types is done in the near-IR, using true near-IR spectral template stars which were measured before and tabulated. Then two spectra are superimposed, starting from an initial guess, and iterated, until the observed spectrum of the binary system is matched (in a chi-square sense). A list of interesting young spectroscopic binaries (SB's) is given in Table 1.

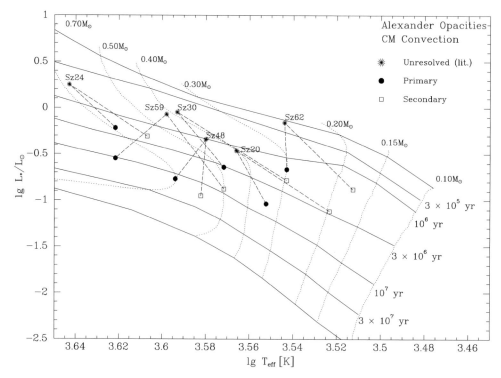

Figure 2. Comparison of literature values for the unresolved binaries with results for the individual binary components (from Brandner & Zinnecker 1997, using D'Antona & Mazzitelli 1994 tracks). Unresolved binaries lead to an underestimate of the age of a T Tauri star population.

As an aside, it is interesting to check, in a kind of a thought experiment, how wrong one can go in terms of pre-MS masses and ages when the binary nature of a young spectroscopic or close visual binary system is unrecognized as opposed to be recognized and resolved (see Figure 2). One can see that pre-MS masses and ages can change by more than 100 % (e.g., Sz 48, Sz 59 in Figure 2). Indeed, overestimating the stellar luminosity in an unresolved pre-MS binary whose components are on Hayashi tracks produces an underestimate of the system age, as already noticed by Simon, Ghez, & Leinert (1993). As a consequence, as these authors have pointed out, the average age of a young group of stars (such as the T Tauri stars in Taurus, whose binary frequency is close to 100 %) is 2 – 3 times longer than it would be if all its binaries were regarded as unresolved objects!

4. Results

4.1. *Previous results*

It was at the IAU Symposium 200 in Potsdam six years ago that the first significant number of dynamical mass measurements of pre-Main Sequence stars were reported, although the very first measurement (the pre-MS secondary of the SB1 eclipsing binary system EK Cep) goes back to Popper (1987). At IAU-S200, M. Simon gave a review on dynamical masses of young stars that focussed primarily on Keplerian rotation measurements of spatially resolved circumstellar disks around *single* pre-MS T Tauri stars (his own work in collaboration with Dutrey & Guilloteau at IRAM/PdBI using ^{12}CO J$=2-1$ maps). Masses of the following six stars were reported: MWC 480, LkCa 15, DL Tau, GM Ori, DM Tau, and CY Tau, with values ranging from $1.65\,M_\odot$ down to $0.55\,M_\odot$. Errors on the masses were $3-15\,\%$, except for the latter where the error exceeds $50\,\%$. These errors do not include the imprecision of the distance to the sources, which may add an extra $10-20\,\%$. [BP Tau's mass was also listed, but later an improved value was published (Dutrey *et al.* 2003). Also, UZ Tau E is a spectroscopic binary with a circumbinary disk, so that the binary system mass could be determined (Prato *et al.* 2002).]

At the same IAU meeting, E. Covino *et al.* announced the first pre-MS eclipsing binary among double-lined T-Tauri stars in Orion (RXJ 0529.4+0041) and gave preliminary masses ($M_1 = 1.25\,M_\odot$, $M_2 = 0.91\,M_\odot$), while L.P. Vaz discussed the TY CrA system, a Herbig Ae/Be star and hierarchical triple with an eclipsing pair and wider tertiary star. The component masses of the eclipsing pair are $3.16\,M_\odot$ (primary, near the ZAMS) and $1.64\,M_\odot$ (secondary, pre-MS, estimated age 3 Myr). The mass of the third member of the system is unknown, its orbit may be highly inclined w.r.t. the inner pair (Casey *et al.* 1998).

Furthermore, R. White described how to use the hierarchical quadruple system GG Tau (consisting of two close visual pairs, about $10''$ apart, with individual component separations of $0\rlap{.}''25$ and $1\rlap{.}''45$, respectively) to test pre-MS evolutionary models, especially under the assumption of coeval formation. *Higher order multiple systems are in principle more powerful test objects than pure binary systems, if all components (including brown dwarfs) can be placed on the HR-diagram.*

Last but not least, F. Palla confronted pre-MS models with observations of four intermediate-mass pre-MS spectroscopic eclipsing binaries (1 SB2: RS Cha; 3 SB1s: EK Cep, BM Ori, TY CrA) and found rather good agreement with his own models (Palla & Stahler 1999). (We omit here the non-eclipsing young SB2s: V773 Tau, NTTS 162814, and P1540, for which only mass ratios but no individual masses can be determined).

Taking all the above information from IAU-S200 together, there have been more than a dozen dynamical masses of young stellar objects gathered up to the year 2000, but the constraints on pre-MS evolutionary tracks from them were not very strong mainly because of a lack of accuracy.

4.2. *Recent results*

A new era of dynamical mass measurements for pre-MS stars began with the publication of the first astrometric-spectroscopic orbit of NTTS 045251+3016 by Steffen *et al.* 2001 (already presented as a poster at IAU-S200). They used a double-hybrid technique: 1) they patiently resolved a good part of the 7 yr SB1 orbit with the HST/FGS and 2) they transformed the optical light SB1 into an SB2 by near-IR spectroscopy. They thus could derive the inclination angle of the orbit, hence the individual masses, and indeed also the orbital parallax, as discussed in Section 3. The measurements of the primary and secondary component masses ($1.45\,M_\odot$ and $0.81\,M_\odot$) were precise enough

($\sim$10%) to indicate that the Baraffe *et al.* (1998) tracks (with mixing-length parameter $\alpha = 1.0$) provided the best fit to the observations. These tracks then predict an age of the system.

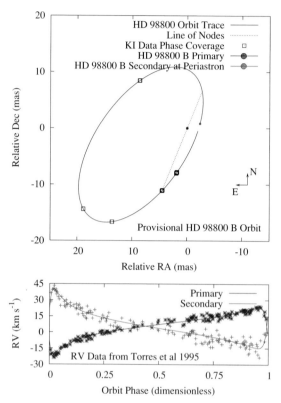

Figure 3. Astrometric-spectroscopic orbit of HD 98800 B (Boden *et al.* 2005).

It took several years until this feat was sort of repeated for another system: HD 98800B in the very nearby TW Hya association (Boden *et al.* 2005). These authors combined FGS data with ground-based K-band Keck interferometric visibility data to derive an astrometric orbital solution for this one binary in a pre-MS quadruple system. Note that the visibility data were compared directly to model predictions in order to constrain the astrometric orbit, without an intermediate determination of the separation. Boden *et al.* (2005) find that the component masses ($M_1 = 0.70 \pm 0.06\,M_\odot$, $M_2 = 0.58 \pm 0.05\,M_\odot$), luminosities and effective temperatures of HD 98800B are inconsistent with solar-metallicity evolutionary tracks; they suggest that a lower metal abundance by a factor of $2-3$ would resolve the discrepancy. The derived orbital parallax of the system is ~ 42 pc.

This forefront observational study bodes well for a significant number of masses being measured via interferometry and spectroscopy in the near future. Indeed, Haro 1-14c has also been studied using the Keck interferometer (Schaefer, Simon, & Prato 2006). The results indicate that the secondary is likely to have a mass of about $0.4\,M_\odot$, the lowest-mass pre-MS tested to date (excluding brown dwarfs). Similarly low-mass secondary components may prove most valuable in the future, because pre-MS models are least constrained in the lowest-mass regime (see Section 2).

Another highlight was the discovery of the eclipsing binary young brown dwarf 2MASS J05352184−0546085 in the Orion Nebula star formation region (Stassun, Mathieu, & Valenti 2006) with component masses of about 54 and 34 Jupiter masses and radii of about 0.67 and $0.51 R_\odot$ (obtained from a combined analysis of the light curve and orbit solution). Radial velocities of the two components were obtained at eight distinct epochs with the Phoenix high-resolution near-infrared spectrograph at the Gemini South telescope. The orbital period is about 10 days. Surprisingly the less-massive brown dwarf is the hotter of the pair, a fact for which no reasonable explanation has been found up to now (but see Wuchterl & Tscharnuter 2003).

Stassun *et al.* (2004) had earlier discovered a pre-MS stellar eclipsing binary (V1174 Ori with $M_1 = 1.01 M_\odot$ and $M_2 = 0.73 M_\odot$). Eclipsing pre-MS binaries offer of course an extra advantage over astrometric-spectroscopic binaries: a measurement of stellar radii which are also a predicted outcome from pre-MS evolutionary models and hence help to further confront theory with observations (for details see Mathieu *et al.* 2006).

Duchêne *et al.* (2006) recently published stellar masses in the T Tau triple system, especially those of T Tau Sa and T Tau Sb (2.73 ± 0.31 and $0.61 \pm 0.17 M_\odot$, respectively). Taking advantage of T Tau N as a reference, absolute astrometric monitoring of the Keplerian motion of the T Tau S binary system was possible, using K-band infrared speckle interferometry at the Keck telescope to fit the orbit.

Further, a 240 AU wide planetary mass young binary (Oph 1622), with model masses of 14 and 7 M(Jupiter), has been identified with the VLT (Jayawardhana & Ivanov 2006).

Finally, another exciting result became known during the present IAU Symposium 240: the star θ^1 Ori E, a member of the Orion Trapezium cluster, has been detected as a double-lined spectroscopic binary (Herbig & Griffin 2006) with estimated masses of 3–$4 M_\odot$ and a period of 9.89 days (see also Costero 2007). This object is the second brightest X-ray source in the Trapezium cluster after θ^1 Ori C, itself resolved as a sub-arcsecond speckle binary (see Schertl *et al.* 2003 for orbital motion).

4.3. *Summary*

Where does that leave us? Most of the results mentioned above have been discussed in two recent comprehensive reviews: Hillenbrand & White (2004) and Mathieu *et al.* (2006, PPV). However, the last 6 years have not been too prolific in providing many new results, and progress has rather stalled, perhaps because the "Golden Age of Astrometry" (quoting M. Simon) is close but hasn't quite arrived yet. Nevertheless VLTI, Keck-I, and CHARA observations of more pre-MS binaries will come in within the next few years. Already a target list of some eight pre-MS spectroscopic binaries has been elaborated for VLTI/AMBER (Guenther *et al.* 2007, see also Melo *et al.* 2001), ready to be observed. An amazing spectroscopic triple system is BS Indi in the young Tucana association: it consists of two eclipsing M0V stars orbiting a K0V star with a period of 3.3 years, readily resolvable with the VLTI. Stay tuned!

Yet, now that dynamical mass measurements will become more frequent with the application of ground–based optical/infrared interferometers, the primary limitations to such tests will be systematic errors in determining the stellar properties necessary for the comparison with evolutionary models, in complete agreement with Mathieu *et al.* (2006).

Acknowledgements

I thank Drs. S. Correia and E. Guenther for insightful discussions about the most recent results in the literature. I also thank Bob Mathieu and Mike Simon for valuable information during the preparation of this talk. Last but not least I thank the IAU and the NOC chairman of the Prague General Assembly, Jan Palouš, for financial support.

References

Baraffe, I., Chabrier, G., Allard, F., & Hauschildt, P.H. 1995, *ApJ* 446, L35 (B98)

Boden, A.F. *et al.* 2005, *ApJ* 635, 442

Brandner, W. & Zinnecker, H. 1997, *A&A* 321, 220

Casey, B.W., Mathieu, R.D., Vaz, L.P.R., Andersen, J., & Suntzeff, N.B. 1998, *AJ* 115, 1617

Costero, R., Poveda, A., & Echevarría, J. 2007, these proceedings, 130

Covino, E., Catalano, S., Frasca, A., Marilli, E., Fernández, M., Alcalá, J.M., Melo, C., Paladino, R., Sterzik, M.F., & Stelzer, B. 2000, *A&A* 361, L49

D'Antona, F. & Mazzitelli, I. 1994, *ApJS* 90, 467

D'Antona, F. & Mazzitelli, I. 1997, *Mem. Soc. Astr. It.* 68, 807 (DM97)

D'Antona, F., Ventura, P., & Mazzitelli, I. 2000, *ApJ* 543, L77

Duchêne, G., Beust H., Adjali F., Konopacky Q.M., & Ghez, A.M. 2006, *A&A* 457, L9

Guenther, E., Esposito M., Mundt R., Covino E., Alcalá, J.M., & Stecklum B. 2007, astro-ph/2268G

Guenther, E.W., Covino, E., Alcalá, J.M., Esposito, M., & Mundt, R. 2005, *A&A* 433, 629

Herbig, G.H. & Griffin, R.F. 2006, *AJ* 132, 1763

Hillenbrand, L.A. & White, R.J. 2004, *ApJ* 604, 741

Jayawardhana, R. & Ivanov, V.D. 2006, *Science* 313, 1279

Kopal, Z. 1986, *Of stars and men. Reminiscences of an astronomer* (Bristol: Hilger)

Mathieu, R.D. 1994, *ARA&A* 32, 465

Mathieu, R., Stassun, K., Baraffe, I., Simon, M., & White, R. 2006, in: B. Reipurth, D. Jewitt, K. Keil (eds.), *Protostars and Planets V*, in press

Mazeh, T., Simon, M., Prato, L., Markus, B., & Zucker, S. 2003, *ApJ* 599, 1344

Melo, C.H.F., Covino, E., Alcalá, J.M., & Torres, G. 2001, *A&A* 378, 898

Palla, F. & Stahler, S.W. 1999, *ApJ* 525, 772 (PS99)

Popper, D.M. 1987, *ApJ* 313, L81

Prato, L., Greene, T.P., & Simon, M. 2003, *ApJ* 584, 853

Prato, L., Simon, M., Mazeh, T., Zucker, S., & McLean, I.S. 2002, *ApJ* 579, L99

Schaefer, G.H., Simon, M., & Prato, L. 2006, in: A. Richichi, F. Paresce (eds.), *The Power of Optical/IR Interferometry*, in press

Schertl, D., Balega, Y.Y., Preibisch, Th., & Weigelt, G. 2003, *A&A* 402, 267

Siess, L., Dufour, E., & Forestini, M. 2000, *A&A* 358, 593 (S00)

Simon, M., Ghez, A.M., & Leinert, Ch. 1993, *ApJ* 408, L33

Stahler, S.W. & Palla, F. 2004, *The Formation of Stars* (Wiley-VCH)

Stassun, K.G., Mathieu, R.D., & Valenti, J.A. 2006, *Nature* 440, 311

Stassun, K.G., Mathieu, R.D., Vaz, L.P.R., Stroud, N., & Vrba, F.J. 2004, *ApJS* 151, 357

Steffen, A.T., Mathieu, R.D., Lattanzi, M.G., Latham, D.W., Mazeh, T., Prato, L., Simon, M., Zinnecker, H., & Loreggia, D. 2001, *AJ* 122, 997

Wuchterl, G. & Tscharnuter, W.M. 2003, *A&A* 398, 1081

Yi, S.K., Kim, Y.-C., & Demarque, P. 2003, *ApJS* 144, 259 (Yi/Yale)

Zinnecker, H. 1984, *Ap&SS* 99, 41

Zinnecker H. & Correia, S. 2004, in: R.W. Hilditch, H. Hensberge, K. Pavlovski (eds.), *Spectroscopically and Spatially Resolving the Components of the Close Binary Stars*, ASP Conference Series, Vol. 318 (San Francisco: Astronomical Society of the Pacific), p. 34

Discussion

G. WUCHTERL: You showed the Boden *et al.* results that show an offset of the Siess tracks relative to the observations. Collapse models (Wuchterl & Tscharnuter 2003) predict such a shift to be a few hundred Kelvin (500 K at $1\,M_\odot$ and $\sim 2\,L_\odot$). How large is the shift in temperature that you discussed?

ZINNECKER: Let's have a look at Figure 6 in Boden *et al.* (2005). One can see that the temperature offset is ~ 200 K for solar metallicity models, both for the primary ($\sim 0.7\,M_\odot$) and for the secondary ($\sim 0.6\,M_\odot$), in the sense that the observed temperatures are hotter than those consistent with the dynamical iso-masses.

Binary Stars as Critical Tools & Tests
in Contemporary Astrophysics
Proceedings IAU Symposium No. 240, 2006
W.I. Hartkopf, E.F. Guinan & P. Harmanec, eds.

© 2007 International Astronomical Union
doi:10.1017/S1743921307004310

Eclipsing Binaries in the LMC: a Wealth of Data for Astrophysical Tests

A. Derekas, L. L. Kiss and T. R. Bedding

School of Physics, University of Sydney, NSW 2006, Australia
email: derekas@physics.usyd.edu.au

Abstract.
  We have analysed publicly available MACHO observations of 6833 variable stars in the Large Magellanic Cloud, classified as eclipsing binaries. After finding that a significant fraction of the sample was misclassified, we redetermined periods and variability class for all stars, producing a clean sample of 3031 eclipsing binaries. We have investigated their distribution in the period-color-luminosity space, which was used, for example, to assign a foreground probability to every object and establish new period-luminosity relations to selected types of eclipsing stars. We found that the orbital period distribution of LMC binaries is very similar to those of the SMC and the Milky Way. We have also determined the rate of period change for every star using the $O - C$ method, discovering about 40 eclipsing binaries with apsidal motion, 45 systems with cyclic period changes and about 80 stars with parabolic $O - C$ diagrams. In a few objects we discovered gradual amplitude variation, which can be explained by changes in the orbital inclination caused by a perturbing third body in the system.

Keywords. binaries: eclipsing, galaxies: individual (Large Magellanic Cloud)

1. Introduction

  The last decade witnessed the birth of a new research field, the large-scale study of variable stars in external galaxies. This has first been made possible by the huge databases of microlensing observations of the Magellanic Clouds, like the MACHO, OGLE and EROS projects (see Szabados & Kurtz 2000 for reviews). These programs (beyond their primary purpose) resulted in the discovery of thousands of new eclipsing binaries with an unprecedented homogeneous coverage of their light curves opening a new avenue of the binary star research. Here we present the first results of an analysis of the publicly available MACHO light curves. The main aim of the project is to measure period changes and discover eclipsing binaries with pulsating components.

2. General properties of the sample

  We have analysed MACHO lights curves for 6833 stars that were originally classified as eclipsing binaries. After re-determining the period and classifying the stars based on their light curve shapes, only 3031 stars remained as genuine eclipsing or ellipsoidal variables (the rest being Cepheids, RR Lyraes and other non-eclipsing variable stars). The period distribution of this binary sample is bimodal, with the strongest peak between 1 and 2 days. Roughly 20% of stars have periods longer than 10 days; many of them show W UMa–like light curve shapes, suggesting ellipsoidal variability of giant componens.

  We classified the binary sample using Fourier decomposition of the phase diagrams. Two coefficients, a_2 and a_4, of the cosine decomposition $\sum_{i=1}^{4} a_i \cos(2\pi i \varphi)$ allow a well-defined distinction between detached, semi-detached and contact binaries (Pojmański 2002). The results show that the sample is dominated by bright main-sequence detached

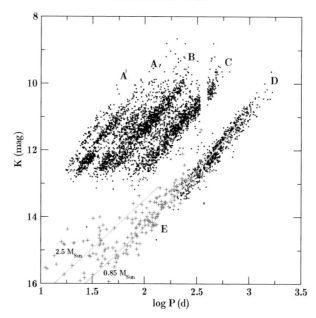

Figure 1. P–L relations of ellipsoidal variables (red pluses) red giant pulsators (black dots). The two lines show a simple model using evolutionary tracks and Roche-geometry.

(50%) and semi-detached (30%) binaries. Contact systems comprise 20% of the sample; the short-period systems are all foreground objects in the Milky Way, while longer periods belong to red giant binaries.

We used the Color-Magnitude Diagram (CMD) to clean the sample of the foreground objects. For this we took evolutionary models of Castellani *et al.* (2003) and calculated the locations of certain minimal orbital period values (where two identical model stars are in contact). The cleaned sample contains about 2800 LMC binaries.

Detached and semi-detached binaries are spread uniformly in the period–K magnitude plane, while there is a well-defined sequence for the contact systems. We found that the widely accepted sequence of eclipsing binaries between Seqs. C and D, known as Seq. E, does not exist. The correct position for Seq. E is at periods a factor of two greater. A simple Roche-model describes Seq. E very well. Although Seq. E seems to merge into Seq. D of the Long Secondary Periods (Wood *et al.* 2004), the two groups are significantly different in their color and amplitude properties (Derekas *et al.* 2006).

3. Period changes and secular amplitude variations

From the 8 years of MACHO observations we measured period changes using the $O - C$ method applied to seasonal subsets of the data. We found about 80 parabolic and 45 cyclic period changes, the rest showing linear $O - C$ diagrams. A significant fraction of the former two groups are candidates for light-time effect in hierarchic triple systems. One example is shown in the top panels of Figure 2. In about 40 eccentric binaries we measured different $O - C$ variations for the primary and the secondary minima, which indicates apsidal motion (bottom panels in Figure 2). With this we double the number of known binaries with apsidal motion in the LMC (Michalska & Pigulski 2005).

In a few objects we discovered gradual amplitude variation, which can be explained by rapid variations in the orbital geometry, most likely in inclination. A third body in the system can perturb the eclipsing pair in such a way that the eclipse depth, as a sensitive

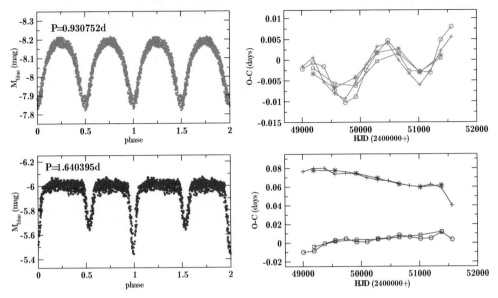

Figure 2. Examples for cyclic $O - C$ diagram and apsidal motion.

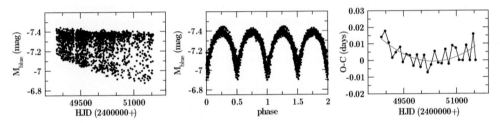

Figure 3. An example for changing minimum depth.

indicator of the inclination variations, follows these perturbations. In Figure 3 we present an example ($P_{ecl} = 0.77$d), for which the large scatter in the $O - C$ diagram may suggest an orbital period of $\leqslant 100$d for the hypothetical third companion.

Acknowledgements

This work has been supported by the Australian Research Council. LLK is supported by a University of Sydney Postdoctoral Research Fellowship. AD is supported by an Australian Postgraduate Research Award. This paper utilizes public domain data obtained by the MACHO Project, jointly funded by the US Department of Energy through the University of California, Lawrence Livermore National Laboratory under contract No. W-7405-Eng-48, by the National Science Foundation through the Center for Particle Astrophysics of the University of California under cooperative agreement AST-8809616, and by the Mount Stromlo and Siding Spring Observatory, part of the Australian National University.

References

Castellani, V., *et al.*, 2003, *A&A*, 404, 645
Derekas, A. *et al.*, 2006, *ApJL*, in press (astro-ph/0608618)
Michalska, G. & Pigulski, A., 2005, *A&A*, 434, 89
Pojmański, G., 2002, *Acta Astron.*, 52, 397
Szabados, L. & Kurtz, D.W., (Eds.), 2000, *ASP Conf. Series*, Vol. 203
Wood P.R., Olivier, E.A., & Kawaler, S.D., 2004, *ApJ*, 604, 800

Binary Stars as Critical Tools & Tests
in Contemporary Astrophysics
Proceedings IAU Symposium No. 240, 2006
W.I. Hartkopf, E.F. Guinan & P. Harmanec, eds.

The Initial–Final Mass Relationship of White Dwarfs in Common Proper Motion Pairs and Open Clusters

S. Catalán[1,2], I. Ribas[1,2], J. Isern[1,2], E. García–Berro[1,3] and C. Allende Prieto[4]

[1]Institut d'Estudis Espacials de Catalunya, c/ Gran Capità 2–4, 08034 Barcelona, Spain
email: catalan@ieec.uab.es

[2]Institut de Ciències de l'Espai, CSIC, Facultat de Ciències, UAB, 08193 Bellaterra, Spain

[3]Departament de Física Aplicada, Escola Politècnica Superior de Castelldefels, Universitat Politècnica de Catalunya, Avda. del Canal Olímpic s/n, 08860 Castelldefels, Spain

[4]McDonald Observatory and Dept. of Astronomy, University of Texas, Austin, TX 78712, USA

Abstract. We have studied white dwarfs in common proper motion pairs (CPMPs) to improve the semi-empirical initial–final mass relationship of white dwarfs. In this contribution, we report new results obtained from spectroscopic observations of both members of several CPMPs composed of an F, G or K type star and a DA white dwarf.

Keywords. stars: white dwarfs, stars: abundances, stars: evolution, binaries: visual

1. Introduction and observations

The initial–final mass relationship connects the properties of a white dwarf with those of its main-sequence progenitor. This function is important for determining the ages of globular clusters and their distances, for studying the chemical evolution of galaxies, and also for understanding the properties of the galactic population of white dwarfs. Despite its relevance, this relationship is still poorly constrained. A promising approach to decrease the uncertainties is to study white dwarfs for which external constraints are available. This is the case of white dwarfs in common proper motion pairs (CPMPs). Important information of the white dwarf can be inferred from the study of the companion, since they were born at the same time and with the same initial chemical composition.

The list of CPMPs under study was selected from the available literature (Wegner & Reid 1991, Silvestri *et al.* 2001). Each CPMP (11 in total) is composed of a white dwarf classified as DA and an F, G or K type star. The observations were carried out using a suite of telescope/instrument configurations. The white dwarf members were observed with the LCS spectrograph of the HJS (2.7-m) telescope at McDonald Observatory (Texas, USA) and with the TWIN spectrograph of the 3.5-m telescope at Calar Alto Observatory (CAHA, Almería, Spain), obtaining a FWHM resolving power of $\sim 4 - 5$ Å. The FGK companions were observed with the FOCES echelle spectrograph of the 2.2-m telescope at CAHA and with the SARG echelle spectrograph at the TNG telescope in La Palma (Canary Islands, Spain) with resolutions of $R \sim 47000$ and $R \sim 57000$, respectively. These spectroscopic observations have revealed that only 5 of the 11 white dwarfs were in fact of the DA type, whereas the rest were misclassified.

Table 1. Stellar parameters derived for the observed FGK stars.

Name	T_{eff} (K)	Z	$\log(L/L_\odot)$	Age (Gyr)
G 158-77[1]	4387 ± 27	–	–	–
BD+44 1847	5627 ± 49	0.006 ± 0.003	-0.188 ± 0.059	–[2]
BD+23 2539[1]	5666 ± 48	–	–	–
BD+34 2473	6268 ± 68	0.015 ± 0.008	0.369 ± 0.109	3.25 ± 2.28
BD−08 5980	5669 ± 52	0.007 ± 0.004	-0.151 ± 0.040	–[2]

[1]Not analyzed because of low S/N. More observations are underway.
[2]The ages obtained from isochrone fits are not reliable (see text).

Table 2. Stellar parameters derived for the observed white dwarfs.

Name	T_{eff} (K)	$\log g$	M_{f} ($M_\odot$)	t_{cool} (Gyr)	t_{ms} (Gyr)	M_{i} ($M_\odot$)
WD0023+109	10377 ± 230	7.92 ± 0.08	0.56 ± 0.03	0.49 ± 0.08	–	–
WD0913+442	8918 ± 111	8.29 ± 0.02	0.78 ± 0.01	1.72 ± 0.06	–	–
WD1304+227	10798 ± 120	8.21 ± 0.05	0.73 ± 0.13	0.73 ± 0.06	–	–
WD1354+340	13650 ± 437	7.80 ± 0.15	0.49 ± 0.10	0.14 ± 0.02	3.11 ± 2.28	$1.450^{+0.215}_{-0.90}$
WD2253−081	7200 ± 170	8.40 ± 0.08	0.87 ± 0.03	3.75 ± 0.19	–	–

2. Analysis and results

To determine the total age of each system, we analyzed the spectra of the FGK stars following this procedure: first, we derived the effective temperatures, T_{eff}, using the available $VJHK$ photometry and following the method of Masana *et al.* (2006). Then, we fitted the observed spectra using the SYNSPEC program (Hubeny & Lanz 1995) and synthetic spectra based on Kurucz atmospheres to derive their metallicities, Z, focusing on spectral windows where unblended lines of FeI, FeII and NiI are present. If the distance and the apparent magnitude are known the calculation of the luminosity, L, is straightforward. Finally, we interpolated the stellar models of Schaller *et al.* 1992 using T_{eff}, Z and L to obtain the ages of these stars, and, consequently, the total ages of the CPMPs — see Table 1. We have found that some of the observed stars are fairly unevolved, so isochrone fits do not provide reliable ages. Other age indicators such as chromospheric activity or X-ray luminosity are currently being considered.

The atmospheric parameters of each white dwarf, T_{eff} and $\log g$, were derived from the fitting of the theoretical models of D. Koester to the observed Balmer lines using the package SPECFIT of IRAF following the procedure described in Bergeron *et al.* (1992) — see Table 2. Then, we derived its mass, M_{f}, and cooling time, t_{cool}, using the cooling sequences of Salaris *et al.* (2000). Since we know the total age of the white dwarf (from the companion), we obtain the main sequence lifetime of the progenitor, t_{ms}, by subtracting its cooling time to the age of the system. Finally, using the stellar models of Domínguez *et al.* (1999) we compute the mass of the progenitor in the main sequence, M_{i}.

In Figure 1 we show the final versus initial mass of the white dwarf in a CPMP for which the analysis has been completed. Also plotted are the results for the white dwarfs in the open clusters M 35 (Williams *et al.* 2004) and M 37 (Kalirai *et al.* 2005). We used the T_{eff} and $\log g$ reported by these authors and then, for internal consistency, we followed the same procedure as for the white dwarfs in our list. By comparing the observational data it can be noted that the error obtained for the M_{i} in the case of the white dwarf in a CPMP is higher than those obtained for white dwarfs in open clusters. In fact, the uncertainty in the determination of the ages is the main drawback of using CPMPs, but it can be minimized by selecting the pairs with evolved companions.

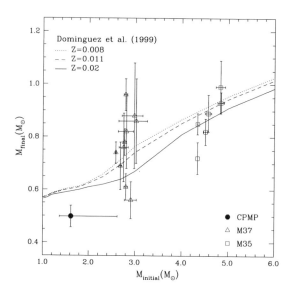

Figure 1. Final masses versus the initial masses obtained for one of the CPMPs in our sample (circle) and for white dwarfs in the open clusters M 35 (squares) and M 37 (triangles).

However, the fact that these systems are potentially more abundant and closer to us makes this a promising approach to improve the semi–empirical initial–final mass relation, since a better spectroscopic study of both members of the pairs can be done. White dwarfs in CPMPs also provide a wider age coverage of this relation, in contrast with open clusters, which are younger and do not contribute to its low–mass range.

A thorough comparison of the semi–empirical data with different theoretical relationships is currently underway and reveals a large scatter in the distribution (Figure 1). This could be due to the fact that maybe the initial–final mass relationship is not a single–valued function. More observational data and improved theoretical stellar models (mainly in the AGB phase) will help to better establish this function. In this sense, to further extend the sample, a cross–correlation of the SIMBAD database and *Villanova White Dwarf Catalog* has provided us with more pairs, which we expect to study shortly.

Acknowledgements

This research was partially supported by the MCYT grant AYA05–08013–C03–01 and 02, by the European Union FEDER funds and by the AGAUR.

References

Bergeron, P., Wesemael, F., & Fontaine, G. 1992, *ApJ*, 387, 288
Domínguez, I., Chieffi, A., Limongi, M., & Straniero, O. 1999, *ApJ*, 524, 226
Hubeny, I. & Lanz, T. 1995, *ApJ*, 439, 875H
Kalirai, J.S., Richer, H.B., Reitzel, D., Hansen, B.M.S., Rich, R.M., Fahlman, G.G., Gibson, B.K., & von Hippel, T. 2005, *ApJ*, 618, L123
Masana, E., Jordi, C., & Ribas, I. 2006, *A&A*, 450, 735
Salaris, M., García–Berro, E., Hernanz, M., Isern, J., & Saumon, D. 2000, *ApJ*, 544, 1036
Schaller, G., Schaerer, D., Meynet, G., & Maeder, A. 1992, *A&AS*, 96, 269
Silvestri, N., Oswalt, T., Wood, M., Smith, J., Reid, I., & Sion, E. 2001, *AJ*, 121, 503
Wegner, G. & Reid, I.N. 1991, *ApJ*, 375, 674
Williams, K.A., Bolte, M., & Koester, D. 2004, *ApJ*, 615, L49

Binary Stars as Critical Tools & Tests
in Contemporary Astrophysics
Proceedings IAU Symposium No. 240, 2006
W.I. Hartkopf, E.F. Guinan & P. Harmanec, eds.

© 2007 International Astronomical Union
doi:10.1017/S1743921307004334

Poster Abstracts (Session 6)

(Full Posters are available at http://www.journals.cambridge.org/jid_IAU)

Analytical Binary Modeling and its Role in Dynamics

T.K.C. Chatterjee

Dept. of Physics & Mathematics, University of the Americas, Puebla, Mexico

Binaries are critical tools that drive stellar systems; which play a major role in galactic dynamics. Internal dynamical evolution of globular clusters (e.g., due to evolution of binaries) can be modulated by external effects (e.g., tidal shocks). This merits a study of the orbital characteristics of binary formation and its applications. We formulate the Hamiltonian treatment of the dynamics of two extended bodies, corresponding to their separation and effective sizes. The internal movement of the constituents of the bodies is taken into account by the hydrodynamic velocities corresponding to their mass distribution and the gravitational potential is considered in the form of a flux. A study of binary motion under spin-orbit synchronization and a condition which favors energy minimization is conducted. Considering binaries to be subsystems in a microcanonical ensemble, we find that for bound binary systems the equipartition distribution of the system parameters favors high eccentricities. Post-collapse evolution of globular clusters are driven by energy output of hard binaries. Our model is of such (hard) type and indicates that most of these eccentric binaries will harden by absorbing energy and lower their eccentricities. Numerical simulation indicate that high eccentricities are maintained for a large timescale; circularization takes place in the adiabatic regimen, but in a very large timescale. In three body encounters the binary is hardened and the single star gains velocity. Thus the core of the cluster is gradually heated; which is a major source of energy input, that drives the post collapse evolution. Preliminary results indicate that this model of hard binaries is likely to stabilize post-collapse oscillations. The model also throws light on the aspects of dynamical evolution of globular clusters that lead to the correlation between their half-mass density and kinetic temperature.

Application of Strong Spin Correlation in Visual Binaries

N. Farbiash and R. Steinitz

Ben-Gurion University, Beer-Sheva, Israel

Using methods to circumvent selection effects, we find correlation between the projected rotational velocities (spins) in binary systems. In visual binaries this correlation is very strong. Moreover, the degree of correlation is independent of component separation. These results indicate the possibility that spin correlation in binaries is the result of evolutionary history, rather than that of tidal interaction. Studies of spin correlation in binaries could thus be an important tool in understanding the evolution of such systems.

The Binary Evolution of the Subdwarf B Star PG 1336−018

H. Hu[1], M. Vuckovic[2], G. Nelemans[1], P. Groot[1], and C. Aerts[1,2]

[1] Department of Astrophysics, Radboud University Nijmegen, The Netherlands,
[2] Institute of Astronomy, Katholieke Universiteit Leuven, Belgium

The formation of subdwarf B (sd B) stars poses several problems for stellar and binary evolution theory. SdB stars are core-helium burning stars ($\sim$0.5 $M_\odot$) with an extremely thin hydrogen envelope ($<$0.02 $M_\odot$). A large fraction of sdB stars are found in binaries. This suggests that Roche lobe overflow can be an effective way to remove almost the entire hydrogen envelope near the tip of the first giant branch.

To test this and other models, a detailed investigation of the sdB interior structure is necessary. Luckily, sdB stars have been observed to pulsate in heat-driven oscillation modes in agreement with theoretical expectations. Thus, asteroseismology provides an excellent tool to test the outcome of sdB formation channels. The work we present here is a first step in this direction and concerns a dedicated study of the range of fundamental parameters of progenitors of the sdB pulsating eclipsing binary PG 1336−018 from a binary evolution code. Our results will constitute a fruitful starting point for our future seismic work on this star which will be based on high-precision VLT photometry and spectroscopy of this target star.

Testing Models of Interacting Binaries Against Observations

S.E. de Mink

Astronomical Institute, Utrecht, Netherlands

One of the big uncertainties in evolution of binary stars is the efficiency of mass transfer. Can it be described as a conservative process or is a significant amount of mass and angular momentum lost? As the uncertainties in hydrodynamical simulations are still very large, currently the best way to address the problem is to compare observations with stellar evolution models.

Recently Hilditch et al. (2005) presented the fundamental parameters of 50 double lined eclipsing binaries in the Small Magellanic Cloud (SMC) with orbital periods between 1 and 5 days. 28 systems are currently transferring mass. This sample is the largest and the most unbiased sample of OB binaries available in any galaxy and therefore suitable to test stellar evolution models of interacting binaries.

No applicable models are currently available for the metallicity of the SMC. Therefore we present a large grid of conservative and non-conservative binary evolution tracks computed with using the fast detailed stellar evolution code STARS (based on Eggleton, 1971). We compare observations and models by fitting evolution tracks to each individual system and by comparing the whole sample at once to a synthetic population. In the near future we plan to present detailed binary evolution models for a larger range of metallicities and orbital periods.

Calibration of the Pre-Main Sequence Binary System RS Cha: Impact of the Initial Chemical Mixture

E. Alecian, Y. Lebreton, M.-J. Goupil, M.-A. Dupret, and C. Catala

Paris-Meudon Observatory, Meudon, France

Accurate observational data are available for the eclipsing double-lined spectroscopic binary system RS Cha, composed of two stars in the pre-main sequence stage of evolution: masses, radii, luminosities and effective temperatures of each component and metallicity of the system. This allows to build pre-main sequence models representing the components of RS Cha and to constrain them in terms of the physical ingredients, initial chemical composition and age.

We present stellar models we have calculated with the CESAM code for different sets of physical inputs (opacities, nuclear reaction rates...) and different initial parameters (global metallicity, helium abundance, individual abundances of heavy elements). We discuss their ability to reproduce the observational constraints simultaneously for the two components. We focus on the impact on the models of the chemical mixture adopted and we propose a calibration for the RS Cha system providing an estimate of its age and initial helium abundance.

Confronting Stellar Evolution Models for Active and Inactive Solar-Type Stars: the Triple System V1061 Cygni

G. Torres[1], C.H.S. Lacy[2], L.A. Marschall[3], H.S. Sheets[4], and J.A. Mader[5]

[1] *Harvard-Smithsonian Center for Astrophysics, Cambridge, MA, United States,*
[2] *University of Arkansas, Fayetteville, AR, United States,*
[3] *Gettysburg College, Gettysburg, PA, United States,*
[4] *Dartmouth College, Hanover, NH, United States,*
[5] *W.M. Keck Observatory, Kamuela, HI, United States*

We present spectroscopic and photometric observations of the chromospherically active (X-ray strong) eclipsing binary V1061 Cyg ($P = 2.35$ days) showing that it is in reality a hierarchical triple system. We combine these observations with *Hipparcos* intermediate data (abscissa residuals) to derive the outer orbit with a period of 15.8 yr. We determine accurate values for the masses, radii, and effective temperatures of the eclipsing binary components, as well as for the mass and temperature of the third star. For the primary we obtain $M = 1.282 \pm 0.015$ M$_\odot$, $R = 1.615 \pm 0.017$ R$_\odot$, $T = 6180 \pm 100$ K, for the secondary $M = 0.9315 \pm 0.0068$ M$_\odot$, $R = 0.974 \pm 0.020$ R$_\odot$, $T = 5300 \pm 150$ K, and for the tertiary $M = 0.925 \pm 0.036$ M$_\odot$ and $T = 5670 \pm 100$ K. Current stellar evolution models agree well with the properties of the primary star, but show a large discrepancy in the radius of the secondary in the sense that the observed value is about 10% larger than predicted (a 5σ effect). We also find the secondary temperature to be $\sim$200 K cooler than indicated by the models. These discrepancies are quite remarkable considering that the secondary is only 7% less massive than the Sun, which is the calibration point of all stellar models. Similar differences with theory have been seen before for lower mass stars. We identify chromospheric acivity as the likely cause of the effect. Inactive stars agree very well with the models, while active ones such as the secondary of V1061 Cyg appear systematically too large and too cool. Both of these differences are understood in terms of the effects of magnetic fields commonly associated with chromospheric activity.

Early Evolution of the 2006 Eruption of the Recurrent Nova RS Oph

G.S. Stringfellow[1], F.M. Walter[2], G. Wallerstein[3], D.G. York[4], J. Dembicky[5], B. Ketzebach[5], and R.J. McMillan[5]

[1] *University of Colorado, Boulder, CO, United States,*
[2] *SUNY Stony Brook, Stony Brook, NY, United States,*
[3] *University of Washington, Seattle, WA, United States,*
[4] *University of Chicago, Chicago, IL, United States,*
[5] *Apache Point Observatory, Sunspot, NM, United States*

Classical novae occur frequently, but repeated outbursts of a KNOWN nova occur rarely. These observed repeated outbursts render recurrent novae (RN) unique amongst all novae, and important astrophysical laboratories in which to study the outburst physics and chemical processing that occurs in novae (or at least this subclass). The primaries of RN are thought to be massive white dwarfs (1.3–1.4 $M_\odot$, close to the Chandrasekhar limit). Consequently, their outbursts are quite brief, returning back to their quiescent state within just a few months. If the white dwarf (WD) retains a net gain in mass after each accretion-eruption cycle, this could eventually result in a Type Ia supernovae explosion. RS Oph is one of only a few known recurrent novae, itself having been observed to erupt in 1898, 1933, 1958, 1967, and 1985; roughly once a generation! RS Oph is further unique in that the secondary is a late-type $\sim$M2III giant star in a long-period orbit of $\sim$460 days. Thus, the hot WD is enshrouded within the extended wind of the giant companion. On 2006 February 12, RS Oph experienced yet another eruption, rising by about 6 mags to $V\sim5$. A high-dispersion spectroscopic campaign was initiated on RS Oph using the APO 3.5-m telescope using the echelle spectrograph (R$\sim$37,500, continuous coverage from 0.35nm to 0.98nm, $\sim$25 nights) complimented by lower-dispersion spectra obtained on the SMARTS 1.5-m telescope at CTIO, and coordinated with various campaigns conducted with space observatories. We report initial results from the spectroscopic ground-based campaign, and compare with results found during previous eruptions. Though the 2006 eruption behaved qualitatively similar to the 1985 outburst, important differences are observed. For example, the 2006 eruption proceeded more rapidly than the 1985 outburst, likely a result of a larger envelope mass accreted before the thermonuclear runaway ensured. Broad hydrogen lines narrowed as the eruption evolved, while He I lines disappeared, replaced by He II lines as the ejecta became shock heated. Details of the evolving ionization state, kinematics, and equivalent widths will be discussed.

Eta Carinae and the Homunculus: an Astrophysical Laboratory

T.R. Gull[1], K.E. Nielsen[1,2], M.F. Corcoran[3,4], D.J. Hillier[5], and K. Hamaguchi[3,4]

[1] *Astrophysical Science Division, NASA/GSFC, Greenbelt, MD, United States,*
[2] *Dept. of Physics & Astronomy, Catholic University of America, Washington, DC, United States,*
[3] *Universities Space Research Association, 10211 Wincopin Circle, MD, United States,*
[4] *Astrophysics Science Division, NASA/GSFCr, Greenbelt, MD, United States,*
[5] *Dept. of Physics & Astronomy, University of Pittsburgh, Pittsburgh, PA, United States*

High spatial resolution spectroscopy with HST/STIS between 1998.0 and 2004.2 has provided much exciting information about the central binary system and the physics of its N-rich, C,O-poor ejecta.

Stellar He I profiles, noticeably blue-shifted relative to P Cygni H and Fe II line profiles, originate from the ionized wind region between two massive companions. Changes in profiles of He I singlet and triplet lines provide clues to the excitation mechanisms involved as the hot, UV companion moves in its highly eccentric orbit.

For 90% of the 5.54-year period, the spectra of nearby Weigelt blobs and the Little Homunculus include highly excited emission lines of Ar, Ne, and Fe. During the few month-long spectroscopic minimum, these systems are deprived of Lyman continuum. Recombination, plus cooling, occurs. In the skirt region between the bipolar Homunculus, a neutral emission region, devoid of hydrogen emission, glows in Ti II, Fe I, Sr II, Sc II, etc. We find the ejecta to have Ti/Ni abundances nearly 100 times solar, not due to nuclear processing, but due to lack of oxygen. Many metals normally tied up in interstellar dust remain in gaseous phase.

Much information is being obtained on the physical processes in these warm N-rich gases, whose excitation varies with time in a predictable pattern. Indeed recent GRB high dispersion spectra include signatures of circum-GRB warm gases. This indicates that the early, primordial massive stars have warm massive ejecta reminiscent to that around η Carinae.

The Eclipsing Model for the Symbiotic Binary YY Her

L. Hric[1], R. Gális[2], and L. Šmelcer[3]

[1] *Astronomical Institute of Slovak Academy of Sciences, Tatranská Lomnica, Slovakia,*
[2] *Faculty of Science, P.J.Šaafárik University, Košice, Slovakia,*
[3] *Public Observatory, Valašské Meziříčí; Czech Republic*

Extensive long-term CCD and photoelectric photometric observations of the classical symbiotic star YY Her, covering the period of its post-outburst activity (JD 2 451 823 – 2 453 880), are presented. We explain the periodic variations of the brightness of YY Her by the eclipses of the components in the symbiotic system. The model with a deformed (non-homogeneous) envelope, surrounding the white dwarf is discussed. In addition, we observed a flare in about JD 2 452 440, during the primary minimum, that was later followed by an energetic outburst in JD 2 452 700. The next outburst activity is well detected on the light curve.

Finding the Primordial Binary Population in Sco OB2: on the Interpretation of Binary Star Observations

M.B.N. Kouwenhoven, A.G.A. Brown, S.F. Portegies Zwart, and L. Kaper

Astronomical Institute, Amsterdam, Netherlands

The majority of stars form in of a binary or multiple systems. Detailed knowledge of the properties of binary systems thus provides important information about the process of star formation. It is difficult to study the binary population during star formation due to the large interstellar extinction. However, after the first massive stars have formed, the interstellar gas is quickly removed by stellar winds, and a freshly exposed population, the "primordial binary population" (PBP), is born. OB associations are the prime targets for finding the PBP. Due to their youth and low stellar density, the binary population has only modestly changed since formation. We targeted Sco OB2 for our study, as this is the nearest young OB association.

Finding the PBP in Sco OB2 involves three steps: (1) performing observations and a literature study to obtain as much information on binarity as possible, (2) finding the true binary population by removing the selection effects from the binarity dataset, and (3) going back in time, to the PBP, by correcting for the effects of stellar and dynamical evolution that have changed the binary population over time.

We present the results of our study. We have performed two binarity surveys with the adaptive optics instruments ADONIS and NACO (in order to fill up the gap between the close spectroscopic and wide visual binaries), and combine our results with literature data. We simulate associations and study the selection effects using "synthetic observations" of visual, spectroscopic, and astrometric binaries in our simulated association. By varying the association properties and performing synthetic observations, we find which properties of the binary population correspond best to the true observations, and find the true binary population. We present the results of detailed N-body simulations (including stellar and binary evolution), which we use to derive the primordial binary population in Sco OB2.

The Eclipsing Binary System V2154 Cyg: Observations and Models

J. Fernandes[1], E. Oblak[2], and M. Kurpinska-Winiarska[3]

[1] *Observatório Astronómico, Coimbra, Portugal,*
[2] *Observatoire, Besançon, France,*
[3] *Obserwatorium, Krakow, Poland*

V2154 Cyg is one of the nearest eclipsing binary systems, located at 88 pc and discovered by the HIPPARCOS mission. Based on photometric and spectroscopic observations carried out respectively at the Nevada Observatory, Spain (*uvby* photometry) and by CORAVEL-ELODIE, at the Haute-Provence Observatory, we determine the masses, effective temperature and radius for both components of the eclipsin system. On the other hand we analyse the HR Diagram position of V2154 Cyg A and B by means of the theoretical models. This system is very interesting thanks to the fact that, being composed of coeval stars, the different masses the eclipsing binary components ($1.27 M_\odot + 0.76$ $M_\odot$) allow a comparison between two stars with clearly different internal structures and evolutionary regimes, mainly on convection processes. We present preliminary results on the determination of the age and helium of the system.

How to Measure Gravitational Aberration?

M. Křížek[1] and A. Solcova[2]

[1] *Mathematical Institute, Academy of Sciences, Prague, Czech Republic,*
[2] *Faculty of Civil Engineering, Czech Technical University, Prague, Czech Republic*

In 1905, Henri Poincaré predicted the existence of gravitational waves and assumed that their speed c_g would be that of the speed of light c. If the gravitational aberration would also have the same magnitude as the aberration of light, we would observe several paradoxical phenomena. For instance, the orbit of two bodies of equal mass would be unstable, since two attractive forces arise that are not in line and hence form a couple. This tends to increase the angular momentum, period, and total energy of the system. This can be modelled by a system of ordinary differential equations with delay. A big advantage of computer simulation is that we can easily perform many test for various possible values of the speed of gravity [1].

In [2], Carlip showed that gravitational aberration in general relativity is almost cancelled out by velocity-dependent interactions. This means that rays of sunlight are not parallel to the attractive gravitational force of the Sun, i.e., we do not see the Sun in the direction of its attractive force, but slightly shifted about an angle less than $20''$. We show how the actual value of the gravitational aberration can be obtained by measurement of a single angle at a suitable time instant T corresponding to the perihelion of an elliptic orbit. We also derive an *a priori* error estimate that expresses how acurately T has to be determined to attain the gravitational aberration to a prescribed tolerance.

[1] Křížek, M. 1999, "Numerical experience with the finite speed of gravitational interaction", *Math. Comput. Simulation*, 50, 237-245

[2] Carlip, S. 2000, "Aberration and the speed of gravity", *Phys. Lett. A*, 267, 81-87

On the Evolutionary History of EHB Objects in Binary Systems with Hot Subdwarf Companions

I.B. Pustylnik and V.-V. Pustynski

Tartu Observatory, Tõravere, Estonia

It has been shown quite recently (Morales-Rueda *et al.* 2003) that dB stars, extreme horisontal branch (EHB) objects in high probability all belong to binary systems. We study in detail the mass and angular momentum loss from the giant progenitors of sdB stars in an attempt to clarify why binarity must be a crucial factor in producing EHB objects. Assuming that the progenitors of EHB objects belong to the binaries with initial separations of a roughly a hundred solar radii and fill in their critical Roche lobes when being close to the tip of red giant branch, we have found that considerable shrinkage of the orbit can be achieved due to a combined effect of angular momentum loss from the red giant and appreciable accretion on its low mass companion on the hydrodynamical time scale of the donor, resulting in formation of helium WD with masses roughly equal to a half solar mass and thus evading the common envelope stage.

Near-Contact Binaries (NCB): Close Binary Systems in a Key Evolutionary Stage

L.Y. Zhu and S.B. Qian

[1] *National Astronomical Observatories, Yunnan Observatory, Kunming, China,*
[2] *Graduate School of the Chinese Academy of Sciences, Beijing, China*

Short-period eclipsing binary systems with EB-type light variations are interesting objects for understanding the evolutionary changes undergone by close binaries. As investigated by many authors (J. Kalužny, A. Yamasaki, D.S. Zhai, X.B. Zhang, R.W. Hilditch, T.M. McFarlane, D.J. King, J.S. Shaw, R.G. Samec, P.G. Niarchos, Kyu-Dong Oh, etc.), a majority of them belong to an important subclass of close binaries called near-contact binaries (NCBs). According to the geometric definition of this subclass, NCBs actually comprise semi-detached, marginal-contact, and marginal-detached systems. They can be in the intermediate stage between detached or semi-detached state and contact state. Therefore, NCBs are the important observational targets which may be lying in key evolutionary states.

In this paper, we observed and investigated several NCBs (BL And, GW Tau, RU UMi, GSC 3658−0076, UU Lyn, AS Ser, IR Cas, EP Aur). Our results show that the orbital periods of BL And, GW Tau, RU UMi and UU Lyn are decreasing, while that of IR Cas is decreasing and oscillating. The mechanisms that could explain the period variations are discussed. Combining the photometric solutions with period variations of these systems, we divide them into four types: BL And is a semi-detached system with a lobe-filling primary; RU UMi and EP Aur are semi-detached systems with lobe-filling secondaries; GW Tau, UU Lyn and AS Ser are marginal contact systems; and GSC 3658-0076 is a marginal detached system. Finally, the evolutionary stage of each system is discussed and some statistical relations of NCBs are presented.

Subdwarf B Stars: Tracers of Binary Evolution

L. Morales-Rueda[1], P.F.L. Maxted[2], and T.R. Marsh[3]

[1] *Radboud University Nijmegen, Nijmegen, Netherlands,*
[2] *Keele University, Keele, United Kingdom,*
[3] *Warwick University, Coventry, United Kingdom*

Subdwarf B stars are a superb stellar population to study binary evolution. In 2001, Maxted *et al.* (MNRAS, 326, 1391) found that 21 out of the 36 subdwarf B stars they studied were in short-period binaries. These observations inspired new theoretical work that suggests that up to 90% of subdwarf B stars are in binary systems with the remaining apparently single stars being the product of merging pairs. This high binary fraction added to the fact that they are detached binaries that have not changed significantly since they came out of the common envelope, make subdwarf B stars a perfect population to study binary evolution. By comparing the observed orbital period distribution of subdwarf B stars with that obtained from population synthesis calculations we can determine fundamental parameters of binary evolution such as the common envelope ejection efficiency. Here we give an overview of the fraction of short period binaries found from different surveys as well as the most up to date orbital period distribution determined observationally. We also present results from a recent search for subdwarf B stars in long period binaries.

Orbits of Post-AGB Binaries with Dusty Discs

T. Lloyd Evans[1] and H. Van Winckel[2]

[1] *School of Physics & Astronomy, University of St. Andrews, United Kingdom,*
[2] *Instituut voor Sterrenkunde, K.U. Leuven, Belgium*

Introduction: Post-AGB stars with RV Tauri-like spectral energy distributions have been studied to establish a connection between binarity and this type of SED.

Methods: Stars were selected from those with the appropriate IRAS colours by their F-type optical spectra and the existence of an infrared excess at K and L. These stars are hotter than classical RV Tauri stars and have pulsations of smaller amplitude, making it simpler to disentangle orbital from pulsational radial velocity variations. Radial velocity measurements were made with CORALIE on the Swiss telescope on La Silla, and spectroscopic orbits were obtained for all six stars. Optical photometry from SAAO, the Flemish Mercator Telescope and the ASAS-3 data bank enabled the determination of pulsation periods for these stars.

Results: These six stars have orbits with periods of several hundred days; several have significantly non-zero eccentricities. The dust may be contained in a circum-binary disc, which in the past has acted as the site where refractory elements have been sequestered in dust grains, leaving depleted gas, which has been returned to the star and has modified the elemental abundances in the photosphere.

Discussion: These observations confirm the close relation between binarity and the presence of a Keplerian dusty disc. The mode of formation of the disc is still poorly understood: dust is commonly observed in the form of an expanding shell about AGB stars, but the presence of a companion provides the possibility of it forming a disk.

The non-zero eccentricities of several of the orbits are surprising, as circularising interactions might have been expected when the primary star was on the AGB and may have filled its Roche lobe.

Binary Stars as a Probe for Massive Star Evolution: the Case of Delta Ori

A.F. Kholtygin[1], T.E. Burlakova[2,3], S.N. Fabrika[2], G.G. Valyavin[2,3], and M.V. Yushkin[2]

[1] *Astronomical Institute of Saint Petersburg University, Russia,*
[2] *Special Astrophysical Observatory, Nizhniy Arkhyz, Russia,*
[3] *Bohyunsan Optical Astronomy Observatory, Korea*

Variability of line profiles in spectra of the triple system δ Ori A with massive components Aa1, Aa2 and Ab is studied. A variability amplitude on the level of 0.5-1% in the continuum units is revealed. The detected variability is probably cyclical with a period about of 4^{h}. In dynamical wavelet spectra of these line variations we have detected large-scale (25-50 km/s) components moving in a band from $-V \sin i$ to $+V \sin i$ for the main star Aa1. A crossing time of the band is $4\text{-}5^{\mathrm{h}}$. However, some variable components are found outside this band, what could be related to non-radial pulsations of the fainter component Ab. The nature of this component is investigated. Evidence that it is a binary star with massive components is found.

This work has been supported by RFBR grant 05-02-16995-a.

Probing the Low-Luminosity X-ray LF of LMXBs in Normal Elliptical Galaxies

D.-W. Kim[1], G. Fabbiano[1], V. Kalogera[2], A. R. King[3], S. Pellegrini[4], G. Trinchieri[5], S.E. Zepf[1], A. Zezas[1], L. Angelini[1], R.L. Davies[1], and J.S. Gallagher[1]

[1] *Harvard-Smithsonian Center for Astrophysics, Cambridge, MA, United States,*
[2] *Northwestern University, Evanston, IL, United States,*
[3] *University of Leicester, Leicester, United Kingdom,*
[4] *University of Bologna, Bologna, Italy,*
[5] *INAF-Osservatorio, Milan, Italy,*
[6] *Michigan State University, East Lansing, MI, United States,*
[7] *NASA GSFC, Greenbelt, MD, United States,*
[8] *Oxford University, Oxford, United Kingdom,*
[9] *Universityof Wisconsin, Madison, WI, United States*

We present the first low luminosity ($L_X > 5$–10×10^{36} erg s^{-1}) X-ray luminosity functions (XLFs) of low-mass X-ray binaries (LMXBs) determined for two typical old elliptical galaxies, NGC 3379 and NGC 4278. Because both galaxies contain little diffuse emission from hot ISM and no recent significant star formation (hence no high-mass X-ray binary contamination), they provide two of the best homogeneous sample of LMXBs. With 110 and 140 ks *Chandra* ACIS S3 exposures, we detect 59 and 112 LMXBs within the D_{25} ellipse of NGC 3379 and NGC 4278, respectively. The resulting XLFs are well represented by a single power-law with a slope (in a differential form) of 1.9 ± 0.1. In NGC 4278, we can exclude the break at $L_X \sim 5 \times 10^{37}$ erg s^{-1} that was recently suggested to be a general feature of LMXB XLFs. In NGC 3379 instead we find a localized excess over the power law XLF at $\sim 4 \times 10^{37}$ erg s^{-1}, but with a marginal significance of $\sim 1.6\sigma$. Because of the small number of luminous sources, we cannot constrain the high luminosity break (at 5×10^{38} erg s^{-1}) found in a large sample of early type galaxies. While the optical luminosities of the two galaxies are similar, their integrated LMXB X-ray luminosities differ by a factor of 4, consistent with the relation between the X-ray to optical luminosity ratio and the globular cluster specific frequency.

Session 7: Binary Stars as Critical Tests:

Binary stars as probes of our Galaxy

Figure 1. (top left to bottom right) Speakers Dany Vanbeveren, Christine Allen, Helmut Abt, and Arcadio Poveda.

Figure 2. Arne Hendon, speaking at an evening public outreach session.

Figure 3. (top) River bank of Vltava with the building of Restaurant Bellevue (from about 1850). Painter Oscar Kokoschka had an atellier in the Bellevue hotel, beginning in 1934. (bottom) Krešimir Pavlovski talking to Ingacio Ribas, Bill Hartkopf watches from behind.

Binary Stars as Critical Tools & Tests
in Contemporary Astrophysics
Proceedings IAU Symposium No. 240, 2006
W.I. Hartkopf, E.F. Guinan & P. Harmanec, eds.

© 2007 International Astronomical Union
doi:10.1017/S1743921307004358

Close Pairs as Probes of the Galaxy's Chemical Evolution

Dany Vanbeveren and Erwin De Donder

Astrophysical Institute, Vrije Universiteit Brussel, Pleinlaan 2, 1050 Brussels, Belgium

Abstract. Understanding the galaxy in which we live is one of the great intellectual challenges facing modern science. With the advent of high quality observational data, the chemical evolution modeling of our galaxy has been the subject of numerous studies in the last years. However, all these studies have one missing element which is 'the evolution of close binaries'. Reason: their evolution is very complex and single stars only perhaps can do the job. (Un)Fortunately at present we know that a significant fraction of the observed intermediate mass and massive stars are members of a binary or multiple system and that certain objects can only be formed through binary evolution. Therefore galactic studies that do not account for close binaries may be far from realistic. We implemented a detailed binary population in a galactic chemical evolutionary model. Notice that this is not something simple like replacing chemical yields. Here we discuss three topics: the effect of binaries on the evolution of ^{14}N, the evolution of the type Ia supernova rate and the effects on the G-dwarf distribution, the link between the evolution of the r-process elements and double neutron star mergers (candidates of short gamma-ray burst objects).

Keywords. Stars: binaries, stars: evolution, Galaxy: evolution

1. Introduction

The ensemble of the physics of galaxy and star formation, the dynamics of stars and gas clouds in a galaxy and the details of stellar evolution from birth till death as function of chemical composition makes the chemo-dynamical evolutionary model (CEM) of a galaxy. During the last four decades many groups all over the world have constructed CEMs with varying degree of sophistication. All these CEMs have one property in common: although most of them account in some parameterized way for the effects of supernova (SN) of Type Ia, the effect of binaries on galactic stellar populations in general, on the chemical yields in particular is largely ignored without justification. Since 2000 our group in Brussels is studying in a systematic way the effects of intermediate mass and massive binaries on populations of stars and stellar phenomena and on CEM. Our CEM has been described in detail in an extended review (De Donder & Vanbeveren, 2004, further DV2004) and references therein.

Summarizing:

• To describe the formation and evolution of the Galaxy we use the model discussed by Chiappini *et al.* (1997) which is supported by recent results of hydrodynamical simulations (Sommer-Larsen *et al.* 2003).

• The star formation rate is a function of the local gravitational potential and thus also depends on the total surface mass density. We use the prescription of Talbot & Arnet (1975), updated by Chiosi (1980). The values of the different parameters that enter the formalism and that we use in our CEM were discussed in DV2004.

Stellar evolution:

• The chemical yields of single stars with an initial mass = [0.1 ; 120] $M_\odot$ are calculated from detailed stellar evolutionary computations where, in the case of massive stars, the most recent stellar wind mass loss rate prescriptions are implemented in the evolutionary code. Of course, the stellar evolutionary dataset depends on the initial metallicity Z and as far as massive stars are concerned not in the least on the effects of Z (read Fe) on the stellar wind mass loss rates. Our dataset corresponds to the case where the Luminous Blue Variable (LBV) mass loss rate is independent from Z whereas the red supergiant star, pre-LBV OB-type star and Wolf-Rayet star mass loss rates depend on Z as predicted by the radiatively driven wind theory (Kudritzki et al. 1989).

• Similarly as for the single stars, the chemical yields of close (=interacting) binary stars with an initial primary mass = [1 ; 120] $M_\odot$ are based on an extended data set of detailed stellar evolutionary calculations of binaries where the evolution of both components is followed simultaneously. Notice that the yields are function of primary mass, binary mass ratio, binary period, the physics of the Roche lobe overflow (RLOF) and/or common envelope process, the physics of mass accretion, the effects of the supernova explosion on binary parameters. Obviously for the massive binaries the stellar wind effects are included similarly as in massive single stars.

• We explore the effects of different initial binary mass ratio (q) distributions, i.e. one that peaks at small q-values (Hogeveen, 1992, we use the letter H), at large q-values (Garmany et al. 1980, we use the letter G) or we assume a flat q-distribution.

• The initial binary period distribution is flat in the Log.

• A Salpeter type IMF for single stars and for binary primaries. We assume that the IMF-slope does not vary in time, which is more than sufficient in order to illustrate the basic conclusion of the present paper.

• To simulate the effects of SN Ia, we calculate the SN Ia rate from first population synthesis principles using the single-degenerate (SD) model of Hashisu et al. (1996, 1999) and/or the double degenerate (DD) model of Iben & Tutukov (1984) and Webbink (1984) in combination with the merger timescale due to gravitational wave radiation. We like to remind the reader that consequently performing binary population synthesis which is linked to a physical SN Ia model is the ONLY method that gives scientifically reasonable results.

• The initial binary formation frequency enters the model as a free parameter. The binary frequency may not be confused with the number of stars in binaries. Suppose that we have 3 O-type stars, 1 single and the two others in 1 binary. Then the binary frequency is 50% but 2/3 of the O-type stars are binary members. Furthermore, the observed binary frequency is not necessarily the same as the binary frequency at birth (on the ZAMS). Stars which are born as binary components may become single stars during the evolution of the binary (due to the merger process or due to the SN explosion of the original companion). Any observed star sample contains the binaries, the single stars which are born as single and the single stars which became single but which were originally binary members. This means that the observed binary frequency is always a lower limit of the binary frequency at birth. The observationally confirmed O-type spectroscopic (thus close) binary frequency in young open clusters has been reviewed by Mermilliod (2001). The rates are as high as 80% (IC 1805 and NGC 6231), and can be as low as 14% (Trumpler 14). The massive close binary (MCB) frequency in Tr 16 is at least 50% (Levato *et al.* 1991) and in the association Sco OB2 it is at least 74% (Verschueren et al., 1996). The results listed above can be considered as strong evidence that the MCB frequency may vary among open clusters and associations. Mason *et al.* (1998) investigated the bright Galactic O-type stars and concluded that the observed spectroscopic binary frequency in clusters and associations is between 34% and 61%, among field stars between 20% and 50% and among the O-type runaways between 5% and 26% (a runaway is defined as a star with a peculiar space velocity larger than 30 km s^{-1}). However, as argued by the authors, these percentages may underestimate the true MCB frequency due to selection effects. The close systems that are missing in most of the observational samples are systems with periods larger than 40 days and mass ratios smaller than 0.4 which are obviously harder to detect. Mason *et al.* conclude that it cannot be excluded that we are still missing about half of all the binaries. Accounting for the observed percentages given above, we therefore conclude that the O-type MCB frequency may be very large. Vanbeveren *et al.* (1998a, 1998b) investigated the bright B0-B3 stars (stars with a mass between 8 M$_\odot$ and 20 M$_\odot$) in the Galaxy and concluded that at least 32% are a primary of an interacting close binary. Again due to observational selection, using similar arguments as for the O-type binaries, the real B0-B3 binary frequency could be at least a factor 2 larger. Let us remark that the observations discussed above give us a hint about the MCB frequency in the solar neighborhood. However, whether or not this frequency is universal is a matter of faith. Indirect evidence about the binary frequency comes from population number synthesis (PNS) studies. Using all we know about single star and binary evolution, we can calculate the number of binaries of some type and compare it with the observed number. It is obvious to realize that predicted numbers depend on the adopted initial binary frequency; it can be concluded that in order to obtain general correspondence with observed numbers, the initial MCB frequency in the simulations must be very large ($> 50\%$). Interestingly, De Donder and Vanbeveren (2002, 2003b) compared observed and theoretically predicted SN type Ia rates with PNS of intermediate mass single stars and binaries, and also in this case the adopted initial intermediate mass close binary frequency must be very large in order to obtain correspondence (see also section 2).

• The stable Roche lobe overflow phase in binaries is assumed to be quasi-conservative (all mass lost by the loser is accreted by the gainer). Notice that in most of the binaries with an initial primary mass larger than 40 M$_\odot$ the Roche lobe overflow phase is avoided due to the very large stellar wind mass loss during the Luminous Blue Variable phase of the primary (the LBV scenario as it was originally introduced by Vanbeveren, 1991).

• The common envelope and/or spiral-in phase is treated according to the prescription of Iben and Tutukov (1984) and Webbink (1984). In some cases these processes lead to the merger of the two binary components; in DV2004 we critically discussed possible consequences of the merger process of different types of binaries on CEM simulations.

• The effect of the SN explosion on the binary and binary orbital parameters is included by adopting a distribution of SN-asymmetries (expressed as the kick velocity that the compact SN remnant gets depending on the SN-asymmetry) and integrating over this distribution; the distribution of SN-asymmetries is based on the observed space velocity distribution of single pulsars which let us suspect that the distribution is χ^2-like (or Maxwellian but with a tail extending to very large values) with an average kick-velocity = 450 km s^{-1}.

• Black hole formation happens when the initial mass of a single star $\geqslant 25$ M$_\odot$ and $\geqslant$ 40 M$_\odot$ for an interacting binary component. Notice that the minimum-mass-difference is due to the fact that the evolution of the helium core in a single star differs from the one of a binary component which lost its hydrogen rich layers due to RLOF during hydrogen shell burning on the thermal timescale (for more details see DV2004). From first physical principles it is unclear whether or not BH formation in massive stars is accompanied by significant matter ejection. De Donder and Vanbeveren (2003) investigated the consequences on CEM simulations of the early Galaxy of the latter uncertainty. It was concluded that a CEM where it is assumed that all massive stars with an initial mass larger than 40 M$_\odot$ form BH's without matter ejection predicts a temporal ^{16}O evolution which is at variance with observations. These observations are much better reproduced when during the core helium burning a star loses mass by stellar wind (a Wolf-Rayet type wind) which depends on the stellar Fe-abundance as predicted by the radiatively driven wind theory whereas prior to or during BH formation on average 4 M$_\odot$ of ^{16}O should be ejected.

Scope of the present paper:

The effect of binaries on the temporal evolution of the galactic chemical elements has been discussed in extenso in DV2004. Here we will focus on the effect of binaries on the SNIa rate, on the G-dwarf metallicity relation, on the temporal evolution of the double neutron star binary mergers and of the r-process elements and on the temporal evolution of ^{14}N.

2. The supernova rates and the G-dwarf metallicity relation

Table 1 gives SNIa/SNIb ratios predicted with our CEM for different values of the initial binary frequency, different mass ratio distributions and for the two SNIa models (SD or DD). We prefer number ratios rather than absolute rates since they do not depend on the adopted model parameters of star formation and Galaxy formation. The observed ratio for spiral Galaxies $\approx$1.6 (Cappellaro et al. 1999) and we conclude that there is a factor 2-3 discrepancy between the predicted and observed SNIa/SNIbc ratio independent from the adopted SNIa progenitor scenario, SD or DD. Since all SNIa's and a major part of the SNIbc's are formed by binaries, increasing the binary frequency doesn't help, at least if we assume that the binary frequency among massive and intermediate mass stars is the same. From present observations it cannot be concluded that SNIa's are exclusively produced by one type of binary systems and therefore SNIa's may form via the SD and the DD channel together. Assuming the latter significantly increases the SNIa rate and consequently brings the number ratio much closer to the observed value. If we look at the G-dwarf disk metallicity distribution which is shown in Figure 1, we also find a better agreement with the observed distribution in case that both the SD and DD channel produce SNIa's. This result can be understood by the fact that the G-dwarf disk metallicity distribution is very sensitive to the iron evolution which is according to our simulations primarily determined by the SNIa rate (about 60-70 % of the total iron content in the Galaxy comes from SNIa's when they are formed by the SD and DD model and an initial binary frequency $\geqslant 50\%$ is adopted).

3. The merger rate of double neutron star binaries and the evolution of r-process elements

Two sites have favorable physical conditions in order to be a major r-process source: the supernova explosion of a massive star (SN II or SN Ibc) and the binary neutron star

Table 1. The predicted supernova number ratios SNIa/SNIbc. The used symbols have the following meaning: f_b=initial total binary frequency, two mass ratio distributions labelled H and G (see text), SD = single degenerate model for SN Ia, DD = double degenerate model for SN Ia

f_b	$q-distr.$	$SNIa model$	$SNIa/SNIbc$
40%(70%)	H	SD	0.5(0.8)
40%(70%)	H	DD	0.4(0.6)
40%(70%)	G	SD	0.2(0.2)
40%(70%)	G	DD	0.7(0.9)

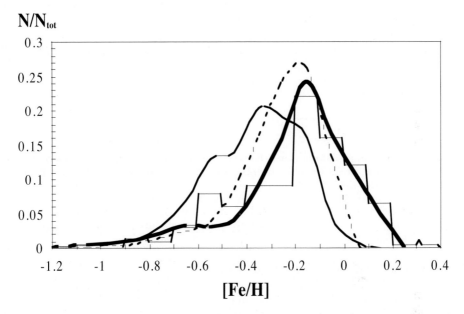

Figure 1. The predicted G-dwarf disk metallicity distribution when only the SD model is applied (full thin line), only the DD model (dashed thin line) and when both the DD and SD model produce SNIa's (thick line). The observed distribution (=histogram) is from Wyse and Gilmore (1995).

merger (NSM). A discussion of the pro's and the contra's of both sites is given in Qian (2000) and Rosswog *et al.* (2000). At present, neither of the two can be promoted as the main enrichment source without reasonable doubt. The main argument against the NSM scenario is the following. Figure 2 shows the temporal evolution of the element europium of which the behavior can be considered as typical for the r-process elements. As can be noticed, europium is observed in stars which were born only a few 10 Myr after the formation of the Galaxy, when the Fe abundance [Fe/H] $\approx$ -3.1. Mathews *et al.* (1992) predicted the temporal evolution of the NSM-rate in the Galaxy and concluded that double neutron star merging starts too late, much later than [Fe/H] = -3.1. However, Mathews *et al.* used knowledge that was available at that time whereas they did not perform detailed binary population synthesis in combination with chemical evolution. We therefore repeated this study with our CEM (De Donder and Vanbeveren, 2003; see also DV2004). Our CEM calculates in detail the temporal evolution of the NS+NS binary population. The corresponding evolution of the NSM-rate is then determined using the theory of gravitational wave radiation. We predict a significant population of double NS

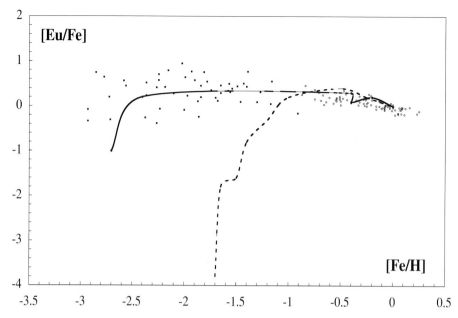

Figure 2. The temporal evolution of [Eu/Fe]. The squares are observations for halo stars from Burris *et al.* (2000), the circles are disk star observations from Woolf *et al.* (1995). On the same figure we also show the theoretically predicted NSM-rate as function of [Fe/H] (for the Solar neighborhood) (full line). The dashed line is calculated with the assumptions of Mathews *et al.* (1992). For both NSM-rate relations we used a normalization to assure that both lines explain the Solar abundance of Eu. We therefore had to assume that during an NSM-event on average 0.004 $M_{\odot}$ of Eu is ejected.

binaries with small periods and very large eccentricities which merge within 1 Myr. This short merging timescale explains why at present no NS+NS binaries are observed with very large eccentricity, but they play a crucial role in the temporal evolution of the NSM-rate. This is demonstrated in Figure 2 where we compare the observed temporal evolution of europium with our predicted temporal evolution of the NSM rate. We conclude that it can not be excluded that NSMs are important sites of r-process element enrichment of the Galaxy.

4. Very massive close binaries and the puzzling temporal evolution of ^{14}N in the Solar neighborhood

Reproducing the observed temporal evolution of ^{14}N during the early evolutionary phase of the Galaxy has been a problem since the beginning of chemical evolutionary modeling of Galaxies. This is illustrated in Figure 5 where we compare the observed ^{14}N evolution in the Solar neighborhood with our CEM prediction. In most of the CEMs it is assumed that the presently observed maximum stellar mass of stars in the Solar Neighborhood (=100-120 $M_{\odot}$) was the same in the past, i.e., independent from Z. The results of the Wilkinson Microwave Anisotropy Probe (WMAP) provide observational evidence that the low metallicity stellar initial mass function was top heavy with the possible presence of stars which were significantly more massive than 100-120 $M_{\odot}$ (the term *very massive star* is used). The optical depth along the line of sight to the last scattering surface of the Cosmic Microwave Background is interpreted in terms of the existence of a population of very massive stars in the early Universe (Kogut *et al.* 2003; Sokasian *et al.* 2003).

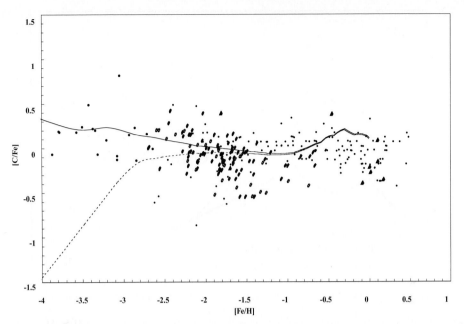

Figure 3. The observed (dots) versus the theoretically predicted temporal evolution of ^{12}C. The thin line (resp. dashed line) corresponds to the case where very massive single stars with a mass $\leqslant 200$ M$_\odot$ (resp. $\leqslant 260$ M$_\odot$) are included. The observations are from various sources which are discussed in De Donder & Vanbeveren (2004).

Vanbeveren and De Donder (2006) investigated the possibility whether or not very massive stars can solve the ^{14}N problem. We first considered the following question: how massive can the maximum stellar mass be in order not to conflict with observations? Using the chemical yields of Heger and Woosley (2002) we adapted our CEM so that the effects of very massive stars during the early phases of Galaxy evolution can be studied. In Figure 3 we compare the observed temporal evolution of ^{12}C with CEM predictions when it is assumed that the maximum stellar mass = 200 M$_\odot$ and when it is 260 M$_\odot$. With the latter value correspondence is very poor and we conclude that the maximum stellar mass during the early evolutionary phase of the Galaxy could have been 200 M$_\odot$ but not much larger. Figure 5 illustrates that very massive single star (with mass up to 200 M$_\odot$) can not solve the ^{14}N discrepancy.

Could very massive interacting binaries provide a way out? Let us consider a typical very massive binary consisting of a 90 M$_\odot$ He star (the post Roche lobe overflow remnant of a 180 M$_\odot$ very massive star) with a 140 M$_\odot$ companion. Low metallicity stars with a mass larger than 140 M$_\odot$ have nearly equal core hydrogen burning timescales. This means that when the He star explodes, our 140 M$_\odot$ is at the beginning of its core helium burning phase. Defining R as the radius of the companion, A as the binary separation, it is easy to understand that roughly

$$\frac{1}{4}90\frac{R^2}{A^2} \tag{4.1}$$

M$_\odot$ of the supernova matter (with a chemical composition which can be deduced directly from the yields published by Heger & Woosley, 2002) will be accreted by the companion (Figure 4 illustrates what happens). The evolutionary consequences are very similar to those during the Roche lobe overflow of a massive binary where nuclearly processed

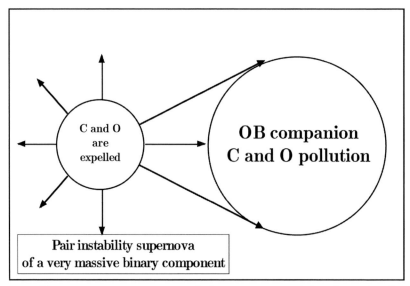

Figure 4. The very massive primary explodes as a pair instability SN and pollutes its companion.

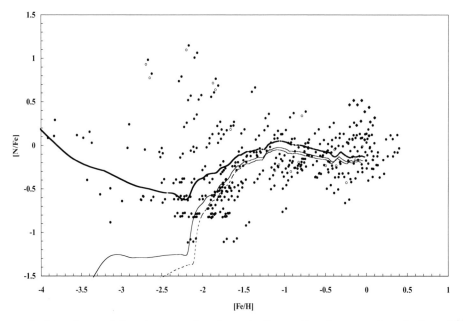

Figure 5. The observed (dots) versus the theoretically predicted temporal evolution of ^{14}N. The dashed line is a CEM simulation without very massive stars. The thin line corresponds to the case where very massive single stars with a mass $\leqslant 200$ M$_\odot$ are included but no very massive binaries; the thick line corresponds to the case where we account for a 20% very massive binary frequency. The observations are from various sources which are listed and discussed in De Donder and Vanbeveren (2004).

matter is transferred from the loser towards the gainer. The accreted matter has a molecular weight μ larger than the μ of the underlying layers which is an unstable configuration that initiates thermohaline mixing (Kippenhahn *et al.* 1980). Thermohaline mixing is a very rapid process and is capable to mix all layers outside the He-core (Braun, 1997).

For the purpose of the present paper we treat the process as an instantaneous one and we mix the accreted mass in a homogeneous way with all the layers outside the He core. The foregoing instantaneous mixing process is strengthened by the fact that very massive core helium burning stars are supra-Eddington and most of the layers outside the He-core are convectively unstable. The core helium burning evolution of the companion after this mixing process is quite interesting. Depending on the period of the binary (thus on the value of A) the layers outside the He-core typically contain 0.3-0.5 % (mass fraction) of ^{12}C (and about a factor of 10 more of ^{16}O), thus also the layers where hydrogen burning takes place. The CN cycle in these layers rapidly transforms the available ^{12}C into ^{14}N which is ejected as primary nitrogen during the SN explosion. Our stellar evolutionary calculations for the 140 M$_\odot$ companion reveal that about 0.02-0.05 M$_\odot$ (depending on the binary period, thus on A) of ^{14}N are ejected, large enough to hope that it could solve the discrepancy between the observed and the theoretically predicted galactic temporal evolution of this element. To calculate whether or not very massive close binaries could explain the ^{14}N discrepancy, we adapted our standard CEM and we assumed that during the early evolutionary phase of the Galaxy 20% of all very massive stars are very massive close binaries which typically eject 0.05 M$_\odot$ of ^{14}N, as discussed above. It can readily be understood that the ^{14}N yield predicted by the binary model discussed above scales linearly with the adopted binary frequency. A CEM where a 10% (resp. a 50%) very massive binary frequency is adopted would predict a ^{14}N yield that is a factor 2 lower (resp. a factor 2.5 larger) that the simulation with a 20% very massive binary frequency. Figure 5 also compares the observations with our prediction. We conclude:

within the uncertainties of the content of a population of very massive stars during the early evolution of our Galaxy, very massive close binaries can produce the observed early temporal evolution of ^{14}N, if the binary component is polluted by the pair instability supernova ejecta of its companion and explodes.

5. Summary

In the present paper we discussed the effects of binaries on galactic supernova rates, on the G-dwarf metallicity distribution, and on the temporal evolution of ^{14}N.

We conclude:

- the observed SN Ia rate in spiral galaxies and the G-dwarf metallicity distribution of the Solar neighborhood is best reproduced by a CEM where SNIa are the result of the SD and DD together

- it can not be excluded that double neutron star mergers are important sites of r-process element enrichment of the Galaxy

- very massive close binaries with primary mass between 140 M$_\odot$ and 200 M$_\odot$ may be important production sites of ^{14}N during the early evolutionary phase of the Galaxy.

References

Braun, H., 1997, Ph.D. thesis, Ludwig-Maximilians-Univ. Mnchen
Burris, D.L., Pilachoski, C.A., Armandroff, T.E., Sneden, C., Cowan, J., & Roe, H., 2000, *ApJ* 386, 206

Cappellaro, E., Evans, R., & Turatto, M., 1999, *A&A* 351, 459

Chiappini, C., Matteucci, F., & Gratton, R., 1997, *ApJ* 477, 765

Chiosi, C., 1980, *A&A* 83, 206

De Donder, E. & Vanbeveren, D., 2003, *New Astron.* 8, 415

De Donder, E. & Vanbeveren, D., 2002 , *New Astronomy*, 7, 55

De Donder, E. & Vanbeveren, D., 2004, *New Astron. Rev.* 48, 861 (DV2004)

Garmany, C.D., Conti, P.S., & Massey. P., 1980, *ApJ* 242, 1063

Hachisu, I., Kato, M., & Nomoto, K., 1996, *ApJ* 470, L97

Hachisu, I., Kato, M., & Nomoto, K., 1999, *ApJ* 522, 487

Heger, A. & Woosley, S.E., 2002, *ApJ* 567, 532

Hogeveen, S. J., 1992 *Ap&SS* 196, 299

Iben, Jr. I. & Tutukov, A. V., 1984, *Ap&SS* 54, 335

Kippenhahn, R., Ruschenplatt, G., & Thomas, H.C., 1980, *A&A* 91, 175

Kogut, A., Spergel, D. N., & Barnes, C., 2003, *ApJS* 148, 161

Kudritzki, R.P., Pauldrach, A., Puls, J., & Abbott, D.C., 1989, *A&A* 219, 205

Levato, H., Malaroda, S., Morell, N., Garcia, B., & Hernandez, C., 1991, *ApJS* 75, 869

Mason, B.D., Gies, D.R., & Hartkopf, W.I., *et al.* 1998, *AJ*, 115, 821

Mermilliod, J.-C., 2001, in *The Influence of Binaries on Stellar Population Studies*, ed. D. Vanbeveren, Kluwer Academic Publishers, Dordrecht, p. 3

Qian, Y.Z., 2000, *A&A* 357, 84

Rosswog, S., Davies, M.B., Thielemann, F.-K., & Piran, T., 2000, *A&A 360*, 171

Sokasian, A., Abel, T., Hernquist, L., & Springel, V., 2003, *MNRAS 344*, 607

Sommer-Larson, J., Gtz, M., & Portinari, L., 2003, *ApJ* 596, 478

Talbot, R.J. & Arnett, W.D., 1975, *ApJ* 197, 551

Vanbeveren, D., 1991, *A&A* 252, 159

Vanbeveren, D. & De Donder, E., 2006, *New Astron.* (in press)

Vanbeveren, D., Van Rensbergen, W., & De Loore, C., 1998a, monograph *The Brightest Binaries*, eds. Kluwer Academic Publishers, Dordrecht

Vanbeveren, D., Van Rensbergen, W., & De Loore, C., 1998b, *A&A Rev.* 9, 63

Verschueren, W., David, M., & Brown, A. G. A., 1996, in *The origins, evolution, and destinies of binary stars in clusters*, ASP Conference Series, Vol. 90, p. 131

Webbink, R, F., 1984, *ApJ* 277, 355

Woolf, V.M., Tomkin, J., & Lambert, D.L., 1995, *ApJ* 453, 660

Wyse, R.F.G. & Gilmore, G., 1995, *AJ* 110, 2771

Discussion

BOB WILSON: Can you explain how the SD and DD curves are combined to form the SD+DD curve (in the Fe fraction versus [Fe/H] diagram)?

VANBEVEREN: We link a detailed stellar population synthesis code (that accounts for the evolution of single stars and binaries) to a galactic and star formation model. SD+DD means that we calculate the galactic temporal evolution of the binary populations that produce SN Ia through the single degenerate channel and through the double degenerate one. Each SN Ia enriches the galaxy with the chemical elements (primarily Fe) corresponding to the observed chemistry of SN Ia.

Binary Stars as Critical Tools & Tests
in Contemporary Astrophysics
Proceedings IAU Symposium No. 240, 2006
W.I. Hartkopf, E.F. Guinan & P. Harmanec, eds.

© 2007 International Astronomical Union
doi:10.1017/S174392130700436X

Halo Wide Binaries and Moving Clusters as Probes of the Dynamical and Merger History of our Galaxy

Christine Allen[1], Arcadio Poveda[1] and A. Hernández-Alcántara[1]

[1]Instituto de Astronomía, Universidad Nacional Autónoma de México, Ciudad Universitaria,
04510 México D.F.
email: chris@astroscu.unam.mx

Abstract. Wide or fragile pairs are sensitive probes of the galactic potential, and they have been used to provide information about the galactic tidal field, the density of GMC and the masses of dark matter perturbers present in both the disk and the halo. Halo wide binaries and moving clusters, since they are likely to be the remains of past mergers or of dissolved clusters, can provide information on the dynamical and merger history of our Galaxy. Such remnants should continue to show similar motions over times of the order of their ages. We have looked for phase space groupings among the low-metallicity stars of Schuster *et al.* (2006) and have identified a number of candidate moving clusters. In several of the moving clusters we found a wide CPM binary already identified in our catalogue of wide binaries among high-velocity and metal-poor stars (Allen *et al.* 2000a). Spectroscopic follow-up studies of these stars would confirm the physical reality of the groups, as well as allow us to distinguish whether their progenitors are dissolved clusters or accreted extragalactic systems.

Keywords. Galaxy: halo. Galaxy: kinematics and dynamics. Binaries, wide

1. Introduction

In recent times evidence has accumulated indicating that the galactic halo formed at least in part by accretion of material from smaller extragalactic systems. This view is in accordance with the hierarchical theory of structure formation in the Universe, in which the first objects to form are small galaxies, which then merge into the larger structures observed today. A consequence of this theory is that substructures should still be present within the old stars of the galactic halo and disk. Thus, the merger history of our galaxy should show up in the distribution of old stars, as coherent tidal streams, tails, moving clusters, etc. (Majewski 2004, 2005 and references therein).

But quite apart from the merger history of our galaxy, the stellar halo we now observe is also the result of the dynamical disruption of many globular clusters. Such disrupted clusters should also leave their signatures in the motions of the halo stars and should show up as dynamically coherent groupings in phase space. Poveda *et al.* (1992) searched for such groupings among the nearby halo stars of the first Schuster *et al.* catalogues (Schuster & Nissen 1988, 1989a, 1989b, henceforth SN1) and they were able to identify seven moving groups, characterized by the similarity of their integrals of motion (the energy E and the z-component of the angular momentum h) as well as by their similar metallicities [Fe/H]. In particular, the Kapteyn and the Groombridge 1830 moving groups were re-discovered.

On the other hand, there exists a small number of old and extremely wide binaries ($a > 10000$ AU) that are very difficult to understand dynamically (Allen *et al.* 2000b, Carney *et al.* 1997). In Table 1 we provide a few examples. The first six rows list parameters

Table 1. Properties of some extremely wide, old binaries

	W 1828 (HIP 15396)	G 15-10 (HIP 74199)	G 40-14 (−)	LDS 519 (HIP 74235)	BD+80°245 (HIP 40068)
E (100 km^{2}s^{-2})	−1163.1	−847.3	−1153.7	−524.6	−878.3
h (10 kpc km s^{-1})	−143.1	−41.8	−69.6	−238.1	−118.5
R_{min} (kpc)	7.6	1.4	1.5	5.0	6.0
R_{max} (kpc)	9.6	30.2	15.1	69.2	25.8
$|z_{max}|$ (kpc)	5.2	24.1	0.2	7.1	20.5
Eccentricity	0.12	0.91	0.81	0.87	0.62
[Fe/H]	−2.05	−2.42	−2.46	−1.28	−2.07
s (AU)	39590	89366	23030	8846	6513
$\langle a \rangle$ (AU)	55410	125073	32283	12380	9115

of the galactic orbits of these binaries. The metallicity, [Fe/H], is given in row 7. The last two rows list the observed (projected) separations of the binaries and the expected major semiaxes, calculated according to the theoretical formula given by Couteau (1960). The values for the major semiaxes are quite large, and since these binaries are old, they should have been already dissociated. One possibility is that they have survived because they spend only a small fraction of their lives in the galactic disk, and they cross it at high speeds (Allen *et al.* 2000a). But perhaps there is another way for them to survive to the present day as binaries. They may, in fact, have been part of bound moving halo clusters, now dissolving or already dissolved.

The wide binaries and other substructures observed in the stellar halo of our galaxy may be the remains of either past mergers or disrupted clusters. An interesting question arises: how can we distinguish between both possibilities? Detailed chemical analyses can provide answers in some cases (King 1997, Ivans *et al.* 2003, Venn *et al.* 2004).

In the present paper we look for moving groups with the same technique used in Poveda *et al.* (1992), but we expand the SN1 sample with the more recent catalogue of Schuster *et al.* (2006), henceforth referred to as SN3. The total number of stars contained in the expanded sample is 1451, out of which 483 turn out to be halo stars according to the V_{rot} - [Fe/H] criterion advocated by Schuster *et al.* to distinguish between halo, thick disk and disk stars.

The outline of this paper is as follows. In Section 2 we define our sample of stars and discuss the technique we used for identifying moving groups. Section 3 contains examples of possible halo moving groups, and compares these groupings with those of other workers who have used similar techniques. Section 4 explores the connection between our moving groups and some of the old and very wide binaries. Sections 5 and 6 study the connection between chemically peculiar stars and wide binaries or moving groups. Section 7 summarizes our main results.

2. Identification of moving groups

To identify moving groups we made use of the existence of two integrals of motion in an axisymmetric potential, namely the energy, E, and the z component of the angular momentum, h. We computed these quantities for the halo stars of the combined sample SN3. We next looked for groupings in the (E, h) plane, considering only stars having similar values of the metallicity [Fe/H]. We will refer to these groupings with similar E, h, and [Fe/H] as "phase space clumps". We choose the (E, h) plane as the most straightforward way of plotting quantities that are strictly conserved in a time-independent axisymmetric potential. Recent work (Pichardo *et al.* 2004, Allen *et al.* 2006) has shown that even in

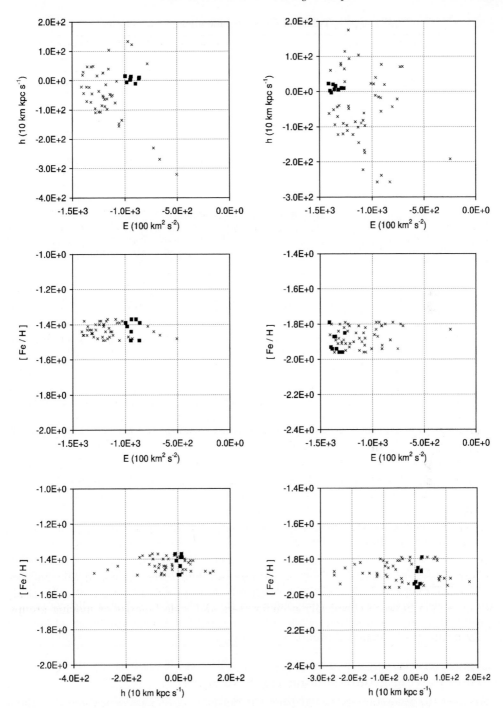

Figure 1. Left column. Plots of the two integrals of motion, E and h, and of the metallicity, [Fe/H] for the candidate moving cluster SN8. The stars belonging to this group are plotted as filled squares. The Schuster *et al.* (2006) stars with metallicities similar to those of the group members are plotted as crosses. Right column. Same plots for the candidate moving cluster SN11

Table 2. Group 8 - members

Leading star name	E 100 km^{2}s^{-2}	h 10 km kpc s^{-1}	[Fe/H]	$|z_{max}|$ kpc	R_{max} kpc	R_{min} kpc	e
G 192-043	−895.8	−10.8	−1.37	17.4	27.8	0.16	0.99
G 088-023	−995.4	14.7	−1.39	17.8	21.8	0.25	0.98
G 202-065	−879.9	−6.4	−1.41	13.7	22.8	0.09	0.99
G 018-024	−946.4	2.0	−1.49	22.9	23.1	1.4	0.89
G 179-054	−942.7	8.1	−1.44	20.8	24.9	0.2	0.99
G 215-047	−939.6	13.5	−1.37	18.2	24.9	0.2	0.98
G 072-006	−864.4	6.4	−1.49	23.7	29.9	0.1	0.99
G 126-063	−860.0	10.1	−1.39	23.1	30.1	0.2	0.99

Table 3. Group 11 - members

Leading star name	E 100 km^{2}s^{-2}	h 10 km kpc s^{-1}	[Fe/H]	$|z_{max}|$ kpc	R_{max} kpc	R_{min} kpc	e
LP 720-028	−1399.9	2.3	−1.93	6.7	9.1	0.04	0.99
HD 116064	−1320.2	4.9	−1.96	8.3	10.7	0.08	0.99
LP 770-071	−1359.0	6.5	−1.87	7.4	9.9	0.11	0.98
LHS 2969	−1267.3	9.6	−1.85	8.2	11.9	0.16	0.97
G 029-071	−1290.4	9.8	−1.96	7.9	11.4	0.16	0.97
LP 673-106	−1347.1	15.9	−1.94	7.4	8.9	0.83	0.83
G 088-027	−1370.0	20.3	−1.87	6.3	9.6	0.38	0.92
HD 160617	−1387.8	−2.9	−1.94	7.4	9.3	0.05	0.99
G 130-007	−1414.0	22.4	−1.79	5.4	8.8	0.41	0.91

the presence of a bar, those quantities are also conserved on the average, except for orbits residing entirely within the bar, which is certainly not the case for relatively nearby stars such as the SN3 sample. Therefore, E and h will keep memory for long times of the values they had when they formed a group. We also require similar values of the metallicity, since the stars within a former group are likely to have but a small spread in their chemical compositions. A further requirement to identify group members is the similarity of their galactic orbital parameters, especially the maximum $|z|$-value.

As a criterion to define "similar values" we take the case of the well-studied common proper motion binary LDS 519 (= HIP 74234/74235 = HD 134439/134440). For this pair, we can compute the difference of the energies and z-components of the angular momenta of the galactic orbit of each component, as well as the difference in metallicity. Since there is little doubt that both components of this wide pair are physically connected, we can take these differences as representative of the differences we would expect to find among members of a common group.

3. Halo moving groups among the SN3 stars

We used the halo stars extracted from the Schuster *et al.* catalogues selecting them according to the population criterion X advocated by these authors, which is a linear combination of the rotational velocity, V_{rot}, and the metallicity, [Fe/H]. A positive value of X identifies a halo star.

In this way, we ended up with 478 stars belonging to the halo population. For these stars we computed galactic orbits using the Allen & Santillán (1993) axisymmetric model potential. We looked for groupings in this halo star sample in the manner described above,

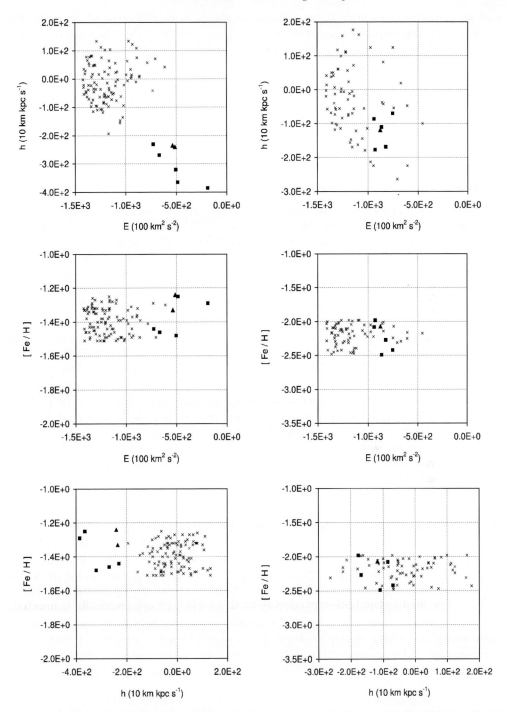

Figure 2. Left column. Plots of the two integrals of motion, E and h, and of the metallicity, [Fe/H] for the candidate moving cluster associated with the wide binary HD 134439/40. The stars belonging to this group are plotted as filled squares. The Schuster *et al.* (2006) stars with metallicities similar to those of the group members are plotted as crosses. Right column. Same plots for the candidate moving cluster associated with the wide binary BD+80°245.

and we were able to identify about a dozen candidate moving clusters which satisfy the criteria of having similar values of E, h, [Fe/H], and similar orbital parameters. A full discussion of these halo moving groups will be given elsewhere. Here, we only show a few examples. Tables 2 and 3 list the members of two candidate groups. Successive rows list the galactic orbital parameters. Figure 1 displays plots of their member stars in three planes containing combinations of E, h, and [Fe/H]. Note that the groupings are much tighter (especially in the planes containing [Fe/H]) than similar groupings proposed in the literature (Eggen 1996, Fiorentin *et al.* 2004, Chiba & Beers 2000, Helmi & White 1999).

4. Wide binaries among moving groups

Interestingly, we find wide binaries in 6 of our candidate moving clusters. These binaries were found by searching among the wide binaries of the catalog of Allen *et al.* (2000a), which was constructed using earlier versions of the Schuster *et al.* catalogues. The binaries were found independently of our search for moving clusters. The fact that some moving groups contain wide binaries supports our conjecture that some of the old wide binaries may be the survivors of former moving groups, which have already dissolved or are in the process of doing so. These binaries are listed in Table 4. The rows and units in this table are the same as in Table 1.

Among the old wide binaries listed in Table 1 there are two well-studied objects that have very peculiar chemical abundances, namely LDS 519 and BD+80°245 (Chen & Zhao 2005, King 1997, Zhang & Zhao 2005, Ivans 2003, Zapatero-Osorio & Martin 2004). Specifically, they show very low ratios of [α/Fe] for their metallicities [Fe/H]. Such chemical peculiarities point to an enrichment history different from that of the bulk of our galactic halo. In fact, it has been suggested that they are reliable indicators of accreted structures. LDS 519 is contained in the SN3 catalog, and its galactic orbit was obtained with the parameters given in that source. For BD+80°245, not contained in the SN3 catalog, we were able to obtain an orbit using data on its distance and kinematics from Hipparcos and Carney *et al.* (1997).

5. Moving groups around chemically peculiar stars

We may ask, conversely, whether there is any evidence of stars with values of E, h and [Fe/H] similar to those of these two chemically peculiar stars. We emphasize that they are both members of wide binaries. Our findings are displayed in Figure 2. We can see that there are indeed other stars with values of E, h, [Fe/H] that are similar to those of LDS 519 and BD+80°245, although the groupings are not as tight as the ones we showed before. No detailed chemical abundances (i.e., [α/Fe] ratios) are available for these candidate moving cluster members, but it would be very interesting if they should also turn out to be peculiar. This would confirm their membership to the halo moving clusters here proposed, and would constitute evidence for their origin by accretion.

6. Binaries or moving cluster members with peculiar chemical composition

Another question we can ask is whether there are other binaries or moving groups with peculiar values of [α/Fe]. To find an answer, we made use of the detailed study of Venn *et al.* (2004). We looked in their Table 2 for information on [α/Fe] for all the wide binaries of our catalog (Allen *et al.* 2000a), as well as for the candidate members of moving

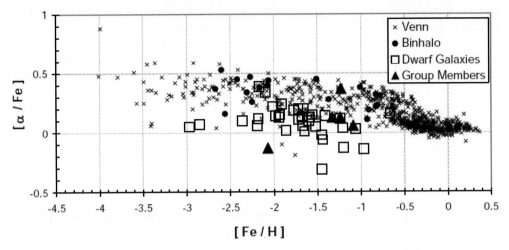

Figure 3. Plot of the abundance of α elements vs metallicity (adapted from Venn *et al.* 2004). The crosses denote stars, the empty squares, dwarf galaxies, as plotted by Venn *et al.*. We have added as dots the wide binaries of Allen *et al.* (2000a) for which α-element abundances were available in Venn *et al.*. Triangles denote binaries that are members of moving clusters.

Table 4. Binaries in moving clusters

	HD 105004	HD 23439	G 18-24	G 40-14	G 15-10	HD 134439/40	BD+80°245
E	−1382.5	−1304.3	−946.4	−1153.7	−847.3	−524.6	−878.3
h	−19.8	99.4	2.0	−69.6	−41.8	−238.1	−118.5
$R_{\min}$	0.3	3.1	1.4	1.5	1.4	5.0	6.0
$R_{\max}$	9.3	9.6	23.1	15.1	30.2	69.2	25.8
$|z_{\max}|$	6.1	1.5	23.0	0.2	24.1	7.1	20.5
e	0.93	0.52	0.89	0.81	0.91	0.87	0.62
[Fe/H]	−1.09	−1.23	1.49	−2.46	−2.42	−1.28	−2.07
s	1620	196	11782	23030	89366	8846	6513
$\langle a \rangle$	2267	274	16489	32233	125073	12380	9115

groups we identified among the SN3 stars. The result of this search is shown in Figure 3, adapted from Venn *et al.* (2004). We show as crosses the stars and as hollow squares the dwarf galaxies studied by Venn *et al.*. The wide binaries of Allen *et al.* appear plotted as thick dots, and the members of candidate moving groups as triangles. The figure shows that four of the stars identified as possible moving group members, namely HD 105004, HD 134439/40 (LDS 519) and BD+80°245 clearly have low [α/Fe] for their [Fe/H]. Recall that BD+80°245 is itself a wide binary, although only the primary is plotted in Figure 3. Such low [α/Fe] stars are chemically similar to the dwarf galaxies, and are therefore probably accreted structures. Figure 3 also shows that at least five of our old wide binaries have such chemical peculiarities, and hence may also be the remains of accreted dwarf galaxies.

7. Summary and conclusions

Old wide binaries may have survived either because they spend little time in the galactic disk, crossing it at high speeds, or else because they were once members of bound clusters or still belong to moving groups. A search for "phase space clumps" among the stars of Schuster *et al.* (2006) yielded about a dozen groups of similar metallicities and galactic orbits. Previously known wide binaries turned out to be present in six of the identified moving groups. We may turn to detailed chemical composition analyses to distinguish whether a grouping had a galactic origin whether it is an accreted structure. Chemical peculiarities, like low [O/Fe] or especially, low [α/Fe] are characteristic of enrichment by type Ia supernovae, and thus of a chemical history different from that of our galaxy. Along with high eccentricities of the galactic orbits, or retrograde motions, these chemical peculiarities are thought to be signatures of accreted structures (dwarf galaxies). At least 5 of our old wide binaries have such peculiarities, and 3 are members of moving groups, which thus may be the remnants of dwarf galaxies accreted onto the galactic halo.

Acknowledgements

Our thanks are due to W.J. Schuster for fruitful discussions and to P. Ronquillo for her help in preparing the typescript. C.A. is grateful to the IAU for a travel grant.

References

Allen, C., Moreno, E., & Pichardo, B. 2006, *ApJ* 652, 1150

Allen, C., Poveda, A., & Herrera, M.A. 2000a, *A&A* 256, 529

Allen, C., Poveda A., & Herrera, M.A. 2000b, in A. Weiss, T. Abel & V. Hill (Eds.) *The First Stars*, Berlin: Springer, p. 66

Carney, B.W., Wright, J.S., Sneden, C., Laird, J.B., Aguilar, L.A., & Latham, D.W. 1997, *AJ* 114, 363

Chen, Y.Q. & Zhao, G. 2006, *MNRAS* 370, 2091

Chiba, M. & Beers, T.C. 2000, *AJ* 119, 2843

Couteau, P. 2000, *J des Observateurs* 43, No.3

Eggen, O.J. 1996, *AJ* 112, 2661

Fiorentin, P.R., Helmi, A., Lattanzi, M.G., & Spagna, A. 2005, *A&A* 439, 551

Helmi, A. & White, S.D.M. 1999, *MNRAS*, 307, 495

Ivans, I., Sneden, C., James, C.R., Preston, G.W., Fulbright, P., Hoflich, A., Carney, W., & Wheeler, J. 2003, *ApJ* 592, 906

King, J.R. 1997, *AJ* 113, 2302

Majewski, S.R. 2004, *Publ. Astr. Soc. Australia* 21, 197

Majewski, S.R. 2005, *ASP Conference Series* vol. 338, p. 240

Pichardo, B., Martos, M., & Moreno, E. 2004, *ApJ* 609, 144

Poveda, A., Allen, C., & Schuster W. 1992, IAU Symp. 149: *The Stellar Populations of Galaxies* p. 471

Schuster, W.J. & Nissen, P.E. 1988, *A&A* 73, 225

Schuster, W.J. & Nissen, P.E. 1989a, *A&A* 221, 65

Schuster, W.J. & Nissen, P.E. 1989b, *A&A* 222, 69

Shuster, W.J., Moitinho, A., Marquez, A., Parrao, L., & Covarrubias, E.B. 2006, *A&A* 445, 939

Venn, K.A., Irwin, M., Shetrone, M.D., Tout, C.A., Hill, V., & Tolstoy, E. 2004, *AJ* 128, 1177

Zapatero Osorio, M.R. & Martin E.L. 2004, *A&A* 419, 167

Zhang, H.W. & Zhao, G 2005, *MNRAS* 364, 712

Discussion

TERRY OSWALT: Do any of the groups and fragile binaries you find coincide with streams found in the "Spaghetti Project" and similar searches for galaxy accretion streams? The Sloan SEGUE project may be an excellent new source of similar groups.

ALLEN: The "Spaghetti" project uses only proper motions (no radial velocities are available), so we cannot compute for them (E, h). We found no CPM binaries in the structures of Chiba & Beers and Fiorenhin *et al.*, but this may be because their proper motions are too small to appear in NLTT (or rNLTT).

Binary Stars as Critical Tools & Tests
in Contemporary Astrophysics
Proceedings IAU Symposium No. 240, 2006
W.I. Hartkopf, E.F. Guinan & P. Harmanec, eds.

© 2007 International Astronomical Union
doi:10.1017/S1743921307004371

Observed Orbital Eccentricities

Helmut A. Abt[1]

[1]Kitt Peak National Observatory, Box 26732, Tucson, AZ 85726-6732, USA
email: abt@noao.edu

Abstract. The eccentricities of catalogued binaries show that tidal interactions extend to periods of at least 1000 days. The maximum periods for complete circularization probably depend on age, rather than mass. For the longest periods all eccentricities are equally probable. For shorter periods the highest eccentricities disappear first. For the same periods, the SB2's have larger mean eccentricities than the SB1's, in accord with Keplers Third Law.

Keywords. binaries: spectroscopic, binaries: visual, stellar: evolution

This is a discussion of the eccentricities in main-sequence spectroscopic and visual binaries with known orbital elements. I collected data on 553 SBs from Dimitri Pourbaix's superb on-line compilation and 616 VBs from the U.S. Naval Observatory's excellent on-line *Sixth Catalog*. Orbital elements of quality 1 were not included.

If one plots the eccentricities as a function of period, one first sees the well-known fact that systems with periods less than a few days have been circularized by tidal interaction. The theory for that has been developed by Zahn (1975, 1977) and Tassoul & Tassoul (1992). But for longer periods the data show a seemingly random distribution. If we fit a mathematical curve to those points, we get a growth curve that goes from zero eccentricity for the shorter periods and then asymptotically approaches 0.5 for the longer periods. But it makes more sense to compute mean eccentricities for bins in the periods because then we can derive error estimates. Figure 1 shows a sample graph, in this case for 221 G dwarfs.

The curves for stars of other spectral types are similar but with various starting points and asymptotic values. Table 1 shows a summary from the fitted curves.

The maximum period for complete circularization seems to vary from about 1.5 days for early-type stars to 4.3 days for GK stars. But is that period a function of primary mass or age? Mathieu & Mazeh (1988) and Mathieu *et al.* (1990) derived numbers like 10–11 days for G stars in old clusters, Latham *et al.* (1988) gave values up to 18 days for halo stars. Mathieu & Mazeh (1988) proposed that the maximum circularization period is a function of age. Therefore I suggest that the periods shown in the table are mostly dependent on age, not mass. This agrees with Zahns calculations.

We see that the asymptotic eccentricity for very long periods averages 0.52±0.02 for all types. If we look at the mean eccentricities for 111 F0–M5 dwarfs with orbital periods greater than 10^5 days (275 yr), we see that all eccentricities are equally probable. That means that when wide binaries are formed, all eccentricities have the same probability of occurring.

When we look at the frequencies of various eccentricities for shorter periods, the higher eccentricities disappear, i.e., for $32 < P < 100$ days, there are no binaries with eccentricities greater than 0.8. For $10 < P < 32$ days, there are no eccentricities greater than 0.7. For $3.2 < P < 10$ days, there are no eccentricities greater than 0.3.

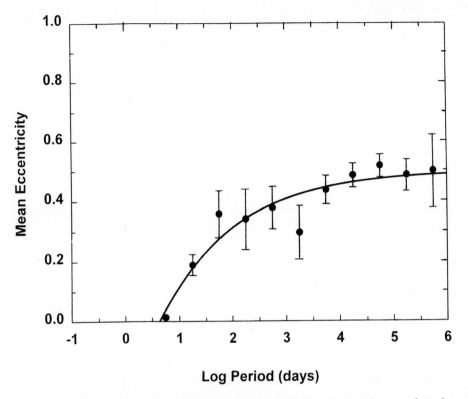

Figure 1. The mean eccentricities of 221 binaries with G dwarf primaries are plotted against orbital periods. The fitted mathematical curve shows a maximum period for total circularization of 4.33 days and an asymptotic eccentricity of 0.50.

Table 1. Results from fitted curves for all spectral types

Spectral Range	Number of Binaries	Max. Circularization Period (days)	Asymptotic Eccentricity
B0–B9.5	131	1.41	0.55 ± 0.02
A0–A5	148	1.60	0.53 ± 0.09
A6–F0	85	2.24	0.55 ± 0.07
F0–F9	363	1.98	0.58 ± 0.06
G0–G9	221	4.33	0.50 ± 0.06
K0–M5	136	4.27	0.44 ± 0.06

We would expect that the tidal interactions should be stronger for binaries having more massive secondaries. We can assume that SB2's have more massive secondaries than SB1's. Note that for binaries with periods up to about 1000 days the tidal effects are substantial because their mean eccentricities have been reduced. We can look at a sample of 143 SB2's and 118 SB1's of primary types F0–M5. For periods between 10 and 1000 days where tidal effects partially reduce the eccentricities, the mean eccentricities for the SB2's are consistently higher. Consider Keplers Third Law. For a fixed period, the separation of the components is larger for SB2's than for SB1's, so the tidal interaction is less. That explains why the eccentricities remain larger for SB2's.

We conclude that:

(*a*) tidal effects in binaries extend to periods of about 1000 days,

(*b*) the maximum period for complete circulations probably depends mostly on age, rather than primary mass,

(*c*) for the longest periods all eccentricities are equally probable,

(*d*) for shorter periods the highest eccentricities disappear first, and

(*e*) for binaries of the same periods, the SB2's have smaller tidal effects than the SB1's.

These results for early-type stars were published in *ApJ* 629, 507, 2005 and for the late-type stars in the November 10, 2006 *ApJ* (*ApJ* 651, 1151, 2006).

References

Latham, D.W., Mazeh, T., Carney, B. W., McCrosky, R.E., Stefanik, R.P., & Davis, R.J. 1988, *AJ* 96, 567

Mathieu, R.D. Latham, D.W., & Griffin, R.F. 1990, *AJ* 100, 1859

Mathieu, R.D. & Mazeh, T. 1988, *ApJ* 326, 256

Tassoul, J.-L. & Tassoul, M. 1992, *ApJ* 395, 259

Zahn, J.-P. 1975, *A&A* 41, 329

Zahn, J.-P. 1977, *A&A* 57, 383

Discussion

JEAN-LOUIS HALBWACHS: I am wondering about the large $< e >$ that you found for SB2. With the Geneva group (Halbwachs *et al.* 2005) we compared the eccentricities of G-K MS binaries with $q \sim 1$ to that of the others, and we found that, for a given P, twins have a smaller median e than the other binaries. Did you also consider q?

ABT: I collected values of q for the SB2's, but did not use them. But since the difference in $< e >$ between SB2's and SB1's is only about 2σ, I hesitated to subdivide the SB2's. But I will look into the difference between Halbwachs *et al.* (2005) and my data. Thank you.

Binary Stars as Critical Tools & Tests
in Contemporary Astrophysics
Proceedings IAU Symposium No. 240, 2006
W.I. Hartkopf, E.F. Guinan & P. Harmanec, eds.

The Frequency Distribution of Semimajor Axes of Wide Binaries: Cosmogony and Dynamical Evolution

Arcadio Poveda[1], Christine Allen[1] and A. Hernández-Alcántara[1]

[1]Instituto de Astronomía, Universidad Nacional Autónoma de México,
Ciudad Universitaria 04510 México D.F.
email: poveda@astroscu.unam.mx

Abstract. The frequency distribution $f(a)$ of semimajor axes of double and multiple systems, and their eccentricities and mass ratios, contain valuable fossil information about the process of star formation and the dynamical history of the systems. In order to advance in the understanding of these questions, we made an extensive analysis of the frequency distribution $f(a)$ for wide binaries ($a > 25$ AU) in various published catalogues, as well as in our own (Poveda *et al.* 1994; Allen *et al.* 2000; Poveda & Hernández–Alcántara 2003). Based upon all these studies we have established that the frequency distribution $f(a)$ is a function of the age of the system and follows Öpik's distribution $f(a) \sim 1/a$ in the range of 100 AU $< a < a_c(t)$, where $a_c(t)$ are the critical semimajor axes beyond which binaries have been dissociated by encounters with massive objects. We argue that the physics behind the distribution $f(a) \sim 1/a$ is a process of energy relaxation, analogous to those present in stellar clusters (secular relaxation) or in the early stages of spherical galaxies (violent relaxation). The existence of runaway stars indicates that both types of relaxation are important in the process of binary and multiple star dynamical evolution.

Keywords. binaries: wide; Galaxy: kinematics and dynamics; stars: proper motions.

---

1. Introduction

The distribution of semimajor axes (separations) of double and multiple stars is a fossil record of the conditions at star formation, as well as of the processes of dynamical evolution, including the dissociation of wide binaries produced by encounters with massive objects: molecular clouds, spiral arms, etc. Our long-standing interest in these topics has led us to investigate the frequency distribution of semimajor axes (separations) of wide binaries as a function of age. In the past, two main distributions of semimajor axes have been proposed:

(1) a power-law frequency distribution $f(a) \sim a^{-\alpha}$

(2) a Gaussian distribution in $\log P$ or $\log a$.

When $\alpha = 1$ in the power-law distribution, we have the well known Öpik (1924) distribution (OD). On the other hand, Kuiper (1935, 1942) proposed the Gaussian distribution, which was further elaborated by Heintz (1969). More recently, Duquennoy & Mayor (1991, DM) again proposed a Gaussian distribution in $\log P$, valid throughout the interval $1 < \log P$ (days) < 10.

Our interest in the subject led us to construct a catalogue of wide binaries in the solar vicinity, based on the catalogue of nearby stars of Gliese & Jahreiss (1991, GJ); see Poveda *et al.* (1994) for details. We have also constructed a list of common-proper-motion binaries in the Orion Nebula Cluster (age 10^6 years), extracted from the Jones & Walker (1988) catalogue of proper motions. (Poveda & Hernández-Alcántara 2003). In our search for evolutionary effects in the observed distributions we have also looked at the oldest stars in the Galaxy. For this purpose we constructed a catalogue (Allen *et al.* 2000) of common-proper-motion companions to the lists of high velocity metal-poor stars of Schuster *et al.* (1988; 1989a; 1989b), with ages of about 10^{10} years.

In all our catalogues, as well as in the *Luyten Double Star Catalogue* (1940–1987, LDS) and in Chanamé & Gould's catalogue (2004, CG), we confirm our previous findings (Poveda *et al.* 1997; Poveda & Allen 2004), that the separations follow Öpik's distribution in an interval that is bounded at the lower end ($a \sim 100$ AU) by the process of close binary and protoplanetary disk formation, and at large separations ($a > 2500$ AU) by the dissociation effects produced by encounters with massive objects.

It can be shown that Öpik's distribution in the plane (N, log P) is a horizontal straight line which is quite consistent (within its error bars) with the DM distribution in the interval $2.44 < \log P$ (years) < 5.44, which corresponds to $53 < a$ (AU) < 5500. (Poveda *et al.* 2004). Since a great number of binaries from many different and largely independent sources confirm the validity of Öpik's distribution, and since there is no stellar formation or single physical process able to produce a Gaussian distribution valid in the interval $1 < \log P$ (days) < 10, we propose to abandon the Gaussian representation for $a > 100$ AU. On the contrary, Öpik's distribution, which is equivalent to a surface density of secondaries $\rho(a) \sim a^{-2}$, has a physical interpretation. In fact, this distribution is similar to the run of surface brightness in globular clusters (King 1962) or to that of elliptical galaxies (Hubble's 1930 or de Vaucouleurs' 1953 distributions). In both cases the physics behind such distributions is well known: it is the result of energy relaxation. The similarity of OD to the surface brightness in clusters and elliptical galaxies indicates that binaries are not born alone; at birth, they must be subject to a process of energy relaxation which cannot be produced by two-body encounters, i.e., stars must be formed in groups of multiplicity $n \geqslant 3$.

This paper is organized as follows. In Section 2 we study the distribution $f(a)$ for a volume–complete sample extracted from our catalogue (Poveda *et al.* 1994). Section 3 examines the distribution $f(a)$ in our catalogue of very young ($T < 10^6$ years), common-proper-motion binaries in the Orion Nebula Cluster (Poveda & Hernández-Alcántara 2003). In Section 4 we examine another sample of wide binaries, namely Chanamé & Gould's (2004) common-proper-motion binaries from the revised NLTT (Salim & Gould 2003; Gould & Salim 2003). Again, the binaries in this catalogue follow OD. Section 5 examines the physics behind Öpik's distribution, and Section 6 presents our conclusions.

2. A Catalogue of Nearby Wide Binaries and a Volume–Complete Sample of these Objects

With the purpose of detecting the effects of dissociation of weakly-bound binaries with the passage of time, we constructed a catalogue of wide binaries (305 double systems, 26 triples and 3 quadruples) with $a > 25$ AU; (Poveda *et al.* 1994).

A sub-sample of this catalogue, i.e., all those systems with primaries of luminosity class V or IV and brighter than absolute magnitude $M_V = 9$ is very important, as it is volume-complete. To show this, in Figure 1 we plot $N(\log r)$ vs. $3 \log r$, where $N(\log r)$ is the number of systems ($M_V < 9$) out to a distance r. As can be seen, with the exception of a few very close systems, the great majority follow the relation $N \sim r^3$ right to the limit of the catalogue ($r = 22.5$ pc), as expected for a volume-complete sample.

Having a volume-complete catalogue, we proceed to investigate the frequency distribution of semimajor axes. It can be shown that an equivalent representation of OD is the cumulative distribution $N(< \log a) \sim \log a$. In the plane $N(< \log a) - \log a$, OD is a

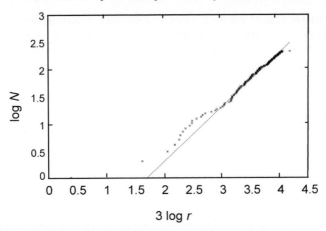

Figure 1. Completeness of the systems in the catalogue of nearby wide binary and multiple systems (Poveda *et al.* 1994). The straight line corresponds to $N(r) \sim r^3$, i.e., a volume-complete sample.

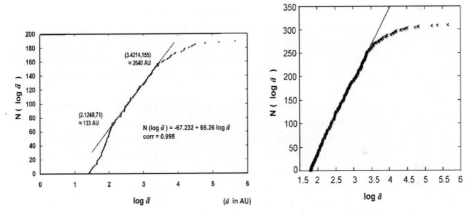

Figure 2. (a)Left. Cumulative distribution of $\log \bar{a}$ for the 189 binaries of the complete sample in Poveda *et al.* (1994), i.e., for systems with $M_V < 9$ and luminosity class IV-V. The straight line is a fit for Öpik's relation. The KS test for this fit gives a value of $Q = 0.99$ in the interval (133AU, 2640AU). (b)Right. Comparison between Öpik's distribution and the cumulative numbers of $\log \bar{a}$ for the binaries in the Poveda *et al.* (2004) catalogue. The straight line is a fit to Öpik's distribution for systems satisfying $60 < \bar{a}$ (AU) < 2965. The KS test gives a value of $Q = 0.96$ for the interval $60 < \bar{a}$ (AU) < 2965.

straight line. In general, we will favor the analysis of the cumulative distribution $f(a)$ in order to reduce the noise introduced by small number sampling fluctuations. Since most of the wide binaries (in our catalogues) do not have reliable orbits because of their long periods, we have to use angular separations s and distances to give projected separations in astronomical units. However, a statistical relation between the average value of a for a given projected separation s, namely $\bar{a} = 1.41s$ (Couteau 1960), can be used to estimate for each binary a value of $\bar{a}$.

Figure 2a shows for our volume-complete sample the run of the cumulative distribution $N(\log \bar{a})$ vs. $\log \bar{a}$. An inspection of Figure 2a shows that this sample closely follows Öpik's distribution in the interval $2.12 < \log \bar{a} < 3.42$, i.e., $133 < \bar{a}$ (AU) < 2640. To evaluate quantitatively how reliably Öpik's distribution represents the data, we shall use the Kolmogorov–Smirnov (KS) test, which was developed precisely for cumulative distributions and where no arbitrary binning of the data is required, (as is the case in the popular χ^2 test). In Figure 2a each binary is plotted, thus $N(\log \bar{a})$ increases one by one. By least-squares we fitted to the data points a number of straight lines, each one defined in the interval ($\log \bar{a}_i$, $\log \bar{a}_j$), until we found the best straight line that minimized the residuals and maximized the interval ($\log \bar{a}_i$, $\log \bar{a}_j$), i.e., the one giving the largest interval ($\log \bar{a}_i$, $\log \bar{a}_j$) in which Öpik's distribution reliably represents the cumulative distribution of separations.

In Figure 2a we give the equation of the straight line that best represents the data, in the interval $\log \bar{a}_i = 2.1248$, $\log \bar{a}_j = 3.4214$ ($\bar{a}_i \approx 133$ AU, $\bar{a}_j \approx 2640$ AU). Having found Öpik's distribution for the wide binaries in the interval $133 < \bar{a}(\text{AU}) < 2640$ we now proceed to test, via KS, what is the level of significance of the theoretical distribution OD. The closer to 1 the estimator Q is, the better the theoretical representation (Press *et al.* 1990). For the present case we find $Q = 0.99$; i.e., we can accept Öpik's distribution at a very high level of confidence in the interval 133–2640 AU. For the interval 133–3100 AU we find $Q = 0.96$, also representing a high level of confidence.

In Figure 2b, we plot in the same plane as in Figure 2a, all the binaries (305) from our 1994 catalogue. Even though this sample is not volume-complete, one can argue that this does not seem to introduce an important bias in the distribution of separations. In fact, as can be seen from Figure 2b, the cumulative distribution $N(< \log \bar{a})$ again defines a straight line. Repeating the statistical analysis for this catalogue of 305 binaries we find that the KS test gives a value $Q = 0.96$ for $\bar{a}$ in the interval between 60 AU and 2965 AU.

We now seek an explanation for the limits of validity of the OD as shown in Figures 2a and 2b, particularly for the volume-complete sample. The following hypothesis is proposed: (1) at the short end of the distribution ($\bar{a} \sim 100$ AU), any primeval OD will be quickly modified by the presence of close binaries and protoplanetary disks; (2) at the wide end ($\bar{a} > 3{,}000$ AU), encounters with massive objects (molecular clouds, spiral arms, black holes, MACHOS, etc.) will gradually dissociate the widest binaries. In fact, in a theoretical paper, Weinberg *et al.* (1987) estimated that a binary with a semimajor axis smaller than 2000 AU has a probability of one to survive (against dissociation by giant molecular clouds) during 10 billion years, yet a binary with $\bar{a} \approx 12{,}600$ AU after two billion years has a probability of survival of only 0.5. The consistency between the results of Weinberg *et al.* and the distributions shown in Figures 2a and 2b gives support to the hypothesis that Öpik's distribution is primeval, valid up to $\bar{a} \approx 45{,}000$ AU (see next section), but with the passage of time it gets truncated at the large separations. To further examine this hypothesis, we analyze the distribution of semimajor axes of the binaries in a very young group ($T < 10^6$ years).

3. Wide Binaries in a Very Young Group

Taking advantage of the Jones & Walker (1988) catalogue of proper motions of the stars in the Orion Nebula Cluster, we have identified 68 candidate common-proper-motion binaries (Poveda & Hernández-Alcántara, 2003). Jones & Walker's determinations of proper motions and infrared magnitudes for 1053 stars of the Orion Nebula Cluster are appropriate for our work, because of the proper motions accuracy ($\sigma_\mu < 0.1$ arcsec/century) and also because the faint limiting magnitude ($I < 13$). Moreover, the one-million year age of the cluster allows us to look into the 'almost' primeval distribution of semimajor

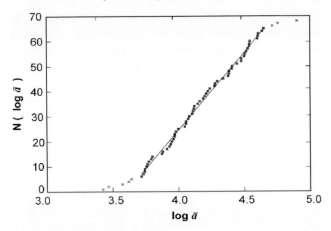

Figure 3. Cumulative distribution of the logarithms of the expected value of a for the common-proper-motion binaries of the Orion Nebula Cluster (Poveda & Hernández-Alcántara 2003). The straight line represents Öpik's distribution. For the interval fitted with a straight line, $5180 < \bar{a}(\text{AU}) < 44800$ (60 binaries) the KS test gives a value of $Q = 0.99$.

axes. The cumulative distribution of semimajor axes for this sample of Orion binaries follows very neatly Öpik's distribution (see Figure 3). The KS analysis of the data plotted in Figure 3 indicates that OD fits the data up to semimajor axes of 45,000 AU with a $Q = 0.99$.

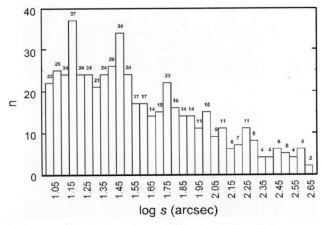

Figure 4. Frequency distribution of the separations s for the disk binaries of Chanamé & Gould (2004) for binaries with $s > 10''$ (523 binaries). Note that in the interval $1 < \log s < 1.55$ ($10'' < s < 36''$) which includes 285 binaries, this distribution is consistent with Öpik's, which in this plane would be a horizontal straight line. For larger separations, the distribution becomes depopulated similarly to what we found to occur for other samples of binaries (see text). We interpret this departure as due to the dissociation of wide binaries by encounters with massive perturbers.

4. Another Sample of Wide Binaries

Chanamé and Gould (2004, CG) have assembled a catalogue of wide binaries based on the revised *Luyten NLTT Catalogue* by Gould & Salim (2003) and Salim & Gould (2003). CG identified 999 common-proper-motion pairs. Good photometry allowed them to construct a reduced proper motion diagram for their binaries. The position of the binary components in this diagram helps to separate disk main-sequence pairs (801) from halo subdwarfs (116). The large number of binaries in the CG Catalogue offers an independent sample to test the validity of OD for main-sequence and halo binaries, respectively.

According to CG, their catalogue is incomplete for separations $s < 10''$. Since we are trying to establish the frequency distribution of separations of binaries in CG we extracted from their list a sample of disk binaries with separations $s > 10''$, i.e., we rejected 276 binaries closer than $10''$ out of 800 disk binaries. For the remaining 524 pairs we plot in Figure 4 the frequency distribution of separations $f(s)ds$. The large number of binaries allows to display this frequency distribution with small sampling fluctuations. In this figure we see clearly that $f(s)ds$ is essentially constant in the interval $1 \leqslant \log s \leqslant 1.55$. For larger separations, i.e., $s > 35''.5$, $f(s)$ is a decreasing function of s. CG noticed this behavior, which they found statistically significant. The constancy of $f(s)$ in the interval $10'' < s < 35''.5$ is just what we expect from OD. Since CG estimate the mean distance of their disk binaries to be 60 pc, we can transform into astronomical units the separations listed by CG. The angular interval where OD holds transforms into: $600 \leqslant s \text{ (AU)} \leqslant 2129$, which is equivalent to $840 \leqslant \bar{a} \text{ (AU)} \leqslant 2981$.

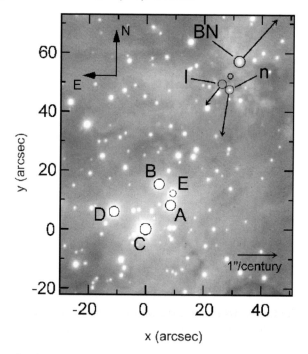

Figure 5. An example of a recent and nearby case of violent relaxation (Rodríguez *et al.* 2005; Gómez *et al.* 2006) Next to the Orion Trapezium, in the Becklin–Neugebauer, Kleinman–Low region we show 3 sources: BN, *I*, *n*, which are moving away from a center with transverse velocities of the order of 26, 8 and 28 km s^{-1}, respectively. These velocities are much larger than those expected for secular relaxation. Indeed, they correspond to violent relaxation.

5. The Physics Behind Öpik's Distribution

The analysis of the various samples of wide binaries, as shown in Figures 2–4, shows convincingly that Öpik's distribution holds for $a > 100$ AU. The distribution

$f(a) \sim a^{-1}$ is equivalent to a surface density of secondaries $\rho(a) \sim a^{-2}$. This surface density distribution is reminiscent of the surface density of stars in globular clusters and elliptical galaxies (King 1962; Hubble 1930; de Vaucouleurs 1953). The physics involved in these distributions is well understood. In the case of globular clusters we have what we may call secular relaxation, i.e., a process in which two-body encounters gradually modify the energy of a given star over a time scale much greater than the crossing time. In the case of elliptical galaxies, what we have is a case of initially violent relaxation, in which the energy of the stars changes rapidly because the potential of the stellar system (galaxy) changes rapidly due to an initial stellar collapse (van Albada 1982) or because of galaxy mergers. In the first case (secular relaxation) there is equipartition of energy and therefore a tendency for the lighter stars to diffuse to the outer parts of the globular cluster (this has been observationally confirmed); on the contrary, in the case of violent relaxation there is no equipartition of energy and hence no segregation of stellar masses, which indeed is the case in elliptical galaxies (no color gradients are observed).

The process of energy relaxation involves the interaction of several close stars ($n \geqslant 3$), very early in the stellar history. In fact, the wide binaries in the Orion Nebula Cluster follow OD even though they are still in the pre-main sequence phase (T $< 10^6$ years). We conclude that the process of energy relaxation is not related to interactions with stars in the cluster environment, but rather it is the result of very early interactions in multiple systems. This suggests that stars are formed in multiple systems that quickly relax and assume OD, with the possible ejection of one or more single stars. The common proper motion binaries in Orion show that there is not enough time for the reverse process to take place, i.e., the formation of double and multiple stars by capture from stars in the cluster.

In the process of star formation, the transition from gas to stellar dynamics may lead to a virialized multiple system where $2T + \Omega \approx 0$, or to a non-virialized one, depending on the initial conditions prevailing in the transition phase. If initial condensations (proto-stars) are not virialized, $2T + \Omega \ll 0$, then the proto-stars will collapse towards the center of mass of the multiple systems. Here we meet the conditions of violent relaxation, i.e., the collapse of the proto-stars will produce a rapid change in the gravitational potential experienced by the stars. In this conditions some stars will be accelerated to velocities larger than those associated with a virialized multiple system. An n-body simulation of this scenario was realized by the present authors many years ago (Poveda *et al.* 1967) with the purpose of finding an alternative explanation for the formation of runaway stars. In those simulations we found not only runaway stars but also that the binaries formed followed Öpik's distribution (Allen 1968; Poveda *et al.* 2004). Figure 5 shows a recent and nearby case of violent relaxation (Rodríguez *et al.* 2005; Gómez *et al.* 2006). Over an image of the Orion Trapezium region (McCaughrean 2001) we have superposed the proper motion vectors of objects BN, I, and n. The proper motions of the heavily obscured B star (Becklin-Neugebauer Object), as well as of the infrared objects I and n imply transverse velocities much larger than those expected to be produced in a virialized multiple system.

6. Conclusions

(1) The study of a large number of wide binaries, from mostly independent sources confirms that the frequency distribution of semimajor axes is $f(a) \sim a^{-1}$, i.e. precisely Öpik's distribution. This distribution is truncated at the short end ($a \approx 100$ AU) by the presence of close binaries and protoplanetary disks; at the large separations the distribution is depopulated by the process of dissociation produced by encounters with

massive objects. Figure 2 clearly exhibits these effects for the volume-complete sample in the solar vicinity.

(2) Öpik's distribution in the plane $N(\log P)$ is a constant, which is entirely consistent (within the error bars) with Duquennoy & Major's Gaussian distribution in the interval $42 < \bar{a}$ (AU) < 4213.

(3) Since there is no single astrophysical process that would generate a Gaussian distribution in $\log P$ (or in $\log a$) over such a wide interval (1 day $< P < 10^{10}$ days) and since, on the other hand, Öpik's distribution can be explained by the process of energy relaxation in few-body interactions, we propose to abandon the Gaussian representation for $\log a$, in favor of Öpik's distribution (for $a > 100$ AU.)

(4) Öpik's distribution suggests that the process of star formation produces multiple stars which evolve towards binaries after ejecting one or more single stars.

Acknowledgement

Our thanks are due to Paola Ronquillo for her help in the preparation of the typescript.

References

Allen, C. 1968 *Thesis* Universidad Nacional Autónoma de México

Allen, C., Poveda, A., & Herrera, M.A. 2000, *A&A* 356, 529

Chanamé, J. & Gould, A. 2004, *ApJ* 601, 289

Couteau, P. 1960, *J des Observateurs* 43, 41

Duquennoy, A. & Mayor, M. 1991, *A&A* 248, 495

Gliese, W. & Jahreiss, H. 1991, in: Brotzman, L.E., & Gessner, S.E. (eds.), *Selected Astronomical Catalogs*, VOL. I, NSSC, NASA, GSFC (GJ91)

Gómez, L., Rodríguez, L., Loinard, L., Lizano, S., Poveda, A., & Allen, C. 2005, *ApJ* 635, 1166

Gould, A. & Salim, G. 2003, *ApJ* 582, 1001

Heintz, W.D. 1969, *JRASC* 63, 275

Hubble, E. 1930, *ApJ* 71, 231

Jones, B. & Walker, M.F. 1988, *AJ* 95, 1755

King, I.R. 1962, *AJ* 67, 471

Kuiper, G.P. 1935, *PASP* 47, 121

Kuiper, G.P. 1942, *ApJ* 95, 201

Luyten, W.J. 1940-87, *Publ. Astr. Obs. Univ. Minnesota III*, part 3, 35; Proper motion survey with the 48-inch Schmidt Telescope, XXI, XXV, XIX, XL, L, LXIV, LV, LXXI, Univ. Minnesota.

Mc Craughrean, M. 2001, ESO PR Photo 03a/01, http://www.eso.org/outreach/press-rel/pr-2001/phot-03-01.html

Öpik, E.J. 1924, *Tartu Obs. Publ.* 25, No. 6

Poveda, A. & Allen, C. 2004, *RevMexAA (SC)* 21, 49

Poveda, A., Herrera, M.A., Allen, C., Cordero, G., & Lavalley, C. 1994, *RevMexAA* 28, 43

Poveda, A. & Hernández-Alcántara, A. in: K. S. Cheng,K. C. Leung, & T. P. Li (eds.), *Stellar Astrophysics: a Tribute to Helmut A. Abt*, Sixth Pacific Rim Conference (Dordrecht: Kluwer), ASSL vol. 298, p. 111

Poveda, A., Allen, C., & Herrera, M. A. in: J.A. Docobo, A. Elipe, & H. McAlister (eds.), *Visual Double Stars: Formation, Dynamics and Evolutionary Tracks (Dortrecht:Kluwer)*, p. 191

Poveda, A., Ruiz, J., & Allen, C. 1967, *Bol. Obs. Ton. Tac* 4, 86

Press, W.H., Flannery, B., Teukolsky, S.A., & Vetterling, W.T. in: *Numerical Recipes in C (Cambridge: Cambridge U. Press)*, p. 490

Rodríguez, L.F., Poveda, A., Lizano, S., & Allen, C. 2005, *ApJ* 627, L65

Salim, S. & Gould, A. 2003, *ApJ* 582, 1011

Schuster, W.J. & Nissen, P.E. 1988, *A&AS* 73, 225

Schuster, W.J. & Nissen, P.E. 1989a, *A&A* 221, 65
Schuster, W.J. & Nissen, P.E. 1989b, *A&AS* 222, 69
Vaucouleurs, G. de 1953, *MNRAS* 113, 134
Van Albada, T.S. 1982, *MNRAS* 201, 939
Weinberg, M.D., Shapiro, S.L., & Wasserman, I. 1987, *ApJ* 312, 367

Binary Stars as Critical Tools & Tests
in Contemporary Astrophysics
Proceedings IAU Symposium No. 240, 2006
W.I. Hartkopf, E.F. Guinan & P. Harmanec, eds.

© 2007 International Astronomical Union
doi:10.1017/S1743921307004395

Poster Abstracts (Session 7)

(Full Posters are available at http://www.journals.cambridge.org/jid_IAU)

Algols Contribute to the Interstellar Mass

Jean-Pierre De Greve, Camiel De Loore, and Walter Van Rensbergen

Astrophysical Institute, Vrije Universiteit Brussel, Brussels, Belgium

We revisited analytical expressions for the distribution of orbital periods and mass ratios for non-evolved binaries with a B-type primary. Selection effects governing the observations were taken into account in order to compare theory with observations. Theory was optimized so as to fit best with the observed P- and q-distributions of non-evolved SB1s and SB2s listed by Pourbaix *et al.* (2004) and the on-line SB9 catalogue of Pourbaix. Van Rensbergen *et al.* (2006) showed that the accuracy of the theoretical mass ratio distribution function is however hindered by the uncertainties on the observations. Our further computations compare statistically the observed distributions of orbital periods and mass ratios of Algols with those obtained from the Brussels binary evolutionary code. Conservative Roche Lobe Over Flow (RLOF) reproduces the observed distribution of orbital periods but fails to explain the observed mass ratios in the range $q \in [0.4\text{-}1]$. In order to obtain a better fit the binaries have to lose a significant amount of matter, without losing too much angular momentum. We tested the following binary evolutionary scenario: The mass acquired by the gainer during RLOF enhances the rotational velocity of the latter, as was shown by Packet (1981) and later refined by Wellstein (2001). Tidal forces counteract this acceleration (Zahn, 1977). Both mechanisms may speed up the equatorial velocity of the gainer close to its critical value leading to an enhanced stellar wind as shown by Langer (1998). The luminosity of the gainer is amplified with the accretion luminosity in the equatorial region, leading to a hot spot in the case of direct hit (e.g., Peters & Polidan, 2004) or a hot line in the case of an accretion disk (e.g., Bisikalo *et al.*, 2005). Only during a short time lapse of rapid RLOF spinning-up and accretion luminosity act together to remove mass and angular momentum from the gainer and blow it into the interstellar space. Consequently also Algols enrich the ISM. Consequently, the binaries persist for a longer time with a larger mass ratio compared to conservative evolution. Since the mass is blown away from the gainer as an enhanced stellar wind not much angular momentum leaves the system; hence the obtained distribution of the orbital periods of Algols differs not very much from what was obtained with the conservative scenario. The proposed time-dependent liberal scenario reproduces the observed distributions of mass ratios and orbital periods of Algols better than conservative evolution would do. As a test case we present the evolution of a binary with a 6 $M_\odot$ primary, a 3.6 $M_\odot$ companion and an initial period of 2.5 d. Only during $\sim$150,000 years of early RLOF A evolution some 1 $M_\odot$ is lost from the system. Later some 0.05 $M_\odot$ is lost during $\sim$30,000 years of RLOF B. Typically, RLOF occurs almost always in a conservative way, but during a short lapse of time the gainer is not capable of grasping the incoming material from the donor.

References

Bisikalo, D., Kaygorodov, P., Boyarchuk, A., & Kuznetzov, O. 2005 *Astron. Rep.* 49, 701

Langer, N. 1998,*A&A* 329, 551

Packet, W. 1981,*A&A* 102, 17

Peters, G. & Polidan, R. 2004 *Astron. Nach.* 325, 225

Pourbaix, D., Tokovinin, A., Batten, A., *et al.* 2004, *A&A* 424, 727

Pourbaix, D. 2006, *The 9th Catalogue of Spectroscopic Binary Orbits*,
http://sb9.astro.ulb.ac.be/

Van Rensbergen, W., De Loore, C., & Jansen, K. 2006, *A&A* 446, 1071

Wellstein, S. 2001, *Ph.D. thesis, Potsdam*

Zahn, J.-P. 1977, *A&A* 57, 383

Candidate Common Velocity Stars from the AGK3 Confirmed with Radial Velocity Measurements

J.-L. Halbwachs[1], M. Mayor[2], and S. Udry[2]

[1] *Observatoire Astronomique de Strasbourg, Strasbourg, France,*
[2] *Observatoire de Genève, Sauverny, Switzerland*

Two set of CPM stars were extracted from the AGK3 a while ago. The first one contains 326 CPM stars with a high probability (99%) to be physical binaries. For the second set (113 pairs), this probability was estimated around 60%.

In order to select the actual physical binaries, a long program of radial velocity (RV) measurements was initiated, using the spectrovelocimeter Coravel. It includes 90 stars from the first set and 177 from the second one. The spectroscopic binaries (SB) that were found were followed during about 15 years.

The RV difference between the components was obtained for 36 pairs from the first set and for 68 pairs from the second set. It appears that the physical wide pairs have components with RV differences less than 1.5 km/s. Only one pair from the first set is *slightly* beyond this limit, with a difference of 2.3 km/s. In the second set, 33 pairs among 68 seem to be optical, in good agreement with our expectations.

Taking into account the parallaxes from Hipparcos, we obtain a sample of 65 wide binaries (WB) with apparent separations more accurate than 25%. The separations range from about 1000 to around 30,000 AU. Among the 130 WB components, 31 are SB, including 24 with periods less than 10 years. The SB frequency is thus similar to that found among solar-type stars in the solar neighbourhood. It is not related to the separation of the WB, neither to the fact that the other component is also a SB. We conclude then that the SB found in WB have the same statistical properties as those found among single stars.

Eclipsing Binary Stars in the Globular Cluster M71

Y.-B. Jeon, S.-L. Kim, and C.-U. Lee

Korea Astronomy & Space Science Institute, Daejeon, South Korea

To search for variable stars in the globular cluster M71, we obtained time-series CCD images for 32 nights from 2000 July 8 to 2004 August 24. The CCD images were obtained with a thinned SITe 2K CCD camera attached to the 1.8-m telescope at the Bohyunsan Optical Astronomy Observatory in Korea. The field of view of a CCD image is $11\overset{'}{.}6 \times 11\overset{'}{.}6$ ($0\overset{''}{.}34$/pixel) at the f/8 Cassegrain focus of the telescope. We adjusted the exposure times from 50 s to 200 s (typically 100 s) in V band depending on the seeing ($1\overset{''}{.}0 - 3\overset{''}{.}3$) and transmission of the night sky. After photometric reduction, we detected 24 eclipsing binary stars. They are classified 8 EAs, 5 EBs and 11 EWs on the base of light curve shape. Ten of them are newly discovered in this research. We also have detected 16 the other type variable stars. Most of them are long-period variable stars.

Monitoring Open Star Clusters

A.S. Hojaev and D.G. Semakov

UBAI, CfSR, UAS, Tashkent, Uzbekistan

Star clusters — especially compact ones (with diameter of few to ten arcmin) — are suitable targets to search of light variability for an orchestra of stars by means of an ordinary Casegrain telescope plus CCD system. A special patrol with short time-fixed exposures and mmag accuracy could also be used to study stellar variability for groups of stars simultaneously. Extra-solar planet transit event detection might be a by-product of long-term monitoring of compact open clusters as well.

We presented a program of open star cluster monitoring with the Zeiss 1-m RCC telescope of Maidanak Observatory (Uzbekistan), which has been automated recently (Hojaev, 2005). In combination with quite good seeing at this observatory (see, e.g., Sarazin, 1999) the automatic telescope equipped with a large-format (2K×2K) CCD camera AP-10 available will allow to collect homogeneous time-series for analysis.

We started this program in 2001 and had a set of patrol observations with the Zeiss 0.6-m telescope and AP-10 camera in 2003. Seven compact open clusters in the Milky Way (NGC 7801, King 1, King 13, King 18, King 20, Berkeley 55, IC 4996) have been monitored for stellar variability and some results of photometry will be presented. A few interesting variables were discovered and dozens were suspected of variability to the moment in these clusters for the first time. Some of them show periodic variability and probably might be eclipsing binaries. It is also shown how observations like these could feasibly be used to look for exo-planets.

Statistical Modeling and Analysis of Wide Binary Star Systems

K.B. Johnston[1], T.D. Oswalt[1], and D. Valls-Gabaud[2]

[1] *Florida Institute of Technology, Melbourne, FL, United States,*
[2] *Paris Observatory, Paris, France*

Post-main sequence (MS) mass loss causes orbital separation amplification in fragile (i.e., common proper motion) binary star systems. Components typically have separations around ∼1000 AU. Such wide pairs experience negligible tidal interactions and mass transfer between companions; they evolve as two separate but coeval stars. In this paper we compute the rate of mass loss during the components' lifetimes and attempt to model how it will statistically distort a frequency distribution of fragile binary separations. Understanding this process provides a robust test of current theories of stellar evolution and sets constraints on the dynamics of the Galactic disk.

The Kinematics and Morphologies of Planetary Nebulae with Close-Binary Central Stars

D.L. Mitchell[1], T.J. O'Brien[1], D. Pollacco[2], and M. Bryce[1]

[1] *The University of Manchester, Manchester, United Kingdom,*
[2] *Queens University Belfast, Belfast, United Kingdom*

A programme is currently underway to study the kinematics and morphologies of PNe known to contain close-binary central stars. Three of the observed PNe have eclipsing binaries: Abell 63, Abell 46 and SuWt2; these objects are important because their physical parameters can be determined in a model independent way. The properties of these PNe can then be used as tests of stellar evolutionary theory.

Long-slit spectra of northern sky targets were obtained in 2004 with the Manchester echelle spectrometer combined with the 2.1-m San Pedro Martir telescope. Southern sky targets were observed in 2005 using UCLES onboard the 3.9-m Anglo-Australian telescope, and the EMMI spectrograph combined with the 3.58-m ESO New Technology Telescope in La Silla.

The test case of the programme is Abell 63 as it contains an almost totally eclipsing binary core. The nebula has two faint, elongated lobes and a tube-like appearance. Two end-caps are visible in [NII] at the tips of the lobes. Slits were positioned along the major and minor axes of the nebula. The longslit spectra from the major axis show clear Hα line-splitting, which can be attributed to receding and approaching sides of an expanding hollow tube. The longslit spectra from the minor axis show a velocity ellipse, which is characteristic of viewing a hollow tube in cross-section.

Soker (1998) predicted that bipolar nebulae with very elongated lobes must be produced by an intrinsically collimated fast wind (or jets) puncturing through a pre-existing spherical AGB wind. Soker proposed that the presence of a low-mass companion star is necessary for the production of jets via an accretion disk around the secondary. We tested this hypothesis for Abell 63 using a hydrodynamic simulation of a low-density 400 km s^{-1} jet blowing into a uniform, static AGB wind. The resulting morphology closely resembled that of Abell 63.

Session 8: Binary Stars as Critical Tests:

Asteroseismology;
Stellar activity

Figure 1. (top left to bottom right) Speakers Conny Aerts,
Katalin Oláh, Heidi Korhonen, and Andrej Pigulski.

Figure 2. Reflecting upon a beautiful city.

Figure 3. (top) Ed Guinan and Ignacio Ribas. (bottom) From left, Mercedes Richards, Katalin Oláh and Styliani Kafka.

Binary Stars as Critical Tools & Tests
in Contemporary Astrophysics
Proceedings IAU Symposium No. 240, 2006
W.I. Hartkopf, E.F. Guinan & P. Harmanec, eds.

© 2007 International Astronomical Union
doi:10.1017/S1743921307004413

Asteroseismology of Close Binary Stars

Conny Aerts

Instituut voor Sterrenkunde, Celestijnenlaan 200D, B-3001 Leuven, Belgium
email: conny@ster.kuleuven.be
and
Department of Astrophysics, Radboud University Nijmegen, P.O.Box 9010, 6500 GL
Nijmegen, The Netherlands

Abstract. In this review paper, we summarise the goals of asteroseismic studies of close binary stars. We first briefly recall the basic principles of asteroseismology, and highlight how the binarity of a star can be an asset, but also a complication, for the interpretation of the stellar oscillations. We discuss a few sample studies of pulsations in close binaries and summarise some case studies. This leads us to conclude that asteroseismology of close binaries is a challenging field of research, but with large potential for the improvement of current stellar structure theory. Finally, we highlight the best observing strategy to make efficient progress in the near future.

Keywords. (stars:) binaries: general, (stars:) binaries (including multiple): close, (stars:) binaries: eclipsing, stars: early-type, stars: evolution, stars: interiors, stars: oscillations (including pulsations), (stars:) subdwarfs, stars: variables: other, stars: statistics

1. Goals and current status of asteroseismology

The main goal of asteroseismology is to improve the input physics of stellar structure and evolution models by requiring such models to fit observed oscillation frequencies. The latter are a very direct and high-precision probe of the stellar interior, which is otherwise impossible to access. In particular, asteroseismology holds the potential to provide a very accurate stellar age estimate from the properties of the oscillations near the stellar core, whose composition, stratification and extent is captured by the frequency behaviour. In general, the oscillations are characterised by their frequency ν and their three wavenumbers (ℓ, m, n) determining the shape of the eigenfunction (e.g., Unno *et al.* 1989). Seismic tuning of interior stellar structure becomes within reach when a unique set of values (ν, ℓ, m, n) is assigned to each of the observed oscillation modes. In practice, the number of identified modes needed to improve current structure models depends on the kind of star. For B-type stars on the main sequence, e.g., even two or three well-identified modes can sometimes be sufficient to put constraints on the internal rotation profile (e.g., Aerts *et al.* 2003, Pamyatnykh *et al.* 2004) and/or on the extent of the convective core (e.g., Aerts *et al.* 2006, Mazumdar *et al.* 2006).

Asteroseismology received a large impetus after the very successful application of the technique to the Sun during the past decade. Helioseismology indeed revolutionised our understanding of the solar structure, including the solar interior rotation and mixing (e.g., Christensen–Dalsgaard 2002 for an extensive review). This remains true today, even though helioseismology is presently undergoing somewhat of a crisis with the revision of the solar abundances (Asplund *et al.* 2005, and references therein). These led to a slightly diminished precision, but still the relative agreement between the observed solar oscillation properties and those derived from the best solar structure models remains below 0.5% for most of the basic quantities, such as the interior sound speed and composition profiles (while it used to be below 0.1% with the old solar abundances).

Helioseismology thus provided us with a *unique* calibrator to study the structure of other stars.

However, the Sun is just one simple star. It is a slow rotator, it is hardly evolved, it does not possess a convective core, it does not suffer from severe mass loss, etc. There are thus a number of effects, of great importance for stellar evolution, that have not yet been tested with high accuracy. Since stars of different mass and evolutionary stage have very different structure, we cannot simply extrapolate the solar properties across the whole HR diagram. Stellar oscillations allow us to evaluate our assumptions on the input physics of evolutionary models for stars in which these effects are of appreciable importance. For the moment, such oscillations are the only accurate probe of the interior physics that we have available.

At present, the interior mixing processes in stars are often described by parametrised laws, such as the time-independent mixing length formulation for convection. Values near the solar ones for the mixing length and the convective overshoot are often assumed, for lack of better information. Similarly, rotation is either not included or with an assumed rotation law in stellar models, while an accurate description of rotational mixing is crucial for the evolution of massive stars (e.g., Maeder & Meynet 2000).

Fortunately, both core overshoot and rotation modify the frequencies of the star's oscillations and they do it in a different way. Core overshoot values can be derived from zonal oscillation modes, which have $m = 0$, because they are not affected by the rotation of the star for cases where the centrifugal force can be ignored, but they are strongly affected if an overshoot region surrounds the well-mixed convective core. On the other hand, the rotational splitting of the oscillation frequencies is dependent of the internal rotation profile. This can be mapped from the identification of the modes with $m \neq 0$ once the central $m = 0$ component of these modes has been fixed by the models. Adequate seismic modelling of core convection and interior rotation is thus within reach, provided that one succeeds in the identification of at least two, and preferably a much larger set of (ν, ℓ, m, n). This observational requirement demands combined high-precision multicolour photometric and high-resolution spectroscopic measurements with a high duty cycle (typically above 50%). For an extensive introduction into asteroseismology, its recent successes, and its challenges, I advise the review papers by Kurtz (2006) and by De Ridder (2006). None of the successful cases so far concerns a close binary ...

2. The specific case of oscillations in close binaries

Close binary stars have always played a crucial role in astrophysics, not only because, besides pulsating stars, they allow stringent tests of stellar evolution models, but also because they are laboratories in which specific physical processes, which do not occur in single stars, take place. Understanding these processes is important because at least half of all stars occur in multiple systems.

Close binaries are subjected to tidal forces and can evolve quite differently than single stars. Seismic mass and age estimates of pulsating components in close binaries of different stages of evolution, would allow to refine the binary scenarios in terms of energy loss and to probe the interior structure of stars subject to tidal effects in terms of angular momentum transport through non-rigid internal rotation.

2.1. *Overview of observational data*

Two excellent review papers on pulsating stars in binaries and multiple systems (including clusters) are available in Pigulski (2006) and Lampens (2006). These are highly recommended to the reader who wants to get a clear overview of the observational status and become familiar with this subfield of binary star research. One learns from these works that numerous pulsating stars are known in binaries, that lots of open questions remain concerning the confrontation between tidal theory and observational data, and that the best cases to monitor in the future are pulsating stars in eclipsing binaries. Eclipsing binaries have indeed revealed values for the core extent in B stars in excess of those found from asteroseismology of single B stars (Guinan *et al.* 2000). A natural thing to do would be to repeat the type of asteroseismic studies that led to the core overshoot value and internal rotation profile of single stars, as discussed in the previous section, but then for pulsating stars in eclipsing binaries. This would allow to disentangle the core overshoot from the internal rotation with higher confidence level than for single stars. In this respect, I refer to Pigulski *et al.*, Golovin & Pavlenko, and Latkovic (these proceedings) for new discoveries of pulsating stars in eclipsing binaries and to Brüntt *et al.* (these proceedings) for the best quality data available of such systems to date.

2.2. *Mode identification through eclipse mapping*

A remark worth giving here is the potential to perform mode identification through the technique of eclipse mapping in eclipsing binaries with a pulsating component. This idea was put forward more than 30 years ago by Nather & Robinson (1974), who interpreted the phase jumps of $360°$ in the nova-like binary UX UMa in terms of non-radial oscillation modes of $\ell = 2$. We now know that this interpretation was premature and that the observed phase phenomenon is far better explained in terms of an oblique rotator model.

Mkrtichian *et al.* (2004) excluded odd $\ell + m$ combinations for the Algol-type eclipsing binary star AS Eri from the fact that the disk-integrated amplitude disappears during the eclipse. Gamarova *et al.* (2005) and Rodríguez *et al.* (2004) made estimates of the wavenumbers for the Algol-type eclipsing binaries AB Cas and found a dominant radial mode, in agreement with the out-of-eclipse identification. By far the best documented version of mode identification from photometric data using eclipse mapping is available in Reed *et al.* (2005). While their primary goal was to search for evidence of tidally tipped pulsation axes in close binaries, they also made extensive simulations, albeit for the very specific case of eclipse mapping of pulsating subdwarf B-star binaries. They find that $\ell > 2$ modes become visible during an eclipse while essentially absent outside of eclipse. Their tools have so far only been applied to the concrete cases of KPD 1930+2752 and of PG 1336−018 (Reed *et al.* 2006) but without clear results.

We must conclude that, still today, more than 30 years after the original idea, mode identification from eclipse mapping is hardly applied successfully in practice, and it certainly has not been able to provide constraints on the wavenumbers for stars which have been modelled seismically. New promising work along this path is, however, in progress (Mkrtichian, private communication).

2.3. *Pressure versus gravity modes*

One thing to keep in mind is that there are two types of oscillations from the viewpoint of the acting forces, and that only one type is relevant in the context of tidal excitation. One either has pressure modes, for which the dominant restoring force is the pressure, or gravity modes, for which the dominant restoring force is buoyancy. Tidal excitation can only occur whenever the orbital frequency is an integer multiple of the pulsation frequency, the integer typically being smaller than ten. This follows from the expression of the tide-generating gravitational potential (e.g., Claret *et al.* 2005, and references therein). Moreover, in that case, one expects only $\ell = 2$ modes to be excited, with an m-value that provides the good combination between the rotation, oscillation and orbital frequencies to achieve a non-linear resonance. Such a situation is much more likely to occur for gravity modes, which have, in main-sequence stars, oscillation periods of order days, than for pressure modes which have much shorter periods of hours. The same holds true for compact oscillators, whose pressure modes have short periods of minutes while their gravity modes have periodicities of hours and these may be of the same order as the orbital periods.

Tidal forces can, of course, alter the free oscillations excited in components of binaries. In that case, one expects to see shifts of the frequencies of the free oscillation modes with values that have something to do with the orbital frequency. This alteration can, but does not need to be, accompanied with ellipsoidal variability. In the latter case, the tides have deformed the oscillator from spherical symmetry, and one needs to take into account this deformation in the interpretation of the oscillation modes.

2.4. Sample studies

Soydugan *et al.* (2006) have presented a sample study of 20 eclipsing binaries with a δ Sct–type component. They came up with a linear relation between the pulsation and orbital period:

$$P_{\mathrm{puls}} = (0.020 \pm 0.002) \, P_{\mathrm{orb}} - (0.005 \pm 0.008). \qquad (2.1)$$

This observational result implies that tidal excitation cannot be active in this sample of binaries, because this would demand a coefficient larger than typically 0.1 as mentioned above, i.e., an order of magnitude larger than the observed one.

A similar conclusion was reached by Fontaine *et al.* (2003), who investigated if the oscillations in pulsating subdwarf B stars could be tidally induced, given that 2/3 of such stars are in close binaries. Tidal excitation can, at most, explain some of the gravity modes observed in some such stars but this is not yet proven observationally. On the other hand, all the formation channels for subdwarf B stars involve close binary evolution (Han *et al.*, Pulstylnik & Pustynski, Morales–Rueda *et al.*, these proceedings). An asteroseismic high-precision mass and age estimate of such a star would imply stringent constraints on the proposed scenarios and on the role of the binarity for the oscillatory behaviour (see Hu *et al.*, these proceedings).

Aerts & Harmanec (2004), finally, made a compilation of some 50 confirmed line-profile variables in close binaries (mainly OBA–type stars). They could not find any significant relation between the binary and variability parameters of these stars.

We come to the important conclusion that we have by no means a good statistical understanding of the effects of binarity on the components' oscillations. This situation can only be remedied by performing several case studies of pulsating close binaries in much more depth than those existing at present.

3. Towards successful seismic modelling of binaries

In general, we have to make a distinction between three different situations when studying oscillations in close binaries with the goal to make seismic inferences of the stellar structure.

3.1. Reduction of the error box of fundamental parameters

In a first case, the binarity is simply an asset for the asteroseismologist, because it allows for a reduction of the observational error box of the fundamental parameters of the pulsating star. This case is relevant whenever we lack an accurate parallax value, and thus a good estimate of the luminosity and the mass. This is mainly the case for OB-type stars, but sometimes also for cooler stars or compact objects. In the absence of a good mass or luminosity estimate, the asteroseismologist cannot discriminate sufficiently between the seismic models fulfilling the observed and identified oscillation modes. It was recently shown that a combined observational effort based on interferometry and high-precision spectroscopy can enhance significantly the precision of luminosity and mass estimates in double-lined spectroscopic binaries with a pulsating component (Davis *et al.*, these proceedings), if need be after spectroscopic disentangling as in Ausseloos *et al.* (2006). Such spectroscopy is in any case also needed to derive the oscillation wavenumbers (ℓ, m) (e.g., Briquet & Aerts 2003, Zima 2006).

The best case study of a pulsating star whose binarity was of great help in the seismic interpretation is the one of α Cen A+B with two pulsating components (Miglio & Montalban 2006, and references therein). The binarity gave such stringent constraints in this case, that in-depth information on the interior of both components was found. In particular, it was found that the components seem to have different values of the mixing length parameter and that the primary, being of the same spectral type as the Sun, is right at the limit of having or not a small convective core. The seismic analysis also provided an accurate age estimate of the system. We refer to Miglio & Montalban (2006), and references therein, for the latest results and an overview of the stellar modelling.

Other well-known, but less successful examples are the δ Sct star θ^2 Tau (Breger *et al.* 2002, Lampens *et al.* these proceedings) and the B-type pulsators β Cen (Ausseloos *et al.* 2006), λ Sco (Tango *et al.* 2006) and ψ Cen (Bruntt *et al.* 2006). For these four stars, which are all fairly rapid rotators, our comprehension of the observed pulsational behaviour is at present insufficient for detailed seismic inference of their interior structure. In particular, we lack reliable mode identification of the detected frequencies. Not being able to identify the oscillation modes properly is the largest stumbling block in asteroseismology of single stars as well.

3.2. *Tidal perturbations of free oscillations*

The second case concerns seismic targets whose free oscillation spectrum is affected by the tides. The first such situation was reported for the δ Sct star 14 Aur A, a 3.8-d circular binary whose close frequency splitting of an $\ell = 1$ mode does not match the one of a single rotating star and was interpreted in terms of a tidal effect by Fitch & Wisniewski (1979). Alterations of the free oscillation modes by tidal effects have also been claimed for the three β Cep stars α Vir (also named Spica) which has an orbital period $P_{orb} = 4.1$d and an eccentricity $e = 0.16$ (Aufdenberg, these proceedings, Smith 1985a,b), σ Sco with $P_{orb} = 33$d, $e = 0.44$ (Goossens *et al.* 1984) and η Ori Aab with $P_{orb} = 8$d, $e = 0.01$ (De Mey *et al.* 1996).

When, besides oscillation frequencies, the orbital frequency and its harmonic is found in the frequency spectrum, one is dealing with oscillations in an ellipsoidal variable. This case occurs for the β Cep stars ψ^2 Ori with $P_{orb} = 2.5$d, $e = 0.05$ (Telting *et al.* 2001) and ν Cen with $P_{orb} = 2.6$d, $e = 0$ (Schrijvers & Telting 2002), as well as for the δ Sct stars θ Tuc with $P_{orb} = 7$d, $e = 0$ (De Mey *et al.* 1998), XX Pyx with $P_{orb} = 1.2$d, $e = 0$ (Aerts *et al.* 2002) and HD 207251 with $P_{orb} = 1.5$d, $e = 0$ (Henry *et al.* 2004). The deformation of the pulsator has so far been neglected in the interpretation of the oscillation frequencies in these binaries. The reason is clear: the complexity of the mathematical description of non-radial oscillations for a deformed star is huge compared to the case of a spherically-symmetric star. Aerts *et al.* (2002) have pointed out that this omission may well be the reason why the extensive efforts to model XX Pyx seismically, as in Pamyatnykh *et al.* (1998), have failed so far.

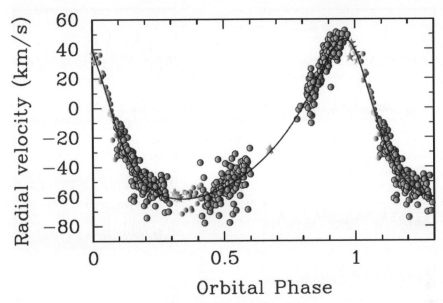

Figure 1. Observed radial velocities of HD 209295 (symbols) phased with the orbital solution (full line). For an explanation of the symbols, we refer to Handler *et al.* (2002), from which this figure was reproduced with permission from MNRAS.

3.3. *Tidally-excited oscillations*

There are at present only two cases known that meet the requirement of having tidally-excited oscillation modes. It concerns the slowly pulsating B star HD 177863 (B8V), for which De Cat *et al.* (2000) found an oscillation mode whose pulsation frequency is an exact multiple of 10.00 times the orbital frequency. The orbit of the star is very eccentric with $e = 0.77$ which provides a good situation to achieve a resonance between an $\ell = 2$ mode with a period of 1 day and the orbit of 10 days. Willems & Aerts (2002) made computations based on tidal oscillation theory for this star and indeed came up with the possibility that it is undergoing a non-linear resonantly excited $\ell = 2$ oscillation mode. This star has only one confirmed oscillation so far, such that seismic modelling is not yet possible since this requires at least two well-identified modes.

By far the most interesting case of tidally-induced oscillations was found by Handler *et al.* (2002). They discovered the star HD 209295 (A9/F0V) to be a binary with $P_{\rm orb} = 3.11$d and $e = 0.352$ (see Figure 1) exhibiting one δ Sct-type pressure mode and nine γ Dor-type gravity modes, of which five have an oscillation frequency which is an exact multiple of the orbital frequency. The frequency spectra of the star, after subsequent stages of prewhitening, are shown in Figure 2. The authors also predicted frequencies of tidally-excited modes for appropriate fundamental parameters of the primary and found several such gravity modes to agree with the observed ones, after having corrected the latter for the surface rotation of the star (see Figure 3). The unfortunate situation of not being able to identify the modes occurs again for this star, such that seismic tuning of its structure was not achieved so far.

4. Conclusions and outlook

We recall that the requirements for successful seismic modelling of a star are stringent. We need accurate frequency values, reliable identification of the spherical wavenumbers

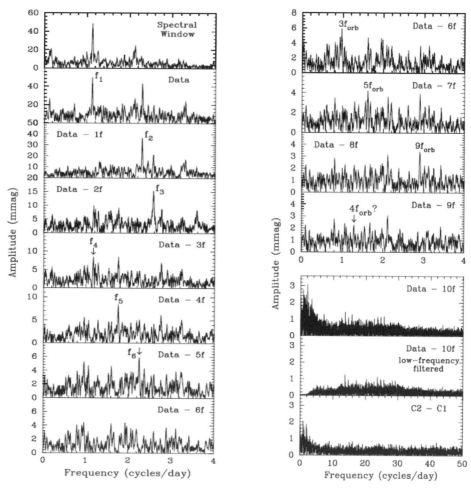

Figure 2. Left: Spectral window and amplitude spectra of the B filter time-series photometry of HD 209295 after subsequent stages of prewhitening. Top Right: amplitude spectra of the combined B and scaled V filter data, after prewhitening with the 6-frequency solution shown in the left panel. Several harmonics of the orbital frequency are found. Bottom Right: amplitude spectra of the differential magnitudes of the comparison stars. Figure reproduced from Handler *et al.* (2002) with permission from MNRAS.

(ℓ, m) and accurate fundamental parameters with a precision better than typically 10% before being able to start the modelling process. These requirements demand long-term high signal-to-noise and time-resolved spectroscopy and multicolour photometry with a duty cycle above, say, 50%. It is important that the data cover the overall beat-period of the oscillations, ranging from a few days for compact oscillators up to several months for main-sequence gravity-mode oscillators. These requirements have been met recently for a few bright single stars, with impressive improvement for their interior structure modelling, but not yet for a pulsating star in a close binary. For those candidates that came close to meeting these requirements, the problem of mode identification occurred and prevented seismic tuning.

We come to the conclusion that the potential of seismic modelling of close binaries is *extremely good*, particularly for eclipsing binaries. At the same time, its application is *extremely demanding* from an observational point of view. It is clear that efficient

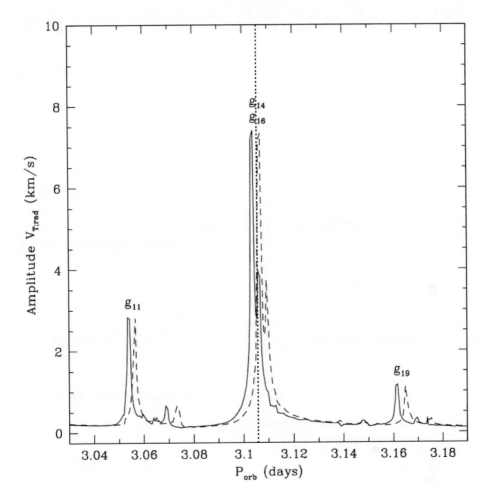

Figure 3. Predicted amplitudes of tidally induced radial velocity variations in a model for HD 209295, for two slightly different values of the rotation frequency ($1.852\,\mathrm{d}^{-1}$: full line; $1.854\,\mathrm{d}^{-1}$: dashed line). The orbital frequency of the star is indicated by the dotted vertical line. Figure reproduced from Handler *et al.* (2002) with permission from MNRAS.

progress in this field can only be achieved from coordinated multisite multitechnique observing campaigns, preferably in combination with uninterrupted space photometry. I strongly encourage the binary and asteroseismology communities to collaborate and take up this challenging project.

Acknowledgements

The author is much indebted to the Research Council of the Catholic University of Leuven for significant support during the past years under grant GOA/2003/04. She is also grateful to the organisers of this meeting for the opportunity to present this work and to Petr Harmanec for inspiring discussions and wise lessons on binary stars.

References

Aerts, C., Handler, G., Arentoft, T., Vandenbussche, B., Medupe, R., & Sterken, C. 2002, *MNRAS* 333, L35

Aerts, C. & Harmanec, P. 2004, in: R. W. Hidlitch, H. Hensberge & K. Pavlovski (eds.), *Spectroscopically and Spatially Resolving the Components of the Close Binary Stars*, ASPC (ASP: San Francisco), vol. 318, p. 325

Aerts, C., Thoul, A., Daszynska, J., Scuflaire, R., Waelkens, C., Dupret, M-A., Niemczura, E., & Noels, A. 2003, *Science* 300, 1926

Asplund, M., Grevesse, N., Sauval, A. J., Allende Prieto, C., & Kiselman, D. 2005, *A&A* 435, 339

Ausseloos, M., Aerts, C., Lefever, K., Davis, J., & Harmanec, P. 2006, *A&A* 455, 259

Breger, M., Pamyatnykh, A. A., Zima, W., Garrido, R., Handler, G., & Reegen, P. 2002, *MNRAS* 336, 249

Briquet, M. & Aerts, C. 2003, *A&A* 398, 687

Bruntt, H., Southworth, J., Torres, G., Penny, A. J., Clausen, J. V., & Buzasi, D. L. 2006, *A&A*, 456, 651

Claret, A., Giménez, A., Zahn, J.-P. (eds.) 2005, *Tidal Evolution and Oscillations in Binary Stars*, ASPC (ASP: San Francisco), vol. 333

Christensen-Dalsgaard, J. 2002, *Rev. Modern Phys.* 74, 1073

De Cat, P., Aerts, C., De Ridder, J., Kolenberg, K., Meeus, G., & Decin, L. 2000, *A&A* 355, 1015

De Mey, K., Aerts, C., Waelkens, C., & Van Winckel, H. 1996, *A&A* 310, 164

De Mey, K., Daems, K., & Sterken, C. 1998, *A&A* 336, 527

De Ridder, J. 2006, in: C. Sterken, & C. Aerts (eds.), *Astrophysics of Variable Stars*, ASPC (ASP: San Francisco), vol. 349, p. 129

Fitch, W. S. & Wisniewski, W. Z. 1979, *ApJ* 231, 808

Fontaine, G., Brassard, P., Charpinet, S., Green, E. M., & Willems, B. 2003, in: M.J. Thompson, M.S. Cunha, M.J.P.F.G. Monteiro (eds.), *Asteroseismology Across the HR Diagram*, ApSS (Kluwer Academic Publishers), vol. 284, p. 517

Gamarova, A. Yu., Mkrtichian, D. E., & Rodríguez, E. 2005, in: A. Claret, A. Giménez & J.-P. Zahn (eds.), *Tidal Evolution and Oscillations in Binary Stars*, ASPC (ASP: San Francisco), vol. 333, p. 258

Goosens, M., Lampens, P., de Maerschalck, D., & Schrooten, M. 1984, *A&A* 140, 223

Guinan, E. F., Ribas, I., Fitzpatrick, E. L., Giménez, A., Jordi, C., McCook, G. P., & Popper, D. M. 2000, *ApJ* 544, 409

Handler, G., Balona, L. A., Shobbrook, R. R., Koen, C., Bruch, A., Romero-Colmenero, E., Pamyatnykh, A. A., Willems, B., Eyer, L., & James, D. J., Maas, T. 2002, *MNRAS* 333, 262

Henry, G. W., Fekel, F. C., & Henry, S. M. 2004, *AJ* 127, 1720

Kurtz, D.W. 2006, in: C. Sterken, & C. Aerts (eds.), *Astrophysics of Variable Stars*, ASPC (ASP: San Francisco), vol. 349, p. 101

Lampens, P. 2006, in: C. Sterken, & C. Aerts (eds.), *Astrophysics of Variable Stars*, ASPC (ASP: San Francisco), vol. 349, p. 153

Maeder A. & Meynet G. 2000, *ARA&A* 38, 143

Mazumdar, A., Briquet, M., Desmet, M., & Aerts, C. 2006, *A&A*, in press

Miglio, A. & Montalban, J. 2005, *A&A* 441, 615

Mkrtichian, D. E., Kusakin, A. V., Rodriguez, E., Gamarova, A. Yu., Kim, C., Kim, S.-L., Lee, J. W., Youn, J.-H., Kang, Y. W., Olson, E. C., & Grankin, K. 2004, *A&A* 419, 1015

Nather, R. E. & Robinson, E. L. 1974, *ApJ* 190, 637

Pamyatnykh, A. A., Handler, G., & Dziembowski, W. A. 2004, *MNRAS* 350, 1022

Pamyatnykh, A. A., Dziembowski, W. A., Handler, G., & Pikall, H. 1998, *A&A* 333, 141

Pigulski, A. 2006, in: C. Sterken, & C. Aerts (eds.), *Astrophysics of Variable Stars*, ASPC (ASP: San Francisco), vol. 349, p. 137

Reed, M. D., Brondel, B. J., & Kawaler, S. D. 2005, *ApJ* 634, 602

Reed, M. D.& Whole Earth Telescope Xcov 21 and 23 Collaborations 2006, *Memorie della Societa Astronomica Italiana* 77, 417

Rodríguez, E., García, J. M., Gamarova, A. Y., Costa, V., Daszyńska-Daszkiewicz, J., López-González, M. J., Mkrtichian, D. E., & Rolland, A. 2004, *MNRAS* 353, 310

Schrijvers, C. & Telting, J. H. 2002, *A&A* 394, 603

Smith, M.A. 1985a, *ApJ* 297, 206

Smith, M.A. 1985b, *ApJ* 297, 224

Soydugan, E., Ibanoğlu, C., Soydugan, F., Akan, M. C., & Demircan, O. 2006, *MNRAS* 366, 1289

Tango, W. J., Davis, J., Ireland, M. J., Aerts, C., Uytterhoeven, K., Jacob, A. P., Mendez, A., North, J. R., Seneta, E. B., & Tuthill, P. G. 2006, *MNRAS* 370, 884

Telting, J. H., Abbott, J. B., & Schrijvers, C. 2001, *A&A* 377, 104

Unno, W., Osaki, Y., Ando, H., Saio, H., & Shibahashi, H. 1989, *Nonradial Oscillations of Stars, 2nd Edition*, University of Tokyo Press

Willems, B. & Aerts, C. 2002, *A&A* 384, 441

Zima, W. 2006, *A&A* 455, 227

Discussion

ROBERT WILSON: Have you looked into the time scale for orbital changes caused by dissipation of orbital energy by excitation of pulsations?

AERTS: I haven't, but other peoples did. In any case, such energy dissipation happens on an evolutionary timescale. This is very different from apsidal motion, with typical timescales between 10 to 100 years (see talk by Gimenez). The computation of timescales of energy dissipation through oscillations (i.e. in the case where there is no mass transfer) is difficult, because it is very dependent on the assumed initial conditions, in particular on the rotation velocity. Analytical computations of energy loss are availabe in, e.g., Willems *et al.* (2003, *A&A*, 397, 973). Numerical computations were made by, e.g., Witte & Savonije (2002, *A&A*, 386, 222). All these works, however, predict too long timescales if we compare them with those derived from data of high-mass X-ray binaries (the only observed cases available). This might be due to the much stronger radiative damping during resonances than anticipated so far. Besides the orbital energy loss, one also has to keep in mind that the oscillations probably imply significant angular momentum loss of the components through non-rigid internal rotation.

PATRICIA LAMPENS: How much time was needed to reach the significant new insight for single stars from the oscillations that you discussed (non-rigid internal rotation and estimate of core overshoot), given that the combination of pulsation and binarity is even more demanding?

AERTS: We managed to derive the non-rigid internal rotation in two main-sequence B stars so far. These stars have multiple oscillation periods of the order of several hours. One star was monitored with one and the same instrument from a single site during 21 years! The other one was monitored during a well-coordinated multisite photometric and spectroscopic campaign lasting 5 months and involving about 50 observers (see references in the paper). It is therefore clear that asteroseismic inference requires long-term monitoring. I think that the binarity does not impose the necessity of even longer runs (at least not for close binaries), because the most stringent demand is the frequency accuracy and the mode identification, while the orbital determination will result naturally and efficiently from a multsite spectroscopic effort.

Binary Stars as Critical Tools & Tests
in Contemporary Astrophysics
Proceedings IAU Symposium No. 240, 2006
W.I. Hartkopf, E.F. Guinan & P. Harmanec, eds.

The Influence of Binarity on Stellar Activity

Katalin Oláh

Konkoly Observatory, 1525 Budapest, P.O. Box 67, Hungary
email: olah@konkoly.hu

Abstract. Activity of late type stars is enhanced by fast rotation, which is maintained in nearly synchronized close binary systems. Magnetic activity originates in the deep convection zones of stars from where magnetic flux tubes emerge to their surfaces. The gravitational forces in binaries help the clustering of activity features giving rise to active longitudes. These preferred longitudes are observed in binaries from dwarfs to giants. Differential rotation is found in many active stars that are components of binary systems. If these binaries are circularized and nearly synchronized, then there will be a corotation latitude in their surfaces, and its position can be determined by observations and by theoretical calculations. Enhanced activity in binaries could have a reverse effect as well: strong magnetism in a binary component can modify the orbital period by the cyclic exchange of kinetic and magnetic energy in its convective envelope.

Keywords. stars: spots – stars: activity – stars: atmospheres – stars: late-type – stars: magnetic fields – stars: binaries: close

1. Introduction

Signatures of stellar activity like spots, plages, flares, enhanced radio and X-ray radiation, activity cycles, are observed on a great variety of single and binary stars with spectral types later than F. The nearest active star to us showing all the mentioned features of magnetic activity is the Sun. In spite of its closeness and the detailed three dimensional picture of its activity drawn from observations across the whole electromagnetic spectrum, the functioning of the solar magnetic dynamo, that produces the activity features, is still not fully understood.

Phenomenologically the observed activity features are the same for both single and binary stars. However, it is obvious that binarity itself has some effect on the activity of stars in binary systems. The reverse situation may work too: stellar activity caused by strong magnetic field may modify the orbits of some binaries.

Statistically, the strength of the activity is higher on stars in binary systems than on single stars. This is well documented by Montes *et al.* (1996) comparing excess CaII H&K and Hε measurements on 73 chromospherically active binaries with those measured on single stars with similar temperature and rotational rate. The reason for this difference is probably the following: the magnetic dynamo working inside the active stars is affected by the internal rotation, which is not the same in single and binary stars due to their different rotational history.

Late-type stars have deep convection zone, which, coupled with fast rotation, excites the hydromagnetic dynamo. Due to the strong tidal interaction in close binaries late type stars in these systems maintain fast rotation and consequently show high level of activity. In this paper the special activity features due to binarity, such as the existence of preferred longitudes and the position of corotation latitudes are summarized. The effect of magnetic cycles on the orbital period changes is briefly discussed.

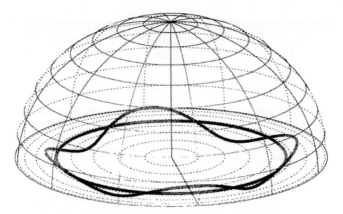

Figure 1. Instability of a toroidal flux tube in the bottom of the convection zone, perturbed by a displacement of wavenumber m=4. From Schüssler (1996).

2. Magnetic dynamo

The origin of the observed stellar activity features is the magnetic dynamo. Unfortunately, the functioning of the solar dynamo is not clearly understood, and our knowledge is even more limited in case of stellar dynamos. Comprehensive reviews and books give detailed picture of our present knowledge on this subject (Covas *et al.* 2005; Moss 2005 and references therein; reviews in: Proceedings of IAU Symp. No. 176, *Stellar Surface Structure*, 1996, eds. Strassmeier & Linsky).

At the beginning we give a brief description of the present idea(s) of the stellar dynamo, which helps one to understand the following sections. The present conception of the strong-field dynamo theory suggests, that the internal rotation changes sharply between the convection zone and the radiative core of stars in the shear layer, which generates a strong toroidal field ($\geqslant 10^5$ G), often called a "cycle dynamo". At the same time throughout the convection zone a turbulent weak-field dynamo operates generating an irregular field from the turbulent flow. These two dynamos are coupled: the weak-field ($< 10^4$ G) dynamo generates some fraction of the field for the cycle dynamo in the overshoot layer, while this large-scale dynamo feeds back the turbulent scale field through different processes (e.g., subsurface reconnections).

The observed activity features (e.g., spots and plages) indicate that the magnetic field in the stellar surface and beyond is not homogeneous; rather, the magnetic flux is concentrated in *flux tubes* surrounded by much less magnetized plasma. The activity features thus originate from the emergence of the magnetic flux tubes from the stellar interior. It is believed that the tubes are formed, intensified and stored in the bottom of the convection zone. After the flux tubes exceed a critical field strength, they lose stability and rise to the stellar surface due to the buoyancy force. More details on the hydromagnetic dynamo are found in the comprehensive book of Schrijver & Zwaan (2000).

Figure 1 shows a scenario with a stable and a perturbed toroidal flux tube in the bottom of the convection zone of a star, where the unstable loop starts to emerge, from Schüssler (1996). The emergence coordinates of flux tubes on the stellar surfaces depend on many parameters: their initial coordinates and magnetic fields, the depth of the convection zone they have to cross, the physical circumstances around (gas pressure, density), the rotational rate of the star and on binarity. In the following, the last two effects are discussed in more details.

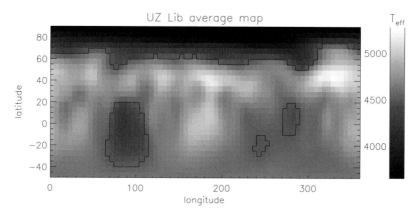

Figure 2. The surface of the non-eclipsing K0III giant UZ Lib, averaging all Doppler maps over 7 years. Note the active longitudes at 90° and at 270°, i.e., the substellar point and opposite. Due to the orbital elements 0° is the time of the quadrature with the primary receding. See Oláh *et al.* (2002b) for more.

3. Preferred longitudes in the binary reference frame

Active stars in eclipsing binaries have been monitored for decades. Earlier, researchers did not assume spots on the components as the cause of the observed light curve distortions outside eclipses, which were found in many cases, and tried to find other explanations (for a comprehensive review of the history of starspots see Hall 1994). Using these old measurements, and with the new knowledge that the distortions may also be caused by starspots, a lot of evidence have been gathered for the existence of preferred longitudes in active close binaries, dating back to almost a century.

One example is the eclipsing binary RT And (types F8V + K0V) which has active components, and has continuously been observed from the early 20th century. The light curves reveal two active longitudes on the quadrature positions all the time (Zeilik *et al.* 1989; Pribulla *et al.* 2000). Other main sequence eclipsing binaries (CG Cyg, BH Vir, WY Cnc, UV Psc) have also two preferred longitudes of spots at the quadrature positions, whereas non-eclipsing giants (IM Peg, HK Lac, UZ Lib) have active longitudes at the substellar points and their antipodes. Binaries with subgiant active components (AR Lac, SZ Psc, RT Lac) show active longitude at the substellar points and at other longitudes as well. Literature of the mentioned binaries is found in Oláh (2006). Figure 2 shows the surface of UZ Lib as an example for active longitudes at the substellar point and opposite, based on Doppler images from spectra averaged over 7 years (Oláh *et al.* 2002b). The figure is a co-added image of the stellar surface of 24 individual images from 7 years. The fact that a definite structure appears (instead of an even surface which would be the result of co-adding images with spots at random positions) means, that indeed, stable active regions are present on the stellar surface throughout the whole observed time interval of 7 years.

A natural explanation, at least partly, for the existence of active longitudes comes from the idea that the gravitational force of the component star affects the emerging flux tubes of the active star in such a way that they cluster around favoured, stable positions. Modelling such scenarios were carried out by Holzwarth & Schüssler (2002, 2003a,b) supposing a hypothetical binary of 1M⊙ + 1M⊙, with an orbital and rotational period of 2 days and solid body rotation. A tidally–deformed solar model was used to describe the internal structure of the primary, whereas the component star was supposed to be a mass point. Beside the tidal effect, the Coriolis force in rapidly rotating stars plays

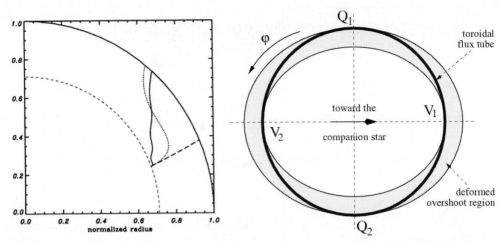

Figure 3. Left: the effect of the Coriolis force on the emerging flux tubes for the initial magnetic field of B=2×10^5 G. Flux tube paths for solar rotational period of 27 days (dashed line), for 9 days rotational period (dotted line) and for 3 days rotational period (solid line) are depicted from Schüssler & Solanki (1992). Right: toroidal flux tube in a tidally deformed overshoot region, from Holzwarth & Schüssler (2003a). The flux ring runs through different environments. V_1 and V_2 mark the substellar points and opposite, where the tube is in a deeper layer than at the quadrature points Q_1 and Q_2.

a significant role by deflecting the flux tubes to higher latitudes (Schüssler & Solanki 1992) in such a way that flux tubes with low initial latitudes (closer to the equator) are much more deflected than the ones originating at high latitudes (Holzwarth & Schüssler 2003b). This was also taken into account in the modelling. In addition to these forces, the strength of the initial magnetic field is a very important factor for finding the emergence place of the flux tubes. The influence of the Coriolis force and the effect of the tidal force are depicted in Figure 3.

The modelling results of Holzwarth & Schüssler (2003a) (see especially their Figure 7) show stable and unstable regions in longitude at the bottom of the convection zone of a binary component for single and double loops (wavenumber equals 1 and 2, respectively), giving the probability that a loop penetrates to the superadiabatic part of the convection zone and rises to the surface. It is found that in the case of single loops, the preferred longitude is the surroundings of the substellar points, whereas for double loops a wide area around the quadratures are favoured by the flux tubes.

The emergence longitude of flux tubes and their clustering depend on many factors. In the quoted literature detailed parameter studies are found assuming different initial magnetic field and initial position (longitude and latitude) of the flux tubes at the bottom of the convection zone. It is discussed, how the accumulation of the tidal effect influences the emergence pattern on the stellar surface during the emergence time of the flux tubes, which can last from months to years. However, these studies suppose solar type stars only, and very few attempts were made to broaden the calculation to evolved systems (see Holzwarth 2004). But the results for solar-type stars are promising and seem to support the observed facts on preferred longitudes of active stars in binary systems. As an example Figure 4 shows resulting flux distributions on the stellar surface with different initial parameter values from the modelling, and an observed example for spots at quadratures of UV Psc by Kjurkchieva *et al.* (2005).

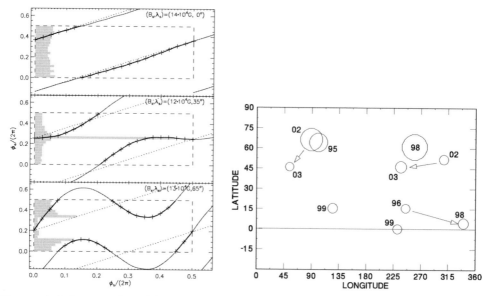

Figure 4. Left: examples of flux distribution on the stellar surface due to tidal effects, from Holzwarth & Schüssler (2003b). Thick lines show the modelled eruption longitudes in the function of initial longitudes at the bottom of the convection zone. The eruption longitude depends also on the initial latitude and magnetic field, given in the top right corners. The grey bars in the left show histrograms of the eruption longitudes on half a stellar surface: (from top to bottom) flat distribution, preferred longitude at quadrature, and two preferred longitude ranges. Right: Preferred longitudes at the quadratures on UV Psc derived from observations by Kjurkchieva *et al.* (2005).

4. Differential rotation - corotation latitude

4.1. *Observing the differential rotation*

The differential rotation of the Sun is evident and naturally, searching for stellar differential rotation started with the advent of studying stellar activity. In principle the task is easy: one should find the latitude of the major spots (spot groups) which cause the light variation, derive the corresponding period of the rotational modulation for several seasons and find the function between them. But in practice this method fails in most cases, because the spot latitudes cannot be derived from photometry with acceptable accuracy (see Kővári & Bartus 1997), and the period determination from a short dataset is also not accurate enough (but during longer time the spots may change their position drastically). Only the most recent photometry from space with accuracy better than a millimagnitude may be useful for this: Croll *et al.* (2006) succesfully deduced differential rotation of ϵ Eri for the first time, from modelling a very accurate light curve (accuracy of the unbinned data is 0.00025 mag.), obtained by the photometric satellite *MOST*.

Migrating waves on the light curves of eclipsing binaries originate from active regions on the surface of a component star producing light variation (usually sinusoidal) on top of the eclipsing light curves, which show continuous shifts measured in the orbital reference frame. Periods of light variability in non-eclipsing binaries usually differ from the accurately known binary periods. Such observational evidence in binaries points toward the presence of differential rotation. (Other reasons of differences between the orbital and rotational periods could be a.) pseudosynchronous rotation in slightly eccentric systems and b.) incorrect orbital period due to unrecognized apsidal motion.)

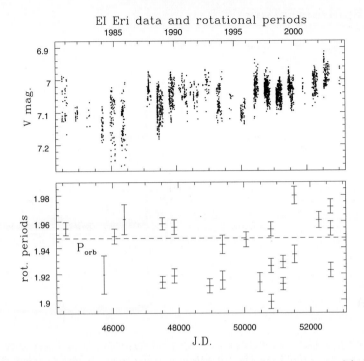

Figure 5. Top: observational data in V colour of EI Eri from the literature and from the Vienna APT (Strassmeier *et al.* 1997). Bottom: rotational periods, sometimes multiple periods, from photometric data for each season, with the errors of the period determinations. Horizontal bars show the lengths of the seasons. The orbital period of the system (P_{orb}) is marked with the horizontal dashed line.

From high resolution spectroscopic data differential rotation for several stars (both for single stars and binary components) were deduced from Doppler imaging, which gives good surface distribution of the starspots. Generally, differential rotation on active stars, both for dwarfs, subgiants and giants in binaries, is found to be much (about an order of a magnitude) weaker than on the Sun (cf., Kővári *et al.* 2004; Oláh *et al.* 2003; Weber & Strassmeier 1998; Weber *et al.* 2005).

4.2. *Differential rotation and the corotation latitude*

The corotation latitude is where the orbital and rotational periods are equal, in case of circularized and nearly synchronized binaries. The only detailed theoretical calculation about the position of the corotation latitudes on binary systems with active components dates back to the early 1980's, by Scharlemann (1981, 1982). He calculated the isorotation surfaces inside the star for various differential rotation laws assuming very deep convection zones. The latitude where the isorotation surface intersects the stellar surface is the corotation latitude.

The rotational periods defined by spots of active stars in binary systems show variability in time. A good example is given in Figure 5 where a long-term dataset of the active binary EI Eri is shown together with the seasonal rotational periods, sometimes multiple periods. The rotational periods were derived using simple Fourier analysis on photometric data for each season separately. If the different periods are due to differential rotation, the spots should belong to distinct active regions at different latitudes. In Figure 5 the orbital period of the system is also marked, and it is well seen that periods shorter and

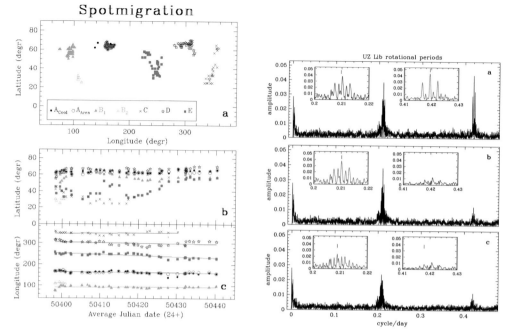

Figure 6. Left: meridional motions of spots on V711 Tau from Strassmeier & Bartus (2000). Note that Vogt *et al.* (1999) put the corotation latitude to about 60° to where the spots tend to move. Right: amplitude spectra of UZ Lib using data from 9 consecutive years. (a): the strongest peak is the first harmonic of the most prominent rotational feature of double humped shape belonging to the spots on the equator. (b): after prewhitening the data with the main frequency and its harmonic, another frequency is revealed caused by high latitude features. (c): removing the second frequency a significant signal still remains, caused possibly by transients. See text.

longer than the orbital period are present throughout the observed time interval. The corotation latitude thus should be somewhere at mid-latitudes on EI Eri.

4.3. *The corotation latitude of HK Lac, V711 Tau and UZ Lib*

The corotation latitude from observations could be deduced directly from the different latitudes of active regions and the corresponding rotational periods. An early study of the active giant HK Lac using photometry was made by Oláh *et al.* (1985), and though the resulting corotation latitude of 32° was uncertain for the reasons detailed in Sect. 4.1, but the value turned out to be reasonable taking into account the theoretical model calculations of Scharlemann (1981, 1982).

Vogt *et al.* (1999) found the corotation latitude of about 60° and antisolar differential rotation (i.e., equator rotates slower than higher latitudes) for V711 Tau using Doppler imaging. A time series analysis of Doppler images on V711 Tau by Strassmeier & Bartus (2000) revealed a poleward latitudinal migration of spots of +0°.41/day. Spots which appear at lower latitudes move upwards to about 60°, i.e., to the latitude which seems to be the corotation latitude and where most spots stay, see Figure 6, left panel. In the case of V711 Tau thus the higher latitudes corotate with the orbit.

A corotation latitude and small antisolar differential rotation were determined for the active giant UZ Lib by Oláh *et al.* (2003) comparing Doppler images (cf. Figure 2) with Fourier analysis of contemporaneous photometric data over 9 years. When differential

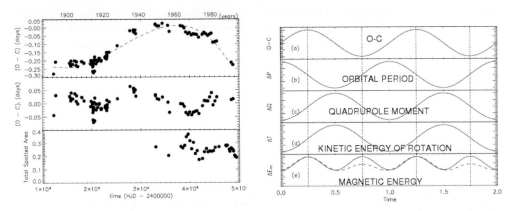

Figure 7. Left: Orbital period variation of RS CVn with a long-term trend (top panel), after removing the long-term trend (middle panel) and the change of the total spotted area on the star (lower panel). Right: Result of theoretical calculations showing the period and phase relations of different quantities of orbital period changes and magnetic cycles. From Lanza & Rodonò (2004).

rotation is present, large spots at different latitudes may result in close but distinct peaks in the amplitude spectrum revealing different periods.

For UZ Lib the rotational frequencies are near 0.21 cycle/day ($\approx$4.76 days), but the highest peak of the amplitude spectrum is found at 0.41944 cycle/day (top of Figure 6, right panel), which is the first harmonic of the most prominent rotational feature of double humped shape, showing that it originates from the two spots placed oppositely on the stellar equatorial zone (see Figure 2), so the main rotational period indicated by these spots is 4.7683 days. After prewhitening the data with the dominant frequency and its harmonic, other two frequencies appear (0.21017 cycle/day = 4.7581 days and 0.20890 cycle/day = 4.7870 days) that should belong to the high latitude features and some transients (middle and lower panels of Figure 6, right panel), which together point toward antisolar differential rotation. See Oláh *et al.* (2003) for more details concerning the identification of the spectral features with spots from Doppler maps. The rotational period belonging to the highest peak, caused by the two equatorial spots agrees with the 4.76824 days orbital period of the system. This means that for UZ Lib the corotation latitude is on the equator.

At present the connection between the corotating latitude and spot clustering in latitude is an open question. The theoretical background of the existence and position of preferred longitudes in binaries is still not well established. The binarity through tidal forces affects spots clustering both in latitudes and longitudes, but many other physical circumstances should also be taken into account. Hopefully, in the future, more and detailed modellings will be carried out to explain the special activity features observed in close binary systems.

5. Orbital period changes due to activity

The above sections deal with the effect of binarity on stellar activity, however, the reverse situation is also interesting: does the stellar activity itself affect the binary system, and if yes, how?

The similarity between timescales of $O - C$ variations and magnetic cycles of certain eclipsing binaries hints the possibility of some connection, as noted first by Hall (1990). An example of such a variation was given by Hall (1991) who showed antiparalel behaviour between the cyclic changes of the $O - C$ diagram and the light variation of CG Cyg. The effect of a third body as a simple explanation for the observed sinusoidal shape of the $O - C$ diagram was ruled out by the fact that these changes were not strictly periodic, whereas the perturbation caused by a third (hierarchical) companion should have to be so.

To gather enough information on the connected orbital period changes and light variability which reflects the magnetic activity, is unfortunately a matter of decades long observations. Data on a few systems (RS CVn, AR Lac, RT Lac) are already available and are summarized in a recent paper by Lanza & Rodonò (2004). As an example in the left panel of Figure 7 the $O - C$ of the orbital period and the spottedness variations of RS CVn are shown. For all systems discussed by Lanza & Rodonò (2004) the cycle timescale of the spottedness variation is about half of the $O - C$ variation.

Different physical mechanisms were proposed to explain the connection between the cyclic magnetic activity and orbital period modulation. Varying magnetised stellar wind cannot account for the observed orbital period variation, since the observed mass loss rates are much smaller than required to explain the observed changes, and the tidal timescale is too slow to couple the spin angular momentum change to the orbital motion. Simple mass transfer would cause not cyclic, but one-way orbital period change.

A possible mechanism to explain the above connection is, that the quadrupole moment of the active component varies during an activity cycle, consequently the outer gravitational potential of the star changes and that induces an immediate change in the orbital period. This model was suggested first by Applegate & Patterson (1987) supposing variation of the stellar radius due to variation of the magnetic pressure gradient, but this idea failed since the required energy for such a change was far too high (Marsch & Pringle 1990). Thus later the model was modified by Applegate (1992) assuming that the internal angular velocity of the star is changing during a magnetic cycle (instead of the radius), which results in the oblateness change, i.e., quadrupole moment change of the star. Recent quantitative calculations by Lanza & Rodonò (2004) show that it is possible to explain the observed connection between the orbital period variations and magnetic cycles, though the observed amplitudes of the $O - C$ diagrams are smaller than that predicted by the model. The phase and period relations between the orbital period, $O - C$ and magnetic cycle of the model agree with the observations and are depicted in the right panel of Figure 7. It is interesting to compare the time and phase behaviour of the observed (RS CVn) and calculated orbital period variation and magnetic cycles. The paper of Lanza & Rodonò (2004) gives much more details for the interested reader.

6. Summary

In the previous sections a brief description of the influence of binarity on stellar activity and *vice versa* has been given. It is known that binarity helps to achieve and maintain fast rotation through spin-orbit coupling. The fast rotation, together with the convective envelope and strong magnetic field of late-type components, results in the operation of hydromagnetic dynamo. Therefore, active close binaries generally show higher level of activity than coeval single stars.

The tidal effects of binaries may organize preferred longitudes and latitudes of activity on the stellar surfaces. However, the efficiency of the tidal force on organizing the emerging flux tubes into certain places depends on several other parameters like the Coriolis force, strength of the magnetic field, depth of the convection zone, initial longitudes and latitudes of the flux tubes at the bottom of the convection zone, etc.

Observations show that in binaries preferred longitudes exist on main sequence stars mostly at quadratures while on giant components at the substellar points. Since the existence of differential rotation is proven for many active stars in binaries, in case of circularized and nearly synchronized systems a corotation latitude should exist on the stellar surface, where the rotational and orbital periods are the same. For the active giant UZ Lib the corotation latitude is the equator (Oláh *et al.* 2003) whereas for the subgiant V711 Tau it is at about 60° (Strassmeier & Bartus 2000), and in both cases spots seem to cluster near these corotating latitudes.

The similarity of the timescales of cyclic behaviour of the orbital period changes and magnetic cycles indicates a connection between them. It is shown by Lanza & Rodonò (2004) that the orbital period modulation and magnetic cycles are connected by the exchange of kinetic and magnetic energy and back, in the convective envelope of stars.

As a conclusion we can say, that the *interaction* between stellar activity and binarity is a very interesting problem, and its exploration has just begun. In the future, hopefully, this special area of stellar activity will attract more interested colleagues who plan, and have the possibility, to work in this field *longer than the PhD timescale*.

Acknowledgements

Thanks are due to J. Jurcsik, Zs. Kővári and A. Prša for their critical reading of the manuscript. Supports from the Hungarian Research Grants OTKA T-043504 and T-048961 is acknowledged.

References

Applegate, J.H. 1992, *ApJ* 385, 621, 99
Applegate, J.H. & Patterson, J., 1987, *ApJ* 322, L
Covas, E., Moss, D., & Tavakol, R. 2005, *A&A* 429, 657
Croll, B., Walker, G.A.H., Kuschnig, R., & Matthews, J.M. 2006, *ApJ* 648, 607
Hall, D.S. 1990, in: *Active Close Binaries*, NATO ASI Series C, Kluwer, ed.: C. Ibanoglu, Vol. 319, p. 95
Hall, D.S. 1991, *ApJ* 380, L85
Hall, D.S. 1994, *IAPPP Comm.* No. 54, 1
Holzwarth, V. 2004, *AN* 325, 408
Holzwarth, V. & Schüssler. M. 2002, *AN* 323, 399
Holzwarth, V. & Schüssler. M. 2003a, *A&A* 405, 291
Holzwarth, V. & Schüssler. M. 2003b, *A&A* 405, 303
Kjurkchieva, D.P., Marchev, D.V., Heckert, P.A., & Ordway, J.I. 2005, *AJ* 129, 1084
Kővári Zs. & Bartus J. 1997, *A&A* 323, 801
Kővári Zs., Strassmeier, K.G., Granzer, T., Weber, M., Oláh, K., & Rice, J.B. 2004, *A&A* 417, 1047
Lanza, A. & Rodonò, M. 2004, *AN* 325, 393
Marsch, T.R. & Prongle, J.E., 1990, *ApJ* 365, 677
Montes, D., Fernández-Figueroa, M.J., Cornide, M., & De Castro, E. 1996, *A&A* 312, 221
Moss, D. 2005, *A&A* 432, 249
Oláh, K. 2006, *AP&SS* 304, 145
Oláh, K. Eaton, J.A., Hall, D.S., & Henry, G.W. *et al.* 1985, *AP&SS* 108, 137
Oláh, K., Strassmeier, K.G., & Granzer, T. 2002a, *AN* 323, 453
Oláh, K., Strassmeier, K.G., & Weber, M. 2002b, *A&A* 389, 202
Oláh, K., Jurcsik, J., & Strassmeier, K.G. 2003, *A&A* 410, 685
Pribulla, T., Chochol, D., & Milano, L. 2000, *A&A* 362, 169
Ransom, R.R., Bartel, N., Bietenholz, M.F., Lebach, D.E. *et al.* 2002, *ApJ* 572, 487
Schrijver, C.J. & Zwaan, K., 2000, *Solar and Stellar Magnetic Activity*, Cambridge Astrophysical Series No. 34, Cambridge University Press
Scharlemann, E.T. 1981, *ApJ* 246, 292
Scharlemann, E.T. 1982, *ApJ* 253, 298

Schüssler, M. 1996, in Proceedings of IAU Symp. No. 176: *Stellar Surface Structure*, eds. Strassmeier & Linsky, p. 269

Schüssler, M. & Solanki, S. 1992, *A&A* 264, L13

Strassmeier, K.G. & Bartus, J. 2000, *A&A* 354, 537

Strassmeier, K.G., Boyd, L.J., Epand, D.H., & Granzer, Th. 1997, *PASP* 109, 697

Vogt, S.S., Hatzes, A.P., Misch, A.A., & Krster, M. 1999, *ApJS* 121, 547

Weber, M. & Strassmeier, K.G. 1998, *A&A* 330, 1029

Weber, M., Strassmeier, K.G., & Washuettl, A. 2005, *AN* 326, 287

Zeilik, M., Cox, D.A., de Blasi, C., Rhodes, M., & Budding, E. 1989, *ApJ* 345, 991

Discussion

MERCEDES RICHARDS: Results of Doppler tomography suggest that the distribution of gas flows can be influenced by coronal mass ejections. So there are direct implications of stellar activity on the mass transfer process.

STYLIANI KAFKA: What would be the expected activity signatures in the case where one or both components of the binary are M stars, especially when they are fully convective?

OLÁH: The activity signatures like spots, flares, etc. of M stars in binaries are similar to those observed on earlier type stars; see, e.g., the cases of YY Gem and CM Dra. If cyclic behaviour of the activity is found, than it is the signature of a large-scale dynamo in the bottom of the convection zone. Inside the convection zone, a turbulent, weak-field dynamo operates which may produce activity signatures as well, but not cycles. The mass limit for fully convective stars is generally thought to be about 0.3–$0.4\,M_\odot$, but some calculations show that it might be as low as 0.1–$0.2\,M_\odot$ (Mullan & McDonald 2002, *ApJ* 559, 353).

Binary Stars as Critical Tools & Tests
in Contemporary Astrophysics
Proceedings IAU Symposium No. 240, 2006
W.I. Hartkopf, E.F. Guinan & P. Harmanec, eds.

Active Longitudes and Flip-Flops in Binary Stars

Heidi Korhonen[1] and Silva P. Järvinen[1,2]

[1]Astrophysikalisches Institut Potsdam, An der Sternwarte 16, D-14482 Potsdam, Germany

[2]Astronomy Division, P.O.Box 3000, FI-90014 University of Oulu, Finland

Abstract. We present results from an investigation where the long-term photometry of several magnetically active RS CVn binaries is studied to see whether or not they show permanent active longitudes and the flip-flop phenomenon. We confirm that it is very common for the active regions to occur on permanent active longitudes. Many of our target stars also show clear flip-flop phenomenon, but often the data set is not long enough for reliable determination of the flip-flop period.

Keywords. stars: activity, stars: late-type, stars: spots

1. Introduction

In many active stars the spots concentrate on two permanent active longitudes which are 180° apart. In some of these stars the dominant part of the spot activity changes the longitude every few years. This so-called flip-flop phenomenon was first reported in the early 1990's in the single, late-type giant FK Com (Jetsu *et al.* 1993). Since then flip-flops have been reported also on binary stars (e.g., Berdyugina & Tuominen 1998), young solar type stars (e.g., Järvinen *et al.* 2005) and the Sun itself (Berdyugina & Usoskin 2003). Even though this phenomenon has been detected on many different kinds of active stars, still only about ten stars are known to exhibit this effect. Therefore no statistically significant correlation between the stellar parameters and the flip-flop phenomenon can be carried out.

2. Sample and methods

Korhonen & Elstner (2005) investigated the long-term light-curve behaviour caused by the flip-flop phenomenon using dynamo models. In the current investigation we have studied in detail stars showing similar patterns as the flip-flop signatures seen by Korhonen & Elstner. The observations are mainly old, already published photometry, but we have also received new data. We have up to now investigated 14 stars, 9 of them binaries, using the light-curve inversion techniques.

In this technique the phased light curves are used for obtaining spot filling factor maps of the stellar surface with an inversion technique using the Maximum Entropy Method. The exact formulation follows closely the one by Lanza *et al.* (1998). Figure 1 shows an example of a light-curve for UZ Lib and a corresponding spot filling factor map.

3. Results

The phases of the spots can be determined from the spot filling factor maps obtained with light-curve inversions. In Figure 2 the results for XX Tri and UZ Lib are shown.

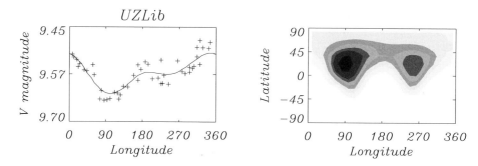

Figure 1. Results from the light-curve inversion for UZ Lib in June-July 1998. The V band observations are given by crosses and the inversion result is marked with a line. In the maps the darker areas mean larger spot filling factor.

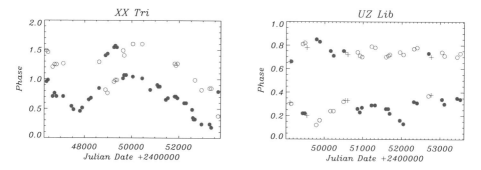

Figure 2. Phases of the spots in XX Tri and UZ Lib determined from the light-curve inversions plotted versus the Julian Date. Filled circles denote the larger spots, open circles the secondary spots and the cross a case where the two spots are equally strong.

The major spots are given as filled circles, minor ones as open circles and spots of equal size are marked with crosses.

Our investigation confirms that many active stars show permanent active longitudes, and that also the flip-flop phenomenon itself is fairly common. But unfortunately in many cases the length of the flip-flop cycle cannot be reliably determined, due to the fact that for pinpointing the length of the flip-flop cycle we would need quite a long sequence of observations and this is often not the case. Table 1 contains the binary star targets and also gives information on active longitudes and flip-flop phenomenon on these stars. A plus sign in the active longitude column means that permanent active longitudes are detected in this star. In the last column information on the flip-flop phenomenon is given: $-$ sign means no flip-flops were detected, $+$ sign means that flip-flops were detected, but no reliable estimate on the cycle length could be obtained. If a flip-flop cycle was detected, the full cycle length in years is given.

From the 14 stars we have investigated, we have managed to obtain an estimate of the flip-flop period for 4 binaries and 1 single star. In Figure 3 we plot the radius against the flip-flop period for these 5 stars and the previously known flip-flop stars. It seems that the long flip-flop periods (> 10 years) tend to occur in large stars (radius > 5 $R_\odot$), whereas the shorter flip-flop periods do not seem to depend on the radius. Also, it seems that the possible range of flip-flop periods for the binary stars (approximately 3.5–20 years) is much larger than for the single stars (approximately 3.5–7 years).

Table 1. The star sample of binary stars investigated in this study. The name of the star, spectral and variability type, stellar radius and rotation period are given together with our results on active longitudes and flip-flops.

Star	HD	Spectral type	Variability type	Radius [R$_\odot$]	P_{rot} [days]	Active longitudes	flip-flops [years]
XX Tri	HD 12545	K0 III	RS CVn	8.0	23.98	+	+
UX Ari	HD 21242	K0 IV	RS CVn	5.78	6.44	+	+
IL Hya	HD 81410	K1 III	RS CVn	8.1	12.73	+	10.1
HU Vir	HD 106225	K1 IV	RS CVn	5.7	10.39	+	11.4
IN Com	HD 112313	G5 III-IV	RS CVn	7.8	5.92	?	−
HK Lac	HD 209813	K0 III	RS CVn	9.5	24.43	+	−
V833 Tau	HD 283750	K5 V	BY Dra	0.77	1.79	−	−
V1355 Ori	HD 291095	K2 VI-V	RS CVn	4.1	3.86	+	3.4
UZ Lib		K0 III	RS CVn	6.3	4.77	+	16

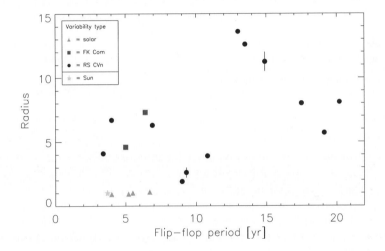

Figure 3. Stellar radius plotted against the flip-flop period for the newly discovered and previously known flip-flop stars

Acknowledgements

We would like to thank K.G. Strassmeier for giving us some of his unpublished data to be used in this study. This project has been supported by German *Deutsche Forschungsgemeinschaft, DFG* grant KO 2320/1-2. SPJ acknowledges support from the Väisälä Foundation, Finland.

References

Berdyugina, S.V. & Tuominen, I. 1998, *A&A* 336, 25
Berdyugina, S.V. & Usoskin, I.G. 2003, *A&A* 405, 1121
Järvinen, S.P., Berdyugina, S.V., Tuominen, I., Cutispoto, G., & Bos, M. 2005, *A&A* 432, 657
Jetsu, L., Pelt, J., & Tuominen, I. 1993, *A&A* 278, 449
Korhonen, H. & Elstner, D. 2005, *A&A* 440, 1161
Lanza, A.F., Catalano, S., Cutispoto, G., Pagano, I., & Rodonò, M. 1998, *A&A*, 332, 541

Binary Stars as Critical Tools & Tests
in Contemporary Astrophysics
Proceedings IAU Symposium No. 240, 2006
W.I. Hartkopf, E.F. Guinan & P. Harmanec, eds.

Discovery of Four β Cephei Stars in Eclipsing Systems

Andrzej Pigulski[1] and Grzegorz Pojmański[2]

[1]Instytut Astronomiczny Uniwersytetu Wrocławskiego, Kopernika 11, 51-622 Wrocław, Poland
email: pigulski@astro.uni.wroc.pl

[2]Obserwatorium Astronomiczne Uniwersytetu Warszawskiego,
Al. Ujazdowskie 4, 00-478 Warszawa, Poland
email: gp@astrouw.edu.pl

Abstract. Using the ASAS-3 photometry, we find the components of four eclipsing binary systems — V916 Cen, HD 101838, V4386 Sgr and HD 168050 — to be β Cephei-type pulsators. The first two systems are members of the young open cluster Stock 14. The pulsating stars are presumably the primary, more massive components in all these systems. The components are detached and for at least two systems, V916 Cen and HD 168050, we may suspect that they will appear to be double-lined spectroscopic ones. In consequence, these stars become very attractive targets for studying pulsations in β Cephei stars by means of asteroseismology.

Keywords. eclipsing stars, pulsating stars

1. Introduction

There is a growing interest in studying stellar interiors of pulsating stars by means of asteroseismology. A general requirement for the method is the detection of several modes which are correctly identified in terms of the geometry of the pulsation and accurate stellar parameters needed in modeling. In this context, the fact that a pulsating star is a component of an eclipsing binary is a great asset because it provides a direct way of obtaining masses and radii of the components through a combination of the double-lined spectroscopic orbit and the analysis of the light curve.

Most types of pulsating stars are known to occur in binaries, but still such cases are rare. Among over a hundred presently known β Cephei-type stars, there are only three that are components of the eclipsing systems. These are 16 (EN) Lac (Jerzykiewicz *et al.* 1978, Pigulski & Jerzykiewicz 1988), V381 Car in NGC 3293 (Engelbrecht & Balona 1986, Jerzykiewicz & Sterken 1992, Freyhammer *et al.* 2002) and λ Sco (Shobbrook & Lomb, De Mey *et al.* 1997, Uytterhoeven *et al.* 2004, Bruntt & Buzasi 2006).

2. The data

The data we used consist of the *V*-filter photometry obtained in the third phase of the All Sky Automated Survey (ASAS-3) (Pojmański 1997, 2001, Pojmański *et al.* 2005). They span the interval of over five years, 2000–2006.

3. The systems with pulsating components

The general data for the four eclipsing systems under consideration are summarised in Table 1. Their orbital periods range from about 1.463 d for HD 101794 up to 10.798 d for HD 167003. Accidentally, two of the four stars, V916 Cen and HD 101838, are members of

Table 1. Data for the four new *β* Cephei stars in eclipsing systems.

HD	ASAS name	V [mag]	P_{orb} [d]	Short period(s) [d]	Remarks
101794	114225−6228.6	8.68	1.46324	0.22465, 0.54362	V916 Cen, Be star member of Stock 14
101838	114249−6233.9	8.42	5.41166	0.31973	member of Stock 14
167003	181442−3308.5	8.45	10.79824	0.14765, 0.13252 0.14253, 0.18593	V4386 Sgr
168050	181839−1906.2	9.81	5.02335	0.1802(var), 0.19044	

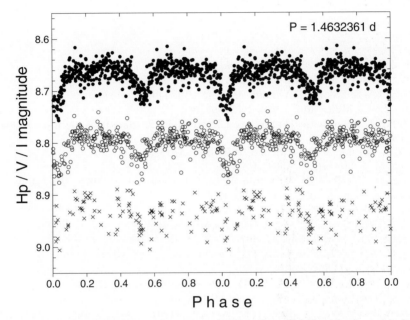

Figure 1. The ASAS-3 eclipsing light curve of V916 Cen folded with the orbital period of 1.4632361 d. The filled circles, open circles and crosses stand for the ASAS-2, ASAS-3 and Hipparcos data, respectively.

the same open cluster, Stock 14 (Moffat & Vogt 1975, FitzGerald & Miller 1983, Peterson & FitzGerald 1988). In addition, HD 101794 is known as a Be star (Garrison *et al.* 1977). In this star, one of the small-amplitude periodic variations has a period of 0.54362 d. This is slightly too long for a *β* Cephei star, but can be attributed either to a *g* mode or *λ* Eri-type of variability.

As far as the number of detected modes are concerned, only HD 101838 was found to be monoperiodic. The largest number of modes – four – was detected in HD 167003. In HD 168050 we detect two modes. One of them shows very fast period changes. HD 168050 is therefore one of the very few *β* Cephei stars which are known to exhibit secular period changes. Since the period of the other mode is constant within errors, the most plausible explanation for the observed period change is some kind of resonant coupling between modes.

From the point of view of the future application of asteroseismology, all four stars are very attractive targets. The deepest secondary eclipses are observed in HD 101794 and HD 168050, which indicates that these systems will be observed as a double-lined spectroscopic ones. This could allow precise determination of masses and radii of the components. They are therefore obvious targets of the follow-up study.

As an example, we show in Figure 1 the eclipsing light curve of HD 101794. The star was found to show erratic changes of γ Cas-type by Hipparcos, but the eclipses in this system were discovered by Pojmański (2000) in the ASAS-2 I-filter data. We show the light freed from the contribution of pulsations and long-term changes. They originate probably in a Be star.

The full study of the four systems will be published elsewhere.

Acknowledgements

This work was supported by the MNiI grant No. 1 P03D 016 27.

References

Bruntt, H., & Buzasi, D.L. 2005, in *Mem. Soc. Astron. Ital.* 77, 278

De Mey, K., Aerts, C., Waelkens, C., *et al.* 1997, *A&A* 324, 1096

Engelbrecht, C.A. & Balona, L.A. 1986, *MNRAS* 219, 449

FitzGerald, M.P. & Miller, M.L. 1983, *PASP* 95, 361

Freyhammer, L.M., Hensberge, H., Sterken, C., Pavlovski, K., Smette, A., & Ilijić, S. 2005, *A&A* 429, 631

Garrison, R.F., Hiltner, W.A., & Schild, R.E. 1977 *ApJS* 35, 111

Jerzykiewicz, M. & Sterken, C. 1992, *MNRAS* 257, 303

Jerzykiewicz, M., Jarzębowski, T., Musielok, B., & le Contel, J.-M. 1978, *IBVS* 1508

Moffat, A.F.J. & Vogt, N. 1975, *A&AS* 20, 125

Peterson, C.J. & FitzGerald, M.P. 1988, *MNRAS* 235, 1439

Pigulski, A. & Jerzykiewicz, M. 1988, *Acta Astron.* 38, 88

Pojmański, G. 1997, *Acta Astron.* 47, 467

Pojmański, G. 2000, *Acta Astron.* 50, 177

Pojmański, G. 2001, *PASPC* 246, 53

Pojmański, G., Pilecki, B., & Szczygieł, D. 2005, *Acta Astron.* 55, 275

Shobbrook. R.R. & Lomb. N.R. 1972, *MNRAS* 156, 181

Uytterhoeven, K., Willems, B., Lefever, K., Aerts, C., Telting, J.H., & Kolb, U. 2004, *A&A* 427, 581

Binary Stars as Critical Tools & Tests
in Contemporary Astrophysics
Proceedings IAU Symposium No. 240, 2006
W.I. Hartkopf, E.F. Guinan & P. Harmanec, eds.

© 2007 International Astronomical Union
doi:10.1017/S1743921307004450

Poster Abstracts (Session 8)

(Full Posters are available at http://www.journals.cambridge.org/jid_IAU)

Possible Distortion Effect on the Pulsation of Eclipsing Binary Components

Burak Ulas and Osman Demircan

Canakkale Onsekiz Mart University, Physics Department, Canakkale, Turkey

Possible effect of tidal and rotational oblateness of a star on the pulsational period in pulsating components of eclipsing binaries is investigated. The angle of foreshortening on the surface of the pulsating component becomes important on account of estimating local pulsating periods. By assuming the linear adiabatic radial oscillation, the pulsation period change was found between $0.947\,P_0$ and $1.048\,P_0$ from equator to the pole for a triaxial ellipsoid star having the axis as a=0.46, b=0.43, c=0.45 as fractional radii. This effect may not be observed due to experimental errors. It was also noted that in the $P_{pulse} \cdot \rho^{1/2}$ equation, the factor of Q changes from equator to pole.

EUVE J0825−16.3 and EUVE J1501−43.6: Two dMe Double-Lined Spectroscopic Binaries

D. Montes, I. Crespo-Chacón, M.C. Gálvez, and M.J. Fernández-Figueroa

Dpto. Astrofísica, Universidad Complutense de Madrid, Madrid, Spain

High-resolution echelle spectroscopic observations taken with the FEROS spectrograph at the ESO 2.2-m telescope confirm the binary nature of the dMe stars EUVE J0825−16.3 and EUVE J1501−43.6, previously reported by Christian & Mathioudakis (2002).

In these binary systems, emission of similar intensity from both components is detected in the Na I D_1 & D_2, He I D_3, Ca II H & K, Ca II IRT and Balmer lines. We have determined precise radial velocities by cross-correlation with radial velocity standard stars, which have allowed us to obtain for the first time the orbital solution of these systems. Both binaries consist of two near-equal M0V components with an orbital period shorter than 3 days. We have analyzed the behaviour of the chromospheric activity indicators (variability and possible flares). In addition, we have determined its rotational velocity and kinematics.

A Hunt for Binaries with Pulsating Components

O. Latković

Astronomical Observatory of Belgrade, Yugoslavia

In the context of the collaboration of several European observatories, we are examining a number of eclipsing binary systems in search for evidence of pulsation phenomena. We hope our candidate systems will be suitable for subsequent asteroseismological studies.

As a first step towards this end, we are analyzing spectroscopic observations of several such binary stars; these observations have been made at National Astronomical Observatory Rozhen, Bulgaria, in the period from 2002 to 2005, and the reduction was done recently at Astronomical Observatory of Belgrade, Serbia. Measurments of radial velocities and RV curve analysis are in preparation.

Only Binary Stars Can Help Us Actually SEE a Stellar Chromosphere

R.E.M. Griffin

Herzberg Institute for Astrophysics, Victoria, Canada

Binary stars of the ζ Aurigae-type (eclipsing systems containing a cool giant plus a hot main-sequence star) offer our only method of probing a stellar chromosphere. Close to occultation, the dwarf acts as a light probe behind the giant's chromosphere, enabling an observer to detect changing conditions in that chromosphere along the line of sight. The technique is powerful, the effects dramatic (as will be illustrated).

However, presently known eclipsing systems number only about 10, and a much greater sample is required for meaningful statistics of the properties of stellar chromospheres. New surveys of fainter binaries should be investigated for eclipses in order to extend the sample size and gain more information on chromospheres in general. Such information is clearly vital for modelling stellar photospheres, from which abundances are derived. The paper describes the very different behaviour of chromospheric material in two $3^{\rm rd}$-magnitude binaries.

Orbital Period Variation in the Chromospherically Active Binary FF UMa (2RE J0933+624)

M.C. Gálvez, D. Montes, M.J. Fernández-Figueroa, E. De Castro, and M. Cornide

Depto. Astrofísica, Universidad Complutense de Madrid, Madrid, Spain

We present here a detailed study of FF UMa (2RE J0933+624), a recently discovered, X-ray/EUV selected, active binary system with strong Hα emission. By using high resolution echelle spectroscopic observations taken during five observing runs from 1998 to 2004, we have derived precise radial velocities that allowed us to determine the orbital solution of the system at different epochs.

Analyzing these orbital solutions and a previous one in 1993, determined by other authors, we have found a change in the heliocentric Julian date of conjunction (T_{conj}) that can be explained by a change with time in the orbital period of the system. The relative amplitude of the orbital period variation derived from these data was $\Delta P/P \approx 5 \times 10^{-4}$, which results to be larger than the variations found in other similar chromospherically active binaries like AR Lac and HR 1099. This orbital period variation can be related (Applegate 1992; Lanza *et al.* 1998, 2006) with the modulation of the gravitational quadrupole moment of its magnetically active secondary star produced by angular momentum exchanges within its convective envelope. In addition, using these observations, we have determined the stellar parameters of the components and we have carried out a study of the chromospheric activity using all the optical indicators from Ca II H & K to Ca II IRT lines.

Starspot activity of V711 Tau from November to December, 2005

Sheng-hong Gu[1], A.C. Cameron[2], J.R. Barnes[3], and Jia-yan Yang[1]

[1] *National Astronomical Observatories/Yunnan Observatory, CAS, Kunming, Yunnan Province, China,*
[2] *SUPA, School of Physics and Astronomy, University of St. Andrews, North Haugh, St. Andrews, Fife, United Kingdom,*
[3] *School of Physics, Astronomy and Mathematics, University of Hertfordshire, Hatfield, United Kingdom*

The active binary V711 Tau was observed using the high-resolution echelle spectrograph of the 2.16-m telescope at Xinglong station of NAOC during November 18-24 and December 17-22, 2005. Based on the least-squares deconvolution method, we have derived high signal-to-noise ratio profiles for V711 Tau, which have been used to reconstruct the starspot patterns for the two datasets by means of Doppler imaging code DoTS. Finally, we discuss the starspot feature and evolution in this observing season.

Surface Imaging of Late-Type Contact Binaries: Hα Emission in AE Phoenicis and YY Eridani

O. Vilhu and C. Maceroni

[1] *Observatory, Helsinki University, Helsinki, Finland,*
[2] *INAF, Rome Astronomical Observatory, Monteporzio C., Italy*

We present and discuss the Hα observations of the contact (W UMa type) binaries AE Phoenicis and YY Eridani, obtained in 1989, 1990 and 1995 with the CAT/CES telescope of the Southern European Observatory (ESO). In particular, we compare the intrinsic equivalent widths of both components with the NextGen theoretical models and the saturation limit. We find that the average Hα equivalent widths are close to the saturation border and that the primary components have excess Hα emission, indicating enhanced chromospheric activity. This is compatible with both theoretical and observational suggestions that the primary is the more magnetically active component and is filled with (mostly unresolvable) dark spots and associated chromospheric plages.

Non-Linear Resonance Model of High-Frequency Quasiperiodic Oscillations (QPO) in X-ray Sources: Theory vs. Observation

G. Török, M.A. Abramowicz, Z. Stuchlik, and E. Sramkova
Silesian University in Opava, Opava, Czech Republic

There is an impressive agreement between theoretical prediction of the high frequency twin peak QPO resonance model and the black-hole as well as neutron-star observational data.

The model explains QPOs as two modes of accretion disk oscillations that are weakly coupled, non-linear and resonant. It predicts that there should be 1/mass scalling of frequencies, that the two frequencies ν_{high}, ν_{low} should be nearly linearly correlated for one source, $\nu_{high} = A\nu_{low} + B$, that the A, B coefficients should be anti correlated in a sample of neutron star sources, and that the difference of amplitudes in the two modes should vanish close to the resonance. All these predictions are confirmed by observations. No any other QPO model is able to explain all these observations.

Summaries and Thanks

Figure 1. Summary speakers Colin Scarfe and Virginia Trimble

Figure 2. Vltava River, between the Charles Bridge and the National Theatre. The Bellevue Hotel is seen at upper right.

Figure 3. The volunteers who provided excellent support throughout the meeting: (upper left) Jana Drbohlavová, Anna Vernerová, and Tereza Krejčová; (upper right) Miloslav Zejda; (middle left, staring at computer) Pavel Chadima.

Binary Stars as Critical Tools & Tests
in Contemporary Astrophysics
Proceedings IAU Symposium No. 240, 2006
W.I. Hartkopf, E.F. Guinan & P. Harmanec, eds.

Summary of Observational Aspects

C. D. Scarfe

Department of Physics and Astronomy, University of Victoria,
Victoria, B.C. V8W 3P6, Canada
email: scarfe@uvic.ca

Abstract. This summary covers the first two days of the Symposium, in which the oral papers were primarily observational in nature. It also mentions several of the many observational posters associated with those discussions.

Keywords. (stars:) binaries: general

1. Introduction

First, let me thank my colleagues on the SOC for giving me the opportunity to attempt to summarize the first portion of this meeting, and with it the incentive to listen carefully to several interesting invited papers, and also to read numerous posters covering a wide range of topics relevant to binary and multiple stars. So much information has been presented at this meeting that I am very grateful not to have to cover it all, but to leave part of it in the very capable hands of Virginia Trimble, who will speak next. We have considered how we might best share our pleasant duty, and have settled on the following: she will cover the primarily theoretical papers of the final day and a half, and I am to deal with the earlier portion, as stated above.

One of the few advantages of growing older is that one can remember a goodly number of interesting meetings. For myself, I can declare that the most interesting of all have been IAU-sponsored, and concerned primarily with binary stars. Some of those that I have enjoyed include Colloquium 5 in Nice, France, in 1969, Colloquium 33 on *Multiple Stars*, in Oaxtepec, Mexico, in 1976, Colloquium 62 on *Current Techniques*, in Flagstaff, U.S.A. in 1981, Colloquium 107 on *Algols*, in Victoria, Canada, in 1988, Colloquium 135 on *Complementary Approaches*, near Atlanta, U.S.A., in 1992, Symposium 200 on *Formation of Binary Stars*, in Potsdam, Germany, in 2000, Colloquium 191 on their *Environment and Evolution*, in Merida, Mexico, in 2003, and finally the present one, Symposium 240. Most of those meetings have been Colloquia, a series whose discontinuance is regrettable because it leaves a gap in the range of meeting size and of breadth of topic which the IAU offers its members.

2. Oral Papers

Most of the above–mentioned meetings have included papers on a range of topics scattered around the stated theme of the meeting, and this one is no exception. However it seemed to me that at least early in the meeting the 'contemporary astrophysics' addressed by the speakers was mainly that of the binaries themselves, rather than wider applications.

Nevertheless the presentations contained much of interest, as presaged in the 'trailer' presented very lucidly at the beginning by Guinan (see Guinan *et al.* 2007). We began by learning about major developments in instrumentation for studying all sorts of binaries.

465

Developments in long-baseline interferometry in the northern and southern hemispheres were summarized with great clarity by McAlister (2007) and Davis (2007) respectively, and Shao (2007) envisaged doing even better from space. Then Balega described the history of speckle interferometry. He reminded us of the vast flood of good speckle observations that began in the 1970s; we await eagerly a similar flood from long-baseline interferometry.

Next, Hatzes reminded us of the rapid progress in the precision of relative radial-velocity data, with wonderfully stable instruments that have left absolute radial-velocity accuracy far behind. But for many of us this is of little concern. Absolute velocities have limited uses for binary-star studies. Stability from night to night and over the long term is of greater value.

The use of large telescopes, and concentration upon spectroscopic observations, was urged by Ribas (2007), and Bonanos (2007) described the vast data banks derived from microlensing programs. In a comment, Wilson warned about the poor quality of some mass-produced light curves. Another speaker claimed that masses can be obtained from light curves without resort to spectroscopic data, but Leung (2007) indicates that this may not be true except in favourable cases, e.g., total eclipses in contact systems.

The prodigious efforts that have been made to extend the *Washington Double Star Catalog*, a valuable resource indeed, were described by Mason (2007). Then Hartkopf (2007) discussed the continued development of multiplicity nomenclature. However, it may not be easy always to apply that nomenclature in a way that does not require changes with time, as a brief consideration of the discovery sequence described by Harmanec *et al.* (2007) will reveal.

New and sophisticated methods of obtaining radial velocities were described by Hensberge & Pavlovski (2007), but Parimucha & Škoda (2007) found that the old method of matching profiles of individual lines with their mirror images produces better results in some situations.

On the following day, Peters (2007) described some remarkable spectroscopic observations of Algols, mostly in the far ultraviolet, and their interpretation in terms of models of the circumstellar material around the mass-gaining stars in those systems. Some information may be gleaned on the systems' past history by determination of the chemical composition of those materials. Then Kafka (2007) indicated how the pattern of accretion in cataclysmic variables depends upon their magnetic properties, and went into detail about what may be observed in the low states of VY Scl and AM Her systems, when the bright light of obscuring material is greatly reduced.

A beautifully clear introduction to tomography was presented by Richards (2007). She discussed the importance of eliminating stellar light to see more clearly the structure of disks (or is 'disk' a too simple description?!), and outlined the difficulties of interpreting the data in terms of models of that structure. Then Strassmeier (2007) gave a fine account of the challenge of mapping spotted stars, with numerous examples, and of the complexities arising if one tries to model the related magnetic fields by Zeeman–Doppler imaging. Then followed a most entertaining account by ten Brummelaar (2007) of the snags and pitfalls of long-baseline interferometry.

Wilson (2007) described the incorporation of various effects into modelling codes. This has been enabled by spectacular increases in computer speed and memory capacity, and have made the models' properties resemble those of real stars increasingly closely. Moreover, one can now archive very many model light curves, to provide rapid matching to the multitude of observed light curves that are becoming available, and to the even greater multitudes that are promised before long.

The six final oral papers that fell under my purview were all concerned with large-scale surveys. Pourbaix described several potential automatic methods of identifying binaries and discovering some of their characteristics. Then Prša & Zwitter (2007) discussed pipeline reduction of binary light curves, which will become essential if large-scale surveys from space discover binaries in the huge numbers expected. Koch *et al.* (2007) followed this with a very practical and straightforward account of the Kepler mission, whose purpose is to observe a single area repeatedly, to seek transits of earth-sized planets across their stars' disks, no small task considering that such an event may be manifested by a 0.1 mmag decrease in light, that lasts for perhaps twelve hours and occurs once a year. It seems likely that objects such as eclipsing binaries will be found much more easily. Niarchos *et al.* (2007) gave us details of some important simulations, preparatory for handling the vast amounts of data anticipated from GAIA. Of course, if the numbers of binaries found by Mazeh *et al.* (2007) prove to be normal, those data will not be so numerous as supposed, by perhaps one order of magnitude. But that will still be a huge amount of data! Finally, Zwitter described a ground–based project to obtain a million radial velocities. This alone will keep the bibliography team busy for some time. One gets an impression from these various projects of something of a trade-off between wonderfully precise data in large but manageable quantities, and an overwhelming quantity of less precise data. The first of these — the manageable quantities of excellent data — makes me wish I were young again. The other makes me not so sure of this.

[Editors' note: Unfortunately the talks by Balega, Hatzes, Pourbaix and Zwitter mentioned above were not submitted to these proceedings.]

3. The Posters

I would now like to say a little about the posters, only a very few of which I have mentioned so far. They are wonderfully numerous and diverse, even though not all of the 181 listed in the abstract book actually appeared — I counted about 140. I spent four interesting lunch periods trying to read them, and although I did not read every word (some of them being very lengthy and detailed for posters), I did at least look at them all. Even so I cannot hope to mention them all here and I hope the authors of those I omit will forgive me. But it is important that I spend some time on them because I saw only a few participants up in the poster hall reading them for themselves. The poster hall's remote location was partly to blame for this, of course.

Some people said to me that the talks were too much on instruments, techniques and projects, and too little about science, and I have to agree, at least to some extent. However, that defect, if it was one, could be remedied by reference to the posters, where science was more predominant. That science covers a broad spread of objects, from common-proper-motion pairs (Kiselev *et al.* 2007, Sinachopoulos *et al.* 2007, Halbwachs *et al.* 2007, Johnston *et al.* 2007) to ultra-compact binaries in which a white dwarf is losing matter to its neutron-star companion (Nelemans 2007). It also ranges from massive O-type systems (Fernández Lajús & Niemela 2007, Iping *et al.* 2007, Nesslinger *et al.* 2007) to brown dwarf pairs (Golimowski *et al.* 2007, Gómez Maqueo Chew *et al.* 2007), with total masses respectively 100 and 90 times that of Jupiter.

To me it seems not long since the discovery of V471 Tau, one of the first known binaries containing a non-interacting white dwarf and main-sequence star. But now several are known, and we heard in one of the posters selected for oral presentation, van den Besselaar *et al.* (2007), about a new object of that type, DE CVn. It is worth noting, however, that a second poster, Goker & Tas (2007), presents similar but by no means identical parameters for the same system.

As editor-in-chief of Commission 42's semiannual bibliography I am very much aware that certain binaries attract a lot of attention, and are the subject of many papers. That is also true among the posters at this meeting. For example, Cygnus X-1 is the subject of at least three, (Karitskaya *et al.* 2007a, 2007b, 2007c), by different members of the same group. As well, η Carinae is the subject of at least two, Falceta-Gonçalves *et al.* (2007) and Gull (2007). Several old friends among the stars are also the subject of new work, including RZ Cas (Golovin & Pavlenko 2007), ϵ Aur (Stencel 2007), AM Her (Kalomeni *et al.* 2007) and β Lyr (Chadima *et al.* 2007). It is nice to see that V1500 Cyg, once known as the fastest nova in the Galaxy, is still misbehaving almost 30 years after it first attracted widespread attention. Also, of course, RS Oph is active once again (Stringfellow *et al.* 2007).

Some of the posters describe quite basic results, which might be regarded as not quite at the cutting edge. But in most cases they are authored by people who do not have access to modern equipment, or who lack financial support in their home countries, or both. And in general they serve a useful purpose in filling in gaps in our knowledge. Their authors are to be congratulated on their heroic efforts, and applauded for their contributions. (Those of us who do have better support may well not appreciate how lucky we are!) So too should we congratulate the amateurs, who have long contributed much to the study of variable stars, including eclipsing binaries, and are now contributing to visual binary work as well. Examples include Agati *et al.* (2007) and Rica *et al.* (2007).

A few further posters that I would like to mention include those of Liu & Qian (2007), who propose an imaginative design for a telescope, Kövári *et al.* (2007), who warn of aliases that may interfere with the interpretation of Doppler imaging, Scardia *et al.* (2007), who describe some beautiful work on γ Vir, which has just passed its only periastron in any of our lifetimes, and Lampens *et al.* (2007), who present new work on θ^2 Tau, a binary both of whose components pulsate, and which lies at the top of the Hyades' main sequence.

There are more posters, in addition to those already mentioned, on a wide variety of topics, such as:

- 1. Chromospherically active binaries (Korhonen & Järvinen 2007, Gálvez *et al.* 2007 and Gu *et al.* 2007),
- 2. Symbiotic stars (Hric *et al.* 2007),
- 3. Dwarf novae (Hourihane *et al.* 2007, Viallet & Hameury 2007),
- 4. Binaries with pulsating components (Latković 2007 and Evans *et al.* 2007),
- 5. Binaries in clusters (Southworth 2007, as well as Hourihane *et al.* 2007, mentioned above), and
- 6. Data mining (Lee *et al.* 2007 and Devor & Charbonneau 2007).

One poster that particularly interests me is that of Amado (2007), who describes the use of binaries to extend a modified version of the Barnes-Evans relationship to low masses.

Many of the posters are straightforward descriptions of interesting systems, but a significant number do indeed live up to the Symposium's title, providing applications to astrophysics in general. The range of topics is breathtaking. This could be regarded as a measure of the Symposium's lack of focus, or more charitably of its inclusiveness. The choice is yours.

4. Some Further Thoughts

Quite a lot has been said, and written, here about obtaining reliable masses and radii from double-lined eclipsing binaries, and of getting indirect distance estimates for such objects, by incorporating effective temperature information. I say indirect, because this always requires a calibration of effective temperature against colour index or spectral type.

Several other contributors, e.g., Fekel & Tomkin (2007), have mentioned deriving masses and distances from visual–spectroscopic systems, which yield so-called orbital parallaxes. These have proved to be much more accurate than Hipparcos parallaxes, and have indeed served as useful checks on the latter. Very often they agree well, but in some cases there is pronounced disagreement, which is usually interpreted in terms of vitiation of the Hipparcos result by the binarity itself. It is curious that this does not happen more consistently. The region containing binaries for which such combined observations are possible is still a small one near the sun; long-baseline interferometry should make it much larger.

However, remarkably little has been said about the pleasant prospect of over-determining the astrophysical quantities, by interferometric resolution of binaries that are also double-lined and eclipsing. One such object is γ Per (Griffin 2007), whose eclipses, discovered in 1990 (Griffin 1992), were of sufficient interest at Colloquium 135 that the system was featured on the cover of that meeting's Proceedings, although the interferometric aspect of the system did not feature prominently in this meeting's poster presentation.

Such a combination of observations would provide distances for eclipsing systems independently of any calibration of colour against temperature, and with the use of observed spectral energy distributions, would permit direct deduction of effective temperatures from absolute bolometric luminosities and radii. We could then provide other astrophysicists with an absolute calibration of colour indices and bolometric corrections against effective temperatures, surely a worthy goal in view of this Symposium's avowed theme.

It will not be easy to achieve that goal. Of course one needs to choose well-separated eclipsing pairs, of which many are known. But those on the lower main sequence are faint and inaccessible to interferometry, and those on the upper main sequence are distant, and hence also inaccessible. So we will have to build outward from solar-type stars towards those both hotter and cooler, as the capability of interferometry increases.

The most stringent tests of stellar models also require chemical information for each component of a binary. Spectral disentangling is now giving us a way forward to reach this goal. It is the subject here of several posters, e.g., Pavlovski & Tamajo (2007), Hadrava (2007), and Gerbaldi & Faraggiana (2007), as well as of the talk by Hensberge & Pavlovski (2007). Ages are also valuable, and these can sometimes be obtained for field main-sequence stars in wide binaries if their companion is a white dwarf, as discussed by Oswalt (2007).

I could go on, and mention yet more interesting posters, but I think it is now time to turn the floor over to Virginia Trimble.

References

Amado, P.J. 2007, these proceedings, 327

Agati, J.L., Caille, S., Debackere, A., Durand, P., Losse, F., Mantle, R., Mauroy, F., Mauroy, P., Morlet, A.G., Pinlou, C., Salaman, M., Soule, E.J., Thorel, Y., & Thorel, J.C. 2007, *these proceedings*, 117

Bonanos, A.Z. 2007, *These Proceedings*, 79

Chadima, P., Ak, H., Harmanec, P., Demirçan, O., Yang, S., Koubský, P., Škoda, P., Šlechta, M., Wolf, M., Božić, H., Ruždjak, D., & Sudar, D. 2007, *These Proceedings*, 205

Davis, J. 2007, *These Proceedings*, 45

Devor, J., & Charbonneau, D. 2007, *These Proceedings*, 208

Evans, N.R., Schaefer, G., Bond, H.E., Nelan, E., Bono, G., Karovska, M., Wolk, S., Sasselov. D., Guinan, E., Engle, S., Schlegel, E., & Mason, B. 2007, *These Proceedings*, 102

Falceta-Gonçalves, D., Abraham, Z., & Jatenco-Pereira, V. 2007, *These Proceedings*, 198

Fekel, F.C. & Tomkin, J. 2007, *These Proceedings*, 59

Fernández Lajús, E. & Niemela, V. 2007, *These Proceedings*, 120

Gálvez, M.C., Montes, D., Fernández-Figueroa, M.J., De Castro, E., & Cornide, M. 2007, *These Proceedings*, 461

Gerbaldi, M. & Faraggiana, R. 2007, *These Proceedings*, 120

Goker, U.D., & Tas, G. 2007, *These Proceedings*, 128

Golimowski, D.A., Minniti, D., Henry, T.J., & Ford, H.C. 2007, *These Proceedings*, 329

Golovin, A. & Pavlenko, E. 2007, *These Proceedings*, 330

Gómez Maqueo Chew, Y., Stassun, K.G., Mathieu, R., & Vaz, L.P., 2007, *These Proceedings*, 330

Griffin, R.E.M. 2007, *These Proceedings*, 333

Griffin, R.F. 1992, in: H.A. McAlister & W.I. Hartkopf (eds.) *Complementary Approaches to Double and Multiple Star Research*, IAU Colloquium 135 (San Francisco: ASP Conf. Ser.), vol. 32, p. 98

Gu, S.-H., Cameron, A.C., Barnes, J.R., & Yang, J.-Y. 2007, *These Proceedings*, 461

Guiman, E., Harmance, P. & Hartkopf, W. 2007, *These Proceedings*, 5

Gull, T.R. Nielsen, K.E., Corcoran, M.F., Hillier, D.J., & Homaguchi, K. 2007, *These Proceedings*, 387

Hadrava, P. 2007, *These Proceedings*, 111

Halbwachs, J.-L., Mayor, M., & Udry, S. 2007, *These Proceedings*, 427

Harmanec, P., Mayer, P., Božić, H., Eenens, P., Guinan, E.F., McCook, G., Koubský, P., Ruždjak, D., Sudar, D., Šlechta, M., Wolf, M., & Yang, S. 2007, *These Proceedings*, 64

Hartkopf, W.I. 2007, *These Proceedings*, 97

Hensberge, H. & Pavlovski, K. 2007, *These Proceedings*, 136

Hourihane, A.P., Callanan, P.J., & Cool, A.M. 2007, *These Proceedings*, 117

Hric, L., Gális, R., & Šmelcer, L. 2007, *These Proceedings*, 387

Johnston, K., Oswalt, T., & Valls-Gabaud, D. 2007, *These Proceedings*, 429

Kafka, S. 2007, *These Proceedings*, 154

Kalomeni, B., Yakut, K., & Pekunlu, E.R. 2007, *These Proceedings*, 123

Karitskaya, E.A., Agafonov, M.I., Bochkarev, N.G., Bondar, A.V., & Sharova, O.I. 2007, *These Proceedings*, 65

Karitskaya, E.A., Lyuty, V.M., Bochkarev, N.G., Shimanskii, V.V., Tarasov, A.E., Galazutdinov, G.A., & Lee, B.-C. 2007, *These Proceedings*, 122

Karitskaya, E.A., Shimanskii, V.V., Bochkarev, N.G., Sakhibullin, N.A., Galazutdinov, G.A., & Lee, B.-C. 2007, *These Proceedings*, 130

Kiselev, A.A., Kiyaeva, O.V., & Izmailov, I.S. 2007, *These Proceedings*, 129

Koch, D., Borucki, W., Basri, G., Brown, T., Caldwell, D., Christensen–Dalsgaard, J., Cochran, W., De Vore, E., Dunham, E., Gautier, T., Geary, J., Gilliland, R., Gould, A., Jenkins, J., Kondo, Y., Latham, D., Lissauer, J., & Monet, D. 2007, *These Proceedings*, 236

Korhonen, H. & Järvinen, S.P. 2007, *These Proceedings*, 453

Kövári, Zs., Bartus, J., Oláh, Strassmeier, K.G., Rice, J.B., Weber, M., & Forgács-Dajka, E. 2007, *These Proceedings*, 212

Lampens, P., Frémat, Y., De Cat, P., & Hensberge, H. 2007, *These Proceedings*, p. (poster 125)

Latković, O. 2007, *These Proceedings*, 460

Lee, K.-W., Lee, B.-C., & Park, M.-G. 2007, *These Proceedings*, 266

Leung, K.C. 2007, *These Proceedings*, 124

Liu, Z. & Qian, S.B. 2007, *These Proceedings*, 64

Mason, B.D. 2007, *These Proceedings*, 88

Mazeh, T., Tamuz, O., & North, P. 2007, *These Proceedings*, 230

McAlister, H.A. 2007, *These Proceedings*, 35

Nelemans, G. 2007, *These Proceedings*, 124

Nesslinger, S., Drechsel, H., Lorenz, R., Harmanec, P., Mayer, P., & Wolf, M. 2007, *These Proceedings*, 326

Niarchos, P.G., Munari, U., & Zwitter, T. 2007, *These Proceedings*, 244

Oswalt, T.D., Johnston, K.B., Rudkin, M., Vaccaro, T., & Valls-Gabaud, D. 2007, *These Proceedings*, 300

Parimucha, S. & Škoda, P. 2007, *These Proceedings*, 62

Pavlovski, K. & Tamajo, E. 2007, *These Proceedings*, 209

Peters, G.J. 2007, *These Proceedings*, 148

Prša, A. & Zwitter, T. 2007, *These Proceedings*, 217

Ribas, I. 2007, *These Proceedings*, 69

Rica, F.M., Benavides, R., & Masa, E. 2007, *These Proceedings*, 121

Richards, M. 2007, *These Proceedings*, 160

Scardia, M., Argyle, R.W., Prieur, J.-L., Pansecchi, L., Basso, S., Law, N., & Mackay, C. 2007, *These Proceedings*, 132

Shao, M. 2007, *These Proceedings*, 54

Sinachopoulos, D., Gavras, P., Medupe, Th., Ducourant, Ch., & Dionatos, O. 2007, *These Proceedings*, 264

Stencel, R.E. 2007, *These Proceedings*, 202

Southworth, J. 2007, *These Proceedings*, 264

Strassmeier, K.G. 2007, *These Proceedings*, 170

Stringfellow, G.S., Walter, F.M., Wallerstein, G., York, D.G., Dembicky, J., Ketzebach, B., & McMillan, R.J. 2007, *These Proceedings*, 386

ten Brummelaar, T. 2007, *These Proceedings*, 178

van den Besselaar, E.J.M., Greimel. R., Morales-Rueda, L., Nelemans, G., Thorstensen, J.R., Marsh, T.R., Dhillon, V., Robb, R.M., Balam, D.D., Guenther, E.W., Kemp, J., Augusteijn, T., & Groot, P.J. 2007, *These Proceedings*, 105

Viallet, M. & Hameury, J.M. 2007, *These Proceedings*, 212

Wilson, R.E. 2007, *These Proceedings*, 188

Binary Stars as Critical Tools & Tests
in Contemporary Astrophysics
Proceedings IAU Symposium No. 240, 2006
W.I. Hartkopf, E.F. Guinan & P. Harmanec, eds.

Applied Binarology: Theoretical Aspects

Virginia Trimble

University of California, Irvine, Irvine, CA, USA
and
Las Cumbres Observatory, Goleta, CA
email: vtrimble@uci.edu

Abstract. The winding up of a conference like this provides the opportunity to look (1) backwards at how we reached the present stage of understanding of binary star behavior and its relationship to the rest of astronomy, (2) around at the garden of unsolved problems, and (3) cautiously forward at what might come next.

Keywords. binary stars

1. Introduction

Like Dr. Scarfe, the previous speaker, I attended my first binary star conference nigh on to 40 years ago. It was IAU Colloquium 6, *Mass Loss and Evolution in Close Binaries*, held in Elsinore, Denmark, 15-19 September 1969. The cover of the proceedings shows a classic evolutionary sequence, taken from the talk of Bohdan Paczyński, in which a primary fills its Roche lobe, dumps material onto the secondary on a thermal time scale until the mass ratio is reversed, and then continues more gentle donation on the nuclear time scale (conservatively, of course).

This is clearly a generation ago, since, those present at S240 included, of the Elsinore participants, only Alan Batten and Robert Wilson of the relatively senior people, Petr Harmanec (who, with the late Jiří Horn represented the Prague group and has somehow in the interim grown from a callow postdoc to a distinguished director), Johannes Andersen (then a graduate student, helping with logistics), and the present writer (another new postdoc who wasn't actually invited to Colloq. 6 but came anyway). The topic, as shown in the poster presentation by Pustylnik & Pustynski and by de Mink & Pols, remains, however, a focus of on-going research. Incidentally, Daniel Popper presented the class of RS CVn stars (near the main sequence, with emission lines in one or both components) there, in a table with 22 examples, so anyone who tries to tell you the class was discovered later by someone else should be referred to the proceedings (Gyldenkerne & West 1970).

2. Golden Moments in Theory of Binary Stars

That most close pairs of stars (as well as triples and clusters) are physical systems was a theoretical discovery, reported by the Rev. John Michell, B.D.F.R.S. (1768) in 1767. The paper, which also obtained stellar distances from the assumption that other stars were really about as bright as the sun, considered β Cap as a possible accidental juxtaposition of stars at different distances, and by the time the author had considered other doubles, triples, and richer groups, he reached the conclusion that the odds were many million millions to one against the chance hypothesis. The arithmetic is essentially the same as that used to show that, with 20-some people in a room, two of them will probably have the same birthday. The cause was left open, "...their mutual gravitation or some other law or appointment of the Creator." Well, that 'B.D.' is, after all, bachelor of divinity.

In 1782, Goodricke and Pigott recognized the periodicity of Algol and suggested eclipses by an opaque, less luminous orbiting body as the cause. They backed off from their correct hypothesis after failing to be able to account for the light curve of δ Cephei the same way, and rotating, spotted stars were the "best buy" hypotheses for periodic variables through most of the 19th century.

Peeking for a moment over the fence at the observers, we find Christian Mayer publishing his first double star catalog in 1778 and remarking that some of the close pairs seemed to have experienced relative motion since the time of Flamsteed. Credit for recognizing actual orbital motion is customarily assigned to William Herschel in an 1803 paper. Several of these items, incidentally, come from the incipient *Biographical Encyclopedia of Astronomers* (Hockey *et al.* 2007).

Pickering revived the hypothesis of eclipsing stars for periodic variability in the 1880s, but by 1914, Shapley had shown that the Cepheids, for instance, would have to have a sin i less than the stellar radii and f $(M) \leqslant 0.001$. He proposed pulsation as the solution and we now, of course, recognize that star spots, eclipses, and pulsation can all yield periodic light curves. The lesson of "all of the above" should perhaps be carried forward to other phenomena to be mentioned later.

Stars try to expand as they age, and Kuiper in 1941 appears to have been first to point out that this would get them into trouble with the interior Lagrangian surfaces (Roche lobes) if two stars orbited with semi-major axis less than the desired final radius. Crawford in 1955 applied this idea to resolve the then-worrisome Algol paradox (the less massive star the more evolved), and the first detailed calculations came from Morton in 1960, neither of them names primarily associated with binary star evolution (but see Paczyński 1971 if you doubt). And in 1964 Fred Hoyle, enthusiastic as ever, proposed a dog-eat-dog series of processes, in which material spilled back and forth between the stars many times.

The field took off in 1966, with independent but similar computations of binary evolution in three places, Munich under Kippenhahn, Ondřejov under Plavec, and Warsaw under Paczyński. Whether the primary filled its Roche lobe before hydrogen core exhaustion, while a red giant, or still later in life made major differences to the outcomes. And it was right here in Prague at the 1967 General Assembly that Anne Underhill most forcefully disagreed that Wolf-Rayet stars might be one of those outcomes, saying that "there are more models that aren't stars than there are stars that aren't models." She lost the WR battle, but the general point is worth remembering!

Some early computations in which mass and angular momentum were allowed to leave the system seemed to show no qualitative differences. This changed in the mid 1970s, and Paczyński, speaking at the first all-European meeting on astronomy, in Leicester in 1975, pointed to systems like V471 Tauri that simply could not have reached their current values of mass and size without very significant removal of angular momentum as well as mass. This was the meeting during which the transient X-ray source, A0620−00 rose to be the brightest thing in the sky in the Ariel-5 (partly a Leicester project) energy band. And the name common envelope binary quickly came to describe the — very short lived — process during which most of the losses occur.

On the assumption that the next truly golden moment has been the present symposium, we now fast-forward to significant issues that seemed ripe for discussion here and now.

3. Further Current Concerns

This is the list I arrived with of items about which something is known but more is desired, only slightly re-ordered. It is given with NASA bullets because, while the thoughts are in some sense complete, the sentences are not.

• Calibration of stellar structure and evolution calculations from observations of L, T, and R vs. mass, composition, and age (where information is available); the core reason why our colleagues tolerate binary star astronomers!

• Input for computations of chemical evolution and stellar population synthesis, from which binaries tend to be systematically excluded, sometimes leading to wrong answers (e.g., age if you leave out the blue stragglers).

• Tests of star formation theory from percentage of binaries vs. primary mass and the distribution of separations, mass ratios, and eccentricities.

• Clarification of scenarios leading to cataclysmic variables, type Ia supernovae (important for cosmology as well as nucleosynthesis), some gamma ray bursts, some sdOBs, and some blue stragglers.

• Disentangling the effects of rotation from age *per se* on stellar activity.

• Discovery of stellar-mass black holes, Cyg X-1 in 1972, two of the discoverers, C.T. Bolton and Paul Murdin having come to Prague in 1967, but not to S240.

• Determination of the equation of state of dense nuclear matter from the maximum mass and M/R ratios of neutron stars in X-ray binaries.

• Discovery of other relativistic effects, including dragging of inertial frames in BHXRBs and gravitational radiation from double neutron-star binary pulsars, although we ought to remember that the effect on the orbits of cataclysmic variables was recognized first, by a decade or more (depending on whom you credit with the recognizing).

• The radial distribution of density, with implications for rotation, convection, and mixing from apsidal motion.

• Limb darkening, convection mapping, spotting, and other deviations from uniformly bright spherical surfaces from tomography and the Rossiter-McLaughlin effect. (look it up!)

• Response of the secondary to receiving material on a time scale shorter than its own thermal one at the onset of rapid mass transfer (apparently not quite so devastating these days as Benson 1970 found).

• Continuity, or absence thereof, of brown dwarfs with low-mass M dwarfs, based on percentages of binaries and distributions of separations, etc.

• Calibration of distance scales with clean (detached) double line SBs that also eclipse.

• Probes of the current and past gravitational potential of the Milky Way from survival of wide binaries etc.

4. Recent progress and Future Prospects

These get paragraphs rather than bullets, because many of them have verbs. Many were also précised in Guinan's introductory talk. And the ordering would be difficult to defend. The posters were identified by abstract numbers during my talk. These have been replaced by surnames of the first author, on the grounds that neither of the readers is likely to have the Prague abstract book, which, being bound in signatures, defied disassembly and partial return, requiring participants to adopt an "all or nothing" approach to bringing it back.

The most striking single number in my view was the extreme deficiency of early-type eclipsing binaries in the LMC (Mazeh, using an OGLE sample). This is, admittedly, an observation, but it prompts several quasi-theoretical questions besides "What did he do wrong?" (OGLE cannot have missed 10 times as many EBs as it reported, but could the parent population somehow have been greatly overestimated?) More to the point, the LMC is known to have lots of supersoft X-ray binaries, which are advertised as coming from not too much further down the main sequence. Can these numbers be reconciled?

Formation ought to come first. Very little was said about binary formation at S237 (star formation in general), but we had a fine talk from Clarke, who emphasized the need for more and better statistics of binary populations in various contexts. Well, I tried long ago, but gave it up when a distinguished colleague (now deceased) began a review by referring to "the incorrect methods of Trimble." We made it up in the cafeteria line at the Patras GA, but belated thanks, Helmut, for fielding that fly ball long ago. The posters by Köhler *et al.* and Kouwenhoven *et al.* provided some of the requested statistics for Orion and Sco OB2. A real surprise was the seemingly flat distribution function of angular momenta (Zwitter).

Next one might put the effects on formation and dynamical evolution of planets. Definitely non-negligible according to posters by Neuhaeuser *et al.* and Fabrycky. And the ratio of "big" to "little" semi-major axis in a hierarchical triple remains out there somewhere around the 7:1 that I learned from R.S. Harrington long ago — 10:1 according to Lane, and one of the implications is that some systems must have arisen by star exchange, because, although the system is stable now, the planet could not have formed there.

Since core uses include calibration of stellar structure and evolution calculations and, more recently, of the extra-galactic distance scale, it is essential to know just how accurate masses (etc.) from eclipsing SB2's can be made and whether most star pairs can be fitted with somebody's isochrones with a consistent age and composition for the two stars. It is, I think, tempting to focus on moderate discrepancies (posters by de Mink, Kholtygin, Lacy, Richichi, Kiyaeva, and Lee and co-authors). My own prejudice is that we should trumpet this area as one of the major triumphs of modern astrophysics: to a very considerable extent, the stars we calculate are the stars we see, although, of course, the calculating and the seeing can always be done better.

The final white dwarf mass produced by a given main sequence star mass is not as firmly established as you might suppose (Catalan), partly because the real range of WD masses is much narrower than one might have expected.

Duplicity can interact with stellar pulsation ("asteroseismology") by driving particular resonant frequencies (a surprise to me) and in other ways (talk by Aerts, posters by Ulas, Pigulski, and Latkovic). The interaction between duplicity and stellar activity goes both ways. Oláh reported preferred longitudes for spots correlated with binary phase; and Kafka showed that mass transfer in CVs can nearly turn off when there is a large spot under L1.

The role of binaries in chemical evolution of galaxies and population synthesis, and its wide neglected by evolvers and synthesizers, was superbly reviewed by Vanbeveren. I don't know what to do about it except to holler periodically, "Don't forget the binaries!" The posters by Han *et al.* and Eggleton *et al.* addressed related issues, although we all heard, at the end of oral presentation of some of the latter work the announcement by R.F. Griffin that the first mistake in the table occurred in line three (a companion that has not been confirmed).

Within the binary systems themselves, there is a rapid mass transfer phase, of which β Lyrae remains the classic example (Chadima *et al.* poster). It is followed by a slow mass transfer phase and, perhaps sometimes, by back transfer to the primary of material flowing off the secondary or a disk (Qian *et al.* poster). And somewhere in the poster forest was a previously-advertised β Lyrae star that is probably just a contact system (as if we understood those properly).

Dynamical issues within systems include circularization, synchronization, and apsidal motion (posters by Roman, Chatterjee, and Farbiash). Some processes, especially disruption, are very sensitive to the presence of a surrounding star cluster, even a modest one (Kroupa). The role of binaries in encouraging or preventing core collapse in globular clusters belonged to JD14 on dense stellar systems, which was summarized by Douglas Heggie, a, perhaps *the*, pioneer of such studies.

Some astronomical phenomena happen only in binaries and, if you believe most of the published papers, each can happen only one way (that, of course, proposed by the authors). Following the Gell-Mann dictum that "everything that isn't forbidden is compulsory" (he meant particle physics processes), I suspect that multiple channels to a given result may be the norm. A few examples from the symposium included W UMa stars (Li poster); CVs (Bisikalo talk, Skopal and Unda-Sanzana posters); blue stragglers (not the focus of any presentation but one of my personal favorites to come from mass transfer; binary and triple mergers, and even perhaps binary–induced mixing); and, especially, Type Ia supernovae (Vanbeveren talk and several of the presentations at JD09), if only because each of about four major suggested progenitor classes presents problems (excess hydrogen for main sequence donors right on down to non-existence for double degenerates of large total mass and short period).

On the more cheerful side, processes in binaries can typify ones that occur across the range of astrophysics, including accretion and disks (Torok and Eze posters) and magnetized jets (Viallet and Lopez posters), as emphasized in Zhang's Invited Discourse (which, like the proceedings of the various JDs mentioned here should appear eventually in Highlights, van der Hucht 2007).

That wide binaries have survived in the Galactic halo population, given all the unpleasantness of mergers and such, is perhaps a bit of a surprise, but survived some of them have (talks by Chanamé, Bisikalo, Poveda, posters by Oswalt and Orlov, the latter actually on binary pulsars).

That binaries are, one way or another, relevant to the acceleration of run-away stars was addressed by McSwain (who was forced to conclude that very few OB runaways have carried their neutron star partners with them), and it is a special pleasure to note that the originators of both of the main hypotheses, Arcadio Poveda (cluster processes) and Adriaan Blaauw (liberation by a supernova) were in Prague, although the latter only during the first week.

Finally come tests of general relativity. I continue to find it remarkable that GR effects (for both "advance of the perihelion" and for "loss of angular momentum in gravitation radiation") should be detectable in cataclysmic variables, but Giménez left us in no doubt that it is so, and that the effects are more or less as expected. And the Holley-Bockelmann poster said that CVs will be a significant background for LISA. I think that the paper by Wana *et al.* (which was displayed, although not with most of the S240 ones) was advocating something other than GR. And the Křížek poster made the very important point that what is sometimes being tested is not the GR but our ability to do relativistic calculations correctly!

5. The Antic Adjective

My normal method of note-taking (write it all down and sort it out later), invariably yields a certain number of thoughts on the borderline between science and sociology. Here are a few of the items from S240.

Harmanec in his introduction pointed out that the regimes of double stars and spectroscopic binaries overlap considerably and increasingly, a point reinforced in a number of the other talks. Might this eventually lead to a merger of the commissions? This was suggested more than a decade ago and opposed by both commissions, especially the vice presidents, one of whom would necessarily have been forced out of the succession. Since the two commissions collaborated successfully in the present organization, the issue is clearly not urgent.

Aristotle, remarked Richards, was the first tomographer, deducing the spherical shape of the earth from its invariably circular shadow on the moon. Perhaps one should say the first scientific tomographer, since I am reasonably certain that our remote Zinjanthropan ancestors, seeing the shadow of a large, fierce animal in the jungle, did not require three-dimensional information before running like hell.

No one now living has split Procyon with the naked eye, reported Mason, meaning with the eye as a detector, we deduced, not "without a telescope" (the unaided eye). In either case, it suggests organization of a star party, perhaps at Lick, where Schaeberle did it first with the 36″ to remedy this defect.

Not too many time assignment committees give speckle astronomers time to observe binaries on 8–10 meter telescopes, noted Balega. And a critical look at the frequency with which binary star papers are cited compared to other hotter topics (Trimble & Zaich 2005) does not lead one to think this is likely to change in the foreseeable future.

The HR diagram really stands for Hans Rosenberg, claimed speaker Valls-Gabaud. And indeed he (a student of Karl Schwarzschild just before the latter moved to Potsdam) did make the first plot of apparent brightness vs. spectral type, for stars in the Pleiades, shortly before Hertzsprung and Russell had the same sort of idea (Rosenberg 1910). I have not attempted to find out what happened to Rosenberg over the next 30-plus years.

"You are lucky if you can find even a weak hydrogen line" in a white dwarf spectrum, according to Oswalt. The majority of white dwarfs are type DA, and the lines are not, in an equivalent–width sense, at all weak. They are, however, very wide, so that perhaps wavelength resolution beyond the 200 Å/mm of the hundreds of prime focus 200-in plates I measured long ago is not an improvement.

The entries in the *Washington Double Star Catalogue* are known to be bound (concordant parallaxes or common proper motion) in about 2% of the cases, unbound (discordant proper motions or distances) in another 2% and undetermined for the rest, said Mason. Clearly one could repeat something like Michell's analysis (but using a better description of the distribution over the sky of stars of various apparent magnitudes) and decide which of those 2%'s is the tip of the iceberg. I cannot begin to guess whether this would take an afternoon or a decade for someone skilled in the use of globes.

The description of image restoration methods by ten Brummelaar has to be seen to be fully appreciated, but, in outline, speckles flash past your eyes; adaptive optics is a form of forcible flattening; and interferometry tears off a tiny bit of the information and discards the rest. The radio version, according to P.A.G. Scheuer, was like being led blindfolded up the side of a mountain by a single path, and then being asked to describe not only the entire mountain but its Fourier transform.

"It's worth rereading for anyone who hasn't read it before," was the description of the Popper (1980) article on stellar masses provided by Henry. Don't miss it if you can, in other words. Popper came from a family that had been in Northern California for a number of generations, but they were Jewish, and the presence of the name (an unlikely one somehow) on the walls of one of the Prague synagogues suggests that they may have come earlier from this part of the world.

"Like so many of us, he fell in love with astronomy as a little boy," was McAlister's description of his undergraduate mentor, who had attended the IVth General Assembly in Cambridge in 1932. This was the earliest GA to have had a daily newspaper as far as I have been able to determine. It was edited by Cecilia Payne (later Payne Gaposchkin), who had fallen in love with science when...

6. Last Thoughts

Several participants toward the end of the discussions expressed the view that the program had been a bit disconnected, perhaps a predictable result of the merging of two proposals and the concerns of both C26 and C42, and one might invoke the view of Santa Barbara cartoonist Ashleigh Brilliant that "the only requirement for getting there is to keep going in the right direction." This presupposes that one is sure which is the right direction (there being no one of whom one can ask directions in this context). The wide range of problems discussed and the high probability of some interesting binary products arising in more than one way more nearly recalls the claim of Peter De Vries that every novel should have a beginning, a muddle, and an end, and Carl Hansen's gloss that "we are now in the stellar muddling stage" (Hansen *et al.* 2004). If so, then the correct approach is that of Lord Ronald in the short story "Gertrude the Governess," who, according to author Stephen Leacock, "jumped on his horse and rode madly off in all directions.

It is the traditional prerogative of the last speaker at a meeting to offer thanks to the hosts and organizers, which I did and do, most heartily, and to say the last, slightly sad, word, in the form of farewell until the next conference, *au revoir, auf weidersehn, do vidzenia, arrividerci, hasta la vista, tot ziens*, or, in the language of our hosts, *Na shledanou*. In the end, however, Colin Scarfe had the last word, to remind us that Leacock was a Canadian, not an English, author!

References

Benson, R.S. 1970, unpublished Ph.D. thesis, U. California, Berkeley
Gyldenkerne, K. & West, R.M. (Eds.) 1970, *Mass Loss and Evolution in Close Binaries*. Copenhagen University Publications, p. 25
Hansen, C.J. *et al.* 2004, *Stellar Interiors*. Springer p. 329
Hockey, T. *et al.* 2007, *Biographical Encyclopedia of Astronomers*. Springer (in press)
Michell, J. 1767, *Phil. Trans. Roy. Soc.* 57, 234
Paczyński, B. 1971, *ARA & A* 10, 183
Rosenberg, H. 1910, *Astron. Nach.* 186, 71
Trimble, V. & Zaich, P. 2005, *PASP* 117, 111
van der Hucht, K. (ed.) 2007, *Highlights of Astronomy*. Vol. 14, Cambridge Univ. Press

Binary Stars as Critical Tools & Tests
in Contemporary Astrophysics
Proceedings IAU Symposium No. 240, 2006
W.I. Hartkopf, E.F. Guinan & P. Harmanec, eds.

© 2007 International Astronomical Union
doi:10.1017/S1743921307004498

Closing Comments

William I. Hartkopf

Astrometry Department, U.S. Naval Observatory, Washington, DC, United States
email: wih@usno.navy.mil

So at last we come to the end of our meeting! I have the honor of being the last speaker at Symposium 240, and apparently the last speaker of the GA itself! I will keep my comments very brief, so that you all may experience a few more minutes of sightseeing or perhaps another fine Czech beer before beginning your travels home.

In matters monetary and logistical: I first wish to thank the IAU for its decision to support our symposium proposal and for the travel grant which accompanied that decision. Surprisingly, the United States Office of Naval Research Global graciously agreed to match those IAU funds; together these two grants enabled us to provide at least partial support for 50 astronomers — mostly students — who wished to attend. Remaining funds will also enable us to send copies of the proceedings to several people who were unable to attend.

Our sincere thanks to the Czech National Committee for providing us with this most pleasant meeting facility, and to the many technicians and volunteers who kept our equipment running smoothly throughout the meeting, handled all our absurd demands, and still managed to keep smiles on their faces through it all. [*A hearty round of applause for the technicians and volunteers followed.*]

In matters scientific: Thank you to our Scientific Organizing Committee, who assisted us tremendously in merging our two original proposals into one coherent program, suggested many of our speakers, and read through all the submitted proposal abstracts to vote for our "poster highlight" talks.

Thank you to some 220 speakers and poster presenters, who over the last four days gave us interesting talks and displays, kept for the most part to our tight schedule, and allowed us glimpses into the many exciting topics that comprise the field of modern binary star research. Thank you, Virginia and Colin, for tackling the daunting task of trying to summarize such a wide range of topics. Thank you as well to the audience members for their intelligent questions and comments, which further stimulated the discussions.

Special thanks must go to my co-chairs, Petr Harmanec and Ed Guinan. Before this symposium was born I knew Ed only slightly, and had never met Petr. Over the past two years and 2,000 emails, however, we have worked together remarkably smoothly, discussed and compromised on thousands of details major and minor, and even gotten to the point where we understand and tolerate each others' senses of humor!

Finally, a most heartfelt thank you to the people of Prague, both those we see on its busy streets today and those who walked those cobblestones centuries ago, for giving us such an amazing, breathtaking city in which to gather together. The memories of this place will remain with us long after the details of particular talks are forgotten, and we are most grateful to have had this once-in-a-life opportunity.

I wish you all safe journeys home.

Author Index

Object Index